Sustainability Sciences in Asia and Africa

Sustainable Agriculture and Food Security

Series Editor

Rajeev K. Varshney, Centre for Crop and Food Innovation, Food Futures Institute, Murdoch University, Perth, WA, Australia

This book series support the global efforts towards sustainability by providing timely coverage of the progress, opportunities, and challenges of sustainable food production and consumption in Asia and Africa. The series narrates the success stories and research endeavors from the regions of Africa and Asia on issues relating to SDG 2: Zero hunger. It fosters the research in transdisciplinary academic fields spanning across sustainable agriculture systems and practices, post- harvest and food supply chains. It will also focus on breeding programs for resilient crops, efficiency in crop cycle, various factors of food security, as well as improving nutrition and curbing hunger and malnutrition. The focus of the series is to provide a comprehensive publication platform and act as a knowledge engine in the growth of sustainability sciences with a special focus on developing nations. The series will publish mainly edited volumes but some authored volumes. These volumes will have chapters from eminent personalities in their area of research from different parts of the world.

Sudesh Kumar Yadav • Avnesh Kumari
Editors

Nanobiotechnology in Agri-food Sector

Springer

Editors
Sudesh Kumar Yadav
Biotechnology Division
CSIR - Institute of Himalayan
Bioresource Technology
Palampur, Himachal Pradesh, India

Avnesh Kumari
Phytofarming and fermentation Division
CSIR - Institute of Himalayan
Bioresource Technology
Palampur, Himachal Pradesh, India

ISSN 2730-6771 ISSN 2730-678X (electronic)
Sustainability Sciences in Asia and Africa
ISSN 2730-6798 ISSN 2730-6801 (electronic)
Sustainable Agriculture and Food Security
ISBN 978-981-96-7755-9 ISBN 978-981-96-7756-6 (eBook)
https://doi.org/10.1007/978-981-96-7756-6

This Springer imprint is published by the registered company Springer Nature Singapore Pte Ltd.
The registered company address is: 152 Beach Road, #21-01/04 Gateway East, Singapore 189721, Singapore

Series Preface

The UN's second sustainable goal (SDG) sets out to end hunger, achieve food security, improve nutrition, and promote sustainable agriculture within the next seven years. Yet despite the tireless efforts of the global scientific community, NGOs, and national agricultural research systems the world over, more than 800 million people still suffer from undernourishment. To address this food security gap, it is estimated that global food production systems will need to expand their capacities and produce almost twice their existing levels to feed the world's burgeoning population.

It is a momentous challenge that is further exacerbated by rising weather extremities, unpredictability in rainfall patterns, and pest-pathogen dynamics associated with global climate change, which continues to have a profoundly negative impact on agricultural productivity and farm incomes worldwide. To address these challenges, we need sustainable innovations in our agri-farming systems that lead to improved food production in resource-constrained agricultural settings, and more equitable, gender-diverse farming practices that encourage the participation of women in agriculture.

Efforts to implement these sustainable innovations have resulted in improvements in food access, farmer autonomy, and the stabilization of agriculture's global carbon footprint, yet the challenge to reduce hunger remains pronounced in the Global South, particularly Asia and Africa, which house the largest proportion of people suffering from malnutrition and other nutrition-related issues. Furthermore, the agri-food systems in Asia and Africa are severely constrained by the subsistence nature of farming, declining land and other agricultural resources, and increasing environmental pollution, soil and biodiversity degradation, and climate change.

This book series, Sustainable Agriculture and Food Security, aims to support global efforts toward sustainability by providing timely coverage of the progress, opportunities, and challenges of sustainable food production and consumption in Asia and Africa. The series narrates the success stories and research endeavors from the regions of Africa and Asia on issues relating to SDG 2: Zero Hunger. It fosters research in transdisciplinary academic fields spanning sustainable agriculture systems and practices, pre-seed and post-harvest innovations, and food supply chains. The focus of the series is to provide a comprehensive publication platform and to act as a repository of knowledge in sustainability sciences with a particular focus on developing nations.

The global population is projected to reach 9.7 billion people by 2050. To successfully meet growing demand without exacerbating the negative impacts of climate change, it is essential that we embrace new technologies such as Nanotechnology. This nascent field of science offers tremendous benefits across the food production value chain thanks to the unique properties that are not achievable. From utilizing nanoparticles to fortify and improve seed germination and growth, to reducing food wastage and spoilage through smart food packaging systems, Nanotechnology has the potential to transform how we grow, produce, transport, and store food. However, before we can scale up this technology for widespread adoption, we need a clearer understanding of its long-term health and environmental impacts. This will require multidisciplinary collaboration between researchers and policymakers, buy-in from consumers through improved awareness campaigns, and a significant effort to deliver the infrastructure and technical expertise requirements to the regions that need innovative technologies such as this the most.

With that in mind, this book, *Nanobiotechnology in Agri-Food Sector*, edited by Sudesh Kumar Yadav and Avnesh Kumari, serves as a timely and important resource for nanotechnology applications in agriculture. The book presents 16 chapters written by experts and practitioners from across the globe, each containing up-to-date and important insights into the topic of Nanotechnology. I commend the Editors, Sudesh Kumar Yadav and Avnesh Kumari, for including nanotechnology applications throughout the agricultural life cycle and would like to congratulate all the authors for their valuable contributions.

I wish to extend my sincere thanks and gratitude to the Springer staff, particularly Sanchi Bhimrajka, Associate Editor (Books, Medicine and Life Sciences), and Nick Melchior, Editorial Director (Books, Medicine and Life Sciences, Asia), for their constant support for the accomplishment of this compendium. The cooperation received from my supervisors and research leaders such as Prof. Andrew Deeks (Vice Chancellor), Prof. Peter Eastwood (Deputy Vice Chancellor- Research), Prof. Peter Davies (Pro Vice Chancellor), and my laboratory colleagues- Rhys G. R. Copeland, Anu Chitikineni, and Vanika Garg from Murdoch University (Australia) is also gratefully acknowledged. I would like to thank my family members—Monika Varshney, Prakhar Varshney, and Preksha Varshney, for their love and support in discharging my duties as Series Editor.

Perth, WA, Australia Rajeev K. Varshney

Preface

With the world's population growing and natural resources like soil, water, and production land becoming depleted, agricultural development needs to be sustainable, economical, and efficient. The demographics of the population have changed significantly throughout the next few decades. Furthermore, the increase of farmed land may jeopardize sustainability in all regions by promoting deforestation due to its effects on greenhouse gas emissions and biodiversity loss. Additionally, agriculture and food security are extremely vulnerable to changing climatic conditions. The United Nations (UN) has established 17 attainable goals (SDGs) to achieving agricultural sustainability. Nanotechnology has a lot of applications and might completely change the agri-food sector. Thus, it is possible to improve agrochemical efficiency as well as agricultural production. Crops have been genetically improved with the application of NPs. In animals and plants under stress, nanoarrays have been utilized to enhance gene expression. Compared to their native counterparts, the activity of agrochemicals enclosed in NMs was noticeably higher. Essential oil-infused NMs showed better physical stability and a greater reduction in the rate of oil evaporation. Moreover, NMs have been employed to deliver genes, DNA, and other beneficial substances to particular plant areas. NMs have a tendency to enter plants through their roots and travel to other parts of the plant through the circulatory system, where they might further affect the many physiological processes of plants. Additionally, NM-based kits have been used to identify viruses and other plant pathogens as well as to manage plant diseases in crops.

Additionally, nanotechnology has reduced postharvest losses of grains, fruits, and vegetables. The food sector has profited from both the nanoencapsulation of nutraceuticals and the incorporation of NPs to food matrices. Smart packaging materials enabled by nanotechnology have been used to increase the shelf life of stored food items. Biosensors based on nanotechnology have been used to detect foodborne infections and toxins. It may be possible to use NMs as bioactive components in functional meals. Globally, functional foods and dietary supplements containing NP have begun to emerge.

Despite the fact that nanotechnology has the potential to revolutionize the agriculture sector, several challenges need to be addressed before NMs can be widely used. These include insufficient field research on the effectiveness of NMs, a lack of knowledge about the hazards that NM application poses to the environment and human health, and a lack of knowledge about the interactions at the NM-plant-soil interface as well as internalization and translocation in plant tissues and organelles. NMs may accumulate in seeds; however, it is unknown how they contribute to transgenerational consequences. NMs also undergo chemical, biological, and physical changes in plants and soils. These changes alter the properties of NMs that were developed for particular purposes.

Controlling the material chemistry throughout the design and synthesis phases is the first step toward making the most of NP transformations for the desired uses. On the one hand, NPs that yield beneficial crop transformation products may develop into controlled-release fertilizers with moderate and safe release dosages, ensuring long supply durations. On the other hand, until they can be recycled, NPs that generate dangerous transformation products should be administered in safe packaging or replaced with safer NPs. To control the rate and substance of release, it is always necessary to precisely modify the material size, aspect ratio, morphology, crystal phase, and multi-element compositions. Additionally, comprehensive transformation studies of NPs in crop systems should be carried out before to their usage in the field.

Discussing the applications of nanobiotechnology in the food and agriculture sector is the aim of this book. We have talked about the use of nanotechnology in crop protection, abiotic stress management, plant function enhancement, hydroponics and aeroponics, plant genetic engineering, and climate change mitigation. The Agritech Revolution, the use of nanotechnology and artificial intelligence to support precision and sustainable agriculture, post-harvest loss reduction, and edible nanoparticles and miRNAs generated from plants have also been discussed. In addition, we have discussed the use of nanomaterials in the dairy industry, food processing, edible films and coatings, and food packaging. This book also discusses nanomaterials employed in phytoremediation.

Our objective is to satisfy the need for a reference book for scientists, researchers, and academicians working in the nutraceutical, agricultural, food, environmental, and material sectors, in addition to being a great resource for undergraduate and graduate students studying agriculture, nutraceutical sciences, food sciences, and nanotechnology.

Palampur, Himachal Pradesh, India

Sudesh Kumar Yadav
Avnesh Kumari

Contents

Series Editor and Editorial Board Members

About the Series Editor

Rajeev K. Varshney, FRS, is a renowned agricultural scientist with over two decades of experience in genomics and molecular breeding. Currently, he holds multiple prestigious positions at Murdoch University in Australia. He has made a significant impact on global food security by creating genomic resources for major "orphan" tropical crops, using these resources to identify genetic loci and candidate genes for drought and pest tolerance in staple crops in Sub-Saharan Africa and India, and spearheading international programs that deliver superior crop varieties to some of the world's most impoverished farmers. He is an accomplished author, with more than 600 publications to his name, and has been recognized as a Clarivate Highly Cited Researcher for ten consecutive years (2014–2023).

Food Futures Institute, Centre for Crop and Food Innovation, State Agricultural Biotechnology Centre, Agriculture and Food Security, Murdoch University, Perth, WA, Australia.

Editorial Board Members

Trilochan Mohapatra Chairman, Protection of Plant Varieties and Farmers' Rights Authority, Ministry of Agriculture and Farmers Welfare, Government of India, New Delhi, India.

Soumya Swaminathan Chairperson, MS Swaminathan Research Foundation, Chennai, India.

Wendy Umberger Chief Executive Officer, Australian Council for International Agricultural Research (ACIAR), Canberra, ACT, Australia.

Florence Wambugu Chief Executive Officer, Africa Harvest, Kenya.

About the Editors

Sudesh Kumar Yadav obtained his MSc in 1999 and PhD in 2002 from the Department of Biochemistry, CCS Haryana Agricultural University, Hisar. Subsequently, he worked as post-doctoral fellow at ICGEB, New Delhi, India. He joined CSIR—Institute of Himalayan Bioresource Technology, Palampur, in 2004, and worked in Centre of Innovative and applied bioprocessing (CIAB)-Mohali (2016–2023), and he took over the charge of Director Council of Scientific and Industrial Research - Institute of Himalayan Bioresource Technology (CSIR-IHBT), Palampur, on June 9, 2023. His research focuses on metabolic engineering of plants and microbes for value addition, synthetic biology and omics approaches to improve metabolic pathways and use of nanobiology for better efficacy as well as improved enzyme-based bioprocesses. He has extensively worked on understanding the regulation of key metabolic pathways of crops like tea, stevia, rice, and horsegram. He has developed several processes based on synthetic biology and nanobiology for the production of high-value functional molecules from natural bioresources and agri-residues. He has published more than 220 research articles in peer-reviewed journals, 20 patents and 20+ book chapters. He is a Fellow of The National Academy of Sciences India, Allahabad, and Fellow of the National Academy of Agricultural Sciences, New Delhi. He is also recipient of S. - Ramachandran-National Bioscience Award, and NASI-Reliance Industries Platinum Jubilee Award.

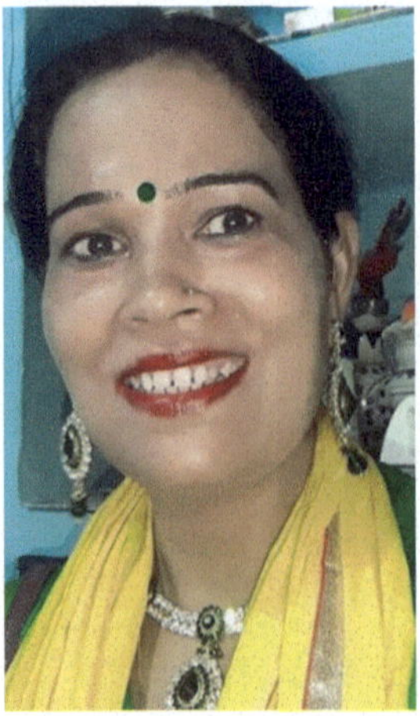

Avnesh Kumari received master's in Chemistry in 2001 and doctorate in 2007 from the Department of Chemistry, Panjab University, Chandigarh. She then spent 2008–2010 as a postdoctoral fellow at CSIR-IHBT, Palampur. She joined CSIR-Institute of Himalayan Bioresource Technology, Palampur, in 2011 and is presently working there as Assistant Professor and Senior Technical Officer (3). She has been working in the area of nanobiology and electron microscopy. Her primary focus in nanobiology is on enhancing the solubility, bioavailability, and effectiveness of significant phytomolecules with antibacterial, anticancer, and antioxidant properties. She received JRF and SRF grants from CSIR, India, early in her career. She also received funding as a CSIR Quick Hire Scientist Fellow. Later, she was awarded a postdoctoral fellowship through the DST-Women Scientist Scheme. She is a life member of numerous scientific associations and has reviewed articles for a number of periodicals. Her exemplary contributions to science and research have earned her significant recognition. She was listed in the Stanford list of Indian researchers in the top 2% most influential list for the year 2020, a testament to her impactful work. Her global rank of 179,475 and Indian rank of 2157 reflect her prominence in the field. With a total of 6923 citations and impressive h-index and i10-index scores, Dr. Kumari's research continues to influence and inspire the scientific community.

Contributors

Sachit Aggarwal Department of Botany, Central University of Punjab, Ghudda, Bathinda, Punjab, India

Amit Crop Genetics & Informatics Group, School of Biotechnology, Jawaharlal Nehru University, New Delhi, India

Ansar Anjum Department of Applied Sciences and Humanities, Galgotias College of Engineering & Technology, Greater Noida, UP, India

Ankur Barwal Plant Biotechnology & Genetic Engineering Lab, Department of Biotechnology, DAV University, Jalandhar, Punjab, India

Rebeca Rocha Batista Department of Chemistry, Biotechnology, and Bioprocess Engineering (DQBIO), Federal University of São João del-Rei (UFSJ), Ouro Branco, MG, Brazil

Ashutosh Bhaik Corteva Agriscience, Hyderabad, Telangana, India

Sakshi Chopra Plant Biotechnology & Genetic Engineering Lab, Department of Biotechnology, DAV University, Jalandhar, Punjab, India

Srishti Chopra Plant Biotechnology & Genetic Engineering Lab, Department of Biotechnology, DAV University, Jalandhar, Punjab, India

Brunna D'Onofre Couto Department of Chemistry, Biotechnology, and Bioprocess Engineering (DQBIO), Federal University of São João del-Rei (UFSJ), Ouro Branco, MG, Brazil

George Lucas da Silva Magalhães Department of Chemical Engineering (DEQUI), Federal University of São João del-Rei (UFSJ), Ouro Branco, MG, Brazil

Lucimara Gaziola de la Torre Department of Materials and Bioprocess Engineering, School of Chemical Engineering, University of Campinas, Campinas, SP, Brazil

Jhonatan Rafael de Oliveira Bianchi Department of Materials and Bioprocess Engineering, School of Chemical Engineering, University of Campinas, Campinas, SP, Brazil

Liliane de Paula Gonçalves Department of Chemical Engineering (DEQUI), Federal University of São João del-Rei (UFSJ), Ouro Branco, MG, Brazil

Marcos Silva de Sousa Department of Chemistry, Biotechnology and Bioprocess Engineering, Federal University of São João del-Rei, Ouro Branco, MG, Brazil

Jane Sélia dos Reis Coimbra Department of Food Technology, Federal University of Viçosa (UFV), Viçosa, MG, Brazil

Nnabuk Okon Eddy Department of Pure and Industrial Chemistry, University of Nigeria, Nsukka, Nigeria

Felipe Ramalho Estanislau Department of Chemistry, Biotechnology, and Bioprocess Engineering (DQBIO), Federal University of São João del-Rei (UFSJ), Ouro Branco, MG, Brazil

Courtney Gallup Corteva AgriScience, Indianapolis, USA

Rajni Garg Department of Applied Sciences and Humanities, Galgotias College of Engineering & Technology, Greater Noida, Uttar Pradesh, India

Rishav Garg Department of Civil Engineering, Galgotias College of Engineering and Technology, Greater Noida, UP, India

Marília Magalhães Gonçalves Department of Chemistry, Biotechnology, and Bioprocess Engineering (DQBIO), Federal University of São João del-Rei (UFSJ), Ouro Branco, MG, Brazil

Praveen Guleria Plant Biotechnology & Genetic Engineering Lab, Department of Biotechnology, DAV University, Jalandhar, Punjab, India

Poonam Jyoti Biotechnology Division, CSIR-IHBT, Palampur, India

Mohd Kashif Department of Applied Sciences and Humanities, Galgotias College of Engineering & Technology, Greater Noida, UP, India

Sukhdeep Kaur Plant Biotechnology & Genetic Engineering Lab, Department of Biotechnology, DAV University, Jalandhar, Punjab, India

Adarsh Kumar Department of Botany, Central University of Punjab, Ghudda, Bathinda, Punjab, India

Amit Kumar Biotechnology Division, CSIR-Institute of Himalayan Bioresource Technology, Palampur, HP, India

Alka Kumari Department of Botany, University of Lucknow, Lucknow, UP, India

Avnesh Kumari Fermentation and Phytofarming Division, CSIR-Institute of Himalayan Bioresource Technology, Palampur, HP, India

Vinay Kumar Department of Botany, Central University of Punjab, Ghudda, Bathinda, India

Vineet Kumar Department of Biotechnology, School of Bioengineering and Biosciences, Lovely Professional University, Phagwara, Punjab, India

Andy Leader Corteva AgriScience, Indianapolis, USA

Dimpal Mehla Crop Genetics & Informatics Group, School of Biotechnology, Jawaharlal Nehru University, New Delhi, India

Ana Luiza Schuenck Alberice Monteiro Department of Chemistry, Biotechnology and Bioprocess Engineering, Federal University of São João del-Rei, Ouro Branco, MG, Brazil

Shashi Rani Biotechnology Division, CSIR-Institute of Himalayan Bioresource Technology, Palampur, HP, India

Patrícia Avelar Ribeiro Department of Chemistry, Biotechnology and Bioprocess Engineering, Federal University of São João del-Rei, Ouro Branco, MG, Brazil

Igor José Boggione Santos Nanotec Research Group-Nanotechnology in Bioprocesses, Chemistry, Biotechnology, and Bioprocesses Engineering Department, Universidade Federal de São João del Rei (UFSJ), Ouro Branco, Brazil

Priya Saxena Department of Botany, University of Lucknow, Lucknow, India

Andersen Escobar Schlog Postgraduate Program in Physics and Chemistry of Materials, Department of Natural Sciences (DCNAT), Federal University of São João del-Rei (UFSJ), São João del-Rei, Brazil

Komal Sharma Department of Botany, University of Lucknow, Lucknow, India

Manoj K. Sharma Crop Genetics & Informatics Group, School of Biotechnology, Jawaharlal Nehru University, New Delhi, India

Tahsin Shoala Environmental Biotechnology Department, College of Biotechnology, Misr University for Science and Technology, Giza, Egypt

Muskan Tiriya Department of Botany, Central University of Punjab, Ghudda, Bathinda, Punjab, India

Ashish R. Warghat Biotechnology Division, CSIR-Institute of Himalayan Bioresource Technology, Palampur, Himachal Pradesh, India

Anurag Yadav Department of Microbiology, C.P. College of Agriculture, Sardarkrushinagar Dantiwada Agricultural University, Sardarkrushinagar, Gujarat, India

Kusum Yadav Department of Biochemistry, University of Lucknow, Lucknow, UP, India

Sudesh Kumar Yadav Biotechnology Division, CSIR-Institute of Himalayan Bioresource Technology, Palampur, HP, India

Chapter 1
Plant-Derived Edible Nanoparticles and miRNAs

Muskan Tiriya, Adarsh Kumar, and Vinay Kumar

Abstract Plant-derived edible nanoparticles (PDNPs) and microRNAs (miRNAs) represent a new emerging tool for boosting human health and nutrition. Here, the properties and sources of PDNPs along with their application, and further the biology and importance of miRNAs in plants and humans, will be discussed. PDNPs are nanovesicles obtained from edible plants with the capability of encapsulating bioactive compounds that include miRNAs. They are biocompatible, stable, and have the feature of targeted delivery. miRNAs participate in the regulation of gene expression and act as mediators of stress responses in plants under various environmental stressors. The interactions between PDNPs and miRNA facilitate cross-kingdom communication and have prospects for new therapeutic strategies. Key applications of PDNPs and miRNAs are antioxidant and anti-inflammatory effects, gut health maintenance, tissue regeneration, and cancer therapy. Despite these challenges, the potential for PDNPs and miRNAs in furthering personalized nutrition and medicine is significant. There is a lot more interdisciplinary research needed to bring this potential fully into human health benefits.

Keywords Plant-derived edible nanoparticles (PDNPs) · MicroRNAs (miRNAs) · Cross-kingdom communication · Target specificity · Stability · Low toxicity · Sustainable

1.1 Introduction

The problem of worldwide health is an issue of increasing global plagues with chronic diseases like cardiovascular disease, cancer, and diabetes. Conventional drugs are limited to systemic toxicity, resistance to drugs, and poor targeting of diseased tissue. All these aspects highlight the critical need for novel and advanced techniques that can optimize drug delivery systems with fewer side effects. Against

M. Tiriya · A. Kumar · V. Kumar (✉)
Department of Botany, Central University of Punjab, Ghudda, Bathinda, Punjab, India
e-mail: vinay@cup.edu.in

S. K. Yadav, A. Kumari (eds.), *Nanobiotechnology in Agri-food Sector*, Sustainability Sciences in Asia and Africa, https://doi.org/10.1007/978-981-96-7756-6_1

this background, scientists have turned to nature, drawing inspiration from the foods that make up the basis of healthy eating, i.e., fruits and vegetables. Diets comprising these foods are often associated with reduced rates of various diseases, including cardiovascular disease, diabetes, several cancers, chronic diseases, and some age-related functional impairments. These bioactive compounds found in plants, i.e., flavonoids, phenolic acids, and carotenoids, are postulated to be responsible for such protective effects. This is where the gap can be filled by increasing the bioavailability and delivery of these bioactive compounds to the target tissues in plant-derived edible nanoparticles (PDNPs). With these characteristics, PDNPs in this context have surfaced as a viable solution since they can efficiently deliver therapeutic agents through the natural benefits of plant material.

PDNPs are nanovesicles originating from edible plants such as *Zingiber officinale* (ginger), *Citrus* × *paradisi* (grapefruit), *Brassica oleracea* (broccoli), and many more fruits and vegetables. These nanoparticles are minute particles that have characteristics that make them superior to synthetic nanoparticles. They are tissue-targeting, non-toxic, and can be produced in bulk with environmentally friendly techniques. Studies have shown that PDNPs offer a high rate of internalization and low immunogenicity, implying that they are effective in delivering drugs without causing many side effects associated with the traditional systems of drug delivery (Iravani et al. 2019). Furthermore, PDNPs remain stable in the alimentary canal and cross the blood-brain barrier (BBB) without inducing the blood-placenta barrier to react. Such features have been adopted by the medical profession to develop drug delivery carrier nanotechnology systems to improve therapeutic specificity and efficacy.

The importance of PDNPs is not just limited to their physical characteristics, they also have a crucial role in improving the bioavailability of bioactive molecules. By encapsulating drugs and nutrients, scientists create nanoparticles to have better stability as well as improved absorption within the body. A good example for this case would be the encapsulation of PDNPs carrying microRNAs, small RNA molecules, as part of the regulation in genes; encapsulation can open up further possibilities to affect gene expression as it may interact with health outcomes (Karnwal et al. 2024). This use is highly applicable in medicine and agriculture, where miRNAs may contribute to the tolerance of plants against stress factors and have therapeutic value in human health (Jain et al. 2022). Furthermore, recent research emphasizes the promise of PDNPs as natural drugs against inflammatory disorders and cancers. The studies reveal that the nanoparticles can stimulate intestinal regeneration by facilitating the growth of intestinal stem cells when taken orally (Chen et al. 2022). This property makes PDNPs a new drug delivery system for treating other digestive tract disorders like inflammatory bowel disease (IBD). Moreover, PDNPs from different plant sources are anti-cancerous as they induce apoptosis and inhibit the growth of the tumor (Iravani and Varma 2019). Their multi-functionality also highlights their value as next-generation drugs.

The study of PDNPs presents a groundbreaking approach in fighting severe health problems posed by chronic diseases. With the use of natural plant properties, PDNPs present a viable, safe, and eco-friendly way to enhance drug delivery systems. In the future, studies will continue to reveal their working methods and applications in agriculture and medicine, keeping PDNPs at the forefront of new approaches for enhancing human health while also promoting environmental sustainability. This would revolutionize how we interact with treatment modalities in modern health care. Incorporating these nanoparticles into therapeutic regimens can open up new doors for the healthcare sector. This chapter discusses the application of PDNPs in health and nutrition, concentrating on their properties and sources. It further delves into the biology of microRNA and its importance to both plant and human health. Some of the key topics include how PDNPs increase the delivery of miRNA and therapeutic intervention. The book concludes with a citation of research gaps and future prospects, giving prominence to the importance of interdisciplinary research to gather the optimum benefit of PDNPs and miRNAs for health and nutrition.

1.2 Fundamentals of Edible Nanoparticles (ENPs)

PDNPs are a recently discovered class of nano-sized vesicles derived from various edible plants, such as ginger, broccoli, and grapefruit. These PDNPs refer to membrane-bound structures that commonly range between 100 and 400 nm in diameter. Nano-sized vesicles have unique features that distinguish them from synthetic nanoparticle counterparts: their small dimensions allow for efficient bioavailability, cellular uptake, and effective drug delivery while decreasing the potentially toxic and immunogenic effects of the synthetic nanoparticles (Jain et al. 2022). The lipid bilayer structure of PDNPs accounts for their distinct properties because it allows for the incorporation of active substances like microRNAs, proteins, and phytochemicals. Such encapsulation safeguards the bioactive materials from being degraded and boosts their stability and activity in biological systems. One of the unique features of PDNPs, which renders them suitable for use in food and drug manufacture, is their high degree of biocompatibility. The modification of the surface of PDNPs through functionalization allows for their targeting of specific tissues or cells, which enhances the potency of treatment applications (Karnwal et al. 2024; Nandan et al. 2024). Besides their structural and functional properties, PDNPs are known to be scalable and economical in terms of the production process. The methods of extraction applied for PDNPs are most often eco-friendly protocols that suit the principles of green chemistry. Additionally, their stability within gastrointestinal environments provides them with suitable applications for oral delivery. Several studies have reported that PDNPs can be resistant to harsh conditions such as acidic environments and proteolytic enzymes found within the digestive tract (Jain et al. 2022). The incorporation of PDNPs into therapeutic applications highlights their promise as natural drug delivery systems that could enhance sustainable treatment approaches. As research progresses in understanding their mechanisms and refining their formulations, PDNPs are likely to become crucial in the future of nanomedicine and functional foods (Karnwal et al. 2024; Nandan et al. 2024).

1.2.1 Composition

Plant-derived edible nanoparticles (PDNPs) were first discovered in 1967. PDNPs comprise lipid, protein, nucleic acid, and other cytoplasmic constituents (Fig. 1.1).

A. Lipid

Lipids are significant to maintain the framework and operation of plant-derived exosome-like nanoparticles (PDENs). The base structural framework of PDENs is the glycolipid and phospholipid bilayer; however, their lipid composition may vary greatly. In contrast to mammalian exosomes that contain mostly PC and cholesterol, the lipid composition of PDENs shows greater diversity. For example, 19% of monogalactosyldiacylglycerol, 27% of digalactosyldiacylglycerol, and 42% of phosphatidic acid are found in ginger-derived nanoparticles. The lipids in grapefruit-derived nanoparticles contain 13% phosphatidylinositol, 23% PC, and 24% phosphatidyl ethanolamine (PE), while grape-derived nanoparticles contain 26% PE and 53% PA.

Functionally, these lipids play a key role in cellular activity. PA from ginger-derived nanoparticles enhances retention in the duodenum and is absorbable by gut flora, whereas the same from grape-derived nanoparticles promotes cell growth in the colon. PS is a vital phospholipid in the biofilm that forms apoptotic cell membranes, whose externalization signals to macrophages for phagocytosis and thus inflammatory responses. PC is also essential for various physiological activities, including the self-assembly of lipid bilayers, cell membrane fusion, activation of various enzymes, the facilitation of cholesterol solubility in bile, and as a surfactant in alveoli (Chen et al. 2022).

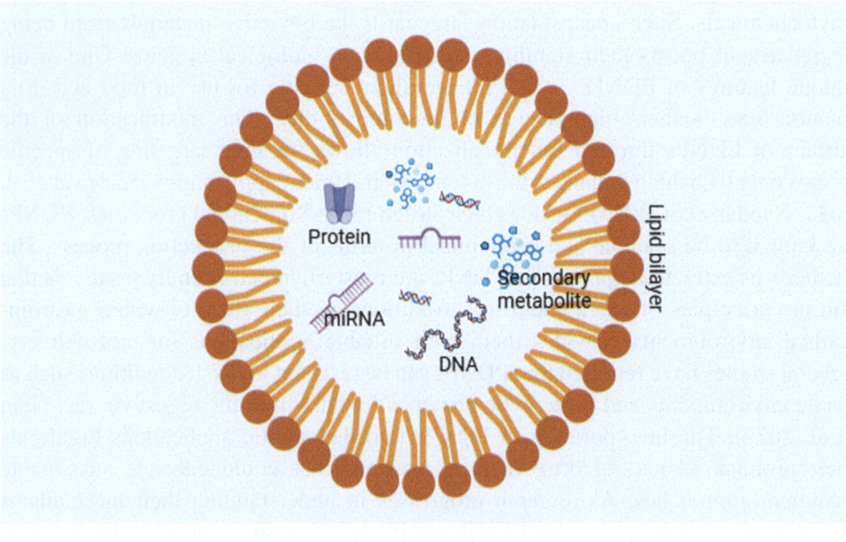

Fig. 1.1 Composition of a PDNPs

B. Protein

Unlike lipids, Gioia et al. (2020) stated that there are very few incongruent studies regarding the proteins of plant-derived nanovesicles, and it has been established that the general protein content in these is less compared to those found in the proteins of mammalian exosomes. 28 proteins were identified by Mass spectrometry (MS) in ginger-derived exosome-like nanoparticles (GELNs), while there were 137 proteins found in grapefruit-derived nanovesicles (GFDNs). The grape-derived nanovesicles, on the other hand, are comprised mostly of cytosolic proteins with some membrane proteins identified through UPLC-MS/MS. It was found that a total of 580 proteins were identified in *Citrus limon L.* juice-derived nanovesicles (CDNs), which overlay by 57% with the mammalian exosome proteins. However, in more recent research, 1018 proteins in *Citrus clementina* juice-derived nanovesicles (CFDNs) using MS-based organelle proteomics and 3129 proteins in grape seed-derived nanovesicles (GSDNs) have been identified through MS.

C. Other Cytoplasmic Components

In addition to lipids and proteins, PDNPs also contain microRNAs (miRNAs) and secondary metabolites, of which miRNAs will be further discussed in a subsequent section. Microanalysis of plant exosomes revealed a high concentration of chlorophylls and curcuminoids. Nevertheless, the analysis of secondary metabolites in exosomes is very much dependent on the basic or acidic conditions of the extraction buffer used. Therefore, the synthesis of such secondary metabolites might be induced in response to bacterial infection, giving antimicrobial attributes to PDNPs. Presumably, this might be critical to the physiological roles of the biological molecules associated with PDNPs (Woith et al. 2021).

1.2.2 Advantages of PdNPs over Mammalian-Derived Nanoparticles

For years, extracellular vesicles (EVs) from mammals have been increasingly utilized in biochemical, genetic, and pharmacological research. A significant number of EVs like cancer cells, immune cells, and stem cells have been identified from mammals. Mammalian EVs have been identified in various body fluids like blood, saliva, and urine (Li et al. 2022). These vesicles transport many cargo molecules, such as lipids, proteins, and nucleic acids, and are involved in cell-to-cell communication (Thakur et al. 2022). The pharmacological properties are beneficial and show compatibility with biological systems; mammalian EVs are recognized as significant regulatory agents in many pathological and physiological processes. Injecting exosomes that were derived from dendritic cells into the myocardial infarction of a mouse was found to boost the infiltration of M2 macrophages and Treg cells into the damaged area, leading to an enhanced cardiac function. Additionally, exosomes from gefitinib-sensitive cancer cells have been shown to effectively bring an end to gefitinib resistance by transferring microRNA-7 (Chen et al. 2021a, b). Further research by Cui et al. (2023) has demonstrated that bioactive molecules can be transported through mammalian exosomes across cellular barriers, e.g., the target glioma cells and the BBB. Despite the efficiency of mammalian EVs for biological applications, several significant challenges occur during their clinical translation, including difficulties in obtaining high-quality EVs, time-consuming processes, and low yield (Kim et al. 2021a, b).

Primarily, using mammalian EVs often triggers host immune responses, which can lead to some adverse side effects. However, identifying plant-derived nanoparticles (PDNPs) and understanding their operational mechanisms may offer a more efficient method to deal with these challenges. Recently, several research teams have proposed the existence of nanovesicles in plant materials that are involved in cell-to-cell communication. While the secretion mechanisms of EVs from plant and animal cells have some similarities (Karamanidou and Tsouknidas 2021), there are distinct biological characteristics that set them apart across various of species. PDNPs can be produced in large quantities from various affordable plant sources, such as fruits and vegetables found in local markets (Kim and Kim 2022), as well as from in vitro plant cells. The PDNP production can be improved under controlled conditions by using food additives (Woith et al. 2021). Being derived from edible plants, PDNPs tend to have lower toxic levels and do not provoke any kind of immune responses in the host. They can easily move past biological barriers such as the BBB through processes like phagocytosis, endocytosis, macropinocytosis, and membrane fusion, enabling absorption by recipient cells (Liu et al. 2020; Stotz et al. 2021). Considering these advantages, it is essential to conduct further research to explore the properties of PDNPs to find new ways to regulate human health. Comparison of PDNPs and mammalian-derived nanoparticles is listed in Table 1.1.

Table 1.1 Comparison of PDNPs and mammalian-derived nanoparticles

Feature	(PDNPs)	Mammalian-derived EVs
Source	Edible plants (e.g., grapefruit, ginger)	Mammalian cells (e.g., immune, cancer cells)
Size range	100–400 nm	Varies, generally similar
Biocompatibility	High	Moderate, can trigger immune responses
Production	Scalable, eco-friendly methods	Low yield, complex isolation processes
Stability in GI tract	High	Variable, often less stable
Targeting ability	Tissue-specific through functionalization	Limited targeting capabilities
Applications	Drug delivery, nutrition enhancement	Cell communication, therapeutic delivery

1.3 miRNAs: Role and Significance

MicroRNAs (miRNAs) are single-stranded, small non-coding RNAs present in both animals and plants. They are usually 19–25 nucleotides long. These molecules play an important role in various physiological and biological processes, making them a deciding factor for normal growth and development. They use post-transcriptional mechanisms like silencing of messenger RNA (mRNA) and translational repression to modulate gene expression. Because of this, a single microRNA (miRNA) can target and modulate hundreds of different mRNA transcripts, thereby allowing control over several pathways and functions in the organism.

1.3.1 Biogenesis and Stability

Plant miRNAs are made up of 22-nucleotide RNAs. The differences between the generation and operational mechanisms of these plant miRNAs and animal miRNAs are less defined compared to those of animals. The stem-loop structures of plant and animal miRNAs vary, with plant-predicted foldbacks being longer than those of the animal. Plant miRNAs emerge from particular stem areas characterized by single-stranded hairpin precursors; these are important for virtually all biological and physiological processes, including stress-responsive regulation. There are a substantial number of miRNA genes in the plant genome, of which RNA polymerase II transcription is responsible for a majority of primary miRNAs (pri-miRNAs). *DICER-LIKE 1* (*DCL1*) present within the nucleus processes pri-miRNAs into stem-loop pre-miRNAs. The plant homolog of exportin 5, called *HASTY*, along with yet-to-be-identified additional factors, helps export pre-miRNAs into the cytoplasm. To generate mature miRNAs, the strands of such pre-miRNAs undergo methylation by the small RNA methyltransferase HUA Enhancer 1 (*HEN1*). MiRNA-mediated silencing involves AGO-containing RISC proteins acting on the guide strand of miRNA; 3′ termini of mature animal miRNAs are devoid of 2′-O-methyl groups. miRNA forms in plants rely on nearly ideal or ideal base-pairing to identify targets, whereas such association with "seed rule" is not so prevalent among animal miRNAs (Saiyed et al. 2022; Bajczyk et al. 2023).

The stability of miRNAs is critical for maintaining their intracellular balance and responding to environmental changes. HEN1 stabilizes miRNAs, which are expressed during light conditions; when the expression is boosted in the light, it can increase photomorphogenesis by altering specific miRNAs. Mutations in HEN1 knock out the stability of miRNAs. Moreover, enzymes like the nuclear exosome targeting complex (NEXT) and XRN degrade pri-miRNA, while proteins like HYL and MAC5 counteract degradation. PAP enhances the stability of pri-miRNA through the further inhibition of XRN. The interaction of miRNA biogenesis with miRNA stability is central to determining the levels of miRNAs in plants, emphasizing the significance of the mechanisms that regulate miRNA production and stability in possible applications to enhance secondary metabolite production (Jeena et al. 2022).

1.3.2 Impact of miRNAs on Growth and Development

miRNAs are indispensable for a multitude of plant developmental facets (such as regulating meristem features, establishment of leaf polarity, facilitation of flowering patterns, and many others), adaptability, abiotic and biotic stress responses, and symbiotic or parasitic interrelations with plant-ecological interactions. Mutation events in genes regulating miRNA transcription or processing complexes were found to produce pleiotropic effects in determining plant form and function, thus attesting to the importance of miRNAs in plant coordination.

A. Role of miRNA in Root Elongation

This dynamic growth allows the plant to establish a solid foundation, anchoring it firmly and thereby enhancing its ability to tap resources necessary for its development. Root elongation is one of the crucial processes of plant growth; it enables the plant to send roots down into the soil, where they search for nutrients and water.

Several studies provide evidence that suggests that miRNAs regulate root elongation. Specifically, miR156 and its target *SQUAMOSA PROMOTER BINDING PROTEIN-LIKE* (*SPL*) genes have a vital role in determining the length of the root. In *Arabidopsis thaliana*, the miR156-*SPL10* module has been described as a central regulator of primary root (PR) growth and length. The main effect of this module on root meristem activity is mainly through the modulation of responses to auxin and cytokinins (CKs), two key hormones governing root development (Barrera-Rojas et al. 2019). By tuning hormonal signals between these pathways, *miR156* and *SPL10* work together to optimize root elongation, thus improving the fertility of the soil. The hormone signaling pathways along with the interplay of miRNAs focus on the role of miRNAs in determining root architecture and plant health.

On the other hand, overexpression of miR163 inhibits primary root (PR) elongation by suppressing the expression of its target gene, *PXMT1*. Overexpression of miR163 has been shown to improve PR lengthening in the *HY5*(LONG HYPOCOTYL 5) mutant. Such a finding offers insights into the modulation of light-influenced root development post-transcriptionally and is mediated by the *HY5-miR163-PXMT1* pathway (Li et al. 2021a, b). In this case, the actual development of a very relevant regulatory axis leads to a better understanding of light signaling of root development through miRNA action. The interplay between HY5, a critical regulator of plant light response, and miR163 also highlights the growing complexity of the hormonal and environmental interactions that govern root growth. This relationship, therefore, reflects how crucial miRNAs in root elongation can create better opportunities to enhance root growth through directed manipulation of such pathways. Understanding these mechanisms may lead to improved methods of optimizing plant growth and plant adaptability under changing environmental conditions.

B. Role of miRNA in Leaf Development

MicroRNAs (miRNAs) shape leaf development by modulating gene expression and directing other developmental processes. Transcription factors like HD-ZIP proteins help regulate, which are distinctively involved in growth and development, specifically in shaping the leaf architecture. *HD-ZIP III* genes regulate leaf polarity, organ development, and vascular tissue patterning through the regulation of the key genes *PHABULOSA (PHB)*, *REVOLUTA (REV)*, and *PHAVOLUTA (PHV)*, which are involved in adaxial-abaxial patterning and vascular differentiation in leaves. The *HD-ZIP III-miR165/166* signaling provides insights into vascular development, maintenance of the meristem, and establishment of leaf polarity. Interaction between *HD-ZIP III* proteins and *KANADI1 (KAN1)* genes is extensively involved in leaf development. Thus, while *HD-ZIP III* proteins affect adaxial development, and *KAN* regulates abaxial polarity, an antagonistic relation between cotyledon development and new shoot apical meristem (SAM) structure development is formed. Additionally, the function of SERRATA, a zinc-finger protein, modulates the activities associated with HD-ZIP III genes on activities of the meristem about leaf polarity. The families of HD-ZIP II and III genes have also been shown to mediate shade avoidance responses and interact with the auxin biosynthesis and transport pathways. Such modulations of auxin production and its polar transport control are key to proper leaf architecture. Therefore, these miRNAs, through HD-ZIP proteins and their interactions with the auxin signaling pathways, comprise the core working of these complex regulatory networks that control leaf development and influence essential processes involved in plant growth and adaptation (Yadav et al. 2021).

To further look into the role of *GRF* and the *miR396* regulatory mechanisms in Chinese cabbage, a study was done where they developed *bra-miR396* and *BrGRF* overexpression vectors, using an enhanced *CaMV 35S* promoter, and then transformed them into *Arabidopsis*. The transgenic plants overexpressing *BrGRF* showed a leaf length increment in the transgenic plants of about 27%, a width increase of approximately 28%, a weight of 39%, and a plant fresh weight of 32% in comparison to WT control leaves. However, bra-miR396 overexpressors' leaves were around 20% narrower and 10% shorter than the control, with reduced plant fresh weight and leaf weight by nearly 33% and 23%. The number of leaves seems unaffected in terms of number by the transgenic lines compared to the control. This result indicates that miRNAs are interfacing with other regulatory pathways for leaf development and morphology (Li et al. 2021a, b).

C. Role of miRNA in Flowering

Flowering is divided into three stages: induction, initiation, and floral organ development. The development of flowers is indeed a very complex phenomenon that includes numerous genes and is a salient event in the developmental evaluation of higher plants. miRNA plays a very vital role in flowering, as shown by many studies undertaken.

One of the key interactions involves *miR156* and *miR172,* which regulate the timing of flowering and patterning of flowers. Some examples of particular interactions of miRNAs can influence developmental processes and underscore their important role in shaping plant architecture and flower patterning (Yadav et al. 2021). Likewise, in *Arabidopsis thaliana*, *SBP/SPL* proteins help in the transformation of the vegetative phase into the reproductive phase. The expression of *SBP/SPL* proteins is hampered in the juveniles by *miR156* and *miR157*. *SBP/SPL* proteins accumulate due to the decrease in the level of *miR156/miR157*, and the plant transitions from the vegetative phase to the reproductive phase (Fouracre et al. 2021).

1.3.3 Responses to Environmental Stress

Plant development and maturation can be impaired or influenced by various environmental factors. The development and maturation of plants are inhibited if plants are subjected to stress from antagonistic activities. The plant adaptation to environmental stress involves the modulation of development and growth using different external or internal processes. These include stresses from biotic factors, such as pathogens, viruses, and pests, and abiotic factors, such as drought, salinity, and low temperatures.

A. Biotic Stress

In the natural environment, biotic stress factors are more prevalent and arise from a wide variety of pathogens such as bacteria, fungi, viruses, and pests. If a pathogen attacks the host plant, miRNA expression responds promptly and changes the expression of its target genes; thus, in turn, this helps regulate the resistance of the plant against pathogenic infection by modifying various processes, such as hormone transfer, secondary metabolism, and the expression of genes that are disease resistance. Zhang et al. (2021) reported *miR394* to be negatively regulating biotic stress because of its increased expression in tomatoes, which subdues the expression of its concerned gene, *LCR*, which then leads to decreased expression of genes related to the JA synthesis and thus increases the disease-causing mildew. In contrast, Yang et al. (2023) observed that overexpression of *miR159a* in transgenic poplar trees enhanced resistance to the necrotrophic fungus *C. chrysosperma*, indicating that JA-inducible signaling or ethylene (Et)-inducible signaling participates in the adaptive response to stress caused by biotic components in both *miR159a* and wild-type overexpressing lines. Moreover, citrus plants can withstand *Candidatus Liberibacter asiaticus* (CLas) by activating the chain reaction of *miR2977–SAMT*, which also controls phytosome defense by depending on methyl salicylate (MeSA) (Cheng et al. 2023). However, CLas manipulate core host functions by hijacking critical miRNAs, whereas host miRNAs synchronize with the regulation of defense-related genes. In total, 54 miRNAs were found in resistant *Cladosporium fulvum* plants to be involved in the regulation of hormone signaling (Liu et al. 2023a, b), with downregulation of *miR9472*, which can lead to upregulation of *SARD1*, an important positive regulator of *ICS1* induction and SA synthesis. This situation leads to elevated SA levels in the resistant plants. Furthermore, overexpression of *miR530* decreases plant resistance to *Verticillium dahliae*, but it degrades *GhSAP6* mRNA to cause susceptibility. Similarly, silencing *GhSAP6* lowers resistance, mainly due to the downregulation of genes associated with SA, such as *GhNPR1* and *GhPR1* (Hu et al. 2023).

NLR proteins play a core role in the defense response between plants and pathogens, and 22 nt miRNAs are responsible for the modulation of *NLR*s to improve plant responsiveness to numerous pathogens (López-Márquez et al. 2023). *NLR* is the largest family of R proteins in plants (Sun et al. 2020). For example, *miR2118* in rice negatively provides resistance to *Xanthomonas oryzae pv.* This overlaps with the modulation of three nucleotide-binding *NLR* genes, *LOC_Os12g37290.1, LOC_Os01g05600.1*, and *LOC_Os08g42700.1,* which are targeted for inhibition as part of the miRNA-NLR relationship, revealing further intricacy in plant defense against pathogens (Zhu et al. 2023). These findings contribute greatly to our understanding of intricate gene circuits regulating resistance to pathogens in plants, indicating some key potential targets to increase resistance against disease.

B. Abiotic Stress

The detrimental impact of inanimate components on a plant in a specified environment is called abiotic stress. Several physiochemical components adversely affect the maturation of plants, including high temperature, low temperature, high salinity, excess or deficiency of water, ultraviolet radiation, and heavy metals. Several miRNAs have been found to mediate the expression of genes under different abiotic stress conditions and thus play significant roles in various key plant models. A recent machine-learning model demonstrated that *miRNA159*, *miRNA169*, *miRNA393*, and *miRNA396* are vital in responding to harsh conditions such as salinity, drought, cold, and heat stress, respectively (Vakilian 2020). It was also found that miRNAs, such as *spi-miRNA166g-3p_osa*, *spi-miRNA166c-3p*, and *spi-miRNA6300_gma*, respond to heat stress and regulate the induction of heat shock proteins (HSPs) (Ding et al. 2020). Different miRNAs are differentially regulated upon exposure to arsenic, cadmium, and mercury, resulting in altered expression profiles of these miRNAs, which may regulate cancer progression and other diseases. For example, exposure to arsenic has been shown to increase miR-21 expression, which induces angiogenesis and regulates the jasmonic acid (JA) pathway, leading to increased levels of different miRNAs that alter cellular functions. Changes in the expression of some miRNAs related to toxicity response are caused by mercury exposure (Aalami et al. 2022; Talukder et al. 2023).

1.4 Interlinkage Between PDNPs and miRNAs

In recent times, studies have been insisting on finding a solution to the complex associations between miRNAs and PDNPs, derived from plants, which not only highlight potential therapeutic applications but also critical roles for miRNAs in cross-kingdom communication. Also referred to as PENs, PDNPs are small vesicles that result from the formation of exosomes from edible plants. Such nanoparticles carry several bioactive substances like lipids, proteins, secondary metabolites, and most importantly, plant-derived miRNAs. These vesicles carry miRNAs, which protect them from being degraded in the digestive process and facilitate their absorption in the gastrointestinal tract of mammals, allowing them to regulate gene expression in mammalian systems.

This notion of plant miRNAs that regulate cross-kingdom interactions has been a matter of utmost interest. Plant miRNAs, referred to as xenomiRs, have been shown to hybridize with certain mRNAs in mammals, where they influence protein synthesis and hence alter gene expression. For example, there are miRNAs like miR156e, miR159, and miR162 found in different edible plants. These miRNAs resist digestion and conditions of cooking in the gut post-plant-based food intake, therefore increasing in the human fecal matter. Bioinformatics also revealed that such miRNAs may target some human genes related to metabolism and cell signaling, adding to the potential role this system may play in interspecies communication and possible health impacts on the host (Díez-Sainz et al. 2024).

These plant-derived miRNAs are encapsulated in PDNPs to enhance their stability and bioavailability. The lipid bilayer of PDNPs prevents miRNAs from enzyme-mediated breakdown in the alimentary canal and permits their absorption by the intestinal epithelium. When taken up, these miRNAs can be systemically delivered and affect different organs through their expression of genes. This mechanism thus points to the possibility of using PDNPs as natural carriers of plant miRNAs with a view to their possible application in therapy. The therapeutic value of PDNPs is also evident in cancer biology and regenerative tissue science. Rahmati et al. (2023) presented a wide overview of the potential of PDNPs as nanocarriers in cancer treatment in a systematic review. PDNPs are biologically compatible and low immunogenic and thus suitable for application as nanocarriers in drug delivery. Moreover, the natural cargo of plant miRNAs in PDNPs suppresses cancer development by targeting genes that promote tumor growth and metastasis. This dual role of PDNPs as a carrier of therapeutic agents and a source of bioactive miRNAs may be useful for cancer therapy.

Despite these results, the area continues to face various challenges and controversies. Other research has challenged the extent to which dietary plant miRNAs could influence human gene expression. These include questions on the stability, absorption efficiency, and complexity of miRNA-mediated gene regulation. Campbell (2020) has, therefore, continued the debate over the bioavailability and physiological relevance of plant miRNAs to human health. It emphasized further research to enlighten how the plant miRNAs are taken up and functionally integrated within the mammalian system.

The interactions between edible nanoparticles derived from plants and microRNAs are an exciting new field of research that holds great promise for cross-kingdom communication and therapeutic uses. The capacity of these engineered nanoparticles to safeguard and transport plant miRNAs into mammalian cells paves the way for deeper insights into the intricate relationship between diet and gene regulation. Nonetheless, additional research is crucial to thoroughly understand the mechanisms at play in these interactions and to leverage their potential in clinical applications.

1.5 Applications

In order to extract medicinal properties from plants, many researchers have started exploring PDNPs for their remedial potential. PDNPs have great potential in medicine because of their natural lipid bilayer, as this layer makes them highly compatible with the body, easy to absorb, and resistant to breakdown. It also helps them avoid getting detected by the immune system and pass through barriers like the BBB without causing any inflammatory response or tissue degeneration. Their low chance of triggering immune reactions, ability to target specific tissues, and useful physical and chemical traits make them excellent carriers for medicines. PDNPs boost medical efficacy by combining the benefits of bioactive molecules with DNA vectors, proteins, miRNAs, and drugs and transferring them to the area concerned. PDNPs are widely available, affordable, and sustainable. Reflecting on the qualities of their source plants, this section discusses the studies on antioxidant, antibacterial, and anti-inflammatory benefits of PDNPs, showing promise in cancer treatment and tissue healing.

1.5.1 PDNPs as an Antioxidant

Reactive oxygen species (ROS) are products of living organisms' gaseous diffusion and exchange that take part in cellular signaling pathways (Snezhkina et al. 2019). An imbalance between the cellular antioxidant response and ROS production creates oxidative stress (Kim et al. 2023). It has been comprehensively proven that oxidative stress is one of the major factors in the onset and evolution of various diseases, including neurodegeneration, cardiovascular disease, and diabetes (Chen et al. 2020). Research by Kim et al. (2021) reported that EVs derived from Aloe vera peels (A-EVs) showed good compatibility and low toxicity levels with human keratinocytes (HaCaT). The superoxide dismutase (SOD) activity and cellular antioxidant activity (CAA) assays showed that A-EVs have antioxidant properties that help reduce intracellular ROS levels in H_2O_2-treated HaCaT cells in a dose-dependent manner. Besides, the scratch assay showed that A-EVs improved the migration potential of both HaCaT cells and HDF. In addition, A-EVs greatly increased the mRNA expression of *CAT, HO-1*, *Nrf2*, and *SOD* genes in H_2O_2-treated HaCaT cells. In conclusion, A-EVs activated the ROS scavenging systems, and activation of the *Nrf2* gene intensified the wound healing process.

1.5.2 PDNPS as Anti-inflammatory

Inflammation is the body's way of defending itself against external stimuli, and it can play a part in the onset of various chronic or acute diseases. Studies lately emphasize the anti-inflammatory effects of plant-based edible NPs, especially in their relation to intestinal macrophages and stem cells. For instance, ginger-derived NPs (GDNP) have displayed selectivity toward the colon, where they are absorbed by the intestinal lining, an active site in inflammation associated with IBD. GDNPs, when taken orally, have been observed to enhance the survival and growth of IECs through a reduction in pro-inflammatory cytokines, such as IL-1β, and TNF-α, while enhancing anti-inflammatory cytokines, such as IL-10 and IL-22. This dual mechanism implies that GDNPs reduce harmful inflammatory factors but also assist in healing processes. Similarly, grape exosome-like nanoparticles (GELNs) were observed to be capable of migrating through the gut and accumulating selectively in intestinal stem cells, leading to significantly improved organoid formation and supporting intestinal tissue regeneration. Importantly, when anti-inflammatory drugs such as methotrexate (MTX) are incorporated into NPs derived from ginger, the MTX loaded within these NPs has lower levels of toxicity and increased therapeutic potency in mouse models against Dextran Sodium Sulfate (DSS)-induced colitis. Furthermore, NPs produced from GDNP-derived lipids and loaded with siRNA-CD98 were also found to be effective in targeting colon tissues, causing a significant downregulation of CD98 expression, thus opening an avenue for the safe delivery of siRNA for IBD therapy (Del Pozo-Acebo et al. 2021).

Moreover, the synthesis of pro-inflammatory cytokines was prevented by broccoli exosome-like nanoparticles (BELN), as well as enhanced cell regeneration in a colitis mouse model, caused by DSS. Similarly, GFDNs also help reduce colitis by targeting the intestinal macrophages by turning down the infiltration of neutrophils that cause inflammation and guarding the epithelial junction proteins in the colon (Chen et al. 2022). Likewise, Garlic ELNs are highly prevalent in *miR-396e*, which critically plays a role in the metabolic reprogramming of macrophages by targeting 6-phosphofructo-2-kinase/fructose-2, 6-bisphosphatase 3 (PFKFB3), and obesity through macrophage-adipocyte cross-talk (Bian et al. 2022). Other studies indicate that garlic ELNs may play a similar role by improving lipid metabolism in hepatocytes by macrophage-hepatocyte cross-talk (Liu et al. 2023a, b). Mulberry bark-derived ELNs help prevent mouse colitis by inducing the AhR/COPS8 pathway through the heat shock protein HSP8 that comes from Mulberry bark ELNs (Sriwastva et al. 2022).

In conclusion, the emerging evidence surrounding plant-based edible nanoparticles (NPs) underlines their potential for inflammation mitigation and therapeutic efficacy in the treatment of inflammation-related disorders, especially gastrointestinal tract-related disorders. The specific targeting of stem cells and intestinal macrophages by nanoparticles acquired from ginger, grapes, broccoli, and garlic shows that these particles can indeed modulate inflammation effectively. By reducing the pro-inflammatory cytokines and enhancing anti-inflammatory factors, these nanoparticles not only aid in lowering inflammation but also facilitate the healing processes of regenerating tissue. Additionally, incorporating anti-inflammatory medications into these plant-based nanoparticles boosts their effectiveness and minimizes toxicity, paving the way for innovative treatments for a range of diseases.

1.5.3 Maintenance of Gut Flora

The relationship of gut flora is complex with various physiological functions in the body like immune response and nutrient uptake. They play a crucial role in metabolic control, protect the integrity of the barrier, and stimulate mucosal immunity. Focusing on gut microbiota using dietary modulation opens up the potential for personalized nutrition. Although many studies are centered on plant exosomes as they are utilized as delivery systems for drug development targeted to specific cells, only a little research has been done on how gut microbiota are delivered and how they react. Liu et al. (2022), by applying bioinformatics, found that PELN miRNAs from Tartary buckwheat can specifically target functional genes in *Lactobacillus rhamnosus,* and *Escherichia coli* functional biological responses. This targeting enhances the growth of these bacteria. Moreover, the co-fermentation of fecal microorganisms with exosomes of Tartary buckwheat revealed that the composition of gut microbiota can be modulated by these exosomes and thus, enhance the production of short-chain fatty acids.

Recent studies have recognized that PDNP obtained from the strawberry juice of *Fragaria x ananassa acts* as nano shuttles possessing health benefits. These bioactive nano nutraceuticals reduce oxidative stress concentration-dependently in human mesenchymal stromal cells and do not harm cellular viability (Perut et al. 2021). Another study by Chen et al. (2021a) states that TFENs-eco-friendly nano therapeutics from tea flower exosome-like nanoparticles have been used to treat metastatic breast cancer. These TFENs are abundant in different bioactive compounds, such as lipids, polyphenols, and functional proteins, and they also possess the ability to influence the flora. Furthermore, the ability of PDNPs to modulate gut microbes (Fig. 1.2) and maintain gut health—by enhancing intestinal barrier functions and targeting specific microbes—makes them potential therapeutic agents for diabesity. However, this promising potential has not yet taken off because the field is still in its infancy. Therefore, preclinical as well as clinical studies with standardized extraction methods and clearly described physical and functional activities are recommended.

1.5.4 Repair Skin Trauma and Improve Skin Condition

The skin being the largest organ is the initial defense against exogenous factors including ultraviolet radiations, oxidative stress, and physical injuries; however, during exposure to the surrounding ecosystem, skin is susceptible to injury. By

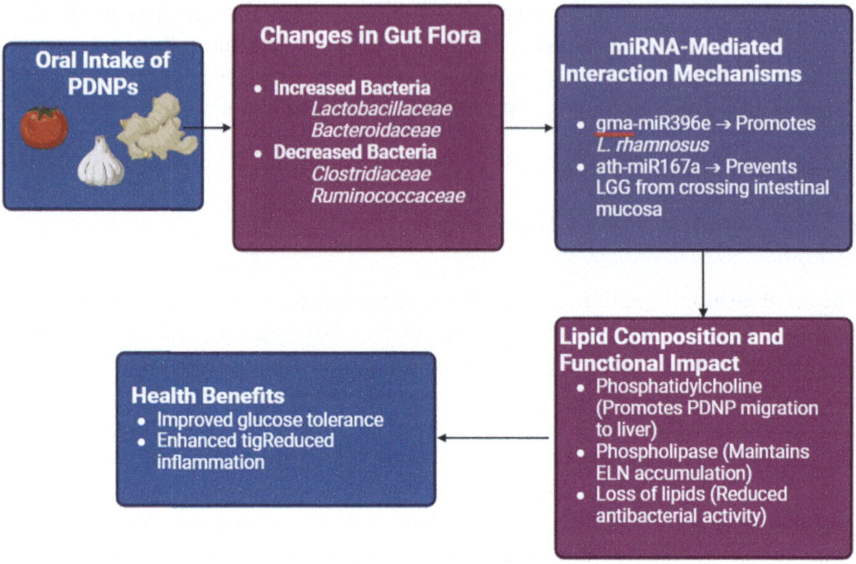

Fig. 1.2 Graphical representation of effect of PDNPs on gut flora

carrying various critical bioactive molecules essential for cellular functions and tissue repair, PDNPs have become crucial players in tissue regeneration. They support cells in their proliferation, migration, and differentiation, which may help regenerate damaged tissues. Beyond their role as delivery systems, plant ELNs have natural immunomodulatory properties, affording them the ability to associate with immune cells such as T cells and macrophages that can moderately modulate the inflammatory response. This dual action thus speeds up the healing process and creates a supportive microenvironment for tissue repair. The understanding of the plant ELNs through continuous progress in their regenerative mechanisms may show significant promise in developing new approaches in regenerative medicine and tissue engineering. The aging-related pigment phenotype in human dermal fibroblasts (HDFs) can be improved by ginseng-derived nanoparticles or by ginseng cell culture supernatants. Their anti-aging activities are further supplemented when there is prolonged exposure (Kumar et al. 2022). Moreover, ginseng-derived nanovesicles not only enhanced the regenerative healing by synergistically combining with C-X-C motif chemokine ligand 12 (CXCL12) in a scaffold but also delivered their miRNAs to BMSCs, which in turn enhanced the neuronal differentiation. These outcomes are due to the regeneration of skin appendages and the stimulation of newly developed tissue, including sweat glands and hair follicles, in a wounded rat model. Moreover, they encourage blood vessel regeneration, leading to a decrease in inflammation (Yang et al. 2022), and recover sensory perception by enhancing regeneration of nerve tissue and neuronal migration. Similarly, the levels of melanin can be reduced by the nanovesicles derived from ginseng, which can surpass the whitening agent melasolv (3,4,5-trimethoxy cinnamic acid

4-hydroxyphenethyl ester) at the same protein concentration in human dermal cells (Cho et al. 2021). In addition, some diseases such as obesity and diabetes can induce cutaneous immune responses. The GDNPs have capabilities that can inhibit inflammation and were proven to prevent high-fat diet-induced insulin resistance in mice and the onset of gray hair and damage by such stress to the skin (Kumar et al. 2022). The mechanisms responsible for these results remain unidentified to date.

1.5.5 PDNPs in Cancer Therapy

Cancer remains one of the most severe threats to human health even today. So far, much scientific research has been carried out, and yet, developing an effective treatment is still at a nascent stage. Among these new promising approaches, PDENs present novel prospects for cancer therapy in addressing reduced durability, bioavailability, and safety often associated with liposomes and exosomes that are derived from non-natural sources.

GDNPs possess therapeutic potential toward kidney and lung cancer cells, with no cytotoxic effects on HDFs. These nanovesicles enhanced endurance by lowering the activity of *caspase 3* and raising the expression level of the anti-apoptotic Bcl-2 (Özkan et al. 2021). Another effective NP is Corn-ENs, which is found to constrain colon cancer (Sasaki et al. 2021). Fingerroot-derived nanoparticles diminished the sustainability of colon cancer cells without affecting any normal colon cells. These particles also elevated regulated cell death, which appears to mediate toxicity through alterations in redox balance and accumulation of ROS. In this study, the authors found that the extract obtained from fingerroot caused a diminished survival of normal cells, and these nanoparticles had no impact on such cells. The findings might therefore be considered a bioactive property of these nanoparticles, where the use of such nanoparticles may revolutionize the present approach to treating cancer (Wongkaewkhiaw et al. 2022). Studies have been reported regarding the anti-carcinogenic effects of nanovesicles derived from lemon on stomach cancer. These nanoparticles were said to stop the S phase of the cell cycle, reduce tumor weight, and increase cell apoptosis due to their suppressing effects on the proliferation of carcinogenic cells (Yang et al. 2020). The effect of GFDNs on melanoma (A375 cells) showed that the cells have increased in the G0/G1 phase and thus reduced the cells in S phase. The cyclin B levels declined, thus causing the cell cycle to stop in the G2/M phase (Stanly et al. 2020).

Cancer therapy based on PDENPs from plants has a great future. It is one of the novel approaches to challenges related to the low bioavailability and toxicity that characterize traditional treatment methods. Selectivity in targeting cancerous cells while sparing healthy tissue makes the potential much more suitable as an effective therapeutic agent. With high stability and low toxicity, PDENPs complement the established treatments and set the way to new treatments. Future research about the mechanisms of such nanoparticles and how they might contribute to new or improved clinical approaches will be an essential aspect for fully capitalizing on their impact in revolutionizing the strategy toward cancer treatment.

1.5.6 PDNP-Related Clinical Trials

In the last couple of decades, the research regarding the understanding and its uses has improved to a good margin in treating various human-related diseases. Many such clinical studies have been registered in the database of ClinicalTrials.gov where the benefits have been highlighted. One such trial (NCT01294072) was exploring the use of extracellular vesicles by isolating them from fruits, which improves the bioavailability of oral curcumin in patients who suffer from colon cancer.

Another key study (NCT04879810) evaluated grape-derived extracellular vesicles as naturally occurring anti-inflammatories for irritable bowel disease. The results indicated that GDEVs can be efficient therapeutic nanoparticles for the disease condition. Another completed trial (NCT01668849) was conducted to demonstrate the anti-inflammatory impacts of grape-derived EVs and showed side-effect improvements like the mitigation of oral mucositis pain in patients with neck and head cancer who receive chemoradiation.

In addition, an exploratory trial with NCT03493984 in which EVs from ginger or aloe would be used as a treatment regimen for polycystic ovarian syndrome; however, it is still undergoing the process with no approval in place. However, the promises of these studies prompt further preclinical and clinical works to better unravel the bioactive properties of the PDEV in humans and more importantly to select the appropriate and safe dosing for therapeutic effects. Overall, PDEVs offer an exciting, new area for drug delivery systems with great prospects for future medicine (Xu et al. 2023). Applications of PDNPs in the delivery of drugs are listed in Table 1.2.

1.6 Challenges

These nanoparticles are very important for agriculture, biomedicine, and environmental science applications. Still, the sector is riddled with regulatory, economic, and technical challenges, which have limited the advancement and industrial scale-up of this field.

1. Optimization

The amount of production of PDNPs requires the control of a number of parameters, for instance, the concentration of the metal precursor, volume of extract, temperature, pH, and duration of the reaction. Even minimal alterations in the values of these parameters can considerably complicate the procedure, as they have a lot of influence on both the morphology and stability of the nanoparticles.

2. Scalability Issues

Scaling up from a laboratory to large-scale industrial production is challenging as there are obstacles in heating and uniform mixing that usually result in an inhomogeneous nanoparticle with a lack of standardization and thereby batch-to-batch inconsistency.

Table 1.2 Applications of various PDNPs for effective delivery of drugs

Source	Agent	Target
Ginger (GDNPs)	Lipids, siRNA	Reduces inflammation (colon, intestinal tissues), suppresses TNF-α and IL-1β, enhances anti-inflammatory cytokines IL-10 and IL-22
Broccoli (BELNs)	Exosome-like nanoparticles	Prevents the synthesis of pro-inflammatory cytokines, promotes cell regeneration (colitis treatment)
Grapefruit	Nanoparticles	Targets intestinal macrophages, reduces neutrophil infiltration, and protects epithelial junction proteins
Garlic (ELNs)	miR-396e	Targets metabolic reprogramming in macrophages, affecting lipid metabolism and inflammation pathways.
Aloe vera (A-EVs)	Antioxidant enzymes (SOD)	Reduces intracellular ROS and improves wound healing through activation of Nrf2, HO-1, and related genes
Blueberry	Bioactive compounds	Cancer treatment and immune modulation (specific targets unclear from the text).
Turmeric	Bioactive compounds	Anti-inflammatory and cancer-related targets (details require more context)
Soybean	miRNAG5781	Targets interleukin-17A as an anti-inflammatory
Strawberry	Anthocyanins, flavanols, folic acid, and vitamin C	Prevents oxygenation stress in BMSCs
Tomato	Proteins	Recognizes the purity of tomato ELNs.
Tea leaves	Proteins	Improves inflammatory bowel disease
Ginseng	Proteins	Associated with the modification of macrophage polarization

3. Biocompatibility and Toxicity Concerns

Though PDNPs are known to be biocompatible, the safety profile varies with the biological activities of phytochemicals used for their synthesis. Because of their potentially toxic nature, many nanoparticles will need to be tested rigorously for safety before they can be used for therapeutic purposes.

4. Regulatory and Ethical Challenges

In regard to regulations, it stands out that green nanoparticle methods-based nanoparticles are not explicitly dealt with. This lack of clarity slows down market approval processes, which in turn affects commercialization efforts in regulated sectors like food and medicine.

5. Environmental Concern

Improper disposal of nanoparticles along with leftover materials can result in environmental pollution. While PDNPs are considered more eco-friendly options, their long-term effects on the environment have not been thoroughly studied.

6. Economic Constraints

High costs of manufacturing inputs and significant infrastructure investment for production at large amounts present major economic constraints. This indicates that PDNP can become rather a luxury for the few in the sector that can afford it (Karnwal et al. 2024; Olawade et al. 2024).

1.7 Future Prospective

The future of PDVLNs looks bright, particularly in therapeutic applications. Researchers need to focus on creating standardized protocols for studying PDVLNs so that our understanding of how they work with both the plant and human systems improves. It will lead to novel therapeutic strategies involving PDVLNs as postbiotics to better health outcomes. With their biocompatibility and possible role as drug carriers, the PDVLNs are expected to be very crucial in clinical application.

1.8 Conclusions

Plant-based edible nanoparticles and miRNAs represent an exciting and very promising area in nutrition, medicine, and biotechnology. Their properties are unique and diverse; they definitely offer considerable potential for health improvement in humans. As a natural method for the delivery of drugs or therapeutic compounds as well as miRNAs, PDNPs ensure targeted treatments with reduced side effects. The role of miRNAs in plant-stress responses is also creating new platforms in understanding basic biological processes through cross-kingdom gene regulation.

Some of the main applications of PDNPs and miRNAs in treatments include antioxidants and anti-inflammatories, gut microbiome modulation, tissue regeneration, and cancer treatment. Some of the advantages of PDNPs:

(a) Compared to their synthetic counterparts, PDNPs possess more advantages in biological compatibility, stability, extension of half-life, and internalization of cells.
(b) They are also cost-effective as they can load a variety of biomolecular substances like nucleic acids, recombinant proteins, and small molecules.
(c) They can be produced on a large scale as plants are natural resources that are present in abundance.
(d) They have excellent tissue penetration and physiochemical stability that help in the exchange of information between cells across various species.
(e) They also act as regulatory factors to exert anti-inflammatory and anti-tumor effects directly at the target cell, as they are target specific.

Even though there are tremendous challenges toward optimizing production, consistency, and scaling for industrial uses of plant-derived nanoparticles, the regulations have to catch up with the peculiar properties of these materials. The unlocking of PDNPs and miRNAs in the future will depend upon interdisciplinarity, but more work needs to be done to establish mechanisms and deliver more efficiently and in a more standardized manner. As our knowledge improves, PDNPs and miRNAs will play a central role in the personalized nutrition area, targeted drug delivery, and innovative therapeutic strategies. As these improvements continue to take place, plant-derived nanoparticles and regulatory molecules could soon transform the way we approach health and disease management.

References

Aalami AH, Hoseinzadeh M, Manesh PH, Sharahi AJ, Aliabadi EK (2022) Carcinogenic effects of heavy metals by inducing dysregulation of microRNAs: a review. Mol Biol Rep 49(12): 12227–12238. https://doi.org/10.1007/s11033-022-07897-x

Bajczyk M, Jarmolowski A, Jozwiak M, Pacak A, Pietrykowska H, Sierocka I, Swida-Barteczka A, Szewc L, Szweykowska-Kulinska Z (2023) Recent insights into plant miRNA biogenesis: multiple layers of miRNA level regulation. Plan Theory 12(2):342. https://doi.org/10.3390/plants12020342

Barrera-Rojas CH, Rocha GHB, Polverari L, Brito DAP, Batista DS, Notini MM, Da Cruz ACF, Morea EGO, Sabatini S, Otoni WC, Nogueira FTS (2019) miR156-targeted SPL10 controls Arabidopsis root meristem activity and root-derived de novo shoot regeneration via cytokinin responses. J Exp Bot 71(3):934–950. https://doi.org/10.1093/jxb/erz475

Bian Y, Li W, Jiang X, Yin F, Yin L, Zhang Y, Guo H, Liu J (2022) Garlic-derived exosomes carrying miR-396e shapes macrophage metabolic reprograming to mitigate the inflammatory response in obese adipose tissue. J Nutr Biochem 113:109249. https://doi.org/10.1016/j.jnutbio.2022.109249

Campbell K (2020) Do the microRNAs we eat affect gene expression? Nature 582(7812):S10–S11. https://doi.org/10.1038/d41586-020-01767-x

Chen H, Liang F, Gu P, Xu B, Xu H, Wang W, Hou J, Xie D, Chai X, An S (2020) Exosomes derived from mesenchymal stem cells repair a Parkinson's disease model by inducing autophagy. Cell Death and Disease 11(4). https://doi.org/10.1038/s41419-020-2473-5

Chen Q, Li Q, Liang Y, Zu M, Chen N, Canup BS, Luo L, Wang C, Zeng L, Xiao B (2021a) Natural exosome-like nanovesicles from edible tea flowers suppress metastatic breast cancer via ROS generation and microbiota modulation. Acta Pharm Sin B 12(2):907–923. https://doi.org/10.1016/j.apsb.2021.08.016

Chen R, Qian Z, Xu X, Zhang C, Niu Y, Wang Z, Sun J, Zhang X, Yu Y (2021b) Exosomes-transmitted miR-7 reverses gefitinib resistance by targeting YAP in non-small-cell lung cancer. Pharmacol Res 165:105442. https://doi.org/10.1016/j.phrs.2021.105442

Chen N, Sun J, Zhu Z, Cribbs AP, Xiao B (2022) Edible plant-derived nanotherapeutics and nanocarriers: recent progress and future directions. Expert Opin Drug Deliv 19(4):409–419. https://doi.org/10.1080/17425247.2022.2053673

Cheng B, Xu L, Bilal MS, Huang Q, Niu D, Ma H, Zhou S, Peng A, Wei G, Chen F, Zeng L, Lin H, Baig A, Wang X, Zou X, Zhao H (2023) Small RNAs contribute to citrus Huanglongbing tolerance by manipulating methyl salicylate signaling and exogenous methyl salicylate primes citrus groves from emerging infection. Plant J 116(5):1309–1324. https://doi.org/10.1111/tpj.16426

Cho E, Choi S, Kim H, Choi E, Lee E, Park P, Ko J, Kim KP, Baek HS (2021) Panax ginseng-derived extracellular vesicles facilitate anti-senescence effects in human skin cells: an eco-friendly and sustainable way to use ginseng substances. Cells 10(3):486. https://doi.org/10.3390/cells10030486

Cui J, Wang X, Li J, Zhu A, Du Y, Zeng W, Guo Y, Di L, Wang R (2023) Immune exosomes loading self-assembled nanomicelles traverse the blood–brain barrier for chemo-immunotherapy against glioblastoma. ACS Nano 17(2):1464–1484. https://doi.org/10.1021/acsnano.2c10219

Del Pozo-Acebo L, De Las Hazas ML, Margollés A, Dávalos A, García-Ruiz A (2021) Eating microRNAs: pharmacological opportunities for cross-kingdom regulation and implications in host gene and gut microbiota modulation. Br J Pharmacol 178(11):2218–2245. https://doi.org/10.1111/bph.15421

Díez-Sainz E, Milagro FI, Aranaz P, Riezu-Boj JI, Lorente-Cebrián S (2024) MicroRNAs from edible plants reach the human gastrointestinal tract and may act as potential regulators of gene expression. J Physiol Biochem 80(3):655–670. https://doi.org/10.1007/s13105-024-01023-0

Ding Y, Huang L, Jiang Q, Zhu C (2020) MicroRNAs as important regulators of heat stress responses in plants. J Agric Food Chem 68(41):11320–11326. https://doi.org/10.1021/acs.jafc.0c03597

Fouracre JP, He J, Chen VJ, Sidoli S, Poethig RS (2021) VAL genes regulate vegetative phase change via miR156-dependent and independent mechanisms. PLoS Genet 17(6):e1009626. https://doi.org/10.1371/journal.pgen.1009626

Gioia, S D., Hossain, M. N., & Conese, M. (2020). Biological properties and therapeutic effects of plant-derived nanovesicles. Medical Review, 50(6), 301–311. https://doi.org/10.1515/med-2020-0160.

Hu G, Wang B, Jia P, Wu P, Lu C, Xu Y, Shi L, Zhang F, Zhong N, Chen A, Wu J (2023) The cotton miR530-SAP6 module activated by systemic acquired resistance mediates plant defense against Verticillium dahliae. Plant Sci 330:111647. https://doi.org/10.1016/j.plantsci.2023.111647

Iravani S, Varma RS (2019) Plant-derived edible nanoparticles and miRNAs: emerging frontier for therapeutics and targeted drug-delivery. ACS Sustain Chem Eng 7(9):8055–8069. https://doi.org/10.1021/acssuschemeng.9b00954

Jain N, Pandey M, Sharma P, Gupta G, Gorain B, Dua K (2022) Recent developments in plant-derived edible nanoparticles as therapeutic nanomedicines. J Food Biochem 46(12). https://doi.org/10.1111/jfbc.14479

Jeena GS, Singh N, Shikha N, Shukla RK (2022) An insight into microRNA biogenesis and its regulatory role in plant secondary metabolism. Plant Cell Rep 41(8):1651–1671. https://doi.org/10.1007/s00299-022-02877-8

Karamanidou T, Tsouknidas A (2021) Plant-derived extracellular vesicles as therapeutic nanocarriers. Int J Mol Sci 23(1):191. https://doi.org/10.3390/ijms23010191

Karnwal A, Jassim AY, Mohammed AA, Sharma V, Al-Tawaha ARMS, Sivanesan I (2024) Nanotechnology for healthcare: plant-derived nanoparticles in disease treatment and regenerative medicine. Pharmaceuticals 17(12):1711. https://doi.org/10.3390/ph17121711

Kim, J., Li, S., Zhang, S., & Wang, J. (2021). Plant-derived exosomelike nanoparticles and their therapeutic activities. Asian Journal of Pharmaceutical Sciences, 17(1), 53–69. https://doi.org/10.1016/j.ajps.2021.05.006.

Kim SQ, Kim K (2022) Emergence of edible plant-derived nanovesicles as functional food components and nanocarriers for therapeutics delivery: potentials in human health and disease. Cells 11(14):2232. https://doi.org/10.3390/cells11142232

Kim MK, Choi YC, Cho SH, Choi JS, Cho YW (2021a) The antioxidant effect of small extracellular vesicles derived from Aloe vera peels for wound healing. Tiss Eng Regen Med 18(4):561–571. https://doi.org/10.1007/s13770-021-00367-8

Kim J, Li S, Zhang S, Wang J (2021b) Plant-derived exosome-like nanoparticles and their therapeutic activities. Asian J Pharm Sci 17(1):53–69. https://doi.org/10.1016/j.ajps.2021.05.006

Kim M, Jang H, Kim W, Kim D, Park JH (2023) Therapeutic applications of plant-derived extracellular vesicles as antioxidants for oxidative stress-related diseases. Antioxidants 12(6):1286. https://doi.org/10.3390/antiox12061286

Kumar A, Sundaram K, Teng Y, Mu J, Sriwastva MK, Zhang L, Hood JL, Yan J, Zhang X, Park JW, Merchant ML, Zhang H (2022) Ginger nanoparticles mediated induction of Foxa2 prevents high-fat diet-induced insulin resistance. Theranostics 12(3):1388–1403. https://doi.org/10.7150/thno.62514

Li T, Lian H, Li H, Xu Y, Zhang H (2021a) HY5 regulates light-responsive transcription of microRNA163 to promote primary root elongation in Arabidopsis seedlings. J Integr Plant Biol 63(8):1437–1450. https://doi.org/10.1111/jipb.13099

Li P, Su T, Zhang D, Wang W, Xin X, Yu Y, Zhao X, Yu S, Zhang F (2021b) Genome-wide analysis of changes in miRNA and target gene expression reveals key roles in heterosis for Chinese cabbage biomass. Horticulture Research 8(1). https://doi.org/10.1038/s41438-021-00474-6

Li L, Zhang L, Montgomery KC, Jiang L, Lyon CJ, Hu TY (2022) Advanced technologies for molecular diagnosis of cancer: state of pre-clinical tumor-derived exosome liquid biopsies. Materials Today Bio 18:100538. https://doi.org/10.1016/j.mtbio.2022.100538

Liu Y, Wu S, Koo Y, Yang A, Dai Y, Khant H, Osman SR, Chowdhury M, Wei H, Li Y, Court K, Hwang E, Wen Y, Dasari SK, Nguyen M, Tang EC, Chehab EW, De Val N, Braam J, Sood AK (2020) Characterization of and isolation methods for plant leaf nanovesicles and small extracellular vesicles. Nanomed Nanotechnol Biol Med 29:102271. https://doi.org/10.1016/j.nano.2020.102271

Liu Y, Tan M, Zhu W, Cao Y, Peng L, Yan Z, Zhao G (2022) In vitro effects of tartary buckwheat-derived nanovesicles on gut microbiota. J Agric Food Chem 70(8):2616–2629. https://doi.org/10.1021/acs.jafc.1c07658

Liu J, Li W, Bian Y, Jiang X, Zhu F, Yin F, Yin L, Song X, Guo H, Liu J (2023a) Garlic-derived exosomes regulate PFKFB3 expression to relieve liver dysfunction in high-fat diet-fed mice via macrophage-hepatocyte crosstalk. Phytomedicine 112:154679. https://doi.org/10.1016/j.phymed.2023.154679

Liu G, Liu F, Zhang D, Zhao T, Yang H, Jiang J, Li J, Zhang H, Xu X (2023b) Integrating omics reveals that miRNA-guided genetic regulation on plant hormone level and defense response pathways shape resistance to Cladosporium fulvum in the tomato Cf-10-gene-carrying line. Front Genet 14. https://doi.org/10.3389/fgene.2023.1158631

López-Márquez D, Del-Espino Á, Ruiz-Albert J, Bejarano ER, Brodersen P, Beuzón CR (2023) Regulation of plant immunity via small RNA-mediated control of NLR expression. J Exp Bot 74(19):6052–6068. https://doi.org/10.1093/jxb/erad268

Nandan A, Anil AM, Nagaraj G, Peter D, Tripathi AD, Paul V (2024) Plant-derived edible nanoparticles and their therapeutic utilities. In Comprehensive Analytical Chemistry. https://doi.org/10.1016/bs.coac.2024.05.005

Olawade DB, Wada OZ, Ige AO (2024) Advances and recent trends in plant-based materials and edible films: a mini-review. Front Chem 12. https://doi.org/10.3389/fchem.2024.1441650

Özkan İ, Koçak P, Yıldırım M, Ünsal N, Yılmaz H, Telci D, Şahin F (2021) Garlic (Allium sativum)-derived SEVs inhibit cancer cell proliferation and induce caspase mediated apoptosis. Sci Rep 11(1). https://doi.org/10.1038/s41598-021-93876-4

Perut F, Roncuzzi L, Avnet S, Massa A, Zini N, Sabbadini S, Giampieri F, Mezzetti B, Baldini N (2021) Strawberry-derived exosome-like nanoparticles prevent oxidative stress in human mesenchymal stromal cells. Biomol Ther 11(1):87. https://doi.org/10.3390/biom11010087

Rahmati S, Karimi H, Alizadeh M, Khazaei AH, Paiva-Santos AC, Rezakhani L, Sharifi E (2023) Prospects of plant-derived exosome-like nanocarriers in oncology and tissue engineering. Hum Cell 37(1):121–138. https://doi.org/10.1007/s13577-023-00994-4

Saiyed AN, Vasavada AR, Johar SRK (2022) Recent trends in miRNA therapeutics and the application of plant miRNA for prevention and treatment of human diseases. Future J Pharm Sci 8(1). https://doi.org/10.1186/s43094-022-00413-9

Sasaki D, Kusamori K, Takayama Y, Itakura S, Todo H, Nishikawa M (2021) Development of nanoparticles derived from corn as mass producible bionanoparticles with anticancer activity. Sci Rep 11(1). https://doi.org/10.1038/s41598-021-02241-y

Snezhkina AV, Kudryavtseva AV, Kardymon OL, Savvateeva MV, Melnikova NV, Krasnov GS, Dmitriev AA (2019) ROS generation and antioxidant defense systems in normal and malignant cells. Oxidative Med Cell Longev 2019:1–17. https://doi.org/10.1155/2019/6175804

Sriwastva MK, Deng Z, Wang B, Teng Y, Kumar A, Sundaram K, Mu J, Lei C, Dryden GW, Xu F, Zhang L, Yan J, Zhang X, Park JW, Merchant ML, Egilmez NK, Zhang H (2022) Exosome-like nanoparticles from mulberry bark prevent DSS-induced colitis via the AhR/COPS8 pathway. EMBO Rep 23(3). https://doi.org/10.15252/embr.202153365

Stanly C, Alfieri M, Ambrosone A, Leone A, Fiume I, Pocsfalvi G (2020) Grapefruit-derived micro and nanovesicles show distinct metabolome profiles and anticancer activities in the A375 human melanoma cell line. Cells 9(12):2722. https://doi.org/10.3390/cells9122722

Stotz HU, Brotherton D, Inal J (2021) Communication is key: extracellular vesicles as mediators of infection and defence during host–microbe interactions in animals and plants. FEMS Microbiol Rev 46(1). https://doi.org/10.1093/femsre/fuab044

Sun Y, Zhu Y, Balint-Kurti PJ, Wang G (2020) Fine-tuning immunity: players and regulators for plant NLRs. Trends Plant Sci 25(7):695–713. https://doi.org/10.1016/j.tplants.2020.02.008

Talukder P, Saha A, Roy S, Ghosh G, Roy DD, Barua S (2023) Role of mi RNA in phytoremediation of heavy metals and metal induced stress alleviation. Appl Biochem Biotechnol 195(9):5712–5729. https://doi.org/10.1007/s12010-023-04599-3

Thakur A, Johnson A, Jacobs E, Zhang K, Chen J, Wei Z, Lian Q, Chen HJ (2022) Energy sources for exosome communication in a cancer microenvironment. Cancers 14(7):1698. https://doi.org/10.3390/cancers14071698

Vakilian KA (2020) Machine learning improves our knowledge about miRNA functions towards plant abiotic stresses. Sci Rep 10(1). https://doi.org/10.1038/s41598-020-59981-6

Woith E, Guerriero G, Hausman J, Renaut J, Leclercq CC, Weise C, Legay S, Weng A, Melzig MF (2021) Plant extracellular vesicles and nanovesicles: focus on secondary metabolites, proteins and lipids with perspectives on their potential and sources. Int J Mol Sci 22(7):3719. https://doi.org/10.3390/ijms22073719

Wongkaewkhiaw S, Wongrakpanich A, Krobthong S, Saengsawang W, Chairoungdua A, Boonmuen N (2022) Induction of apoptosis in human colorectal cancer cells by nanovesicles from fingerroot (Boesenbergia rotunda (L.) Mansf.). PLoS One 17(4):e0266044. https://doi.org/10.1371/journal.pone.0266044

Xu, Z., Xu, Y., Zhang, K. et al. (2023) Plant-derived extracellular vesicles (PDEVs) in nanomedicine for human disease and therapeutic modalities. J Nanobiotechnol 21(114). https://doi.org/10.1186/s12951-023-01858-7

Yadav A, Kumar S, Verma R, Lata C, Sanyal I, Rai SP (2021) microRNA 166: an evolutionarily conserved stress biomarker in land plants targeting HD-ZIP family. Physiol Mol Biol Plants 27(11):2471–2485. https://doi.org/10.1007/s12298-021-01096-x

Yang M, Liu X, Luo Q, Xu L, Chen F (2020) An efficient method to isolate lemon derived extracellular vesicles for gastric cancer therapy. J Nanobiotechnol 18(1). https://doi.org/10.1186/s12951-020-00656-9

Yang S, Lu S, Ren L, Bian S, Zhao D, Liu M, Wang J (2022) Ginseng-derived nanoparticles induce skin cell proliferation and promote wound healing. J Ginseng Res 47(1):133–143. https://doi.org/10.1016/j.jgr.2022.07.005

Yang X, Fu T, Yu R, Zhang L, Yang Y, Xiao D, Wang Y, Wang Y, Wang Y (2023) miR159a modulates poplar resistance against different fungi and bacteria. Plant Physiol Biochem 201: 107899. https://doi.org/10.1016/j.plaphy.2023.107899

Zhang, Y., Cai, Z., Shen, Y., Lu, Q., Gao, W.,Zhong, X., Yao, K., Yuan, J., & Liu, H. (2021). Hydrogel-load exosomes derived from dendritic cells improve cardiac function via Treg cells and the polarization of macrophages following myocardial infarction. Journal of Nanobiotechnology, 19(1). https://doi.org/10.1186/s12951-021-01016-x

Zhu X, Kuang Y, Chen Y, Shi J, Cao Y, Hu J, Yu C, Yang F, Tian F, Chen H (2023) miR2118 negatively regulates bacterial blight resistance through targeting several disease resistance genes in rice. Plan Theory 12(22):3815. https://doi.org/10.3390/plants12223815

Chapter 2
Nanotechnology in Crop Protection

Poonam Jyoti and Sudesh Kumar Yadav

Abstract Adverse climate changes and increasing population are negatively impacting agriculture, which is leading to significant crop losses. The excessive use of pesticides and chemical fertilizers to increase plant productivity is further exacerbating the climate problems. It is important to adopt sustainable practices not only to meet future food demands but also to mitigate the harmful effects of chemicals like fertilizers and pesticides. Nanotechnology plays an important role in this context, as this technique helps to reduce pesticide use by enhancing its effectiveness against pests. Though this approach shows promise, more studies are needed to understand the long-term impact of nanoparticles (NPs) on the environment and human health. This book chapter highlights the benefits of nanoparticles in crop protection, as well as the challenges to address in the future.

Keywords Nanoparticles · Nanofertilizers · Nanopesticides · Crop protection · Nanosensors

2.1 Introduction

Increased population and adverse climate effects are the biggest challenges the world is facing today. These factors place immense pressure on agriculture; every year there is huge crop loss because of environmental effects, pests, and diseases. Global

P. Jyoti
Biotechnology Division, CSIR-Institute of Himalayan Bioresource Technology, Palampur, Himachal Pradesh, India
e-mail: poonamjyoti@ihbt.res.in

S. K. Yadav (✉)
Biotechnology Division, CSIR-Institute of Himalayan Bioresource Technology, Palampur, Himachal Pradesh, India

Academy of Scientific and Innovative Research (AcSIR), Ghaziabad, India

Center of Innovative and Applied Bioprocessing (CIAB), Mohali, India
e-mail: sudesh@ciab1.res.in; director@ihbt.res.in

S. K. Yadav, A. Kumari (eds.), *Nanobiotechnology in Agri-food Sector*, Sustainability Sciences in Asia and Africa, https://doi.org/10.1007/978-981-96-7756-6_2

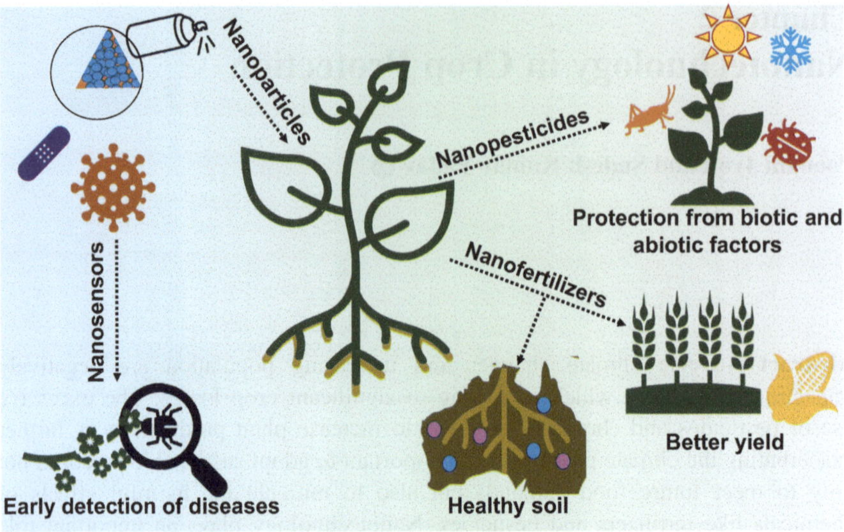

Fig. 2.1 Various applications of nanoparticles for crop protection. Nanoparticles help in the delivery of pesticides and nutrients effectively with a targeted delivery approach, while nanosensors help in the early detection of plant diseases, infections, and concentrations of nutrients and heavy metals or pathogens in the rhizosphere

crop loss because of pests is between 20 and 40%, and around 34% of crop loss happens because of weeds, while diseases and insects lead to a cost of around $220 billion and $70 billion, respectively, to the global economy (Oerke 2006; Gula 2023; Junaid and Gokce 2024). Reports also suggest the bulk use of pesticides (around two million tons) where herbicides, insecticides, and fungicides are approximately 47.5%, 29.5%, and 17.5%, respectively (Sharma et al. 2019). The use of chemical fertilizers has also increased, and the amount reported was 187.92 million tons by 2022, with nitrogen being 58%, phosphate being 23%, and potash being 18% (Statista 2024). The biggest problem with these chemicals is they often wash away into the water sources and contaminate the natural sources of water and soil. These chemicals also harm the beneficial organisms and even possess threat to humans (Pathak et al. 2022). To reduce these negative impacts, it is crucial to develop and implement more efficient methods for delivering these inputs directly to plants and reduce the spillage.

Nanotechnology is one such emerging tool that has the potential to resolve this problem by increasing the efficiency of fertilizers and pesticides through direct delivery to the plants, which also reduces wastage and environmental harm (Jakhar et al. 2022). Nanotechnology works on a nanoscale between 1 and 100 nm where material possesses increased strength and unique chemical and biological properties (Joudeh and Linke 2022) that can be leveraged for various applications in agriculture, such as seed germination, fertilizers for enhanced growth, biosensors, bioremediation, and crop protection (Fig. 2.1) (Shang et al. 2019; Yadav et al. 2023).

Nanotechnology enables highly efficient tools like nanofertilizers (for controlled release of nutrients) and nanopesticides (controlled delivery of encapsulated material and enhanced effectivity against pests) for crop protection (Channab et al. 2024; Batool et al. 2024). Nanotechnology also helps in the early detection of diseases through nanosensors, while some other nanoparticles can be used for increasing resistance in plants against various biotic and abiotic stresses (Giraldo et al. 2019; Ijaz et al. 2023; Imran and Ortas 2024).

Nanotechnology is also helpful in achieving sustainable agriculture by improving water use efficiency, reducing chemical fertilizers and pesticides, and enhancing plant productivity through precision farming. Overall, nanotechnology promises a significant role in crop improvement; here we have listed a few applications of nanotechnology in crop protection along with the challenges and future prospects to achieve food security.

2.2 Nanotechnology Helps Mitigate Biotic and Abiotic Factors Stress in Plants

Nanotechnology is widely explored against biotic and abiotic stresses, and it offers innovative solutions to overcome these stress conditions in the plants. Nanoparticles (NPs), due to their unique size and properties, help plants to generate efficient responses by changing the levels of stress hormones, reactive oxygen species, and other markers to stress conditions like disease and harsh environmental conditions. They can help plants grow better by improving nutrient uptake and also protect plants by enhancing their immunity.

2.2.1 Nanoparticles for Disease Control

Nanotechnology and the use of NPs are excellent approaches to overcoming the biotic stress in plants caused by various pathogens such as bacteria, fungi, and viruses. Many metal-based NPs, for example, silver (Ag), copper (Cu), zinc oxide (ZnO), and titanium oxide (TiO_2), have antimicrobial and antifungal properties. Ag NPs are mostly used against bacterial infections, while Zn NPs play an important role against fungal infections to prevent plants from infections (Dhaka et al. 2023; Jali et al. 2024; Kalia et al. 2020; Lalabadi et al. 2019). Similarly, silicon dioxide (SiO_2)-based NPs are responsible for inducing systemic resistance in plants as they produce pathogenesis-related proteins in plants. These NPs activate the plant immune system, which helps plants to better respond to pathogens. NPs such as Ag, Cu, and ZnO protect plants by interfering with the biofilm formation in the bacterial pathogens and help make the pathogens more susceptible to disease treatments (Afrasiabi and Partoazar 2024; Li et al. 2024).

There are several beneficial microbes, such as rhizobacteria and their active compounds, e.g., primary or secondary metabolites, which have antimicrobial activities, but due to their less stability and efficacy, their use is limited. Nanoformulations or nanocapsulations of such active compounds help overcome these limitations and can be used directly by plants to protect them from diseases or biotic stress (Riseh et al. 2022). For example, chitosan NPs are reported to help plants by stimulating growth and also help them cope with stress by inducing systemic resistance. Chitosan stimulates the production of proteins such as phytoalexins and proteinase inhibitors and the regulation of second messengers and phytohormones, thereby activating the plant's defence mechanisms (Tortella et al. 2023; Rana et al. 2025). Chitosan NPs are also effective against pathogens like *Fusarium andiyazi and F.* oxysporum (Sravani et al. 2023; Tortella et al. 2023). Some examples of microbial nanoencapsulation include *Bacillus velezensis* combined with SiO2 nanoparticles to combat *Phytophthora drechsleri, B. subtilis* encapsulated in sodium alginate to target *Ralstonia solanacearum*, *Pseudomonas fluorescens* nanoparticles for controlling *Ralstonia solanacearum*, and *Streptomyces fulvissimus* Uts22 against *Pythium aphanidermatum* (Saberi et al. 2020; Saberi and Moradi 2021; Moradi et al. 2022; Rana et al. 2025).

2.2.2 Nanotechnology for Abiotic Stress Control

Nanoparticles are also important against abiotic stresses such as drought, salinity, and heavy metals (El-Saadony et al. 2022; Soni et al. 2024; Zhou et al. 2020). Metal NPs and metal oxide-based NPs enhance tolerance through improved water retention, antioxidant defense, photosynthesis, and osmotic regulation in plants and help plants to grow better in adverse environmental conditions (El-Saadony et al. 2022; Soni et al. 2024). NPs also facilitate the uptake of nutrients and scavenging of reactive oxygen species (ROS), which accumulate under stress (El-Saadony et al. 2022). NPs modulate the plant's physiological processes and enhance resilience in harsh environmental conditions.

Taran and co-workers (2017) assessed the impact of Cu and Zn NPs under drought conditions on wheat seedlings. They have observed a positive effect of these NPs on pro-oxidative/antioxidative balance and an increase in carotenoid content along with a change in chlorophyll (Chl *a* to Chl *b*) ratio, which is an adaptive response of plants against drought (Taran et al. 2017). In another study, the effect of iron (Fe), Cu, cobalt (Co), and ZnO nanoparticles was evaluated on soybean plants under drought conditions. Researchers observed the positive effect of these NPs, especially Fe NPs, by improving physiological traits, such as relative water content, drought tolerance index, and biomass reduction rate (Linh et al. 2020).

Semida and co-researchers conducted a field experiment to evaluate the effect of ZnO NPs on eggplant under drought conditions and observed a 12.2% and 22.6% increase in fruit yield when supplemented with 50 and 100 ppm of ZnO NPs, respectively (Semida et al. 2021). ZnO NPs are reported to help alleviate the salt stress in cotton plants. Plants sprayed with nanoparticles resulted in more root biomass, leaf area, and seedling height under salt stress. These NPs significantly reduce the hydrogen peroxide (H_2O_2) and oxygen (O_2) levels, increase the activity of antioxidant enzymes, remove the excess sodium ion (Na^+) toxic effects, and maintain the sodium/potassium (Na^+/K^+) ion ratio to mitigate salt stress (Qian et al. 2024). Zafar and co-workers have used the green synthesis method for selenium nanoparticles using lemon extracts, and they used different concentrations (0.01%, 0.05%, 0.1%) on wheat plants. They observed significantly better yield, biomass, and chlorophyll content and a significant reduction in H_2O_2 and malondialdehyde content with 0.1% selenium NPs (Zafar et al. 2024).

NPs are also reported to mitigate the heavy metal stress in plants; for example, ZnO NPs were tested against heavy metals lead (Pb) and cadmium (Cd). These NPs reduced malondialdehyde content and increased the antioxidative enzymes, total protein content, and photosynthetic pigment to help plants overcome heavy metal stress (Venkatachalam et al. 2017). These studies explain various applications of nanoparticles to help plants overcome stress conditions and produce better growth and yield.

2.3 Nanotechnology for Pesticides

Nanotechnology has paved the way for the development of nanoformulations of traditional pesticides, which provide various advantages like improved solubility, stability, and bioavailability of active ingredients (Kumar et al. 2019; Anjaneyulu et al. 2024). Nanotechnology has also revolutionized the delivery systems for effective release of active molecules in a controlled manner; for example, nanoencapsulated pesticides are useful for controlled release of active molecules for extended duration without degrading them (Kumar et al. 2019). Nanopesticides are mainly of three types: organic, which includes polymers; inorganic, which is metal based; and third includes both organic and inorganic nanoparticles. Most synthesized nanopesticides typically encompass nanocapsules (which are minute capsules of pesticides), nanoemulsions (oil-water or water-oil emulsions of nano-sized pesticides), nanogels (gel-like nanostructures that can carry pesticides), nanospheres (spherical nanostructures capable of carrying pesticides), and metal and metal oxide nanoparticles (Kapeleka and Mwema 2024). A few applications of nanotechnology for crop protection from pests are as follows.

2.3.1 Targeted Application

NPs can ensure that pesticides are applied precisely to the target where they are needed. This reduces the use of pesticides and also protects the non-target organisms. For example, carrier nanoparticles like liposomes (lipid vesicles that can attach to the plant cells and deliver the pesticide inside the cell), polymers (like chitosan and alginate), or nanocapsules that encapsulate the pesticide molecule and deliver it to the target without harming other plants or insects (Rathore et al. 2024). Owing to their size and unique properties, NPs can penetrate easily compared to conventional pesticides, ensuring better efficiency. In addition to the pest-specific, there are other NPs that release the pesticide molecules under the influence of certain temperatures, pH, enzymes, or different environmental conditions (Camara et al. 2019). For example, an alginate-based system for cypermethrin (insecticide) is pH specific; at low pH, there is cross-linking of alginate polymer with calcium ions, which inhibits the release of the active molecules, but as there is an increase of pH, it releases the insecticide compound by the breakdown of polymer links (Patel et al. 2018). Another pH-responsive NP is λ-cyhalothrin with chitosan, which is covered with polyurethane, which releases the pesticide in an alkaline pH (Hou et al. 2023). These NPs can be designed for the combined delivery of plant growth promoters along with the pesticides, which will enhance the overall efficiency of plant productivity, or a single formulation for antifungals and pesticides. Pei and co-workers have developed a similar nanosystem that is effective against *Botrytis cinerea* fungus and aphids. This co-delivery system is based on mesoporous silica and a metal framework and is pH-responsive, containing azoxystrobin fungicide and insecticide thiamethoxam (Pei et al. 2025).

2.3.2 Controlled Release of Pesticides

Control over the release of active ingredients or pesticides through nanopesticides is an innovative approach that enhances the efficacy of pesticides and also helps protect the environment. Encapsulation of the active material in nanopolymers allows their targeted release, and that too in a controlled, slow manner, which reduces the need for frequent applications and also reduces the harmful environmental effects. In nanopesticides, noncovalent interactions such as hydrophobic and electrostatic interactions occur between pesticides and carrier molecules, which significantly influence the controlled release of pesticides (Zhang et al. 2024a, b). These interactions help in the efficiency of nanopesticides by increasing their stability, sensitivity, and the amount of active ingredient loading capacity. For pesticides such as carbaryl and λ-cyhalothrin, hydroxyl and amino groups containing carriers like chitosan and poly(vinyl alcohol) are used where hydrogen bonding is used for holding the pesticide molecules (Wu et al. 2023; Xu et al. 2023; Zhang et al. 2024a, b). For the hydrophobic pesticides, carbon-based NPs encapsulated in chitosan are used for controlled release, which is triggered by low pH and high temperature (Dong et al. 2021). Shen and co-workers have developed a nano-gel-based nanopesticide that is responsive to temperature for the release of active ingredients. They used λ-cyhalothrin pesticide, loaded it on Pickering emulsion, and finally encapsulated it in stable poly(*N*-isopropylacrylamide) gel. This smart delivery system was effective against *Plutella xylostella*, *Aphis gossypii*, and *Pieris rapae* (Shen et al. 2022).

Liu and researchers have developed metal-organic PCN-777 nanoparticles for the smart delivery of large-sized insecticide avermectin. They have used sodium lignosulfonate and chitosan for the controlled delivery of the insecticide, which is triggered by various factors like pH, laccase, phosphate, and phytic acid. This system was tested against *Spodoptera litura* and found to be highly effective with good efficacy (Liu et al. 2023). These studies show that the controlled release of pesticides made them more effective and reduced chemical waste. Additionally, the use of biocompatible and biodegradable materials in NP formulation, such as proteins (zein) and polysaccharides (chitosan), promotes eco-friendly solutions for agriculture (Zhao et al. 2022a, b).

2.3.3 Enhanced Pesticide Activity and Less Effect on Non-target organisms

Due to the unique size and properties of nanopesticides, they penetrate easily inside the insects or pests and release the active ingredient in response to various factors, for example, alkaline pH in the gut of insects, which makes them more effective (Zhao et al. 2022a, b). NPs can be engineered to improve their solubility, stability, and bioavailability; also, encapsulation protects them from degradation because of light, heat, cold, UV, or other environmental factors (Kumar et al. 2019). NPs also help in targeted delivery, as there is a lot of glutathione present in the insect cells, which can break the disulfide bonds. This can be used for the effective release of pesticide or insecticide. Researchers have developed nanocarriers with disulfide bonds, and when these nanocarriers with active ingredients come in contact with the insects, glutathione breaks the disulfide bonds and leads to the release of active ingredients (Liang et al. 2020). Zhao and co-workers have developed an eco-friendly nanopesticide (Fig. 2.2) carrier by using chitosan oligosaccharides with zein protein and bridging them with a disulfide bond from 3,3′-dithiodipropionic acid for the avermectin insecticide delivery. This system was effective against insects; chitosan increases the water dispersibility of the active ingredient and also helps overcome light sensitivity through encapsulation (Zhao et al. 2022a, b). Jiang and co-workers have developed a nanocarrier for the delivery of dinotefuran (pesticide); with the nanocarrier, the pesticide can easily spread on the plant leaves and be more accessible to the aphids. This system is efficient in killing the aphids without having harmful effects on oilseed grapes (Jiang et al. 2022).

Yan and researchers have focused on reducing the toxic effects of pesticides on non-target organisms. They have developed a carbon dots-doped mesoporous silica-based nanopesticide system where mesoporous silica was used as a carrier and thiacloprid was the pesticide. This system was effective against *Aphis gossypii,* and the acute toxicity to other non-target *Chrysoperla pallens* larvae and 4 T1 cells was significantly reduced compared to the conventional use of thiacloprid (Yan et al. 2024). In a similar study, Deng and researchers have developed a

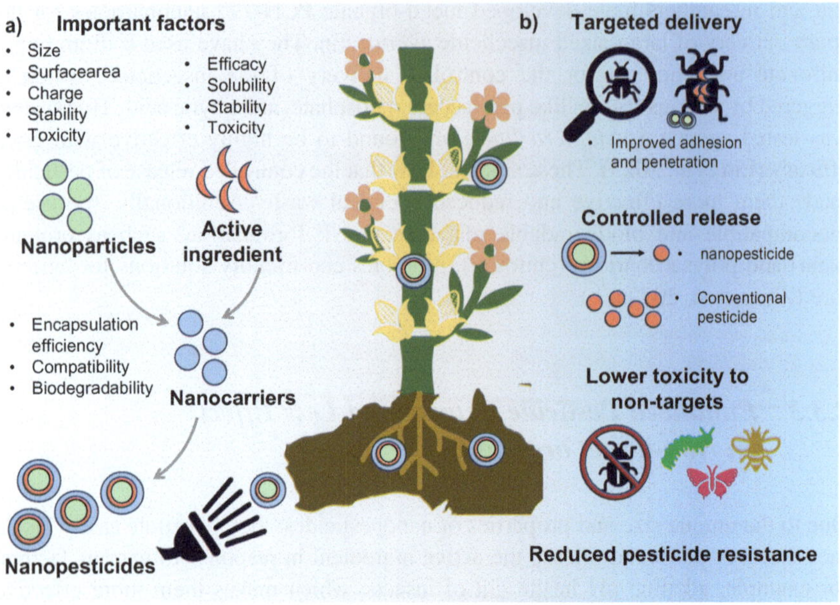

Fig. 2.2 Nanopesticides and their benefits for crop protection along with the key considerations in nanoparticle engineering for nanopesticide formation. (**a**) represent the factors to consider for nanoparticles, active ingredients (pesticides), and nanocarriers in nanopesticide formulation. (**b**) represent the various benefits of nanopesticides in crop protection

nanopesticide delivery system against the pest *Megalurothrips usitatus*, which majorly affects cowpea crops. They have used amino acids γ-aminobutyric acid, aspartic acid, and glutamic acid (GABA, ASP, and GLU)-based modified polysuccinimide to increase the distribution and efficiency of thiamethoxam pesticide. Studies have shown that this nanocarrier system is absorbed by the plant roots and transported to the leaves through the vascular system. This system is effective against *Megalurothrips usitatus* while reducing its toxicity to non-target species, like worker bees, and improving plant growth (Deng et al. 2024).

2.4 Nanocoatings for Crop Protection

Nanocoatings provide potential benefits to plants by improving their health and productivity. Nanocoatings are typically NPs such as metals, metal oxides, carbon, and polymer-based materials that are applied to plants such as seeds, leaves, stems, and roots, where they interact with the plant at the cellular level. These nanocoatings provide various benefits to the plant, mainly helping in water management, providing nutrients, and protecting against pests and diseases (Singh et al. 2024). These can also be utilized as barriers from cold stress (Ag nanoparticles), heat stress, or UV radiations (TiO_2 NPs) (Singh et al. 2024).

2.4.1 *Enhanced Disease Protection*

Nanocoatings, when applied on the seeds or plants, act as barriers against fungi, bacteria, and viruses and provide protection against them. Metal and metal oxide NPs, when applied to the seeds or seedlings, prevent them from pathogens like *Fusarium* or soil-borne pathogens and help in the better survival of plants from the young stage (Zhang et al. 2024a, b). Nanocoating of ZnO NPs on bamboo plants is reported to provide resistance to the plants against *Aspergillus niger* and *Penicillium citrinum* (Li et al. 2017). Lin and co-workers have utilized iron oxide (Fe_2O_3) NPs in aromatic rice varieties and observed improvement in seed germination and antioxidant activity under straw stress with the help of nanoparticles (Lin et al. 2024). Nanocoatings are useful against heavy metal stress; silicon nanocoating prevents the uptake of heavy metal through rice roots (Pang et al. 2024). Feng and co-workers have utilized silica nanocoat on pea (*Pisum sativum*) root tips to provide a barrier and protection against aluminum toxicity (Feng et al. 2019). Nanocoatings are also reported to prevent fruits and vegetables from diseases postharvest or during storage conditions. For example, chitosan-based NPs of mango peel and seed extracts are effective against various fungi such as *Colletotrichum asianum*, *Talaromyces variabilis*, *Fusarium solani,* and *Neofusicoccum oculatum* (Istúriz et al. 2022). Nanocoating of chitosan with thyme essential oil is effective against *Pectobacterium carotovorum* and increases the self life of green bell peppers (Correa et al. 2021).

2.4.2 *Improved Nutrient and Water Retention*

Beneficial microbes can be a great alternative to chemical fertilizers, but they are sensitive to the harsh environmental conditions. Nanocoatings of these microbes can be the solution to this problem. For instance, *Pseudomonas chlororaphis* is a beneficial nitrogen-fixing bacterium, and metal-phenolic network coatings can protect the bacteria from environmental stress conditions, and their application on seeds helps in enhanced germination (Burke et al. 2023). NPs help plants in water retention in drought stress. Graphene oxide was tested on soybean plants under drought stress and observed 127% and 128% higher relative water content in stems and leaves, respectively. This NP treatment has also improved root growth, increased antioxidant enzymes, reduced oxidative stress markers, and increased drought-related hormones such as jasmonic acid, salicylic acid, and abscisic acid in the treated plants (Zhao et al. 2022a, b). Drought stress inhibits the uptake of nutrients like zinc, nitrogen, and phosphorus and also reduces the panicle initiation in wheat.

Dimkpa and co-workers have used ZnO NPs with a urea coating and observed a 51% increase in grain yield (Dimkpa et al. 2020).

2.5 Nanosensors for Crop Protection

Nanosensors are valuable tools in crop protection, offering precise and real-time analysis of various pathogens, disease progression, pests, nutrient deficiencies, environmental conditions, and plant-soil health (Shaw and Honeychurch 2022). For example, nanosensors can provide information about soil conditions like pH, moisture conditions, temperature, etc., which helps in predicting the abiotic stress, if any (Voke et al. 2021). Lu and co-workers have developed a multimodal nanosensor for optical measurement of temperature and humidity using $ZnIn_2S_4$ nanosheets (Lu et al. 2020). Nanosensors are also useful in providing information about harmful pathogens like fungi or bacteria; these can also detect the early symptoms and help overcome the disease and its widespread with proper management (Zia et al. 2022). To detect stress, ROS are the molecules to focus on, and among them, H_2O_2 is considered to be ideal owing to its long half-life and easy transport across the membrane (Lew et al. 2020). Nanosensors like near-infrared carbon nanotubes help predict the stress by analyzing the H_2O_2 (Wu et al. 2020). Similar single-chirality carbon NPs help in the recognition of signaling molecules like nitric oxide and H_2O_2 and form optical signals (Giraldo et al. 2015) for in vivo studies. Lew and co-researchers utilized H_2O_2-based single-walled carbon nanotube nanoparticles to access the H_2O_2 concentration in different plants such as lettuce, sorrel, arabidopsis, spinach, and strawberry under wounding conditions (Lew et al. 2020).

Nanosensors are also applicable to track the presence of various pests by detecting pheromones or unique markers; for example, polyaniline and silver and polyaniline and graphene oxide are explored in the real environment to detect the semiochemicals secreted by southern green stink bug (Dias et al. 2024) and cantilever nanosensors for detecting stink bug (*Euschistus heros*) (Martinazzo et al. 2022), while other rhizosphere parameters like root exudates, carbon flux, quorum sensing, etc., also allow the creation of targeted formulations against different organisms (Dimri et al. 2020). Fluorescent silica nanoparticles and gold NPs-based nanosensors are reported to check infections caused by *Xanthomonas axonopodis* pv. *vesicatoria* and *Pseudomonas syringae*, respectively (Yao et al. 2009; Cebula et al. 2019).

Nanosensors are also significantly important to check the number of pesticides or insecticides present in the rhizosphere or various insects and pests present around the plant; this helps in decision-making on how much pesticide is needed to kill the pest, which avoids excessive use (Abdel and Heikal 2021). Nanosensors are useful to monitor heavy metals and their uptake by plants, for example, Lew and co-workers have used single-walled carbon nanotube-based near-infrared fluorescent nanosensors to recognize arsenite and monitor its uptake through plant roots. This gives an idea about the plants that accumulate heavy metals and the concentration of accumulated heavy metals in vivo without damaging the plants (Lew et al. 2021). Ahmed and co-researchers have developed selenium NP for heavy metal toxicity detection. They have used selenite-reducing bacteria, and in the presence of heavy metals, there is inhibited selenite reduction, which leads to loss of red color and helps monitor heavy metal contamination (Ahmed et al. 2020). By providing real-time information on the nutrients available in the soil (Bharti et al. 2024), nanosensors help in providing required fertilizers, for example, detection of nitrate content (Kundu et al. 2022), potassium content in the soil (Kumar et al. 2020), detecting the soil leachates (Saini and Devi 2022), and monitoring plant and soil health (Parameswari et al. 2024; Arief et al. 2023. Nanosensors enable more accurate and timely decision-making, which helps achieve agricultural sustainability.

2.6 Environmental and Health Benefits

Nanoparticles are valuable tools and present a transformative opportunity to achieve environmental sustainability by offering precise and controlled release of pesticides and nutrients. This reduces the excessive use of pesticides and also the ecological impacts. For instance, nanopesticides utilize NPs that can be engineered for a target pest with high precision (Mustafa and Hussein 2020). This reduces the use of broad-spectrum pesticides and causes no harm or lesser effects on the non-target plants or animals (Prasad et al. 2017). Another advantage of nanoparticles as nanopesticides is the controlled release (Bhattacharyya et al. 2016); for example, nanoencapsulation protects humans and animals from exposure to pesticide residues by controlling their release (Chaud et al. 2021). This allows the effective use of pesticides for a longer time and reduces the multiple sprays or frequent application of pesticides, which ultimately lowers the risk of contamination of natural sources such as soil, water, or ecosystems with the pesticides. Additionally, there is always less amount of active material utilized in the nanopesticides compared to the conventional pesticides, leading to lesser exposure to harmful chemicals (Kah et al. 2018). Another application of NPs in crop protection is the targeted delivery of fertilizers as nanofertilizers, which provide essential nutrients to the plants with the controlled delivery mechanism. There is better absorption of nutrients, limiting the excess use of chemical fertilizers, and helping achieve better productivity (Yadav et al. 2020). With the reduced reliance on chemical fertilizers and conventional pesticides, there can be safer food production without harming the environment.

2.7 Challenges and Risks

Although there are numerous benefits of NPs in almost all sectors, such as agriculture, health, and technology, because of their size and unique properties, they can create environmental and health hazards. They can easily penetrate the cell membranes, which is the best point for nanopesticides, but this can be the biggest problem if they enter inside plants, animals, and humans and cause severe cellular toxicity (Parveen et al. 2021; Forest 2022).

The long-term consequences of the use of NPs in agriculture are also questionable. There is an accumulation of NPs in the soil and plants in the form of nano fertilizers and nano pesticides, which can enter the food chain and disrupt the ecosystem. De Albuquerque and co-workers have reported that even with the targeted delivery of pesticides with nanopesticides, only 0.1% are delivered to the pest while the rest enter the environment, which can lead to biodiversity disruption (de Albuquerque et al. 2020; Forest 2022).

Exposure of NPs to humans can lead to inflammation, tumor, hemolysis, cell toxicity, and organ failure (Parveen et al. 2021; Ettlinger et al. 2022). NPs are also reported to disturb aquatic life; for example, TiO2 nanoparticles were tested to check the toxic effects of *Daphnia magna* and *Gammarus fossarum* and were reported to cause acute to chronic effects (Seitz et al. 2014).

The production of engineered nanoparticles has increased drastically with TiO_2 as the most produced with 10,000 t per year, while iron oxide, aluminum oxide, and zinc oxide are between 100 to 1000 t per year of worldwide production (Piccinno et al. 2012). Most of the NP-based studies and their dosage exposures are studied for a short duration, and it is important to understand their long-term fate, toxicological effects on plants, safety, and regulatory awareness (Mishra et al. 2019). Toxicity concerns and regulatory hurdles are the major challenges in the world of nanotechnology that need more research to utilize their potential for the benefits of agriculture and sustainability rather than creating environmental disasters.

2.8 Conclusion

Nanotechnology has great potential in agriculture and can revolutionize crop protection by providing efficient solutions against pests and other biotic and abiotic stresses. Its ability to improve plant growth by making nutrients easily accessible, increasing water availability, increasing the efficiency of fertilizers, and increasing the precision of pesticide delivery is valuable to achieve agricultural sustainability and reduce the harmful environmental impacts caused by pesticides and other chemicals. Though NPs play a significant role in agriculture and other sectors, their use is limited, and there are a few concerns that need to be addressed. For example, the toxic effects of nanoparticles on plants and the environment, long-term impacts on microbes, plants, animals, and humans, accumulation of NPs in the soil, addition of NPs in the food chain, and effect on human health (Zhao et al. 2017; Paramo et al. 2020). Future research should be on biodegradable nanoparticles and assess their long-term fate in the environment. Nanoparticles are valuable tools and could become great next-generation crop protection approaches; however, to utilize their full potential, more studies are required to address the safety concerns and regulatory and economic challenges.

Acknowledgments PJ and SKY acknowledge the support of the Council of Scientific and Industrial Research (CSIR)-Institute of Himalayan Bioresource Technology (IHBT). Poonam Jyoti also thanks the Department of Science and Technology (DST)-Innovation in Science Pursuit for Inspired Research (INSPIRE) faculty funds (DST/INSPIRE/04/2023/003156/IFA23-LSPA165). CSIR-IHBT Communication No. for this book chapter is 5766.

References

Abdel-Aziz HM, Heikal YM (2021) Nanosensors for the detection of fertilizers and other agricultural applications, vol 1. Nanosensors for Environment, Food and Agriculture, pp 157–168

Afrasiabi S, Partoazar A (2024) Targeting bacterial biofilm-related genes with nanoparticle-based strategies. Front Microbiol 22(15):1387114

Ahmed F, Dwivedi S, Shaalan NM, Kumar S, Arshi N, Alshoaibi A, Husain FM (2020) Development of selenium nanoparticle based agriculture sensor for heavy metal toxicity detection. Agric 10(12):610

Anjaneyulu B, Chauhan V, Mittal C, Afshari M (2024) Innovative nanocarrier systems: a comprehensive exploration of recent developments in Nano-biopesticide formulations. J Environment Chem Eng 29:113693

Arief MA, Kim H, Kurniawan H, Nugroho AP, Kim T, Cho BK (2023) Chlorophyll fluorescence imaging for early detection of drought and heat stress in strawberry plants. Plan Theory 12(6): 1387

Azizi-Lalabadi M, Ehsani A, Divband B, Alizadeh-Sani M (2019) Antimicrobial activity of titanium dioxide and zinc oxide nanoparticles supported in 4A zeolite and evaluation the morphological characteristic. Sci Rep 9(1):17439

Batool A, Nazir M, Zargar SM (2024) Nano-pesticides and nano-fertilizers from natural (plant/animal) wastes. Biocatal Agric Biotechnol 60:103265

Bharti A, Jain U, Chauhan N (2024) From lab to field: nano-biosensors for real-time plant nutrient tracking. Plant Nano Biol 2:100079

Bhattacharyya A, Duraisamy P, Govindarajan M, Buhroo AA, Prasad R (2016) Nano-biofungicides: emerging trend in insect pest control in advances and applications through fungal nanobiotechnology. Springer, pp 307–319

Burke B, Fan G, Wasuwanich P, Moore EB, Furst AL (2023) Self-assembled nanocoatings protect microbial fertilizers for climate-resilient agriculture. JACS Au 3(11):2973–2980

Camara MC, Campos EV, Monteiro RA, do Espirito Santo Pereira A, de Freitas Proença PL, Fraceto LF (2019) Development of stimuli-responsive nano-based pesticides: emerging opportunities for agriculture. J Nanobiotechnol 17:1–9

Cebula Z, Żołędowska S, Dziąbowska K, Skwarecka M, Malinowska N, Białobrzeska W, Czaczyk E, Siuzdak K, Sawczak M, Bogdanowicz R, Nidzworski D (2019) Detection of the plant pathogen pseudomonas syringae pv. Lachrymans on antibody-modified gold electrodes by electrochemical impedance spectroscopy. Sensors 19(24):5411

Channab BE, Idrissi AE, Ammar A, Dardari O, Marrane SE, El Gharrak A, Akil A, Essemlali Y, Zahouily M (2024) Recent advances in nano-fertilizers: synthesis, crop yield impact, and economic analysis. Nanoscale 16:4484–4513

Chaud M, Souto EB, Zielinska A, Severino P, Batain F, Oliveira-Junior J, Alves T (2021) Nanopesticides in agriculture: benefits and challenge in agricultural productivity, toxicological risks to human health and environment. Toxics 9(6):131

Correa-Pacheco ZN, Corona-Rangel ML, Bautista-Baños S, Ventura-Aguilar RI (2021) Application of natural-based nanocoatings for extending the shelf life of green bell pepper fruit. J Food Sci 86(1):95–102 de Albuquerque FP, Preisler AC, Fraceto LF, Oliveira HC, de Castro VL (2020) Overview of nanopesticide environmental safety aspects and regulatory issues: the case of nanoatrazine. In Nanopesticides:281–298. https://doi.org/10.1007/978-3-030-44873-8_9

Deng W, Zhang Y, He L, Xu L, Ye X, Xu H, Zhu L, Jia J (2024) Optimized nanopesticide delivery of thiamethoxam to cowpeas (Vigna unguiculata) controls thrips (Megalurothrips usitatus) and reduces toxicity to non-target worker bees (Apis mellifera). Sci Total Environ 954:176327

Dhaka A, Mali SC, Sharma S, Trivedi R (2023) A review on biological synthesis of silver nanoparticles and their potential applications. Results Chem 11:101108

Dias DA, Fernandes IA, Machado EP, Pedott L, Blassioli-Moraes MC, Borges M, Steffens J, Steffens C (2024) Nanosensors for detecting volatile compounds in pest management: a focus on agricultural sustainability. ACS Agric Sci Technol 4(11):1250–1259

Dimkpa CO, Andrews J, Fugice J, Singh U, Bindraban PS, Elmer WH, Gardea-Torresdey JL, White JC (2020) Facile coating of urea with low-dose ZnO nanoparticles promotes wheat performance and enhances Zn uptake under drought stress. Frontiers in Plant Sci 11:168

Dimri A, Pathak N, Sharma S (2020) Nanosensors for root zone parameters influencing plant growth. In: Nanomaterials for agriculture and forestry applications. Elsevier, pp 387–406

Dong J, Liu X, Chen Y, Yang W, Du X (2021) User-safe and efficient chitosan-gated porous carbon nanopesticides and nanoherbicides. J Colloid Interf Sci 594:20–34

El-Saadony MT, Saad AM, Soliman SM, Salem HM, Desoky ES, Babalghith AO, El-Tahan AM, Ibrahim OM, Ebrahim AA, Abd El-Mageed TA, Elrys AS (2022) Role of nanoparticles in enhancing crop tolerance to abiotic stress: a comprehensive review. Frontiers Plant Sci 13: 946717

Ettlinger R, Lächelt U, Gref R, Horcajada P, Lammers T, Serre C, Couvreur P, Morris RE, Wuttke S (2022) Toxicity of metal—organic framework nanoparticles: from essential analyses to potential applications. Chem Soc Rev 51(2):464–484

Feng Y, Li X, Guo S, Chen X, Chen T, He Y, Shabala S, Yu M (2019) Extracellular silica nanocoat formed by layer-by-layer (LBL) self-assembly confers aluminum resistance in root border cells of pea (Pisum sativum). J Nanobiotechnol 17:1–1

Forest V (2022) Experimental and computational nanotoxicology-complementary approaches for nanomaterial hazard assessment. Nano 12(8):1346

Giraldo JP, Landry MP, Kwak SY, Jain RM, Wong MH, Iverson NM, Ben-Naim M, Strano MS (2015) A ratiometric sensor using single chirality near-infrared fluorescent carbon nanotubes: application to in vivo monitoring. Small 11(32):3973–3984

Giraldo JP, Wu H, Newkirk GM, Kruss S (2019) Nanobiotechnology approaches for engineering smart plant sensors. Nature Nanotechnol 14(6):541–553

Global Fertilizer Consumption 1965–2022, by nutrient. Published by Statista Research Department, Nov 7, 2024. https://www.statista.com/statistics/438967/fertilizer-consumption-globally-by-nutrient/

Gula LT (2023) Researchers helping protect crops from pests. USDA, editor

Hou R, Zhou J, Song Z, Zhang N, Huang S, Kaziem AE, Zhao C, Zhang Z (2023) pH-responsive λ-cyhalothrin nanopesticides for effective pest control and reduced toxicity to Harmonia axyridis. Carbohydr Polym 302:120373

Ijaz M, Khan F, Ahmed T, Noman M, Zulfiqar F, Rizwan M, Chen J, Siddique KH, Li B (2023) Nanobiotechnology to advance stress resilience in plants: current opportunities and challenges. Mater Today Bio 6:100759

Imran OI (2024) Nano biotechnology: exploring the potential of Nano materials in enhancing crop health and combating abiotic stress. Commun Soil Sci and Plant Anal 30:1–20

Istúriz-Zapata MA, Correa-Pacheco ZN, Bautista-Baños S, Acosta-Rodríguez JL, Hernández-López M, Barrera-Necha LL (2022) Efficacy of extracts of mango residues loaded in chitosan nanoparticles and their nanocoatings on in vitro and in vivo postharvest fungal. J Phytopathol 170(10):661–674

Jakhar AM, Aziz I, Kaleri AR, Hasnain M, Haider G, Ma J, Abideen Z (2022) Nano-fertilizers: a sustainable technology for improving crop nutrition and food security. NanoImpact 27:100411

Jali P, Acharya S, Mahalik G (2024) Antimicrobial efficacy of nano-particles for crop protection and sustainable agriculture. Discover Nano 19(1):117
Jiang Q, Xie Y, Peng M, Wang Z, Li T, Yin M, Shen J, Yan S (2022) A nanocarrier pesticide delivery system with promising benefits in the case of dinotefuran: strikingly enhanced bioactivity and reduced pesticide residue. Environ Sci Nano 9(3):988–999
Joudeh N, Linke D (2022) Nanoparticle classification, physicochemical properties, characterization, and applications: a comprehensive review for biologists. J Nanobiotechnol 20(1):262
Junaid MD, Gokce AF (2024) Global agricultural losses and their causes. Bull Biol All Sci Res 1:66
Kah M, Kookana RS, Gogos A, Bucheli TD (2018) A critical evaluation of nanopesticides and nanofertilizers against their conventional analogues. Nature Nanotechnol 13(8):677–684
Kalia A, Abd-Elsalam KA, Kuca K (2020) Zinc-based nanomaterials for diagnosis and management of plant diseases: ecological safety and future prospects. J Fungi 6(4):222
Kapeleka JA, Mwema MF (2024) State of nano pesticides application in smallholder agriculture production systems: human and environmental exposure risk perspectives. Heliyon 10:e39225
Kumar S, Nehra M, Dilbaghi N, Marrazza G, Hassan AA, Kim KH (2019) Nano-based smart pesticide formulations: emerging opportunities for agriculture. J Control Release 294:131–153
Kumar AA, Kumar SN, Fernandez RE (2020) Real time sensing of soil potassium levels using zinc oxide-multiwall carbon nanotube-based sensors. IEEE Trans Nanobioscience 20(1):50–56
Kundu M, Krishnan P, Chobhe KA, Manjaiah KM, Pant RP, Chawla G (2022) Fabrication of electrochemical nanosensor for detection of nitrate content in soil extract. J Soil Sci Plant Nutr 22(3):2777–2792
Lew TT, Koman VB, Silmore KS, Seo JS, Gordiichuk P, Kwak SY, Park M, Ang MC, Khong DT, Lee MA, Chan-Park MB (2020) Real-time detection of wound-induced H2O2 signalling waves in plants with optical nanosensors. Nature plants 6(4):404–415
Lew TT, Park M, Cui J, Strano MS (2021) Plant nanobionic sensors for arsenic detection. Adv Mater 33(1):2005683
Li J, Wu Z, Bao Y, Chen Y, Huang C, Li N, He S, Chen Z (2017) Wet chemical synthesis of ZnO nanocoating on the surface of bamboo timber with improved mould-resistance. J Saudi Chem Soc 21(8):920–928
Li H, Yang Z, Khan SA, Walsh LJ, Seneviratne CJ, Ziora ZM (2024) Characteristics of metallic nanoparticles (especially silver nanoparticles) as anti-biofilm agents. Antibiotics 13(9):819
Liang Y, Gao Y, Wang W, Dong H, Tang R, Yang J, Niu J, Zhou Z, Jiang N, Cao Y (2020) Fabrication of smart stimuli-responsive mesoporous organosilica nano-vehicles for targeted pesticide delivery. J Hazard Mater 389:122075
Lin T, Chen X, Ren Y, Qing B, Zhang M, Mo Z, Wang S (2024) Effects of iron oxide nanocoatings on the seed germination, seedling growth, and antioxidant response of aromatic rice grown in the presence of different concentrations of rice straw extracts. J Nanopart Res 26(4):78
Linh TM, Mai NC, Hoe PT, Lien LQ, Ban NK, Hien LT, Chau NH, Van NT (2020) Metal-based nanoparticles enhance drought tolerance in soybean. J Nanomater 6(1):4056563
Liu J, Xu D, Xu G, Li X, Dong J, Luan X, Du X (2023) Smart controlled-release avermectin nanopesticides based on metal-organic frameworks with large pores for enhanced insecticidal efficacy. Chem Eng J 475:146312
Lu Y, Xu K, Zhang L, Deguchi M, Shishido H, Arie T, Pan R, Hayashi A, Shen L, Akita S, Takei K (2020) Multimodal plant healthcare flexible sensor system. ACS Nano 14(9):10966–10975
Martinazzo J, Ballen SC, Steffens J, Steffens C (2022) Sensing of pheromones from Euschistus heros (F.) stink bugs by nanosensors. Sens Actuators Rep 4:100071
Mishra M, Dashora K, Srivastava A, Fasake VD, Nag RH (2019) Prospects, challenges and need for regulation of nanotechnology with special reference to India. Ecotoxicol Environ Saf 171:677–682
Moradi Pour M, Saberi Riseh R, Skorik YA (2022) Sodium alginate–gelatin nanoformulations for encapsulation of Bacillus velezensis and their use for biological control of pistachio gummosis. Materials 15(6):2114

Mustafa IF, Hussein MZ (2020) Synthesis and technology of nanoemulsion-based pesticide formulation. Nano 10(8):1608

Oerke EC (2006) Crop losses to pests. J Agric Sci 144(1):31–43

Pang Z, Shen S, Ma Q, Wang Y, Peng H, Liang Y (2024) Extracellular silicon (Si) nanocoating induced by cationic polymers enhances heavy metal resistance in both Si-accumulating and non-Si-accumulating plants. Environ Technol Innov 33:103500

Parameswari P, Belagalla N, Singh BV, Abhishek GJ, Rajesh GM, Katiyar D, Hazarika B, Paul S (2024) Nanotechnology-based sensors for real-time monitoring and assessment of soil health and quality: a review. Asian J Soil Sci Plant Nutr 10(2):157–173

Paramo LA, Feregrino-Pérez AA, Guevara R, Mendoza S, Esquivel K (2020) Nanoparticles in agroindustry: applications, toxicity, challenges, and trends. Nano 10(9):1654

Parveen R, Maiti PK, Murmu N, Datta A (2021) Preparation of serum capped silver nanoparticles for selective killing of microbial cells sparing host cells. Sci Rep 11(1):11610

Patel S, Bajpai J, Saini R, Bajpai AK, Acharya S (2018) Sustained release of pesticide (Cypermethrin) from nanocarriers: an effective technique for environmental and crop protection. Process saf and Environ Prot 117:315–325

Pathak VM, Verma VK, Rawat BS, Kaur B, Babu N, Sharma A, Dewali S, Yadav M, Kumari R, Singh S, Mohapatra A (2022) Current status of pesticide effects on environment, human health and it's eco-friendly management as bioremediation: a comprehensive review. Front Microbiol 13:962619

Pei H, Wang M, Yang T, Li J, Sun S, Wang T, Liu Y, Zhang F (2025) A hybrid of bimodal mesoporous silica and metal organic framework for intelligent co-delivery of dual-pesticide for synergistic controlling fungal disease and insect pest. Colloid Surf A: Physicochem Eng Asp 7: 136140

Piccinno F, Gottschalk F, Seeger S, Nowack B (2012) Industrial production quantities and uses of ten engineered nanomaterials in Europe and the world. J Nanopart Res 14:1–1

Prasad R, Bhattacharyya A, Nguyen QD (2017) Nanotechnology in sustainable agriculture: recent developments, challenges, and perspectives. Front Microbiol 8:1014

Qian J, Shan R, Shi Y, Li H, Xue L, Song Y, Zhao T, Zhu S, Chen J, Jiang M (2024) Zinc oxide nanoparticles alleviate salt stress in cotton (Gossypium hirsutum L.) by adjusting Na+/K+ ratio and antioxidative ability. Life 14(5):595

Rana A, Rani A, Nayana KR, Deswal S, Singh AP, Rana S, Chahar M, Singh N, Dhaka RK (2025) Biotic stress alleviation in plant using rhizobacteria: an overview of mechanism of action, antimicrobial compounds production, (Nano) formulations and employment methods. Indian J Microbiol 3:1–27

Rathore A, Hasan W, NM R, Pujar K, Singh R, Panotra N, Satapathy SN (2024) Nanotech for crop protection: utilizing nanoparticles for targeted pesticide delivery. J Zool 45(6):46–71

Riseh RS, Hassanisaadi M, Vatankhah M, Soroush F, Varma RS (2022) Nano/microencapsulation of plant biocontrol agents by chitosan, alginate, and other important biopolymers as a novel strategy for alleviating plant biotic stresses. Int J Biol Macromol 222:1589–1604

Saberi-Rise R, Moradi-Pour M (2020) The effect of Bacillus subtilis Vru1 encapsulated in alginate–bentonite coating enriched with titanium nanoparticles against Rhizoctonia solani on bean. Int J Biol Macromol 152:1089–1097

Saberi-Riseh R, Moradi-Pour M (2021) A novel encapsulation of Streptomyces fulvissimus Uts22 by spray drying and its biocontrol efficiency against Gaeumannomyces graminis, the causal agent of take-all disease in wheat. Pest Manag Sci 77(10):4357–4364

Saini R, Devi P (2022) Nanosensors for detecting nutrient losses from soil (as gaseous ammonia and nitrous oxide, and/or soil leachates as nitrate and phosphate). In Nanosensors for Smart Agriculture Elsevier 1:303–327

Seitz F, Rosenfeldt RR, Schneider S, Schulz R, Bundschuh M (2014) Size-, surface-and crystalline structure composition-related effects of titanium dioxide nanoparticles during their aquatic life cycle. Sci Total Environ 493:891–897

Semida WM, Abdelkhalik A, Mohamed GF, Abd El-Mageed TA, Abd El-Mageed SA, Rady MM, Ali EF (2021) Foliar application of zinc oxide nanoparticles promotes drought stress tolerance in eggplant (Solanum melongena L.). Plan Theory 10(2):421

Shang Y, Hasan MK, Ahammed GJ, Li M, Yin H, Zhou J (2019) Applications of nanotechnology in plant growth and crop protection: a review. Molecules 24(14):2558

Sharma A, Kumar V, Shahzad B, Tanveer M, Sidhu GP, Handa N, Kohli SK, Yadav P, Bali AS, Parihar RD, Dar OI (2019) Worldwide pesticide usage and its impacts on ecosystem. SN Applied Sciences 1:1–6

Shaw DS, Honeychurch KC (2022) Nanosensor applications in plant science. Biosensors 12(9):675

Shen Y, An C, Jiang J, Huang B, Li N, Sun C, Wang C, Zhan S, Li X, Gao F, Zhao X (2022) Temperature-dependent nanogel for pesticide smart delivery with improved foliar dispersion and bioactivity for efficient control of multiple pests. ACS Nano 16(12):20622–20632

Singh A, Sharma A, Singh O, Rajput VD, Movsesyan HS, Minkina T, Alexiou A, Papadakis M, Singh RK, Singh S, Sousa JR (2024) In-depth exploration of nanoparticles for enhanced nutrient use efficiency and abiotic stresses management: present insights and future horizons. Plant Stress. 25:100576

Soni S, Jha AB, Dubey RS, Sharma P (2024) Nanowonders in agriculture: unveiling the potential of nanoparticles to boost crop resilience to salinity stress. Sci Total Environ 6:171433

Sravani B, Dalvi S, Narute TK (2023) Role of chitosan nanoparticles in combating Fusarium wilt (Fusarium oxysporum f. sp. ciceri) of chickpea under changing climatic conditions. J Phytopathol 171(2–3):67–81

Taran N, Storozhenko V, Svietlova N, Batsmanova L, Shvartau V, Kovalenko M (2017) Effect of zinc and copper nanoparticles on drought resistance of wheat seedlings. Nanoscale Res Lett 12: 1–6

Tortella G, Rubilar O, Pieretti JC, Fincheira P, de Melo Santana B, Fernández-Baldo MA, Benavides-Mendoza A, Seabra AB (2023) Nanoparticles as a promising strategy to mitigate biotic stress in agriculture. Antibiotics 12(2):338

Venkatachalam P, Jayaraj M, Manikandan R, Geetha N, Rene ER, Sharma NC, Sahi SV (2017) Zinc oxide nanoparticles (ZnONPs) alleviate heavy metal-induced toxicity in Leucaena leucocephala seedlings: a physiochemical analysis. Plant Physiol Biochem 110:59–69

Voke E, Pinals RL, Goh NS, Landry MP (2021) In planta nanosensors: understanding biocorona formation for functional design. ACS Sensors 6(8):2802–2814

Wu H, Nißler R, Morris V, Herrmann N, Hu P, Jeon SJ, Kruss S, Giraldo JP (2020) Monitoring plant health with near-infrared fluorescent H2O2 nanosensors. Nano Lett 20(4):2432–2442

Wu J, Chang J, Liu J, Huang J, Song Z, Xie X, Wei L, Xu J, Huang S, Cheng D, Li Y (2023) Chitosan-based nanopesticides enhanced anti-fungal activity against strawberry anthracnose as "sugar-coated bombs". Int J Biol Macromol 253:126947

Xu X, Yu Y, Ling M, Ares I, Martínez M, Lopez-Torres B, Maximiliano JE, Martínez-Larrañaga MR, Wang X, Anadón A, Martínez MA (2023) Oxidative stress and mitochondrial damage in lambda-cyhalothrin toxicity: a comprehensive review of antioxidant mechanisms. Environ Pollut 4:122694

Yadav RK, Singh NB, Singh A, Yadav V, Bano C, Khare S, Niharika (2020) Expanding the horizons of nanotechnology in agriculture: recent advances, challenges and future perspectives. Vegetos 33:203–221

Yadav A, Yadav K, Abd-Elsalam KA (2023) Nanofertilizers: types, delivery and advantages in agricultural sustainability. Agrochemicals 2(2):296–336

Yan X, Shan Y, Ma Y, Wang Y, Wu C, Ren X, Song X, Wang D, Hu H, Ma X, Ma Y (2024) Thiacloprid-silica nano-delivery system enhances toxicity against Aphis gossypii and improves non-target biosafety. Chemosphere 367:143596

Yao KS, Li SJ, Tzeng KC, Cheng TC, Chang CY, Chiu CY, Liao CY, Hsu JJ, Lin ZP (2009) Fluorescence silica nanoprobe as a biomarker for rapid detection of plant pathogens. Adv Mater Res 79:513–516

Zafar S, Hasnain Z, Danish S, Battaglia ML, Fahad S, Ansari MJ, Alharbi SA (2024) Modulations of wheat growth by selenium nanoparticles under salinity stress. BMC Plant Biol 24(1):35

Zhang Q, Shi X, Gao T, Xing Y, Jin H, Hao J, Liu X, Liu X, Liu P (2024a) Precision management of Fusarium fujikuroi in rice through seed coating with an enhanced nanopesticide using a tannic acid-ZnII formulation. J Nanobiotechnol 22(1):717

Zhang H, Xu W, Li G, Qu H, Ma C, Noruzi EB, He Q, Cheng J, Li H (2024b) Controlled release system of nanopesticides based on noncovalent interactions. ACS Agric Sci Technol 4(9): 851–871

Zhao Xiang ZX, Cui Hai Xin CH, Wang Yan WY, Sun Chang Jiao SC, Cui Bo CB, Zeng Zhang Hua ZZ (2017) Development strategies and prospects of nano-based smart pesticide formulation. J Agric Food Chem 66(26):6504–6512

Zhao M, Li P, Zhou H, Hao L, Chen H, Zhou X (2022a) pH/redox dual responsive from natural polymer-based nanoparticles for on-demand delivery of pesticides. Chem Eng J 435:134861

Zhao L, Wang W, Fu X, Liu A, Cao J, Liu J (2022b) Graphene oxide, a novel nanomaterial as soil water retention agent, dramatically enhances drought stress tolerance in soybean plants. Front Plant Sci 13:810905

Zhou P, Adeel M, Shakoor N, Guo M, Hao Y, Azeem I, Li M, Liu M, Rui Y (2020) Application of nanoparticles alleviates heavy metals stress and promotes plant growth: an overview. Nano 11(1):26

Zia R, Iftikhar M, Rafiq A, Hakim S, Nawaz MS, Imran A, Bajwa SZ (2022) Nanosensors for microbial detection in soil. In: Nanosensors for smart agriculture, vol 1. Elsevier, pp 367–400

Chapter 3
Exploring the Role of Nanotechnology in the Management of Abiotic Stress

Tahsin Shoala

Abstract Abiotic stresses such as heavy metals, drought, salinity, cold, and heat harm plant development and the production of crops in general. To attain the aim of being sustainable in the face of abiotic environmental limits, crop production systems must employ effective tactics in conjunction with technology innovation. Recently, nanotechnology has emerged as an attractive topic to investigate with potential applications in agriculture, and it requires attention to treat some of the abiotic challenges. The administration of nanoparticles (NPs) as nanofertilizers has huge potential because of their ecological friendliness, cost-effectiveness, specific physicochemical features, and higher plant production, and there has been a lot of attention in the sector lately. Nanomaterials (NMs) are vital in many aspects of plant growth, especially seedling strength, the emergence of seeds, growth, the photosynthesis process, blooming, and root development. Furthermore, these NMs play an important role in plant defense from oxidative stress. NMs stimulate the generation of reactive oxygen species (ROS) in plants and imitate the effects of specific enzymes (antioxidative). As a result, it is vital to understand the cellular, biochemical, and molecular mechanisms of NMs in plants under abiotic stress. Various studies suggest that NPs are used to control abiotic stress. This chapter discusses the effects of abiotic stress on plants, the functions of nanoparticles in abiotic stress management, and future possibilities to establish a feasible and environmentally beneficial approach for ensuring agriculture's long-term viability.

Keywords Abiotic stress · Nanofertilizers · Drought stress · Salt stress · Agriculture

T. Shoala (✉)
Department of Environmental Biotechnology, College of Biotechnology, Misr University for Science and Technology, Giza, Egypt
e-mail: tahsen.shoala@must.edu.eg

S. K. Yadav, A. Kumari (eds.), *Nanobiotechnology in Agri-food Sector*, Sustainability Sciences in Asia and Africa, https://doi.org/10.1007/978-981-96-7756-6_3

3.1 Introduction

Plants evolved as persistent species in a profoundly inhospitable terrestrial habitat around 700 million years ago, eventually making the land suitable for various species (Hotton et al. 2001; Selden and Edwards 1989). Plants have evolved a wide range of adaptive processes that allow them to react appropriately to surrounding stresses (e.g., the water stressful situations: severe drought or flooding; elevated salinity; extreme temperatures: both hot and cold; heavy metal [HM] poisoning) (Bray 2000; Wani et al. 2016). Drought, the most common type of abiotic stress, would severely damage almost 45% of arable lands globally by 2100 (Field et al. 2014). Water is a vital element; provided water is accessible, each biological field, regardless of exactly how extreme, can be adopted by life-forms (Wood 2005). A drought is defined as an extended duration of insufficient rainfall or watering, which is more likely to occur in extremely dry and semiarid environments. The agricultural industry consumes the most water, accounting for more than 70% of subterranean water resources harvested, primarily in poor countries. Rainfall immediately impacts over 90% of agricultural land undergoing agriculture worldwide. Drought will wreak havoc on roughly 45% of the global agricultural area by the end of this century. Salinity is a different prevalent source of significant biotic stress, known for its multifaceted consequences for plant health (Dai 2011; Vibha 2016). With the progressive rise in global temperature of about 1 °C over the previous century, salinity of arable land has become a limiting issue in agriculture. This increase has aggravated the issue by increasing the soil rates of evaporation (Nouman et al. 2018; Zhao et al. 2017), disrupting hydrodynamic paradigms, which is another important cause for concern for the agricultural sector. Salinity may take place naturally or anthropologically because of inadequate handling of water and soil degradation caused by intensive agricultural practices. Salinity affects approximately one million hectares of agricultural land, and the higher production from irrigated farms over rain-fed lands favors salinization (Munns and Gilliham 2015).

The emergence of ecological change-driven influences has increased the prevalence of environmental stresses, especially extreme temperatures and extreme heat, which exacerbates the extent of additional stresses. For example, a 1 °C increase in temperature worldwide leads to a significant decrease in the productivity of crops (Iizumi et al. 2017). Global warming has a comparable impact on chilled stress, additional extreme weather phenomena. Although scientists have primarily focused on extreme temperature stress, stress caused by low temperatures is endangering crop yields on a wide scale due to variations in the climate lately (Budhathoki and Zander 2019).

A further indicator of climate change is atmospheric volatility, which creates uncertainty regarding duration and intensity, and also the substantial concentration of rainfalls, resulting in devastating floods in farmlands following storms (Onyekachi et al. 2019). Regarding the hazardous effects of HMs in the rhizosphere, their buildup can be linked to a number of resources, including areas that experience flooding (such as manganese or iron), bedrock decomposition (such as nickel, cobalt, cadmium, and lead), soil acidification (such as manganese and zinc), and human interference (such as zinc, nickel, cobalt, copper, cadmium, molybdenum, chromium, and lead) (White and Pongrac 2017). As a result of climate change, the sources are expanding and have the potential to drastically change crop productivity (Fageria et al. 2010). More seriously, the presence of one stimulus creates conditions for the development of other stresses, especially elevated temperatures and drought or salinity and drought, which usually happen concurrently (Sahin et al. 2018). To make conditions more severe, global environmental change, as created by human phenomena, is endangering the farming industry by enforcing and raising the probability of the combination of abiotic stresses at an accelerated rate. This book chapter will focus on the use of nanobiotechnology to mitigate the various and harmful effects of abiotic stressors on crop plants.

3.2 Abiotic Stress

Abiotic stress variables that can have an impact on modern farming output globally include low or high temperatures, flooding, drought, salinity, HMs, and ultraviolet (UV) radiation (Badawy et al. 2021). Plants respond to abiotic stress in a variety of ways, according to the crop type, stress kind, and duration of exposure (Rady et al. 2021; Shaaban et al. 2022). As a result, ecologically friendly agricultural practices and higher yield output can potentially improve the nutritional content of water, soil, and various plant-use supplies.

To satisfy the rising world food demand, scientists are working to reduce the negative impacts of abiotic stress, increase crop productivity and food supply, and attain sustainability and food security. Indeed, to solve these pressing global issues, scientists have to keep trying to discover novel innovations or alternatives. Abiotic stresses are those that arise naturally but suddenly and have a subsequent impact on agricultural plants via their surrounding environment. This has a negative influence on crop output, survival, overall biomass, as well as quality output (Grover et al. 2001; Calanca 2017). Food is one of the most important aspects of existence for humans; hence, farming and agribusiness have always been prioritized. Despite this, due to population growth overflow, land for farming is being converted into non-arable areas to accommodate the rapidly growing population of humans. Unconcerned human activities have caused an imbalance in climatic patterns, which has significantly elevated global temperatures. Plants grown under stress have evolved greater mechanisms for survival and versatility than plants grown under common and optimal development conditions. The former's responses to stress are far more advanced. Since the last century, the earth's air temperature has increased substantially, and plant yield is continually constrained, making it difficult to meet global demand. It is becoming a major concern due to its prolonged exposure (Cramer et al. 2011; Kumar et al. 2018). Plant responses to these overlapping abiotic stresses are extremely dynamic and complex systems. The plant community is constantly exploited due to environmental pressures. The advancement of techniques and technologies helped to reduce these exploitations drastically. In today's world, dealing with abiotic stressors is a constant process. Despite improvements in molecular and physiological biology that have increased our understanding of stress tolerance, scientists continue to investigate the mechanism and gene structure using a variety of bio- and nanotechnologies.

3.3 Nanotechnology's Impact on Agriculture

Most developing nations rely heavily on farming as their primary source of employment. According to projections, our planet will need to house approximately 9 billion people by 2050. The rapid increase in population will have major consequences on water, food, and energy supplies (Marchiol 2018). As a result, there is an urgent requirement to implement sustainable development to improve agricultural yield on currently cultivated land while remaining ecologically friendly. With the rapid expansion of the global population, the overuse of limited resources such as the depletion of land, water, and soil necessitates agricultural development that is both economically and environmentally sustainable. Thus, fresh agricultural development strategies are required (Singh et al. 2015; Prasad et al. 2014).

Most recently, prospective uses of nanotechnology for agricultural purposes have been offered as a viable technology for the twenty-first century, opening up new avenues for investigation in academia and industry (Dasgupta et al. 2015). Nanotechnology applied to farming has evolved as an integrated field with important implications for agricultural operations, including diagnosis of diseases, control, nutrition, and tissue engineering (Srilatha 2011). It is concerned with the smallest feasible elements that can improve crop production through enhanced management and preserving agricultural inputs (Kalita et al. 2021). NMs are either synthetic or natural particles with sizes ranging from 1 to 100 nm, such as nanoparticles (NPs), nanocapsules, and nanocrystals, which have the potential to update agricultural techniques. It enhances the biological absorption of herbicides, insecticides, and nutrients, improving agricultural output. As a result, the use of nanotechnology for agricultural purposes will continue to have a significant impact in areas such as disease detection, disease control, nutrition delivery, developing and designing product approaches, and food security and biosecurity monitoring. Consequently, this technology will help to modernize ways of farming on an international level by providing numerous techniques for lowering crop production costs, increasing the profitability of crops, and preserving the environment. Furthermore, the distinctive characteristics of NPs make them ideal for generating and creating novel nanotechnological solutions to promote sustainable agriculture.

3.3.1 Nanotechnology's Function in Managing Abiotic Stress

Unfavorable environmental circumstances, including a variety of biotic and abiotic stressors, are well known to have a sensitive influence on plants. They see themselves as substantial threats to the sustainability of our crop yield. Because of their inevitability and repetition, abiotic pressures are a greater source of worry in modern research than biotic stresses. Drought, flood, salinity, heat and cold, and HM toxicity are the most common contributors. Plants are also continuously exposed to UV rays, which reduces crop development and yield. These pressures lead to an average yield loss of more than 50% (Bray et al. 2000).

Water-associated stress has been found as one of the most common abiotic stressors worldwide, with drought and salinity being close relatives. Stresses consistently reduce plant life and growth, limiting production. According to Ashraf (1994), drought and salinity might cause major salinization of all agricultural areas by up to 30% over the next 25 years, resulting in negative worldwide repercussions. Land loss of up to 50% is anticipated by 2050. Long-term prolonged exposure to multiple kinds of abiotic stress causes numerous morphological, physiological, biochemical, and molecular alterations. The need for crop products continues to rise due to the ongoing influence of abiotic stress on agricultural output (Zhao et al. 2017). Through the previous century, precipitation has increased by around 11% during the winter season, promoting the establishment of several types of diseases that impair crop fertility and productivity. The world population is expected to surpass 9.8 billion by 2050 and 11.2 billion by 2100 (United Nations 2017). Farming is undergoing extensive research and technological improvement to fulfill the demands of an expanding population (Dwivedi et al. 2016; Kou et al. 2018).

The agricultural sector feeds nearly the entire global population and accounts for more than 60% of all livelihoods. Despite being considered the structural foundation of developing countries, the agricultural industry faces several issues. Climate change, rising temperatures, excessive use of resources, the use of chemical fertilizers, and numerous biotic and abiotic pressures are among the significant concerns that require prompt attention (Raliya et al. 2017). Abiotic pressures are more damaging than biotic stresses; hence, they receive a lot of attention in scientific research. Abiotic stress on crop plants has sparked academic attention during the last decade since it is inescapable and repeating in nature.

Despite these unavoidable instances, nanotechnologies have received a lot of interest recently due to their vital role in boosting and enhancing crop yields. Biotechnology and nanotechnology usages in material energy, medicine, medical medicines, and catalysis have received substantial research funding (Ghidan and Antary 2019). Nanotechnology is a multidisciplinary field that includes biological sciences, chemistry, physics, healthcare, and engineering (Abd-Elrahman and Mostafa 2015). Progress in nanotechnology research has shown a number of methods for improving crop plant resilience to abiotic stress.

3.3.2 Application of Nanofertilizers

Agricultural crops are generally grown with synthetic fertilizers to increase production, but they do not absorb all of the synthetic nutrients offered to them, resulting in the development of an excessive quantity of chemical fertilizer in the soil, eventually rendering the soil infertile. As a result, nanofertilizers are now preferred over chemical and biofertilizers since they can give the necessary amount of nutrients in a little quantity. These nanofertilizers help crop plants survive environmental restrictions and decrease toxicity. According to a particular study, the use of Si NPs and Si fertilizer under stressful circumstances (salinity stress) may enhance the physiological and morphological aspects of plants (Janmohammadi et al. 2015). Water, as an essential element of life, can severely impede agriculture if not available in suitable quantities or if overabundant. Drought is one of the most typical conditions that dry region crops experience. This stress from the environment decreases production to about half of the mass, resulting in decreased nutrition and development of crops. NPs, specifically Si NPs, have proven to be effective solutions to such challenges. Zn NPs have also demonstrated good outcomes (Cakmak et al. 1996) (Fig. 3.1).

3.3.3 Nanofertilizers Versus Commercial Fertilizers

Agricultural chemicals can be sprayed in precise doses, and macromolecules can be selectively delivered. Fertilizers and insecticides can be used more efficiently by combining nanoscale transporters and chemicals, resulting in a reduction in overall chemical use while maintaining crop output. Commercial fertilizers, on the other hand, are less beneficial to plants due to both increased particle size and less bioavailability. Furthermore, continuous chemical fertilizer applications cause a hazardous accumulation of hazardous molecules, disrupting the soil's environmental equilibrium. Furthermore, improper chemical fertilizer application can pollute soil via overflow or inadequate use by plants, leading to leftovers being converted into insoluble salts in the soil. Nanoagrochemicals play an important role in enhancing nutrient utilization efficiency and water monitoring in ecologically conscious farming. Nonetheless, biological buildup and prolonged exposure of NPs to plants may affect edible plants and food chains (Rajput et al. 2020). Staro'n et al. (2020) discovered that nanoparticles can be taken in and retained in crop plant consumable tissues. The spontaneous development of NPs or metal ions may disrupt the functioning of plants, affect the viability of cell function and subcellular organelle organizations, and modify the structure of proteins, lipids, and nucleic acids through the creation of hydroxyl radicals (Cota-Ruiz et al. 2018; Rajput et al. 2020) (Fig. 3.1).

In general, various uses of NPs may present several challenges in terms of ecology, ethics, health, and safety (Rajput et al. 2018). To date, the possibility of

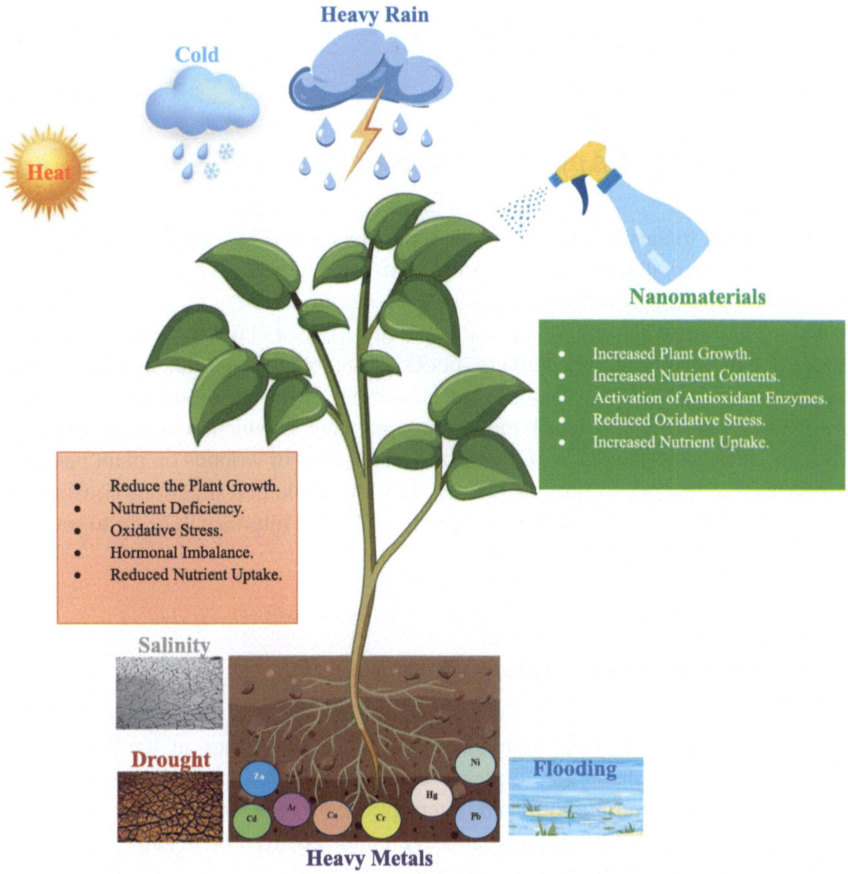

Fig. 3.1 Potential impact of nanoparticles (NPs) on plant growth and development in the presence of abiotic stress

adverse impacts of NPs on human health has been hypothetical and unfounded (Staro'n et al., 2020). Nanotechnology has gained in popularity as a result of the development of different NPs as new agricultural tools. There is an urgent need to expand our expertise and comprehension of the particular advantages and cons of using NPs. The progress of nanotechnology has resulted in considerable volumes of produced NPs in the agricultural environment. Although this technique has various benefits, scientists and others are concerned about the hazardous disposal of NPs in enormous numbers (several hundred tons) per year (Rajput et al. 2020). It is proposed that NPs live in a variety of regulated environments (the atmosphere air, water objects, soils, hydrobionts, algae, fungi, and land plant/animal tissues) (Rajput et al. 2020). In comparison with different resources, there has been little investigation into the fate and transportation of NPs in soil. Additionally, the soil provides essential nutrients to food crops, which can also act as an NP-collecting sink (Rajput et al. 2020). The current research gives insight into the possible effects of NPs on the ecosystem, human health, and food security.

3.4 The Mode of Application of Nanoparticles

NMs are given to agricultural plants via their roots, branches, and leaves. They can be absorbed via stomata, hydathodes, and bark pores (Eichert et al. 2008). Nanomaterials used in nanofertilizers today come in a variety of shapes and sizes, including 2D forms (sheets, disks, rods, and tubes), nanoclays, and nanoemulsions. According to certain studies, delivery modes affect absorption efficiency. These NPs are passively delivered to the targeted areas after being absorbed by the various organs and passing the exterior multimembrane layers (which contain cellulose or hemicellulose, as well as many other proteins). Injuries or wounds on plant parts can also serve as entry points into the plant system. Multiple investigations consistently propose understanding the physiology of the NM-plant interface in order to improve our understanding of their dynamics.

3.4.1 Mechanisms of Action of Nanoparticles in Plants

Although nanoparticles have numerous potential applications in agriculture, the majority of them are damaging to plants when administered in large concentrations. The absorption, accumulation, and interruption of NPs with critical metabolic functions in diverse plant tissues can have either positive or negative effects on plants, depending on their number, mobility, characteristics, and reactivity.

3.4.1.1 Uptake of Nanoparticles

Considerable amounts of NPs can pass through plant cells and the plasma membrane, potentially interfering with essential cellular functions (Mazumdar and Ahmed 2011; Mirzajani et al. 2013). NPs can reach tissues in plants through their root systems or aboveground structures such as root connections and lesions. Being carrier molecules, NPs have to pass several physiological hurdles prior getting taken up by the plant and transmitted. Plant cell walls, made of cellulose, allow small NPs ranging in size from 5 to 20 nm to enter plant cells (Dietz and Herth 2011). Some nanoparticles have been shown to permeate the cell wall via larger pores (Kurepa et al. 2010). NPs can spread to neighboring plant tissues via both apoplastic and symplastic mechanisms (Ma et al. 2010). Wong et al. (2016) proposed a lipid interchange mechanism for NP administration via plant cells. The size, magnitude, and zeta potential of NPs influence their transit in plant cells.

3.4.1.2 Interaction Between Nanoparticles and Plant Pathways

NPs may have an impact on plant metabolism by delivering micronutrients (Liu and Lal 2015), regulating genes (Nair and Chung 2014), and interfering with several plant oxidative processes (Hossain et al. 2015). An oversupply of NPs can cause ROS, disrupting the oxidative process; other mechanisms are unknown. According to Pakrashi et al. (2014) and Cvjetko et al. (2017), nanoparticles can disrupt the electron transportation system in mitochondria and chloroplasts, causing an oxidative burst and high ROS levels. Highly stressful circumstances cause a decrease in the rate of carbon fixation, which heightens the inhibition of photosynthesis and may cause the photosystem to overproduce H_2O_2 and superoxide anion radicals. All biological elements are impacted when NPs produce ROS, which results in lipid peroxidation, DNA damage, and alterations in proteins (Van Breusegem and Dat 2006). Plants that interact with NPs experience increased lipid peroxidation and DNA damage, according to several research (Atha et al. 2012; Saha and Dutta Gupta 2017). Plant cells may die because of necrosis or apoptosis brought on by an increase in ROS levels. ROS contribute to biological processes, such as stress tolerance, despite their damaging character (Sharma et al. 2012a). Whether ROS serves a signaling or destructive purpose depends on how well they are balanced between being produced and scavenged. To carefully regulate the amount of ROS, the cells have evolved a strong antioxidant system. Plant defense mechanisms rely on both enzymatic (SOD, CAT, and guaiacol peroxidase) and non-enzymatic (ascorbate, glutathione, carotenoids, tocopherols, and phenolics) antioxidants (Sharma et al. 2012b). Plants subjected to NPs have been shown in multiple studies to create more antioxidant molecules (Jiang et al. 2014; Costa and Sharma 2016). Phytohormones can also affect plant stress response signaling (Sham et al. 2019). According to Sham et al. (2017), plant growth hormones are naturally occurring chemicals that control how plants grow and respond to stress. Various hormonal pathways can be triggered or repressed in response to abiotic stressors (O'Brien and Benková 2013). When exposed to AgNPs, cytokinin levels rose in red pepper (*Capsicum annuum*); meanwhile, auxin and ABA levels in cotton (*Gossypium* sp.) decreased in response to CuO NPs. This implies that NPs have an impact on the hormonal balance and metabolism of plants. According to several research (Perreault et al. 2014; Tripathi et al. 2017), NPs can also impact the amount and activity of photosynthetic pigments in plants. Plants may experience growth retardation or even die because of high NP concentrations' detrimental effects on photosynthesis (Tripathi et al. 2017) (Fig. 3.1).

3.4.1.3 The Role of Nanoparticles Under Salinity Conditions

Applying NPs is crucial to reducing the adverse impact of salinity on plants as an abiotic stressor. When Ag NPs are given to *Lathyrus sativus* undergoing salty conditions, the rate of the germination process, shoot and length of root, seedling

FW, and dry weight (DW) all increase during the germination process. This helps to regulate osmotic pressure and reduces the negative effects of salinity (Hojjat 2019). Noman et al. (2020) found that adding Cu NPs to the soil boosted plant development and yield but reduced oxidative stress in wheat. According to Shi et al. (2016), adding NPs to wheat not only stimulates plant development but also improves germination under salt stress. When wheat seeds are pretreated with Ag NPs, antioxidant enzyme activity changes, oxidative damage is reduced, and salt-stress tolerance increases in these plants (Kashyap et al. 2015).

Furthermore, during salt stress, ZnO NPs have been reported to raise sunflowers' DW (Torabian et al. 2016). It was discovered that CeO NPs (100 and 200 mg kg 1) improved *Brassica napus*'s physiological reactions to salt stress (100 mM NaCl). It is additionally recognized that CeO NPs increase plant biomass in canola that is stressed by salt (Rossi et al. 2016). When basil seeds are exposed to salt stress, the application of Ag NPs promotes seed germination (Darvishzadeh et al. 2015; Hojjat and Kamyab 2017). When utilized on *Satureja hortensis*, Ag NPs decrease the impact of salt stress on plant shoot length and percentage of germination while increasing plant tolerance to salt stress (Nejatzadeh 2021). Moreover, Ag NP application significantly increases plant resilience to salt in cumin plants under salt stress (Ekhtiyari and Moraghebi 2012). Fe3O4 NPs shield mint plants against oxidative stress brought on by elevated NaCl content, according to research by Askary et al. (2017) (Fig. 3.1).

3.4.1.4 The Role of Nanoparticles Under Drought Conditions

One of the main abiotic stresses that can severely restrict crop productivity is drought (Roy et al. 2021). Applying NPs to plants can effectively mitigate the effects of drought by boosting the activity of antioxidant enzymes, enhancing phytohormone levels, and changing physiological characteristics. It has been shown that adding analcite NPs to soil during hot, dry weather improves wheat sprouting and development of plants (Hossain et al. 2021). Furthermore, introducing ZnO NPs to soybean seeds in arid settings increases germination rates (Sedghi et al. 2013). Cu and Zn NPs applied to drought-stressed wheat plants increase relative moisture and antioxidant enzyme activity, lower thiobarbituric acid levels, modify substance precipitation, stabilize photosynthesis pigment levels in leaves, and reduce stress effects (Semida et al. 2021). Delivering SiO_2 NPs to barley during drought conditions can increase shoot length and relative water content (RWC) while decreasing superoxide radical generation and membrane damage. In arid conditions, applying Cu NPs to maize increased the level of water in the leaf tissue, biomass of the plant, and anthocyanin, chlorophyll, and carotenoid pigments (van Nguyen et al. 2022). According to Jaberzadeh et al. (2013), applying TiO_2 NPs topically to wheat can effectively counteract the yield drop brought on by drought stress. SiO_2 NPs administered to drought-stressed hawthorn decreased stomatal conductance and photosynthesis, according to Ashkavand et al. (2015). Silicon (Si) NPs, on the other hand, have been shown to lessen the effects of drought stress on banana plants.

Si NPs applied topically to coriander produced the highest amount of essential oil output and maximum antioxidant capacity during mild drought circumstances (Afshari et al. 2021). According to Shallan et al. (2016), cotton plants grown in dry environments can experience less detrimental impacts from drought stress when SiO_2 and TiO_2 NPs are applied topically (Fig. 3.1). Si NPs applied to the soil improve the relative water retention of chickpea plants, hence mitigating the adverse impacts of drought (Gunes et al. 2007). According to Badawy et al. (2021), silica (Si) and selenium (Se) NPs can benefit rice under saline circumstances by promoting growth and ion selectivity in roots. According to Khan et al. (2019), ZnO NPs can mitigate both drought stress and the harmful effects of Cd in wheat, even while drought stress amplifies the effects of Cd.

3.5 Future Prospects on the Potential of Nanoparticles for Improving Crop Tolerance to Abiotic Stress

According to Wu and Li (2022), nanobiotechnology has the potential to improve plant strain resilience, stress sensing/detection, controlled release and precise application of agrochemicals, transgenic events, and seed nanopriming. Because of their high dispersion and lack of HMs, these NMs could be synthesized for use in agriculture. Our major priority going forward ought to be investigating the biological impacts of nanozymes, or Mn_3O_4 NPs, in crops experiencing stress. It is necessary to look at the mechanisms behind how nanopriming induces germination in seeds, breaks dormancy in seeds, and interacts with seeds. By comprehending how NPs increase plant stress tolerance, scientists will be able to create customized NMs that address issues in agriculture. Furthermore, NMs undoubtedly have a promising future, particularly in terms of how well they work in plants. For instance, Santana et al. (2020) have created a specific administration strategy utilizing NMs to improve the photosynthesis of plants in low-light environments by turning chloroplasts into a "chloroplast factory." Utilizing NMs for CRISPR-Cas genome editing in cargo delivery would boost the effectiveness of genetic engineering to improve plant stress tolerance (Demirer et al. 2021). Creating laws and regulations could aid in controlling biological safety risks related to using NMs in agriculture. We think that the future of agriculture will be greatly impacted by NMs.

3.6 Conclusion

The scientific community must find suitable environmentally acceptable ways to alleviate agricultural crop losses caused by abiotic stress to attain environmentally friendly farming. Boosting the farming industry, controlling the worldwide hunger crisis, and increasing crop output and quality may all be accomplished by nanotechnology's creative and practical solutions. Extensive research is being done to determine how certain NPs may help mitigate damage caused by abiotic stressors while enhancing crop output and development in plants. By inducing ROS generation and phytotoxicity, NPs like TiO_2, SiO_2, and Ag NPs can activate plant defense mechanisms and mitigate the harmful impacts of abiotic stress. Because of their small size, NPs can also successfully permeate plant tissues. Once they are inside, they have a positive impact on the plant's appearance, and biological and metabolic operations, promoting development and increasing crop yield in plants that encounter a range of abiotic difficulties. Furthermore, the huge wide surface of NPs improves the transmission and uptake of specific nutrients. However, investigation into this field is still limited and, to some extent, incompatible, and NP applications in crop enhancement and ecologically friendly farming are still in their early stages. As a result, further study needs to focus on these topics to lessen the detrimental effects of NPs on plants and ecosystems (before applying NPs in the field, their size and concentration should be optimized): (a) how NPs affect plants and their metabolic processes at the biochemical and intracellular levels; and (b) the effects of NPs and suspected nanotoxicity.

References

Abd-Elrahman SH, Mostafa MAM (2015) Applications of nanotechnology in agriculture: an overview. Egypt J Soil Sci 55(2):197–214

Afshari M, Pazoki A, Sadeghipour O (2021) Foliar-applied silicon and its nanoparticles stimulates physio-chemical changes to improve growth, yield and active constituents of coriander (*Coriandrum sativum* L.) essential oil under different irrigation regimes. Silicon 13:4177–4188. https://doi.org/10.1007/s12633-02101101-8

Ashkavand P, Tabari M, Zarafshar M, Tomášková I et al (2015) Effect of SiO2 nanoparticles on drought resistance in hawthorn seedlings. For Res Pap 76:350–359

Ashraf M (1994) Breeding for salinity tolerance in plants. Crit Rev Plant Sci 13:17–42

Askary M, Talebi SM, Amini F, Bangan ADB (2017) Effects of iron nanoparticles on Mentha piperita L. under salinity stress. Biologija 63:65–67. https://doi.org/10.6001/biologija.v63i1.3476

Atha DH, Wang H, Petersen EJ, Cleveland D et al (2012) Copper oxide nanoparticle mediated DNA damage in terrestrial plant models. Environ Sci Technol 46:1819–1827. https://doi.org/10.1021/es202660k

Badawy SA, Zayed BA, Bassiouni SMA, Mahdi AHA, Majrashi A et al (2021) Influence of nano silicon and nano selenium on root characters, growth, ion selectivity, yield, and yield components of rice (*Oryza sativa* L.) under salinity conditions. Plan Theory 10:1657. https://doi.org/10.3390/plants10081657

Bray EA (2000) Response to abiotic stress. Biochem Mol Biol Plant:1158–1203

Bray EA, Bailey-Serres J, Weretilnyk E (2000) Responses to abiotic stresses. In: Buchanan BB, Gruissem W, Jones RL (eds) Biochemistry and molecular biology of plants. ASPB, Rockville, pp 1158–1203

Budhathoki NK, Zander KK (2019) Socio-economic impact of and adaptation to extreme heat and cold of farmers in the food bowl of Nepal. Int J Environ Res Public Health 16:1578

Cakmak I, Yilmaz A, Kalayci M, Ekiz H et al (1996) Zinc deficiency as a critical problem in wheat production in Central Anatolia. Plant Soil 180(2):165–172

Calanca PP (2017) Effects of abiotic stress in crop production. In: Ahmed M, Stockle CO (eds) Quantification of climate variability, adaptation and mitigation for agricultural sustainability. Springer, Cham, pp 165–180

Cota-Ruiz K, Hernández-Viezcas JA, Varela-Ramírez A, Valdés C, Núñez-Gastélum JA, Martínez-Martínez A, Gardea-Torresdey JL (2018) Toxicity of copper hydroxide nanoparticles, bulk copper hydroxide, and ionic copper to alfalfa plants: a spectroscopic and gene expression study. Environmental Pollution, 243:703–712

Costa MVJD, Sharma PK (2016) Effect of copper oxide nanoparticles on growth, morphology, photosynthesis, and antioxidant response in Oryza sativa. Photosynthetica 54:110–119. https://doi.org/10.1007/s11099-015-0167-5

Cramer GR, Urano K, Delrot S et al (2011) Effect of abiotic stress on plants: a systems biology perspective. BMC Plant Biol 11:163. https://doi.org/10.1186/1471-2229-11-163

Cvjetko P, Milošic A, Domijan AM, Vinkovic-Vrcek I, et al. (2017). Toxicity of silver ions and differently coated silver nanoparticles in *Allium cepa* roots. Ecotoxicol Environ Saf 137: 18–28. https://doi.org/10.1016/j.ecoenv.2016.11.009

Dai A (2011) Characteristics and trends in various forms of the palmer drought severity index during 1900–2008. J Geophys Res Atmos 116. https://doi.org/10.1029/2010jd015541

Darvishzadeh F, Najatzadeh F, Iranbakhsh AR (2015) Effect of silver nanoparticles on salinity tolerance of basil plant in germination stages under laboratory conditions. J Cell Biotechnol Mol 20:63–70

Dasgupta N, Ranjan S, Mundekkad D et al (2015) Nanotechnology in agro-food: from field to plate. Food Res Int 69:381–400

Demirer GS, Silva TN, Jackson CT, Thomas JB, W. Ehrhardt D, Rhee SY, Landry MP (2021) Nanotechnology to advance CRISPR–Cas genetic engineering of plants. Nature Nanotechnology, 16(3):243–250

Dietz KJ, Herth S (2011) Plant nanotoxicology. Trends Plant Sci 16:582–589

Dwivedi S, Saquib Q, Al-Khedhairy AA, Musarrat J (2016) Understanding the role of nanomaterials in agriculture. In: Singh DP, Singh HB, Prabha R (eds) Microbial inoculants in sustainable agricultural productivity. Springer, New Delhi, pp 271–288

Eichert T, Kurtz A, Steiner U, Goldbach HE (2008) Size exclusion limits and lateral heterogeneity of the stomatal foliar uptake pathway for aqueous solutes and water-suspended nanoparticles. Physiol Plant 134(1):151160. https://doi.org/10.1111/j.13993054.2008.01135.x

Ekhtiyari R, Moraghebi F (2012) Effect of nanosilver particles on salinity tolerance of cumin (*Cuminum cyminum* L.). J Plant Biotechnol 25:99–107

Fageria NK, Baligar VC, Jones CA (2010) Growth and mineral nutrition of field crops. CRC Press, Florida

Field CB, Barros VR, Mastrandrea MD et al (2014) Summary for policymakers. Climate change: impacts, adaptation, and vulnerability Part A: global and sectoral aspects contribution of working group II to the fifth assessment report of the intergovernmental panel on climate change. Cambridge University Press, London, pp 1–32

Ghidan AY, Al Antary TM (2019) Applications of nanotechnology in agriculture. In: Applications of nanobiotechnology. IntechOpen

Grover A, Kapoor A, Lakshmi OS et al (2001) Understanding molecular alphabets of the plant abiotic stress responses. Curr Sci (Special Section: Plant Mol Biol) 80:206–216

Gunes A, Pilbeam DJ, Inal A, Bagci EG et al (2007) Influence of silicon on antioxidant mechanisms and lipid peroxidation in chickpea (Cicer arietinum L.) cultivars under drought stress. J Plant Interact 2:105–113. https://doi.org/10.1080/17429140701529399

Hojjat SS (2019) Effect of interaction between Ag nanoparticles and salinity on germination stages of *Lathyrus sativus* L. J Environ Soil Sci 2:186–191. https://doi.org/10.32474/oajess.2019.02.000132

Hojjat SS, Kamyab M (2017) The effect of silver nanoparticle on fenugreek seed germination under salinity levels. Russ Agric Sci 43:61–65. https://doi.org/10.3103/S1068367417010189

Hossain Z, Mustafa G, Komatsu S (2015) Plant responses to nanoparticle stress. Int J Mol Sci 16: 26644–26653. https://doi.org/10.3390/ijms161125980
Hossain A, Skalicky M, Brestic M, Maitra S et al (2021) Consequences and mitigation strategies of abiotic stresses in wheat (*Triticum aestivum* L.) under the changing climate. Agronomy 11:241. https://doi.org/10.3390/agronomy11020241
Hotton C, Hueber F, Detal G (2001) Early terrestrial plant environments: an example from the Emsian of Gaspé, Canada. In: Gensel PG, Edwards D (eds) Plants invade the land: evolutionary and environmental perspective. Columbia University Press, New York, pp 179–212
Iizumi T, Furuya J, Shen Z et al (2017) Responses of crop yield growth to global temperature and socioeconomic changes. Sci Rep 7:1–10
Jaberzadeh A, Moaveni P, Tohidi Moghadam HR, Zahedi H (2013) Influence of bulk and nanoparticles titanium foliar application on some agronomic traits, seed gluten and starch contents of wheat subjected to water deficit stress. Not Bot Horti Agrobot 41:201–207. https://doi.org/10.15835/nbha4119093
Janmohammadi M, Sabaghnia N, Ahadnezhad A (2015) Impact of silicon dioxide nanoparticles on seedling early growth of lentil (*Lens culinaris* Medik.) genotypes with various origins. Poljoprivreda i Sumarstvo 61(3):19
Jiang HS, Qiu XN, Li GB, Li W et al (2014) Silver nanoparticles induced accumulation of reactive oxygen species and alteration of antioxidant systems in the aquatic plant *Spirodela polyrhiza*. Environ Toxicol Chem 33:1398–1405. https://doi.org/10.1002/etc.2577
Kalita R, Saha O, Rahman N, Tiwari S et al (2021) Nanotechnology in agriculture. In: Al-Khayri JM, Ansari MI, Singh AK (eds) Nanobiotechnology. Springer, Cham. https://doi.org/10.1007/978-3-030-73606-4_5
Kashyap PL, Xiang X, Heiden P (2015) Chitosan nanoparticle-based delivery systems for sustainable agriculture. Int J Biol Macromol 77:36–51. https://doi.org/10.1016/j.ijbiomac.2015.02.039
Khan ZS, Rizwan M, Hafeez M, Ali S et al (2019) The accumulation of cadmium in wheat (Triticum aestivum) as influenced by zinc oxide nanoparticles and soil moisture conditions. Environ Sci Pollut Res 26:19859–19870. https://doi.org/10.1007/s11356-019-05333-5
Kou TJ, Yu WW, Lam SK et al (2018) Differential root responses in two cultivars of winter wheat (*Triticum aestivum* L.) to elevated ozone concentration under fully open-air field conditions. J Agron Crop Sci 204:325–332
Kumar A, Sengar RS, Singh A, Dixit R et al (2018) Biotechnological tools for enhancing abiotic stress tolerance in plant. In: Sengar R, Singh A (eds) Eco-friendly agro-biological techniques for enhancing crop productivity. Springer, Singapore
Kurepa J, Paunesku T, Vogt S, Arora H et al (2010) Uptake and distribution of ultrasmall anatase TiO2 alizarin red S nanoconjugates in *Arabidopsis thaliana*. Nano Lett 10:2296–2302. https://doi.org/10.1021/nl903518f
Liu R, Lal R (2015) Potentials of engineered nanoparticles as fertilizers for increasing agronomic productions. Sci Total Environ 514:131–139. https://doi.org/10.1016/j.scitotenv.2015.01.104
Ma X, Geisler-Lee J, Deng Y, Kolmakov A (2010) Interactions between engineered nanoparticles (ENPs) and plants: phytotoxicity, uptake and accumulation. Sci Total Environ 408:3053–3061. https://doi.org/10.1016/j.scitotenv.2010.03.031
Marchiol L (2018) Nanotechnology in agriculture: new opportunities and perspectives. In: Çelik Ö (ed) New vision in plant science. Intech, London, pp 121–141
Mazumdar H, Ahmed GU (2011) Phytotoxicity effect of silver nanoparticles on Oryza sativa. Int J Chem Tech Res 3:1494–1500
Mirzajani F, Askari H, Hamzelou S, Farzaneh M et al (2013) Effect of silver nanoparticles on Oryza sativa L. and its rhizosphere bacteria. Ecotoxicol Environ Saf 88:48–54. https://doi.org/10.1016/j.ecoenv.2012.10.018
Munns R, Gilliham M (2015) Salinity tolerance of crops—what is the cost? New Phytol 208(668): 673
Nair PMG, & Chung IM (2014) Physiological and molecular level effects of silver nanoparticles exposure in rice (Oryza sativa L.) seedlings. Chemosphere, 112:105–113

Nejatzadeh F (2021) Effect of silver nanoparticles on salt tolerance of *Satureja hortensis* L. during in vitro and in vivo germination tests. Heliyon 7:e05981. https://doi.org/10.1016/j.heliyon.2021.e05981

Noman M, Shahid M, Ahmed T, Tahir M et al (2020) Green copper nanoparticles from a native *Klebsiella pneumoniae* strain alleviated oxidative stress impairment of wheat plants by reducing the chromium bioavailability and increasing the growth. Ecotoxicol Environ Saf 192:110303. https://doi.org/10.1016/j.ecoenv.2020.110303

Nouman W, Qureshi MK, Shaheen M et al (2018) Variation in plant bioactive compounds and antioxidant activities under salt stress. In: Vats S (ed) Biotic and abiotic stress tolerance in plants. Springer, Singapore, pp 77–101

O'Brien JA, Benková E (2013) Cytokinin cross-talking during biotic and abiotic stress responses. Front Plant Sci 4:451. https://doi.org/10.3389/fpls.2013.00451

Onyekachi OG, Boniface OO, Gemlack NF et al (2019) The effect of climate change on abiotic plant stress: a review. In: de Oliveira AB (ed) Abiotic and biotic stress in plants. IntechOpen. Available via DIALOG. https://www.intechopen.com/books/abiotic-and-biotic-stress-in-plants/the-effect-of-climate-change-on-abiotic-plant-stress-a-review. Accessed 7 July 2020

Pakrashi S, Jain N, Dalai S, Jayakumar J et al (2014) In vivo genotoxicity assessment of titanium dioxide nanoparticles by *Allium cepa* root tip assay at high exposure concentrations. PLoS One 9:e87789. https://doi.org/10.1371/journal.pone.0087789

Perreault F, Samadani M, Dewez D (2014) Effect of soluble copper released from copper oxide nanoparticles solubilisation on growth and photosynthetic processes of *Lemna gibba* L. Nanotoxicology 8:374–382. https://doi.org/10.3109/17435390.2013.789936

Prasad R, Kumar V, Prasad KS (2014) Nanotechnology in sustainable agriculture: present concerns and future aspects. Afr J of Biotechnol 13(6):705–713

Rady MM, Boriek SHK, Abd El-Mageed TA, Seif El-Yazal MA et al (2021) Exogenous gibberellic acid or dilute bee honey boosts drought stress tolerance in *Vicia faba* by rebalancing osmoprotectants, antioxidants, nutrients, and phytohormones. Plan Theory 10:748. https://doi.org/10.3390/plants10040748

Rajput VD, Minkina TM, Behal A, Sushkova SN, Mandzhieva S, Singh R, Movsesyan HS (2018) Effects of zincoxide nanoparticles on soil, plants, animals and soil organisms: A review. Environmental Nanotechnology, Monitoring & Management, 9:76–84

Rajput V, Minkina T, Mazarji M, Shende S, Sushkova S, Mandzhieva S, Jatav H (2020) Accumulation of nanoparticles in the soil-plant systems and their effects on human health. Annals of Agricultural Sciences, 65(2):137–143

Raliya R, Saharan V, Dimkpa C, Biswas P (2017) Nanofertilizer for precision and sustainable agriculture: current state and future perspectives. J Agric Food Chem 66(26):6487–6503

Rossi L, Zhang W, Lombardini L, Ma X (2016) The impact of cerium oxide nanoparticles on the salt stress responses of Brassica napus L. Environ Pollut 219:28–36. https://doi.org/10.1016/j.envpol.2016.09.060

Roy R, Núñez-Delgado A, Sultana S, Wang J et al (2021) Additions of optimum water spent mushroom compost and wood biochar to improve the growth performance of *Althaea rosea* in drought-prone coal-mined spoils. J Environ Manag 295:113076. https://doi.org/10.1016/j.jenvman.2021.113076

Saha N, Dutta GS (2017) Low-dose toxicity of biogenic silver nanoparticles fabricated by Swertia chirata on root tips and flower buds of *Allium cepa*. J Hazard Mater 330:18–28. https://doi.org/10.1016/j.jhazmat.2017.01.021

Sahin U, Ekinci M, Ors S et al (2018) Effects of individual and combined effects of salinity and drought on physiological, nutritional and biochemical properties of cabbage (*Brassica oleracea var. capitata*). Sci Hortic 240:196–204

Santana I, Wu H, Hu P, Giraldo JP (2020) Targeted delivery of nanomaterials with chemical cargoes in plants enabled by a biorecognition motif. Nature communications, 11(1):2045

Sedghi M, Mitra H, Sahar T (2013) Effect of nano zinc oxide on the germination of soybean seeds under drought stress. Ann West Univ Timi Ser Biol 16:73–78

Selden PA, Edwards D (1989) Colonisation of the land. In: Keith A, Briggs DEG (eds) Evolution and the fossil record. Smithsonian Institution Press, Washington, pp 122–152

Semida WM, Abdelkhalik A, Mohamed GF, Abd El-Mageed TA et al (2021) Foliar application of zinc oxide nanoparticles promotes drought stress tolerance in eggplant (*Solanum melongena* L.). Plan Theory 10:421. https://doi.org/10.3390/plants10020421

Shaaban A, Al-Elwany OA, Abdou NM, Hemida KA et al (2022) Filter mud enhanced yield and soil properties of water-stressed *Lupinus termis* L. in saline calcareous soil. J Soil Sci Plant Nutr 22:1572–1588. https://doi.org/10.1007/s42729-021-00755-y

Shallan MA, Hassan HM, Namich AA, Ibrahim AA (2016) Biochemical and physiological effects of TiO_2 and SiO_2 nanoparticles on cotton plant under drought stress. Res J Pharm Biol Chem Sci 7:1540–1551

Sham A, Moustafa K, Al-Shamisi S, Alyan S et al (2017) Microarray analysis of Arabidopsis WRKY33 mutants in response to the necrotrophic fungus *Botrytis cinerea*. PLoS One 12: e0172343. https://doi.org/10.1371/journal.pone.0172343

Sham A, Al-Ashram H, Whitely K, El-Tarabily KA et al (2019) Meta transcriptomic analysis of multiple environmental stresses identifies RAP2.4 gene associated with Arabidopsis immunity to botrytis cinerea. Sci Rep 9:17010. https://doi.org/10.1038/s41598-019-53694-1

Sharma P, Bhatt D, Zaidi MGH, Saradhi PP et al (2012a) Silver nanoparticle-mediated enhancement in growth and antioxidant status of *Brassica juncea*. Appl Biochem Biotechnol 167:2225–2233. https://doi.org/10.1007/s12010-012-9759-8

Sharma P, Jha AB, Dubey RS, Pessarakli M (2012b) Reactive oxygen species, oxidative damage, and antioxidative defense mechanism in plants under stressful conditions. J Bot 2012:217037. https://doi.org/10.1155/2012/217037

Shi Y, Zhang Y, Han W, Feng R et al (2016) Silicon enhances water stress tolerance by improving root hydraulic conductance in *Solanum lycopersicum* L. Front Plant Sci 7:196. https://doi.org/10.3389/fpls.2016.00196

Singh S, Singh BK, Yadav SM, Gupta AK (2015) Applications of nanotechnology in agricultural and their role in disease management. Res J Nanosci Nanotechnol 5(1):1–5

Srilatha B (2011) Nanotechnology in agriculture. J Nanomedic Nanotechnol 2:123. The European Commission. Commission Recommendation of 18 Oct 2011 on the Definition of Nanomaterial. Brussels, Belgium: Official Journal of the European Union

Staroń A, Długosz O, Pulit-Prociak J, Banach M. (2020). Analysis of the exposure of organisms to the action of nanomaterials. Materials, 13(2):349

Torabian S, Zahedi M, Khoshgoftar AH (2016) Effects of foliar spray of two kinds of zinc oxide on the growth and ion concentration of sunflower cultivars under salt stress. J Plant Nutr 39:172–180. https://doi.org/10.1080/01904167.2015.1009107

Tripathi DK, Singh S, Singh S, Srivastava PK et al (2017) Nitric oxide alleviates silver nanoparticles (AgNps)-induced phytotoxicity in *Pisum sativum* seedlings. Plant Physiol Biochem 110:167–177. https://doi.org/10.1016/j.plaphy.2016.06.015

United Nations (2017) Department of Economic and Social Affairs, Population Division. World population prospects: the 2017 revision, key findings and advance tables. Working Paper No. ESA/P/WP/248 (ed) New York, United Nations

Van Breusegem F, Dat JF (2006) Reactive oxygen species in plant cell death. Plant Physiol 141: 384–390. https://doi.org/10.1104/pp.106.078295

van Nguyen D, Nguyen HM, Le NT, Nguyen KH et al (2022) Copper nanoparticle application enhances plant growth and grain yield in maize under drought stress conditions. J Plant Growth Regul 41:364–375. https://doi.org/10.1007/s00344-021-10301-w

Vibha (2016) Macrophomina phaseolina: the most destructive soybean fungal pathogen of global concern. In: Kumar P, Gupta VK, Tiwari AK, Kamle M (eds) Current trends in plant disease diagnostics and management practices. Springer, Cham, pp 193–205

Wani S, Kumar V, Shriram V (2016) Phytohormones and their metabolic engineering for abiotic stress tolerance in crop plants. Crop J 4:162–176

White PJ, Pongrac P (2017) In: Shabala S (ed) 12 heavy-metal toxicity in plants, 2nd edn. Plant Stress Physiology, pp 301–328

Wong MH, Misra RP, Giraldo JP, Kwak SY et al (2016) Lipid exchange envelope penetration (LEEP) of nanoparticles for plant engineering: a universal localization mechanism. Nano Lett 16:1161–1172. https://doi.org/10.1021/acs.nanolett.5b04467

Wood AJ (2005) Eco-physiological adaptations to limited water environments. In: Jenks MA, Hasegawa PM (eds) Plant abiotic stress. Wiley, New York, pp 1–10

Wu H, & Li Z (2022) Nano-enabled agriculture: How do nanoparticles cross barriers in plants?. Plant Communications, 3(6)

Zhao C, Liu B, Piao S et al (2017) Temperature increase reduces global yields of major crops in four independent estimates. Proc Natl Acad Sci 114:9326–9331

Chapter 4
Nano-Bionics Approach for Augmentation of Plant Functions

Anukul Barwal, Srishti Chopra, Sakshi Chopra, Sukhdeep Kaur, Vineet Kumar, and Praveen Guleria

Abstract Nanotechnology is an emerging field of science and technology that offers a promising future in enhancing plant functions. Nano-bionics is a novel approach that innovatively integrates nanotechnology and plant biology. The recent era of agriculture needs sustainability to fulfill the demands of the growing population. In such cases, nano-bionics prove their worth, as they possess the ability to alter plant growth and development during stressed conditions positively. Moreover, the embedding of nanoparticles/nanomaterials into plant cells upgrades the functioning of plants, such as the induction of chemical sensing to monitor environmental changes and other hazards, promotion of light absorption capacity, and improvement of yield. Thus, nano-bionics can be considered a potential strategy to enhance sustainability in agriculture. However, ethical concerns and prior testing are still under consideration. This article discusses different types of nanomaterials used to design nano-bionics, the mechanism of nanomaterial/nanoparticle interaction, and applications of nano-bionics in detail.

Keywords Nanoparticles · Nano-bionics · Light-emitting plants · Nano-bionics tool kit · Interaction of NPs with plants

4.1 Introduction

The present agricultural era is of green revolution 2.0, focusing on enhanced crop yields through chemical fertilizers and high-yield varieties. However, modern challenges like climate change, resource depletion, and the urgent need for food security still require attention. The current agricultural studies emphasize sustainable

A. Barwal · S. Chopra · S. Chopra · S. Kaur · P. Guleria (✉)
Plant Biotechnology & Genetic Engineering Lab, Department of Biotechnology, DAV University, Jalandhar, Punjab, India

V. Kumar
Department of Biotechnology, School of Bioengineering and Biosciences, Lovely Professional University, Phagwara, Punjab, India

S. K. Yadav, A. Kumari (eds.), *Nanobiotechnology in Agri-food Sector*, Sustainability Sciences in Asia and Africa, https://doi.org/10.1007/978-981-96-7756-6_4

practices to boost productivity and protect the environment for future generations (Tsakok and Guedegbe 2019; Hamdan et al. 2022). So, the studies are now focusing on enhancing the potential of plants to grow and thrive under stressed conditions using innovative techniques like biotechnology and genetic engineering. By improving traits such as abiotic and biotic stress resistance and nutrient uptake, more resilient and productive crops can be created (Rizzo et al. 2023).

Given this, the generation of nano-bionics can be a striking novel approach for the development of high-end crops (Ranjan et al. 2022). Nanotechnology is a fast-growing technology that deals with the manipulation of matter at the atomic level (ranging from 1 to 100 nm). This strategy finds application in the medical, electrical, and agricultural sectors. Nano-bionics also deals with designing novel materials with special abilities at the nanoscale and using nanoparticles/nanomaterials (NPs/NMs) to enhance plant functions, such as improving photosynthesis and nutrient absorption (Tan et al. 2025). NPs possess unique properties of large surface area, high reactivity, conductivity, and magnetic properties. Therefore, these are considered the "magic bullets," as they can serve as herbicides, insecticides, fertilizers, etc. (Alagoz et al. 2022).

The word "bio" means life, and "onics" refers to electronics; thus, the amalgamation of nature and technology offers nano-bionics as a sustainable solution for upcoming agricultural challenges. Plant cells have exceptional potential to generate energy using sunlight through a process called photosynthesis (Barbinta-Patrascu et al. 2024). Besides, the size of plant cells is almost similar to the size of NPs, and major plant organelles also range in nanoscale (Sharma and Kar 2019). The size comparison reveals the importance of the NMs inside the plant cell. Thus, it was documented that by introducing NPs/NMs, plants with exceptional capabilities can be created, such as light-illuminating plants or sensor plants to sense environmental stresses like drought or heavy metal contamination (Faizan et al. 2023). Nano-bionics finds enormous applications in the research fields of artificial intelligence, nanobiotechnology, nanorobotics, medicine, and military purposes.

The main aim of nano-bionics is to integrate NPs into the plant organelles to improve and enhance their physiological mechanisms, thus supporting plant growth (Fig. 4.1). The use of engineered NPs to create nano-bionics can improve essential processes like photosynthesis, nutrient absorption, and stress resistance. For example, three times enhanced photosynthetic activity and electron transport rates under the influence of single-walled carbon nanotubes (SWNTs) and nanoceria in plants have been reported. Their exposure even allowed the plants to function as photonic chemical sensors (Ranjan et al. 2022; Bhaskar et al. 2024). Moreover, continuous studies on such modified plants can also lead to the generation of biosensors with high sensitivity to measure even a small fluctuation in the environment (Kumar et al. 2024). Hence, nano-bionics prioritize agricultural productivity as well as ecological health, which can pave the way for the development of tolerant food crops capable of supporting their growth and productivity under adverse conditions.

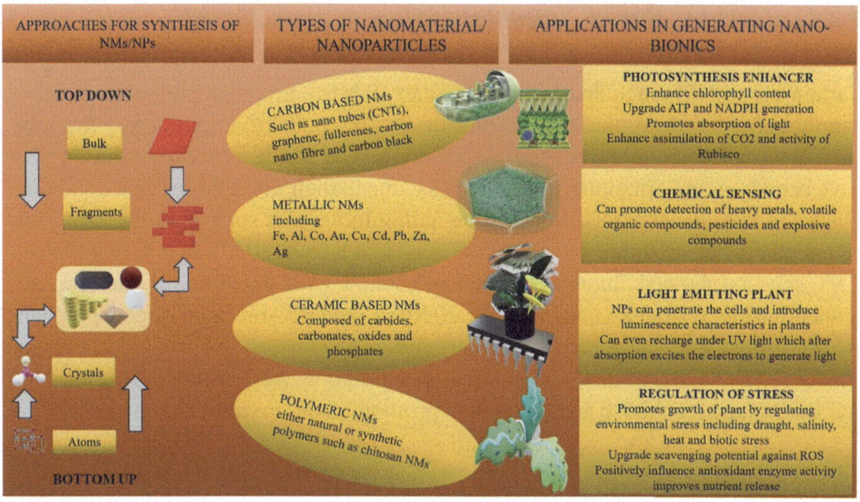

Fig. 4.1 The figure depicts the approaches for the synthesis of NMs/NPs, different types of NMs, and general applications of these NMs in enhancing plant functions

4.2 The Nano-Bionics Toolkit: Materials and Mechanisms

The term nanotechnology deals with materials working at the nanoscale. Nanotechnology possesses multiple applications in various fields such as biomedical, healthcare, environmental remediation, food, pharmaceutical industry, and agriculture (Nasrollahzadeh et al. 2019). There are two basic approaches, "top-down" and "bottom-up," which are used in nanotechnology to generate NMs/NPs. The bottom-up approach involves building up atoms into molecules in order to assemble nanostructures, materials, and devices. However, the top-down approach includes making structures and devices from larger entities down to their atomic levels. Because of different approaches to synthesis, nanotechnology is widely used in drug delivery, tissue engineering, biosensors, dentistry, and nanoscale diagnostic tool kits (Subramani et al. 2019).

As mentioned earlier, the primary element of nanotechnology known as NPs/NMs is less than 100 nm in size, possessing unique physical and chemical properties (Wiederrecht et al. 2023). There are different shapes of NMs, like NPs, nanorods, and nanosheets, which can be distinguished based on their dimensions, where nanomaterials with zero dimensions are said to be NPs, one dimension are nanorods, and two dimensions are films or layers (Kolahalam et al. 2019; Wiederrecht et al. 2023). The classification of NMs is based upon different criteria, but one of them is based on the type of synthesizing materials and is so, broadly classified as follows:

4.2.1 Carbon-Based Nanomaterials

Carbon-based NMs include carbon nanotubes (CNTs), graphene, fullerenes, carbon nanofiber, and carbon black. These are thermally conductive, but across the tube, these are non-conductive (Parvej et al. 2022; Gu et al. 2022). Few carbon-based NMs have been used in agriculture, such as CNTs, due to their good fluorescence stability. These are broadly exploited to detect signaling molecules such as H_2O_2, Ca^{2+}, and NO in plants during biotic and abiotic stress (Huang et al. 2020). In cherry tomatoes, CNT sensors showed fluorescence on interaction with diazinon, glyphosate, and semicarbazide residues (Fatahi et al. 2020). Likewise, nitrogen-doped graphene quantum dots were also used to prepare test strip sensors for pesticide detection (Liu et al. 2018). It was also reported that quantum dots were applied to plastic films and LEDs for greenhouse applications to promote plant growth (Zhang et al. 2020).

4.2.2 Metallic Nanomaterials

Metal-based NMs/NPs can be synthesized through constructive or destructive processes using specific metal materials such as iron (Fe), aluminum (Al), cobalt (Co), gold (Au), copper (Cu), cadmium (Cd), lead (Pb), zinc (Zn), and silver (Ag). These NMs have a large surface area to volume ratio, are highly sensitive to the UV-visible range, and possess significant catalytic and electrical potential and thermal characteristics (Mekuye and Abera 2023). Studies have revealed the potential of metallic NPs like Ag-NPs and ZnO NPs to improve germination and plant growth (Kale et al. 2021; Umavathi et al. 2021). Likewise, Fe_2O_3 NPs, reportedly, improved the amount of Fe in soil and accumulation in plant cells, thus regulating the phytohormone accumulation and antioxidant enzyme activity to improve the growth of peanuts (Zia-ur-Rehman et al. 2018). Furthermore, MgO NPs priming was reported to enhance germination and morphological and biochemical parameters in various plants, including mungbean, radish, mustard, and horse gram (Fernandes et al. 2020; Anand et al. 2020; Sharma et al. 2022; Gautam et al. 2023). Likewise, exposure to the mixture of calcium carbonate and hydroxyl apatite NPs was reported to promote the yield of soybean plants (El-Hady and Hussein 2021). Similarly, seeds treated with TiO_2 NPs resulted in chlorophyll content as well as a photosynthetic rate increment in spinach plants (Dağhan 2018).

4.2.3 Ceramic-Based Nanomaterials

Ceramic-based NMs are inorganic solids made up of carbides, carbonates, oxides, and phosphates formed through heat and successive cooling. Ceramic NMs, reportedly, can improve photocatalysis and can be used for imaging and photodegradation of dyes (Mekuye and Abera 2023). In agriculture, the utilization of calcium carbonate NMs was documented to positively affect plant development (Jayaraj et al. 2024). The exposure of calcium carbide NPs was observed to positively influence the growth and yield of crops such as green pod and okra, maximally up to 37% (Mahmood et al. 2014; Li et al. 2020).

4.2.4 Polymeric Nanomaterials

The polymeric NMs are colloidal systems that include either natural or synthetic polymers. These can also be synthesized through the direct polymerization of monomers or by using the dispersion of preformed polymers (Sarkar et al. 2020). These NMs can act as a carrier system in pest control (Pascoli et al. 2018). It was reported that polymeric NPs, such as chitosan, can modify the plant signaling pathways and promote plant growth and development. Likewise, nitric oxide and salicylic acid-releasing NPs can also be used as protection agents for plants against biotic and abiotic stresses (Pereira et al. 2019).

4.3 Mechanism of Interaction of NPs and Plants

The cell wall is considered a complex, however, an integral part of the plant cell (Koziel et al. 2021). It consists of a network of polysaccharide polymers such as cellulose, hemicellulose, and pectin; glycoproteins; and lignin. The cell wall plays a critical role in determining plant morphogenesis, acts as a mechanical support for plants, helps in transporting water and nutrients, and provides defense against biotic and abiotic stresses (Zhang et al. 2021). Transportation across membranes plays a crucial role in providing source material to plants for their growth and development, as shown in Fig. 4.2 (Klejchova et al. 2021; Cheung et al. 2020).

The cell wall serves as a barrier due to the small pore sizes ranging between 20 and 100 nm, allowing only NPs to penetrate. Thus, cell walls provide scope to be targeted in many studies in order to promote agricultural sustainability. NMs have been reported to penetrate cell walls through apoplastic or symplastic pathways, leading to their entry into cells and organelles. However, the exact mechanism is not completely understood and is still under study. Previous studies have documented that NPs pierce into cell walls through endocytosis, get engulfed by membrane vesicles, and are transported into cells, or NPs can diffuse from their high concentration gradient to low concentration gradient via stomatal openings. The endocytosis-mediated absorption of carbon-based NPs by roots in *Catharanthus roseus* has already been reported (Etxeberria et al. 2006). Moreover, the specialized proteins derived from transmembrane channels present in cell membranes can also act as to-and-fro gateways for NPs passage. The influence of NMs/NPs exposure

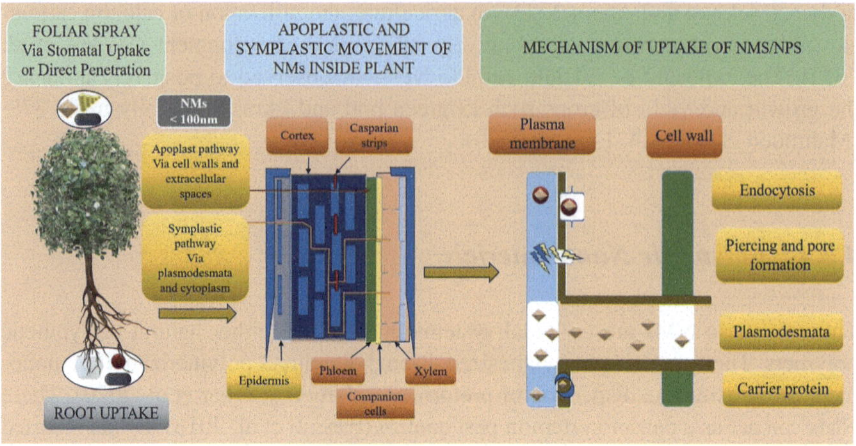

Fig. 4.2 The figure shows the general mechanism of uptake and transportation of NMs/NPs inside the plant system

significantly depends upon their size, type, and concentration. Studies have reported positive as well as negative impacts of NPs exposure on plants (Tenhaken 2015).

Similarly, the root cells have been reported to absorb NPs/NMs and transport them into the plant cells. Further, the absorbed small-sized NPs can easily penetrate via various pathways as mentioned earlier (Mousavi et al. 2007). During nutrient-deficient conditions, plants release root exudates that are organic compounds such as oxalate and malate. NPs interact with these root exudates, and this interaction helps in the uptake of NPs by plants. However, if the size of the NP-root exudate aggregate increases, the uptake process is halted (Cervantes-Avilés et al. 2021; Faizan et al. 2023). Additionally, NPs possess specific surface charges that can influence their absorption via roots. It was observed that positively charged NPs are absorbed more readily in comparison to negatively charged particles by the roots. The root epidermis, cell wall porosity levels, and presence of the Casparian strip act as additional barriers to NPs uptake (Bao et al. 2020). The Fe_2O_3 NPs and Ti-nano conjugates have been studied to enter root epidermis in pumpkin and *Arabidopsis thaliana* via the adsorption process, maintained in an aqueous medium (Munir et al. 2023; Wang et al. 2023; Kurepa et al. 2010).

Moreover, foliar NM exposure has also been reported as a medium of plant treatment. During foliar exposure, NMs are absorbed by two methods: stomatal uptake and direct penetration. In stomatal uptake, the size of the stomata plays a critical role in the regulation and exchange of gases and water in plants. NMs sprayed onto the leaves are absorbed by stomata, from where they get translocated into the whole plant (Yang et al. 2015). The leaf epidermis, composed of wax, cutin, and pectin, possesses two channels on its surface: hydrophilic and lipophilic. These channels allow the absorption of NMs via diffusion and infiltration (Bussières et al. 2014). A study documented that the absorption of ZnO NPs in *Triticum aestivum* via stomata has resulted in the delivery of micronutrients into the plant cells (Zhu et al. 2020).

Post-penetration, the NMs/NPs get translocated into plants/tissues via symplastic and apoplast transportation routes. The movement of water and substances between the cytoplasm of adjacent cells via plasmodesmata and sieve plates occurs in symplastic transport. However, the apoplastic route involves transportation from roots to vascular tissues through xylem vessels, extracellular spaces, and cell walls, leading to upward translocation of NMs into aerial parts (González-Melendi et al. 2008). The apoplastic movement of cerium oxide NMs reported in *Raphanus sativus L.* ultimately influenced their growth and development (Zhang et al. 2017). Another study has demonstrated the symplastic movement of carbon NMs in *Oryza sativa*, thus affecting plant growth (Tombuloglu et al. 2019).

4.4 Applications of Nano-Bionics

4.4.1 Enhancing Photosynthesis

Alterations in photosynthetic activity can increase crop yield while minimizing chemical inputs. Scientists are exploring innovative ways to use different types of NPs in order to optimize light capturing and utilize the capability of plants, which can ultimately improve photosynthesis. Nano-bionics possesses the ability to enhance the light absorption capacity of plants up to the near-infrared region, as chloroplasts can absorb light within the visible spectrum only. Nano-bionic applications can promote chlorophyll content, which can boost crop yield, biomass, and production. A recent study documented an improvement in the light absorption potential of *Arachis hypogaea* chloroplast under the influence of Ag and core-shell silica (AgS) NSs due to increased photoluminescence properties, which improved the electron transport chain and up-regulated the generation of ATP and NADPH (Dhiman et al. 2024).

Plant nano-bionics-based bio-mimetic materials are being widely studied that can efficiently promote photosynthesis by sensing biological alteration in order to harvest light. For instance, carbon nanotubes, such as SWNTs, have the capability to passively transport within the chloroplast, where they get localized. Here, it contributes to promoting light absorption, which in turn increases electron transport rates, thus significantly increasing photosynthesis. Moreover, it boosts plant functions as photonic chemical sensors that can help in monitoring nitric oxide by enabling near-infrared fluorescence monitoring (Giraldo et al. 2014; Kumar et al. 2024). Carbon quantum dots (CQDs) are zero-dimensional NMs that can act as artificial photosynthetic pigments upon foliar application. A study indicated that at concentrations ranging from 0 to 400 ppm, CQDs showed significant enhancement in Rubisco activity and CO_2 assimilation rate depending upon the light conditions. The potential of CQDs as light converters or photosensitizers has been explored under white light and red/blue (R/B) light combinations (Izati Fadzil et al. 2025).

Furthermore, the foliar application of CQDs in indoor hydroponics in *Brassica juncea* has been reported to act as a photosynthesis enhancer (Izati Fadzil et al. 2025; Tian et al. 2024). Similarly, TiO_2 NPs also promoted photosynthesis by enhancing the activity of Rubisco and elevating chlorophyll content in *Brassica napus* and *Solanum lycopersicum* L. (Qu et al. 2021; Ranjan et al. 2022). Zinc acts as an antioxidant parameter to regulate reactive oxygen species (ROS) in plant cells during stress conditions. It helps to modulate the cellular redox state and stress responses. Seed priming and foliar exposure of ZnO NPs have also revealed noteworthy promise as effective enhancers of plant growth and photosynthesis, particularly at lower concentrations (10–50 mg/L). Its exposure has been reported to increase chlorophyll content by 40–49% and improve photosystem efficiency, stomatal conductance, and CO_2 assimilation by up to 35% (Rajput et al. 2021).

4.4.2 Chemical Sensing

Plant-based nano-bionics can potentially promote the competence of plants for chemical sensing, such as the detection of heavy metals, volatile organic compounds, pesticides, and explosive compounds. The introduction of hybrid biosensors in plants is being extensively studied nowadays due to their ability to detect pollutants, amplify signals with accuracy, and act as measurable signal detectors. These biosensors are built with NMs that possess specific molecules or polymers to promote the sensitivity and specificity of the sensors for pollutant detection. On the introduction of these sensors to plants, specific proteins present in plants interact with the target molecule, which triggers a biochemical signal, thus activating the reporter gene. The expression of a gene is then converted into a fluorescence signal or a color change. The intensity of the observed color is directly correlated with the concentration of the detected pollutant (Nikoleli et al. 2018; Ang et al. 2021). Likewise, CNTs can respond to minor changes occurring in the environment. The integration with SWCNTs and metal-organic frameworks (MOFs) can significantly enhance the sensing ability of plants against pollutants and explosives. For instance, SWCNTs conjugated with Bombolitin II successfully detected nitro-aromatic compounds through near-infrared fluorescence in spinach. This allowed for the real-time monitoring of groundwater contaminants using a smartphone app (Wong et al. 2017; Barbinta-Patrascu et al. 2024). A similar approach was followed to create an MOF-plant hybrid sensor for detecting harmful metal ions and organic contaminants. A photoluminescent MOF was introduced in *Syngonium podophyllum,* after which the roots of plants were submerged in disodium terephthalate and terbium chloride solution. The sensor altered the intensity of fluorescence upon interaction with a high concentration of pollutants, making it highly reliable, sensitive, and specific with real-time monitoring (Liang et al. 2020). Likewise, the corona phase molecular recognition (CoPhMoRe) technique also exploited SWCNTs enfolded in polymers that detected the target molecules, leading to a change in fluorescence shifts in the presence of specific pollutants (Ansari et al. 2020).

4.4.3 Light-Emitting Plants

Due to the critical shortage of non-renewable resources and increasing energy demands in daily life, there is an urgent need to explore alternative energy sources. Light-emitting plants/bioluminescent plants might work as a solution and can be generated by genetic engineering approaches. In a study, NPs were used as carriers to carry light-illuminating compounds into the plants. NPs were introduced into the stomata of leaves via high-pressure system, which allowed their penetration into the cells. Luciferase and luciferin were, reportedly, transported using silica NPs and poly (lactic-co-glycolic acid) (PLGA) NPs, respectively. Similarly, chitosan NPs were used to deliver coenzyme A. After the release of luciferin within the cells, a reaction occurred between luciferin and luciferase, which emitted light (Ebrahimi et al. 2015). An exceptional $SrAl_2O_4$ (strontium aluminate) NPs, a phosphor, has shown remarkable luminescence characteristics, making it suitable to generate light-emitting plants. Due to its ability to extend the period of glow and the capability to recharge with both ultraviolet and visible light sources, $SrAl_2O_4$ NPs were injected into the vascular system of plants, where these NPs travel through the xylem and phloem, distributing throughout the plant tissues. $SrAl_2O_4$ NPs absorbed energy during light exposure and became excited. As soon as the excited electrons return back to their ground state, photoluminescence occurs, in which energy is released as visible light, specifically phosphorescence, which allows for prolonged light emission even after the excitation source is removed (Kanjanapokin et al. 2023).

4.4.4 Nano-Bionics in Nutrient Enhancement and Plant Growth

Nano-bionics is a transformative approach in agriculture that contributes to overall plant growth. The potential of NPs to act as nutrient carriers could trigger the growth and viability of plants. It also facilitated water absorption through tough seed coats. For example, the growth of *Brassica juncea* and *Glycine max* was promoted under the influence of MWCNTs, which accumulated within the endosperm of seeds. The controlled release of nutrients by NPs has also been documented to promote plant growth and health (Ranjan et al. 2022; Vithanage et al. 2023). Certain NPs containing nitrogen (N), phosphorus (P), and potassium (K) acted as sources of nutrients, thus significantly enhancing the growth of wheat, including improvements in shoot and root development as well as fruit production (Wohlmuth et al. 2022). A study has indicated that a foliar spray of NPK-based NPs at a concentration of 500 ppm over 21 days could lead to a notable increase in their total saccharide content (Rossi et al. 2019). Additionally, carbon NPs at concentrations between 40 and 60 mg/L were found to promote plant regeneration and improve the root-to-bud ratio in *Vetiveria zizanioides* (Landa 2021). In maize (*Zea mays*), exposure to carbon-based nanodots resulted in reduced root mass and smaller shoots but also enhanced their biochemical enrichment and antioxidant potential (Zhao et al. 2021b).

4.5 Nano-Bionics and Abiotic Stress Tolerance

Due to their small size, NPs can interact with plants by penetrating their cell wall, which enlarges the pores in the wall for efficient uptake. Magnesium (Mg), zinc (Zn), manganese (Mn), copper (Cu), calcium (Ca), iron (Fe), potassium (K), and molybdenum (Mo) are necessary and vital metallic nutrients required for the growth of plants. Hence, these nutrients as elements are exclusively used to generate metallic NPs that have been reported to induce and promote plant stress tolerance against multiple stresses. Various studies have reported that NM exposure can induce the antioxidant system of plants for scavenging ROS generation (Zhu et al. 2022). For instance, the exposure of metal NPs such as Mn_3O_4 NPs and ZnO NPs in tomatoes increased ROS scavenging, which allowed the plant to endure salinity stress. Additionally, NPs based on magnesium, zinc, iron, and copper contributed to making plants more resilient to environmental stressors (Faizan et al. 2021; Bidi et al. 2021). Moreover, it has been reported that from seed germination to final yield, all the physiological processes from start to end, including antioxidant enzyme activity and metabolism, can be positively influenced under the exposure of metallic NPs, carbon NPs, and Ag-NPs (Wang et al. 2022). Ag-NPs, due to significant bioactivity, showed an impact on the germination of *Bacopa monnieri* seeds, promoting their shoot length and chlorophyll pigment by stimulating the production of important metabolites and elevating levels of phenol and catalase activity (Vithanage et al. 2023). Research showed that application of SiO_2 NPs exposure to *Cucurbita pepo* under salt stress can improve transpiration rates, water use efficiency, total chlorophyll content, and antioxidant enzyme activity, thus regulating stress (Faizan et al. 2023).

Drought stress is one of the major abiotic stresses that limit plant growth, development, and productivity, threatening worldwide food security. Plant nanobionics utilizes NMs that can penetrate and get localized within tissue/organelles such as chloroplasts to enhance the capacity of plants to survive during stress (Raza et al. 2023). Some NPs like super-absorbent polymers can be applied to soil, which increases their water-holding capacity. These NPs possess the potential to absorb water for plants, even during drought, and thus cope with the limited water availability (Palma et al. 2024).

Copper is one of the major micronutrients required for the adequate growth and functioning of plants, as it can regulate chlorophyll, photosynthesis, respiration, metabolism of proteins and carbohydrates and seed development. Under drought stress, vascular plants have been reported to be significantly affected by Cu-based NPs in a dose-dependent manner. For example, the Cu-NP treatment was documented to promote the shoot-to-root index in *C. cajan* plants (Bhaskar et al.

2024). Furthermore, soil irrigation including γFe_2O_3 NPs was observed to regulate oxidative stress in *Brassica napus* by decreasing the levels of hydrogen peroxide and lipid peroxidation (Palmqvist et al. 2017; Hasanuzzaman et al. 2023). Further, the potential of nano-bionics to sense and cope with water shortages in plants has also been reported. SWCNTs-based nanosensors, when integrated with spinach, could detect the drought conditions in the plant as well as in the vicinity (Skrzypczak et al. 2017; Verma et al. 2024). An innovative sensor was developed that could detect water scarcity once implanted into plant leaves to identify drought-affected plants. This can also potentially provide early warnings to the farmers for proactive responses (Saxena et al. 2016). These sensors can also be used in soil monitoring and satellite imaging to assess drought conditions.

Due to the large surface area of NPs, toxic metals can bind onto its surface, thus allowing its two-fold role to not only promote plant growth but also protect plants against abiotic stress. Unlike bioremediation, which is time-consuming, nano-bionic-based remediation is effective and environmentally friendly. Transgenic plant nano-bionics can be used effectively to detect and remove metal and metalloid contaminants from soil by exploiting the potential of NPs to transport metals, their ability to enhance metabolism enzymes, and their role as chelators like metallothioneins and phytochelatin to detoxify metals (Kumar et al. 2024). Polyhydroxyfullerene (PHF)-based water-soluble NM was documented to detect and reduce the heavy metal toxicity in plants (Pradhan et al. 2014; Sigala-Aguilar et al. 2024). Likewise, paraquat toxicity was regulated in *Arabidopsis thaliana* with MWCNTs and carbon nitride dot applications (Fan et al. 2018; de Freitas et al. 2024). Moreover, the integrated system of arbuscular mycorrhizal fungi and carbon-based fullerenol NPs was found to regulate lead stress in *Brassica napus L.* The integration matrix improved the photosynthetic performance of plants, promoted their nutrient uptake, and enhanced antioxidant potential, thereby regulating oxidative stress (Shah et al. 2024).

Temperature stress interrupts the photosynthetic machinery and hinders the conversion of light energy into chemical energy. Nano-bionics can also act as thermoregulators under mild heat and cold stress. NP/NMs integrated plants can increase water conductance and transpiration, thus maintaining the integrity of thylakoid membranes and protecting the plant cells from thermal damage (Lew et al. 2020). For instance, CNT exposure improved plant tolerance against heat stress by generating a strong water transport system and enhancing water retention. ZnO NPs and Fe_3O_4 NPs also regulated plant hormonal responses to heat stress (Zhao et al. 2021a, b). Abscisic acid (ABA) is a stress-responsive plant hormone that, on moderate production triggers, stomatal closure to conserve water. ZnO NPs treatment was documented to maintain ethylene and ABA equilibrium to regulate plant responses during heat stress (Gupta et al. 2022). Likewise, cold stress significantly disrupts cell membrane integrity, leading to reduced plant growth (Mony et al. 2022). However, NPs treatment in plants was documented to maintain and enhance the cell membrane stability during cold and heat stress conditions (Zain et al. 2023). In such a study, SiO_2 NPs and graphene oxide NPs were observed to stabilize cell membrane integrity and regulate plant growth during cold stress (Khatri and Rathore 2018).

4.6 Ethical Considerations

Nano-bionics are being looked upon as an innovative approach targeting agricultural sustainability. Nano-bionics focuses on the utilization of NPs/NMs/NSs to support the growth, development, and overall performance of the plants (Faizan et al. 2023; Khan et al. 2023; Bhaskar et al. 2024). However, this innovative approach comprises various ethical and regulatory challenges that need careful investigations to ensure responsible development and application (Barbinta-Patrascu et al. 2024; Meena and Chandel 2024). For example, NPs exposure to a certain limit promotes plant growth and regulates oxidative stress. However, excessive NPs exposure beyond a concentration can be toxic, leading to enhanced ROS production, thus causing severe DNA damage and critical growth reduction. So, the NPs treatment can either be positive or negative depending on the plant species and the type and amount of nano-bionics used. Thus, the addition of NPs into an ecosystem can be a boon as well as a potential risk. Further, the use of nano-bionics exploiting NP/NMs can also affect the soil parameters, water quality, and interacting beneficial bacteria. Moreover, nano-bionics can accumulate in the soil, raising the concerns of biomagnification and bioaccumulation, which can equally be harmful to the consumers. Additionally, the long-term NPs exposure to agricultural fields and workers is still undiscovered and needs attention. Hence, it is essential to decipher strong safety protocols for the use of nano-bionics in agriculture, including toxicological assessments to evaluate the potential hazards associated with specific NPs and their interactions with biological systems. Moreover, clear guidelines must be developed by regulatory bodies before considering nano-exposed products for markets. Thus, sustainable, safe, and effective regulation of nano-based products requires strict implementation of specific guidelines targeting safe applications.

4.7 Future Scope and Conclusion

Plant-based nano-bionics embraces great promise for transforming agriculture into a more sustainable and efficient practice. The market value of nano-bionics is increasing gradually by 1.08 billion US dollars. It is further expected to increase by 2.9 trillion US dollars in industry by 2030 (Hassanzadeh et al. 2022). These have emerged as a technological advancement to transform the agri-food sector, with the potential to increase agricultural productivity, food security, and economic growth for industries. Furthermore, researchers have created polymer-based nanocarriers that release fertilizers only when they detect specific soil conditions, ensuring nutrient availability to plants in need (Zheng et al. 2022; Dhiman et al.

2022). Another cutting-edge promise uses carbon dots and carbon nanotubes that have been shown to enhance photosynthesis by improving the efficiency of electron transport within chloroplasts, as these nanotubes are incorporated into plant tissues (Guirguis et al. 2023; Khan et al. 2023, 2024).

Given this, the integration of biology with nanotechnology can fascinatingly lead to the creation of biohybrid systems. Here, the plants function as living sensors. These living plant-based nanosensors can detect pollutants in the soil or air by changing color or emitting signals when they sense harmful substances, thus providing real-time monitoring of environmental health. These novel innovations perform dual roles: enhancing plant functions and maintaining environmental sustainability. Thus, the generation of nano-bionics is marked as a promising approach in augmenting a variety of plant functions and maintaining agricultural sustainability. However, testing is still required, as these might be toxic to the environment, plants, and humans, thus requiring specific guidelines and a regulatory framework.

References

Alagoz SM, Arvanaghi HR, Dolatabadi N, Khalaki MA, Moameri M, Lajayer BA, van Hullebusch ED (2022) Impact on nutritional status of plants treated with nanoparticles in toxicity of nanoparticles in plants. Academic Press, pp 333–358

Anand KV, Anugraga AR, Kannan M, Singaravelu G, Govindaraju K (2020) Bio-engineered magnesium oxide nanoparticles as nano-priming agent for enhancing seed germination and seedling vigour of green gram (Vigna radiata L.). Mater Lett 271:127792

Ang MCY, Dhar N, Khong DT, Lew TTS, Park M, Sarangapani S, Strano M (2021) Nanosensor detection of synthetic auxins in planta using corona phase molecular recognition. ACS Sensors 6(8):3032–3046

Ansari MH, Lavhale S, Kalunke RM, Srivastava PL, Pandit V, Gade S, Singh AV (2020) Recent advances in plant nanobionics and nanobiosensors for toxicology applications. Curr Nanosci 16(1):27–41

Bao Y, Ma J, Pan C, Guo A, Li Y, Xing B (2020) Citric acid enhances Ce uptake and accumulation in rice seedlings exposed to CeO2 nanoparticles and iron plaque attenuates the enhancement. Chemosphere 240:124897

Barbinta-Patrascu ME, Bita B, Negut I (2024) From nature to technology: exploring the potential of plant-based materials and modified plants in biomimetics, bionics, and green innovations. Biomimetics 9(7):390

Bhaskar R, Pandey SP, Kumar U, Kim H, Jayakodi SK, Gupta MK, Han SS (2024) Nanobionics for sustainable crop production: recent development to regulate plant growth and protection strategies from pests. Open Nano 15:100198

Bidi H, Fallah H, Niknejad Y, Tari DB (2021) Iron oxide nanoparticles alleviate arsenic phytotoxicity in rice by improving iron uptake, oxidative stress tolerance and diminishing arsenic accumulation. Plant Physiol Biochem 163:348–357

Bussières, P. (2014). Estimating the number and size of phloem sieve plate pores using longitudinal views and geometric reconstruction. Scientific Reports, 4(1), 4929

Cervantes-Avilés P, Huang X, Keller AA (2021) Dissolution and aggregation of metal oxide nanoparticles in root exudates and soil leachate: implications for nanoagrochemical application. Environ Sci Technol 55(20):13443–13451

Cheung AY, Qu LJ, Russinova E, Zhao Y, Zipfel C (2020) Update on receptors and signaling. Plant Physiol 182(4):1527–1530

Dağhan H (2018) Effects of TiO2 nanoparticles on maize (Zea mays L.) growth, chlorophyll content and nutrient uptake. Appl Ecol Environ Res 16(5):6873–6883

de Freitas ADS, Fré LV, Rodrigues JS, Amaro SF, de Barros A, Nascimento ME, Ferreira M (2024) Simultaneous electrochemical detection of paraquat and carbofuran using a biodegradable flexible electrode, enhanced with carbon nitride dots and poly (allylamine hydrochloride) via layer-by-layer technique. Top Catal:1–13. https://doi.org/10.1007/s11244-024-02014-7

Dhiman A, Sharma AK, Agrawal G (2022) Polymer based engineered materials for sustainable agriculture. ACS Agric Sci Technol 2(4):693–711

Dhiman S, Debnath N, Bandyopadhyay K, Das S (2024) Novel approach of nanophotonic electron transfer for augmenting photosynthesis in *Arachis hypogaea*: a biophysical rationale behind the plasmonic enhancement of chemical energy transfer. ACS Omega 9(33):35332–35347

Ebrahimi M, Hosseinkhani S, Heydari A, Akbari J (2015) Simple and rapid immobilization of firefly luciferase on functionalized magnetic nanoparticles; a try to improve kinetic properties and stability. Biom J 1(1):104–112

El-Hady A, Hussein H (2021) Effect of foliar nano fertilizers and irrigation intervals on soybean productivity and quality. J Plant Prod 12(9):1007–1014

Etxeberria E, Gonzalez P, Baroja-Fernandez E, Romero JP (2006) Fluid phase endocytic uptake of artificial nano-spheres and fluorescent quantum dots by sycamore cultured cells: evidence for the distribution of solutes to different intracellular compartments. Plant Signal Behav 1(4): 196–200

Faizan M, Bhat JA, Chen C, Alyemeni MN, Wijaya L, Ahmad P, Yu F (2021) Zinc oxide nanoparticles (ZnO-NPs) induce salt tolerance by improving the antioxidant system and photosynthetic machinery in tomato. Plant Physiol Biochem 161:122–130

Faizan M, Karabulut F, Alam P, Yusuf M, Tonny SH, Adil MF, Hayat S (2023) Nanobionics: a sustainable agricultural approach towards understanding plant response to heavy metals, drought, and salt stress. Nanomaterials 13(6):974

Fan X, Xu J, Lavoie M, Peijnenburg WJGM, Zhu Y, Lu T, Qian H (2018) Multiwall carbon nanotubes modulate paraquat toxicity in Arabidopsis thaliana. Environ Pollut 233:633–641

Fatahi Z, Ghasemi SF, Taherian A, Esfandiari N (2020) Ultrasensitive fluorescent detection of pesticides in real sample by using green carbon dots. PLoS One 15(3):e0230646–e0230646

Fernandes, M., Singh, K. R., Sarkar, T., Singh, P., & Singh, R. P. (2020). Recent applications of magnesium oxide (MgO) nanoparticles in various domains. Adv. Mater. Lett, 11(8), 1–10

Gautam A, Sharma P, Ashokhan S, Yaacob JS, Kumar V, Guleria P (2023) Magnesium oxide nanoparticles improved vegetative growth and enhanced productivity, biochemical potency and storage stability of harvested mustard seeds. Environ Res 229:116023

Giraldo JP, Landry MP, Faltermeier SM, McNicholas TP, Iverson NM, Boghossian AA, Strano MS (2014) Plant nanobionics approach to augment photosynthesis and biochemical sensing. Nat Mater 13(4):400–408

González-Melendi P, Fernández-Pacheco R, Coronado MJ, Corredor E, Testillano PS, Risueño MC (2008) Nanoparticles as smart treatment-delivery systems in plants: assessment of different techniques of microscopy for their visualisation in plant tissues. Ann Bot 101(1):187–195

Gu Q, Lv J, Mo X, Jiang X (2022) High aspect ratio metamaterials and their applications. Sens Actuators A: Phys 335:113376

Guirguis A, Yang W, Conlan XA, Kong L, Cahill DM, Wang Y (2023) Boosting plant photosynthesis with carbon dots: a critical review of performance and prospects. Small 19(43):2300671

Gupta N, Jain SK, Tomar BS, Anand A, Singh J, Sagar V, Singh AK (2022) Impact of foliar application of ZnO and Fe3O4 nanoparticles on seed yield and physio-biochemical parameters of cucumber (Cucumis sativus L.) seed under open field and protected environment vis a vis during seed germination. Plants 11(23):3211

Hamdan MF, Mohd Noor SN, Abd-Aziz N, Pua TL, Tan BC (2022) Green revolution to gene revolution: technological advances in agriculture to feed the world. Plants 11(10):1297

Hasanuzzaman M, Zhou M, Shabala S (2023) How does stomatal density and residual transpiration contribute to osmotic stress tolerance? Plants 12(3):494

Hassanzadeh P, Atyabi F, Dinarvand R (2022) Nanobionics: from plant empowering to the infectious disease treatment. J Control Release 349:890–901

Huang X, Zhang J, Zhang L, Su H, Liu X, Liu J (2020) A sensitive H2O2 biosensor based on carbon nanotubes/tetrathiafulvalene and its application in detecting NADH. Anal Biochem 589:113493

Izati Fadzil N, Nulit R, Jusoh M, Abdul Rashid S, Tan TL (2025) Foliar application of carbon quantum dots photosynthesis enhancer for enhancing the growth and development of green mustard (*Brassica Juncea*) in indoor hydroponics. https://doi.org/10.2139/ssrn.5087482

Jayaraj S, Jayasree JN, Murali T (2024) Applications and synthesis processes of biosynthesized calcium oxide nanoparticles with sulfamethoxazole: a comprehensive review. Nano-Struct Nano-Objects 39:101244

Kale SK, Parishwad GV, Patil ASHAS (2021) Emerging agriculture applications of silver nanoparticles. ES Food Agrofor 3:17–22

Kanjanapokin C, Thiravetyan P, Krobthong S, Aonbangkhen C, Yingchutrakul Y, Kittipornkul P, Treesubsuntorn C (2023) Possibility to apply strontium aluminate to produce light-emitting plants: efficiency and safety. Chem Biodivers 20(8):e202300552

Khan MN, Fu C, Li J, Li Y, Khan Z, Zhu L, Wu H (2023) Plant nanobionics: nanotechnology for augmentation of photosynthesis efficiency. In: In nano-enabled sustainable and precision agriculture. Academic Press, pp 119–142

Khan MN, Ali W, Grillo R, Wu H, Nie L (2024) Carbon dots and single-walled carbon nanotubes enhances maize shading stress tolerance. bioRxiv 2024-07

Khatri K, Rathore MS (2018) Plant nanobionics and its applications for developing plants with improved photosynthetic capacity. In: Photosynthesis: from its evolution to future improvements in photosynthetic efficiency using nanomaterials, vol 95, p 1016

Klejchova M, Silva-Alvim FA, Blatt MR, Alvim JC (2021) Membrane voltage as a dynamic platform for spatiotemporal signaling, physiological, and developmental regulation. Plant Physiol 185(4):1523–1541

Kolahalam LA, Viswanath IK, Diwakar BS, Govindh B, Reddy V, Murthy YLN (2019) Review on nanomaterials: synthesis and applications. Mater Today: Proc 18:2182–2190

Kozieł E, Otulak-Kozieł K, Bujarski JJ (2021) Plant cell wall as a key player during resistant and susceptible plant-virus interactions. Front Microbiol 12:656809

Kumar KM, Nadaradjan S, Anand MS, Premkumar A, Hemalatha G (2024) Enhancing agricultural sustainability through plant Nanobionics: innovations, challenges and future directions. Int J Res Agron 7(5):251:259

Kurepa J, Paunesku T, Vogt S, Arora H, Rabatic BM, Lu J, Wanzer MB, Woloschak GE, Smalle JA (2010) Uptake and distribution of ultrasmall anatase TiO2 Alizarin red S nanoconjugates in Arabidopsis thaliana. Nano Lett 10:2296–2302

Landa P (2021) Positive effects of metallic nanoparticles on plants: overview of involved mechanisms. Plant Physiol Biochem 161:12–24

Lew TTS, Koman VB, Gordiichuk P, Park M, Strano MS (2020) The emergence of plant nanobionics and living plants as technology. Adv Mater Technol 5(3):1900657

Li Y, Li Q, Guan G, Chen S (2020) Phosphate solubilizing bacteria stimulate wheat rhizosphere and endosphere biological nitrogen fixation by improving phosphorus content. PeerJ 8:e9062

Liang J, Zulkifli MY, Choy S, Li Y, Gao M, Kong B et al (2020) Metal–organic framework–plant nanobiohybrids as living sensors for on-site environmental pollutant detection. Environ Sci Technol 54(18):11356–11364

Liu Y, Cao N, Gui W, Ma Q (2018) Nitrogen-doped graphene quantum dots-based fluorescence molecularly imprinted sensor for thiacloprid detection. Talanta 183:339–344

Mahmood R, Ali A, Yaseen M, Ahmad M (2014) Role of encapsulated calcium carbide in nitrogen economy, growth and yield of crops: a review. JAPS 24(2):354

Meena LK, Chandel A (2024) Ethical and regulatory perspective on nanotechnology in agriculture. Just Agric 5(2):94–104

Mekuye B, Abera B (2023) Nanomaterials: an overview of synthesis, classification, characterization, and applications. Nano Select 4(8):486–501

Mony C, Kaur P, Rookes JE, Callahan DL, Eswaran SV, Yang W, Manna PK (2022) Nanomaterials for enhancing photosynthesis: interaction with plant photosystems and scope of nanobionics in agriculture. Environ Sci Nano 9(10):3659–3683

Mousavi SR, Galavi M, Ahmadvand G (2007) Effect of zinc and manganese foliar application on yield, quality and enrichment on potato (Solanum tuberosum L.). Asain J Plant Sci 6(8):1256. https://doi.org/10.3923/ajps.2007.1256.1260

Munir N, Gulzar W, Abideen Z, Hasanuzzaman M, El-Keblawy A, Zhao F (2023) Plant–nanoparticle interactions: transcriptomic and proteomic insights. Agronomy 13(8):2112

Nasrollahzadeh M, Sajadi SM, Sajjadi M, Issaabadi Z (2019) An introduction to nanotechnology. In: Interface science and technology, vol 28. Elsevier, pp 1–27

Nikoleli GP, Siontorou CG, Nikolelis DP, Karapetis S, Bratakou S (2018) Prototype biosensing devices: design and microfabrication based on nanotechnological tools for the rapid in the field detection of food toxicants and environmental pollutants. In: Nanotechnology and biosensors. Elsevier, pp 1–28. https://doi.org/10.1016/B978-0-12-813855-7.00001-5

Palma D, Lagos O, Souto C, Pérez A, Quezada L, Hirzel J, Urbano B (2024) Evaluation of a natural superabsorbent polymer on water retention capacity in coarse-textured soils. Water 16(22):3186

Palmqvist NM, Seisenbaeva GA, Svedlindh P, Kessler VG (2017) Maghemite nanoparticles acts as nanozymes, improving growth and abiotic stress tolerance in Brassica napus. Nanoscale Res Lett 12:1–9

Parvej MS, Khan MI, Hossain MK (2022) Preparation of nanoparticle-based polymer composites. In: Nanoparticle-based polymer composites. Elsevier, pp 55–94

Pascoli M, Lopes-Oliveira PJ, Fraceto LF, Seabra AB, Oliveira HC (2018) State of the art of polymeric nanoparticles as carrier systems with agricultural applications: a minireview. Energy Ecol Environ 3:137–148

Pereira AE, Sousa BT, Iglesias MJ, Alvarez VA, Casalongué CA, Oliveira HC, Fraceto LF (2019) Potential use of polymeric particles for the regulation of plant growth. In: Polymers for agri-food applications. Springer, pp 45–66. https://doi.org/10.1007/978-3-030-19416-1_4

Pradhan A, Pinheiro JP, Seena S, Pascoal C, Cássio F (2014) Polyhydroxyfullerene binds cadmium ions and alleviates metal-induced oxidative stress in Saccharomyces cerevisiae. Appl Environ Microbiol 80(18):5874–5881

Qu Y, Sakoda K, Fukayama H, Kondo E, Suzuki Y, Makino A, Yamori W (2021) Overexpression of both Rubisco and Rubisco activase rescues rice photosynthesis and biomass under heat stress. Plant Cell Environ 44(7):2308–2320

Rajput VD, Minkina T, Fedorenko A, Chernikova N, Hassan T, Mandzhieva S, Burachevskaya M (2021) Effects of zinc oxide nanoparticles on physiological and anatomical indices in spring barley tissues. Nanomaterials 11(7):1722

Ranjan A, Rajput VD, Kumari A, Mandzhieva SS, Sushkova S, Prazdnova EV, Chung G (2022) Nanobionics in crop production: an emerging approach to modulate plant functionalities. Plants 11(5):692

Raza A, Charagh S, Salehi H, Abbas S, Saeed F, Poinern GE, Varshney RK (2023) Nano-enabled stress-smart agriculture: can nanotechnology deliver drought and salinity-smart crops? J Sustain Agric Environ 2(3):189–214

Rizzo P, Chavez BG, Dias SL, D'Auria JC (2023) Plant synthetic biology: from inspiration to augmentation. Curr Opin Biotechnol 79:102857

Rossi L, Fedenia LN, Sharifan H, Ma X, Lombardini L (2019) Effects of foliar application of zinc sulfate and zinc nanoparticles in coffee (Coffea arabica L.) plants. Plant Physiol Biochem 135: 160–166

Sarkar S, Ponce NT, Banerjee A, Bandopadhyay R, Rajendran S, Lichtfouse E (2020) Green polymeric nanomaterials for the photocatalytic degradation of dyes: a review. Environ Chem Lett 18(5):1569–1580

Saxena R, Tomar RS, Kumar M (2016) Exploring nanobiotechnology to mitigate abiotic stress in crop plants. J Pharm Sci Res 8(9):974

Shah AA, Usman S, Noreen Z, Kaleem M, Raja V, El-Sheikh MA et al (2024) Fullerenol nanoparticles and AMF application for optimization of Brassica napus L. resilience to lead toxicity through physio-biochemical and antioxidative modulations. Sci Rep 14(1):30992

Sharma SR, Kar D (2019) An insight into plant nanobionics and its applications. In: Plant nanobionics: advances in the understanding of nanomaterials research and applications, vol 1. Springer, pp 65–82

Sharma P, Gautam A, Kumar V, Guleria P (2022) In vitro exposed magnesium oxide nanoparticles enhanced the growth of legume Macrotyloma uniflorum. Environ Sci Pollut Res 29:13635–13645

Sigala-Aguilar NA, López MG, Fernández-Luqueño F (2024) Carbon-based nanomaterials as inducers of biocompounds in plants: potential risks and perspectives. Plant Physiol Biochem 212:108753

Skrzypczak T, Krela R, Kwiatkowski W, Wadurkar S, Smoczyńska A, Wojtaszek P (2017) Plant science view on biohybrid development. Front Bioengin Biotechnol 5:46

Subramani K, Elhissi A, Subbiah U, Ahmed W (2019) Introduction to nanotechnology. In: Nanobiomaterials in clinical dentistry. Elsevier, pp 3–18

Tan TL, Fadzil NI, Nulit R, Jusoh M, Rashid SA (2025) Foliar application of carbon quantum dots photosynthesis enhancer for enhancing the growth and development of green mustard (Brassica Juncea) in indoor hydroponics. https://doi.org/10.2139/ssrn.5087482

Tenhaken R (2015) Cell wall remodeling under abiotic stress. Front Plant Sci 5:771

Tian W, Su C, Zhang N, Zhao Y, Tang L (2024) Simulation of the physiological and photosynthetic characteristics of C3 and C4 plants under elevated temperature and CO2 concentration. Ecol Model 495:110805

Tombuloglu H, Slimani Y, Tombuloglu G, Korkmaz AD, Baykal A, Almessiere M, Ercan I (2019) Impact of superparamagnetic iron oxide nanoparticles (SPIONs) and ionic iron on physiology of summer squash (Cucurbita pepo): a comparative study. Plant Physiol Biochem 139:56–65

Tsakok I, Guedegbe T (2019) From Asian green revolution 1.0 to sustainable green revolution 2.0 (No. 1901). Policy Center for the New South

Umavathi S, Mahboob S, Govindarajan M, Al-Ghanim KA, Ahmed Z, Virik P, Kavitha C (2021) Green synthesis of ZnO nanoparticles for antimicrobial and vegetative growth applications: a novel approach for advancing efficient high quality health care to human wellbeing. Saaudi J Biol Sci 28(3):1808–1815

Verma A, Choudhary R, Singh VJ, Kachhwaha S, Kothari SL, Jain R (2024) Assessment of the effect of multi-walled carbon nanotubes (-OH functionalized) on growth characteristics and biochemical profile of Brassica juncea (L.) Czern. & Coss. Environ Sci Pollut Res 82:1–16

Vithanage M, Zhang X, Gunarathne V, Zhu Y, Herath L, Peiris K, Siddique KH (2023) Plant nanobionics: fortifying food security via engineered plant productivity. Environ Res 229: 115934

Wang C, Yang H, Yue L, Sun W, Chen F, Cao X, Xing B (2022) Physiological and molecular level understanding of advanced carbon dots to enhance maize drought tolerance: modulation of photosynthesis and signaling molecules. Environ Sci Nano 9(10):3821–3832

Wang X, Xie H, Wang P, Yin H (2023) Nanoparticles in plants: uptake, transport and physiological activity in leaf and root. Materials 16(8):3097

Wiederrecht GP, Bachelot R, Xiong H, Termentzidis K, Nominé A, Huang J, Pupek KZ (2023) Nanomaterials and sustainability, pp 3443–3449

Wohlmuth J, Tekielska D, Čechová J, Baránek M (2022) Interaction of the nanoparticles and plants in selective growth stages-usual effects and resulting impact on usage perspectives. Plants 11(18):2405

Wong MH, Giraldo JP, Kwak SY, Koman VB, Sinclair R, Lew TTS et al (2017) Nitroaromatic detection and infrared communication from wild-type plants using plant nanobionics. Nat Mater 16(2):264–272

Yang C, Powell CA, Duan Y, Shatters R, Zhang M (2015) Antimicrobial nanoemulsion formulation with improved penetration of foliar spray through citrus leaf cuticles to control citrus. PLoS One 10:e0133826. https://doi.org/10.1371/journal.pone.0133826

Zain M, Ma H, Nuruzzaman M, Chaudhary S, Nadeem M, Shakoor N, Ahamad T (2023) Nanotechnology based precision agriculture for alleviating biotic and abiotic stress in plants. Plant Stress 10:100239

Zhang W, Dan Y, Shib H, Ma X (2017) Elucidating the mechanisms for plant uptake and in-planta speciation of cerium in radish (Raphanus sativus L.) treated with cerium oxide nanoparticles. J Environ Chem Eng 5:572–577

Zhang X, Sun Z, Zhu Z, Luo J, Wu Z, Wang Z (2020) High-efficient, spherical and thermal-stable carbon dots@ silica fluorescent composite as rare earth-free phosphors for white LED. Ceram Int 46(10):14706–14712

Zhang B, Gao Y, Zhang L, Zhou Y (2021) The plant cell wall: biosynthesis, construction, and functions. J Integr Plant Biol 63(1):251–272

Zhao D, Wang X, Cheng Z, Tang Y, Tao J (2021a) Multi-walled carbon nanotubes prevent high temperature-induced damage by activating the ascorbate-glutathione cycle in Paeonia ostii T. Hong et JX Zhang. Ecotoxicol Environ Saf 227:112948

Zhao F, Xin X, Cao Y, Su D, Ji P, Zhu Z, He Z (2021b) Use of carbon nanoparticles to improve soil fertility, crop growth and nutrient uptake by corn (Zea mays L.). Nanomaterials 11(10):2717

Zheng L, Seidi F, Liu Y, Wu W, Xiao H (2022) Polymer-based and stimulus-responsive carriers for controlled release of agrochemicals. Eur Polym J 177:111432

Zhu J, Li J, Shen Y, Liu S, Zeng N, Zhan X, Xing B (2020) Mechanism of zinc oxide nanoparticle entry into wheat seedling leaves. Environ Sci Nano 7(12):3901–3913

Zhu L, Chen L, Gu J, Ma H, Wu H (2022) Carbon-based nanomaterials for sustainable agriculture: their application as light converters, nanosensors, and delivery tools. Plants 11(4):511

Zia-ur-Rehman M, Naeem A, Khalid H, Rizwan M, Ali S, Azhar M (2018) Responses of plants to iron oxide nanoparticles. In: Nanomaterials in plants, algae, and microorganisms. Academic Press, pp 221–238

Chapter 5
Role of Nanotechnology in Aeroponics and Hydroponics

Shashi Rani, Amit Kumar, and Ashish R. Warghat

Abstract Conventional agricultural techniques must be reassessed to tackle the issues of diminishing groundwater, unprecedented climate changes, soil nutrient depletion, and disrupted ecosystem cycles. These issues significantly diminish crop productivity and quality, directly jeopardizing the food supplies required to satisfy the increasing demands of the expanding global population. Soilless cultivation (hydroponics and aeroponics) integrated with nanotechnology may offer an innovative approach and a more sustainable, profitable alternative to conventional agricultural techniques. Nanomaterials (NMs) such as nano-fertilizers, nano-elicitors, nano-pesticides, and nano-coatings have attracted the interest of researchers due to their distinctive properties, including their specificity and reactivity, which render them potential molecules in the agriculture and food production sector. Green nanotechnology for nanoparticle (NPs) synthesis with plant extracts is beneficial due to its simplicity, convenience, environmental safety, and reduced reaction time. NPs synthesized in environmentally sustainable ways may improve crop nutrient uptake and bioavailability, reduce abiotic stress, boost crop productivity, and promote plant health. By combining soilless cultivation and nanotechnology, plants' quality, purity, and production rate can be enhanced while utilizing limited resources. The increasing worldwide demand for high-quality food and products mandates the implementation of a collaborative strategy, which can lead to resilient agricultural systems, guarantee food availability, and promote environmental sustainability. This book chapter examines the integration of nanotechnology with hydro-aeroponics, focusing on the application of NPs in plant growth and development.

Keywords Aeroponic · Hydroponics · Nanotechnology · Sustainable agriculture · Secondary metabolites

---

S. Rani · A. Kumar · A. R. Warghat (✉)
Fermentation and Phytofarming Technology Division, CSIR-Institute of Himalayan Bioresource Technology, Palampur, Himachal Pradesh, India

Academy of Scientific and Innovative Research (AcSIR), Ghaziabad, India
e-mail: ashishwarghat@ihbt.res.in

S. K. Yadav, A. Kumari (eds.), *Nanobiotechnology in Agri-food Sector*, Sustainability Sciences in Asia and Africa, https://doi.org/10.1007/978-981-96-7756-6_5

5.1 Introduction

Since the commencement of the Green Revolution in the middle of the twentieth century, plant productivity has grown enormously. Nevertheless, over time, this trend has diminished due to several challenges, including substandard farming methods, global warming, declining soil fertility, overgrazing, and reduced water availability (Gomiero 2016). The agriculture sector must be rejuvenated through novel approaches to address these concerns. Initiatives are underway to investigate alternative food and plant production systems to guarantee global food security and meet industry standards for increased production yields (Partap et al. 2021; Jaeger et al. 2024). In response to these problems, soilless cultivation (aeroponics and hydroponics) is a cutting-edge and sustainable growing method that has been developed for the mass production of commercially significant plants under regulated environmental circumstances. In horticulture and agriculture, these systems are extensively promoted for their purity and quality-rich biomass production. This cultivation approach provides a sustainable and dependable framework for the development of several commercially significant crops, including medicinal herbs, spices, aromatics, floral plants, and vegetables (Partap et al. 2021; Sharma et al. 2024).

Nanotechnology offers an enormous opportunity for promoting agricultural sustainability in multiple aspects. The integration of nanotechnology and hydroponics has developed an effective system for plant cultivation, providing accurate regulation of nutrient conditions, thereby ensuring agricultural sustainability, and alleviating the detrimental environmental effects (Srivastava and Bhargava 2022; Jadoon et al. 2024). Elicitors are chemicals derived from biotic and abiotic origins that can provoke plant responses to adverse conditions, resulting in enhanced production of bioactive metabolites for acclimatization in adverse conditions (Partap et al. 2024; Kumar et al. 2024). Nano-elicitation is a nanotechnology method designed to improve plant production, secondary metabolite content, and yield using NPs as elicitors. Nano-elicitation improves plant characteristics to endure environmental diseases and stress, resulting in increased crop yields and production rates, energy conservation, and reduced waste (Jadoon et al. 2024). Hydroponics and aeroponics have become increasingly preferred by plant researchers for examining the effects of NPs on plant health and growth (Movahedi and Moieni 2024). Owing to a larger surface area and significant activity, NPs provide plants with easily accessible nutrients by augmenting their soluble and available forms (Shukla et al. 2024). The unique attributes of NPs, including their modifiable physicochemical characteristics and capability to move through the cell wall, can improve ion and mineral absorption more effectively by the roots (Arora et al. 2020; Jadoon et al. 2024). NPs serve as superior nutrient transporters compared to traditional agrochemicals, rendering them a highly attractive resource, particularly for these systems. Recent research on NPs manufactured by eco-friendly methods has shown their equivalent effectiveness to other materials in enhancing plant growth, agricultural yield, and elevating phytochemical content (Ghidan and Al Antary 2019; Shukla et al. 2024).

Numerous investigations indicated that NPs were used in hydroponics to mitigate abiotic stress (Singh et al. 2022; El-Shal et al. 2022). The collaborative integration of nano-elicitation with hydroponics & aeroponics facilitates accurate monitoring of the delivery of nutrients and water intake, leading to enhanced growth, yield, and abiotic stress alleviation. Creating an environmentally friendly framework that incorporates these two strategies can resolve the previously mentioned challenges. This book chapter will examine the utilization of NPs with hydro-aeroponics to enhance plant production and growth. Additionally, it delineates the prospective applications of this integrated strategy in facilitating the establishment of a sustainable agricultural system.

5.2 Review of Literature

The chapter offers an in-depth examination of the integration of nanotechnology with soilless cultivation (hydroponics & aeroponics) to enhance crop productivity and yield. This chapter's material is significantly important for developing future methodologies that will benefit research groups, business organizations, and the agriculture sectors ultimately enhancing the well-being of people. This chapter presents a comprehensive review of relevant literature using various academic databases, including PubMed, Google Scholar, SciFinder, and Science Direct. The knowledge and conclusions presented in this chapter were sourced solely from educational sources, review papers, book chapters, and publications in the English language.

5.3 Overview of Hydroponics and Aeroponics Cultivation

Hydroponics is a technique in which plants grow in nutrient-rich liquid artificial media (Partap et al. 2020; Sharma et al. 2024). In this high-tech system, liquid nutrients are given to the growing medium to provide the necessary nourishment, such as micro and macro elements, vitamins, and minerals for the optimum and fast growth of plants (Thakur et al. 2019; Rattan et al. 2022). Hydroponics employs several plant support and growth substrates, including rock wool, gravel, pebbles, and perlite (Savvas et al. 2017). Aeroponics is an advanced soilless cultivation method that delivers nutrient-rich mist to roots, enhancing nutrient uptake and oxygenation (Garzón et al. 2023). The sterile environment and optimal oxygen supply in aeroponics promote rapid plant growth and biomass production (Lakhiar et al. 2018). In a soil-based system, nutrients adhere to soil particles, dissolve in the soil's aqueous solution, and are subsequently absorbed by the roots. In hydroponic systems, there is an absence of friction, yet it ensures a more oxygen-rich environment for the roots to enhance growth and development. Hydro-aeroponics is a modern agricultural technique in which all environmental variables, such as temperature, light, humidity, pH, and biofortification for value enhancement, are artificially controlled and monitored (Erekath et al. 2024). This advanced technique facilitates the cultivation of food, spices, herbs, and medicinal plants in urban environments, providing advantages such as improved resistance to diseases, pests, and drought; year-round crop production; and decreased transportation costs for agricultural products (Benke and Tomkins 2017; Erekath et al. 2024). Technological advancements have optimized hydro-aeroponic systems through automated nutrient delivery, environmental regulation, and real-time monitoring, reducing soil-borne diseases and chemical dependency. Several investigations demonstrate that plants grown in an artificial growing medium exhibit enhanced growth and biomass production relative to those cultivated in soil, promoting rapid growth and improved nutrient composition compared to traditional farming (Partap et al. 2020; Fatima et al. 2023; Sharma et al. 2024).

5.4 Green Synthesis of Nanoparticles

Currently, there is an expanding field of research focused on the synthesis of nanomaterials (NMs) such as NPs, nanotubes, nanocrystals, nanofibers, and quantum dots, using chemical, physical, and green synthesis methods (Horwat et al. 2011). Physical and chemical methods are progressively being supplanted by green synthesis techniques (Alsammarraie et al. 2018) due to concerns regarding excessive use of energy (Horwat et al. 2011), the emission of harmful materials (Hoag et al. 2009), and the necessity for intricate instruments and synthesis environments (Saiqa Ikram 2015). Green nanotechnology or green synthesis includes the advancement of efficient methods and methods that are ecologically viable and encourage practices to minimize deleterious impacts on the environment (Tangley 1987). The synthetic fabrication of tailored NPs encounters obstacles due to elevated costs, harmful byproducts, and environmental dangers (Abbas et al. 2020). In response to the increasing worries over nanotechnology, a subfield known as "green nanotechnology" or "green synthesis" has evolved, integrating nanotechnology with green technology (Verma et al. 2019; Soni et al. 2022). The technique incorporates many methodologies that adhere to the concepts of green chemistry for the synthesis and design of ecologically friendly NMs that employ renewable energy sources. Many investigations have emphasized the significance of the green synthesis of NPs in fostering sustainable agriculture and alleviating the eco-friendly consequences of NMs (Verma et al. 2019; Soni et al. 2022). Before commercial utilization, emphasis must be placed on the green synthesis, testing, and optimization of NMs for diverse plant species. Consequently, ongoing progress in NPs for novel and improved synthesis methods, with precise regulation of product composition and delivery systems within a suitable timeframe, would enhance efficacy in the agriculture and food sectors (Balusamy et al. 2023). We must advance toward sustainable technology in the food and agriculture sector to address emerging demands.

Nanotechnology is essential for addressing the requirements of future generations and ensuring agricultural sustainability.

5.5 Molecular Mechanism of Nano-Elicitation Inside the Plant Cell

Plants synthesize a wide array of secondary metabolites with various chemical characteristics. These metabolites serve as a mechanism for interaction with nature, facilitating survival and adaptability to the surroundings (Isah 2019). Notably, the communication between NPs and plants results in the excess production of reactive oxygen species (ROS), oxidative stress, membrane structural damage, modifications in antioxidant enzymes, alteration in secondary metabolic and phytohormone routes, and signal transmission. Secondary metabolic alterations were correlated with elevated concentrations of ROS, phenylalanine ammonia-lyase (PAL), and polyphenol oxidase (PPO) in *Arabidopsis thaliana* subjected to 250 and 1000 mg/L of cerium oxide (CeO_2) and indium oxide (In_2O_3) NPs, respectively (Ma et al. 2016). Furthermore, NPs modify the expression of genes and transcription factors (TFs) associated with cell division, cellular structure, electron transport, and pathways related to biotic and abiotic stress (Landa et al. 2012; Van Aken and Van Breusegem 2015). Elevated levels of ROS and cytoplasmic calcium ions (Ca^{+2}) are the key responses of plants to NPs exposure, leading to the activation of the mitogen-activated protein kinase (MAPK) cascade under subsequent abiotic stress (Marslin et al. 2017; Selvakesavan et al. 2023). The detection of AgNPs by plasma membrane-localized receptors in *Arabidopsis thaliana* elicited a Ca^{+2} surge and ROS activation (Sosan et al. 2016). The proteomic study in rice subjected to AgNP treatment revealed elevated Ca^{+2} levels and an increase in proteins associated with the corresponding signaling pathway (Mirzajani et al. 2014). The authors posited that AgNPs, or the ions they release, obstruct metabolic processes in cells by interacting with Ca^{+2} receptors and channels and Ca^{+2}/Na^{+} ATPases. Calcium-binding proteins (CaBPs) or other NP-specific proteins either mimic Ca^{+2} or act as signaling molecules within the cytosol. Ca^{+2} ions serve as secondary messengers in response to various stresses and facilitate Ca^{+2} ion channels, enabling plants to adjust to adverse conditions (Kang et al. 2023). The phosphorylation of MAPK followed by the stimulation of descending TFs constitutes a typical mechanism via which plants alter secondary metabolic pathways and signaling (Zhang et al. 2022; Sena et al. 2023). NPs elicited single-channel-like conductance, facilitating Ca^{+2} influx, leading to ROS formation due to cytosolic Ca^{+2} levels. This situation induces the expression of MAPK genes, which influences the transcriptional reprogramming of genes associated with secondary metabolites (Rahmawati et al. 2021). This ultimately enhances the activation of genes and TFs responsible for phytochemicals, increasing their accumulation (Fig. 5.1).

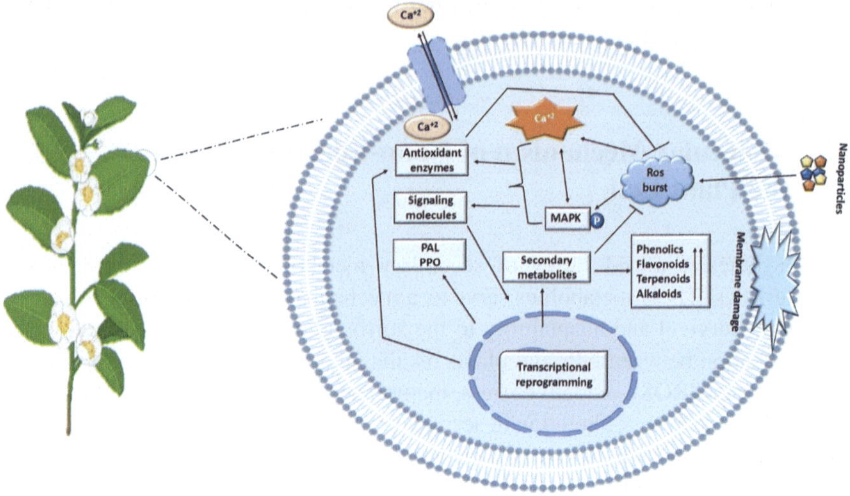

Fig. 5.1 Diagram illustrating the general processes of nanoparticles (NPs) influencing the plant's secondary metabolism

5.6 Role of Nanoparticles in Plant Growth and Productivity

Nanotechnology will significantly contribute to creating a structured, sustainable agriculture and food sector and resolving current issues. NPs play a significant role in plant growth, including enhanced morpho-physiological parameters, phytochemical production, and photosynthetic and antioxidant activity, and provide resistance against abiotic stress due to better nutrient absorption than traditional agrochemicals in hydro-aeroponics (Fig. 5.2). Table 5.1 represents the impact of NPs on hydroponically cultivated crops.

5.6.1 Nanoparticles for Nutrient Uptake and Absorption

The primary objectives of employing NPs in hydroponic and aeroponic systems are to reduce nutrient losses while enhancing yields via improved nutrient uptake efficiency. Owing to the higher relative surface area and sensitivity, NPs may supply plants with more easily absorbed and bioavailable nutrient forms, hence inhibiting the precipitating and insolubilization of nutrients commonly observed with traditional fertilizers (Singh et al. 2024a, b). Furthermore, they enhance nutrient transport across plant cell membranes, enhancing nutrient absorption and use, and improving overall plant health and production. Maluin and Katas (2022) investigated the effects of chitosan nanoparticles (CNPs) on hydroponically grown microgreens compared to mature Bok Choy (*Brassica rapa*). CNPs significantly enhance nutrient absorption and uptake, consequently improving crop growth, quality, yield, and nutritional

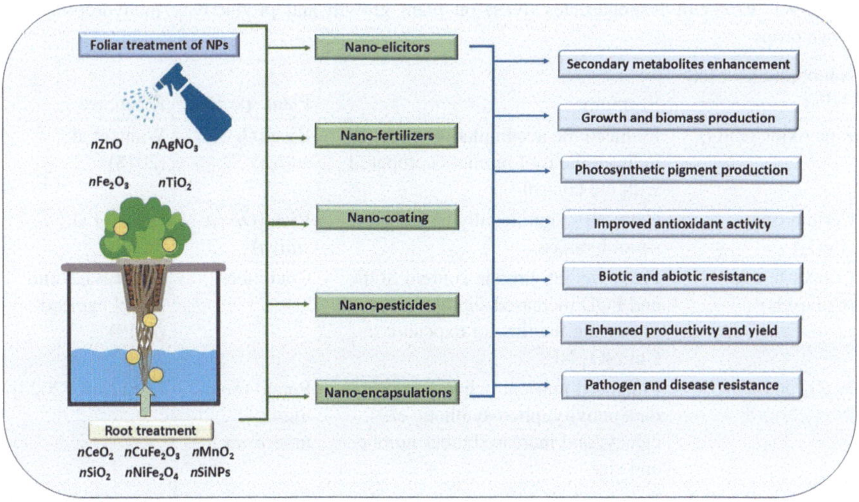

Fig. 5.2 Role of Nanoparticles (NPs) in plant growth and development

content, indicating their promise as an environmentally friendly nutrient composition for hydroponically cultivated horticulture crops. Yadav et al. (2024) synthesize CuO micronutrient NPs using cyanobacterial (*Pseudanabaena foetida* RJ1) biomass extract and extract free of cells to create a nano-formulation. The NPs in the form of nano-formulations were administered to germinating paddy seeds at varied doses (5, 10, 15 mg/L) in hydroponically cultivated rice crops, and results showed a notable enhancement in rice growth. Fe_2O_3 (200 mg/L) markedly reduced both fresh and dry biomass at a concentration of spinach (Jeyasubramanian et al. 2016). In tomatoes, a lower concentration of CeO_2 NPs reduced the vegetative growth and retarded fruit ripening, but CeO_2 at 10 mg/L greatly enhanced plant development (Wang et al. 2012). Recently, Shahid et al. (2024) used nano-bone-char (NBC) in the *Syngonium podophyllum* under hydroponic and soil conditions and observed an enhancement in biomass, shoot length, chlorophyll content, leaf number, total $Ca^{2+,}$ and PO_4^{3-} content, and total organic carbon. Zoufan et al. (2020) demonstrated that all treatments with ZnO NPs (10, 50, and 250 mg L^{-1}) led to a reduction in growth and chlorophyll content while substantially increasing carotenoids, lipid peroxidation, leaf H_2O_2, and leaf electrolyte leakage in *Chenopodium murale*. Nazarideljou et al. (2024) showed beneficial impacts of nano-chelated fertilizers (NCFs) on tomato and cucumber plants in terms of morphological and biochemical characteristics, increased photosynthetic efficiency, antioxidant enzyme capacity, and overall improved plant growth. The plant biomass exhibited a substantial rise at lower doses of Al_2O_3 NPs or NiO NPs application, while higher doses of these NPs resulted in a substantial reduction. Antioxidant enzymes, including ascorbate peroxidase (APX), catalase (CAT), superoxide dismutase (SOD), and peroxidase (POD) in roots and shoot parts, demonstrated an overall enhancement in activity following exposure to Al_2O_3 and NiO NPs in *Nigella arvensis* (Chahardoli et al. 2020).

Table 5.1 Effect of Nanoparticles (NPs) on plant growth and productivity in hydroponically grown crops

Nanoparticles used (NPs)	Response	Plant species	Reference
Zinc oxide (ZnO)	Reduced the accumulation of As(III) reduced the root biomass compared with the control	Rice (*Oryza sativa*)	Wang et al. (2018)
Cerium oxide (CeO_2)	CeO_2 NPs significantly increased the shoot biomass	Rice (*Oryza sativa*)	Wang et al. (2018)
Copper ferrite ($CuFe_2O_4$)	Fresh weight, protein content SOD and POD increased significantly with increasing the time of exposure to $CuFe_2O_4$ NPs	Cucumber	Abu-Elsaad and Abdel hameed. (2019)
Fe_3O_4 and MnO_2	Enhanced nutrient delivery, antioxidant activity, photosynthetic efficiency, and increased tuber number and size	Potato (*Solanum tuberosum*)	Joshi et al. (2024)
Iron oxide (Fe_2O_3)	Beneficial effects on plant growth include enhanced root and shoot length, increased biomass, and elevated iron content in a dose-dependent manner	Spinach	Jeyasubramanian et al. (2016)
Titanium dioxide (TiO_2)	The concentrations of shoot nutrients (N, P, K, Zn, Cu, Mn) were elevated after TiO2 NPs treatment	Maize (*Zea mays*)	Dağhan (2018)
Silver (AgNPs)	Plants treated with AgNPs exhibited higher levels of protein and carbohydrates, but comparably reduced levels of total phenolic content (TPC), catalase (CAT), and peroxidase (POX) enzyme activity	*Bacopa monnieri*	Krishnaraj et al. (2012)
Silica (SiO_2) and calcium chelate (Ca-chelate)	Different concentrations of Ca-chelate and SiO_2 NPs increased flower and flower bud growth	Gerbera (*Gerbera jamesonii*)	Tofighi Alikhani et al. (2021)
(Fe_2O_3)	Presoaking of seeds with Fe2O3 (50 mg/L) NPs has a better growth effect Fe_2O_3 NPs (50 and 200 mg/L) seed has not significantly affected chlorophyll and iron content while increasing antioxidant (SOD, POD, and CAT) enzyme activity	Alfalfa	Yang et al. (2023)
Silver (AgNPs)	Higher proline content and fresh weight compared to control plants	*Malva parviflora*	Oraibi et al. (2022)
Titanium dioxide (TiO_2)	The exposure of TiO_2 NPs (300 ppm) and light spectrum showed a substantial impact on the anthocyanin and TPC in the shoot part	*Carum copticum*	Baharmoghadam et al. (2024)
Titanium dioxide (TiO_2)	The chlorophyll content was decreased with increasing TiO_2 NPs		Daghan et al. (2020)

(continued)

Table 5.1 (continued)

Nanoparticles used (NPs)	Response	Plant species	Reference
	concentration The higher N, P, Zn, and Cu concentrations were recorded in TiO_2 (40 mg/L) NPs treatment	Wheat (*Triticum vulgare*)	
Nickel ferrite ($NiFe_2O_4$)	$NiFe_2O_4$ NPs exposure led to a significant increase in chlorophyll, carotenoids, and soluble protein content Higher dose treatment (1000 mg/L) significantly reduced the plant growth	Barley (*Hordeum vulgare*)	Tombuloglu et al. (2019)
Ceria (CeO_2)	CeO_2 NPs (100 and 500 mg L^{-1}) showed elevated nitrogen levels in roots and shoots in low nitrogen conditions by 6–12% and 22–30%, respectively, and increased nitrogen assimilation enzymes activity including glutamine synthetase (GS), glutamine oxoglutarate aminotransferase (GOGAT), and glutamate dehydrogenase (GDH)	Rice (*Oryza sativa*)	Wang et al. (2020)
CuO and ZnO	Lower concentrations of CuO and ZnO NPs showed improved plant growth Plants treated with CuO and ZnO NPs have increased antioxidant activity, enhancing their resistance to diseases and pests while facilitating growth	*Amaranthus hybridus*	Francis et al. (2022)
Magnetic NPs and Salicylic acid Magnesium oxide (MgO) and hydroxyapatite (HA)	Enhanced phenolics, flavonoids, and aromatic metabolites Improved seed germination rate and root development under ammonium (NH^{4+}) stress in *Ipomoea aquatica* and reduction in catalase and elevation in peroxidase and superoxide dismutase enzyme activity during NH^{4+} stress in *Cucumis melo*	*Ipomoea aquatica* and *Cucumis melo*	Wu et al. (2023)
Ceria (CeO_2)	CeO_2 NPs (100 and 500 mg L^{-1}) showed elevated nitrogen levels in roots and shoots in low nitrogen conditions by 6–12% and 22–30%, respectively, and increased nitrogen assimilation enzymes activity including GS, glutamine oxoglutarate aminotransferase (GOGAT), and glutamate dehydrogenase (GDH)	Rice (*Oryza sativa*)	Wang et al. (2020)

5.6.2 Nanoparticles as a Nano-Fertilizer

Nano-fertilizers have garnered considerable attention in agricultural research to address the rising worldwide food need and introduce novel approaches to boost crop production (Seleiman et al. 2020). Nano-fertilizers can efficiently retain enough nutrients to deliver optimal doses over a prolonged duration (Adhikary and Basak 2024; Bathla et al. 2024). Engineered nano-fertilizers with targeted physical properties and surface changes have arisen as sustainable alternatives to conventional chemical fertilizers. In hydroponic alfalfa seedlings subjected to 50 mg/L Fe_2O_3 NPs exhibit increased root length, enhanced lateral roots, greater fresh weight, and considerably elevated chlorophyll and iron levels compared to the control treatment (Yang et al. 2023). The Fe_2O_3 NPs could harm the plant cell wall and plasma membrane to enable entrance and transport. Consequently, the NPs were ingested into cell walls, incorporated by new cells, and ultimately invaded the vascular tissue. The Fe_2O_3 NPs are absorbed by plant roots, translocate through the stem, and accumulate in leaf tissue. Francis et al. (2022) investigated two distinct methods of fertilizer treatment (hydroponics and foliar) with varying concentrations of CuO (0.06 μM, 0.12 μM) and ZnO (0.12 μM, 0.24 μM) NPs, alongside water control and Hoagland's medium. The hydroponic treatment technology has shown superior efficiency in promoting plant growth and development. It could occur due to the efficient absorption of nutrients and ions by the plants in the hydroponic system. The smaller size of the NPs enhances effectiveness for uptake and transport. This led to the utilization of about 70% reduced concentrations of CuO and ZnO NPs compared to $CuSO_4$ and $ZnSO_4$ salts in the nutritional media. A nano-fertilizer comprising humic acid at a concentration of 1000 mg/L enhances nutritional absorption effectiveness, resulting in elevated macronutrient levels in the root and shoot part of the lettuce (Roosta et al. 2018). Nano Si, in combination with Atonik, is employed to examine its impact on the growth characteristics of hydroponically cultivated barley. The findings regarding yield, water use efficiency, protein content, lipids, and fibers demonstrated a beneficial effect on the treatment (Al-Juthery et al. 2019). Many aeroponically cultivated medicinal plants (valerian, chicory, withania, and purple coneflower) were subjected to ZnO NPs (3 g/L dose) for approximately 60 days to examine its impact on yield (dry weight), root and shoot growth, photosynthetic pigment content, and number of leaves (Rabary et al. 2022). Wang et al. (2019) demonstrated that rice cultivated in hydroponics exhibited decreased biomass when simultaneously treated with sodium chloride and cadmium chloride solutions. In this context, the use of cerium oxide NPs (200 mg/L) reduced DNA damage and improved antioxidant capacity and chlorophyll content. In *Silybum marianum,* coupled administration of silver NPs and MeJA resulted in enhanced concentrations of chlorophyll content, respectively, under hydroponic conditions (Mubeen et al. 2023). Joshi et al. (2024) administered metal oxide NPs (MONPs) to hydroponically cultivated potato plants, namely iron oxide (Fe_3O_4) (4.0 mg L^{-1}) and manganese oxide (MnO_2) at (1.0 mg L^{-1}). The results indicated enhanced biomass, tuber yield, and antioxidant enzyme activity, including SOD, CAT, and POD, alongside diminished levels of malondialdehyde (MDA) and hydrogen peroxide (H_2O_2) content. These data indicate that MONPs improve plant growth and crop productivity in hydroponic systems. Ran et al. (2024) examined the advantageous function and possible detoxifying mechanism of biogenic selenium NPs (Bio-SeNPs) derived from *Psidium guajava* extracts in mitigating antimony (Sb) toxicity in rice seedlings. The findings indicated that the external application of Bio-SeNPs (0.05 g/L) in hydroponically grown rice seedlings significantly improved rice shoot length, shoot fresh weight, and dry biomass under 50 μM Sb(III) stressful conditions. In comparison to sole Sb treatment, the hydroponic administration of Bio-SeNPs significantly enhanced rice photosynthetic activity, augmented cell viability and membrane integrity, diminished ROS levels, and elevated antioxidant capacities. Simultaneously, the application of external Bio-SeNPs markedly diminished Sb accumulation in rice roots and shoots while also decreasing root-to-shoot translocation. Tiwari et al. (2017) administered 0.5, 1, 2, and 4 g/L TiO_2 NPs to tomato plants in a hydroponic system over 2 weeks. Lower doses (0.5–2 g/L) of TiO_2 NPs superior growth by roughly 50% and significantly increased photosynthetic metrics and chlorophyll content, while also inducing the expression of the photosystem I (PSI) gene compared to control. The elevated concentration (4 g/L) adversely impacted these measures. The enzyme activity of CAT and POX in the treated plants increased in a dose-dependent way. Similarly, treated plants demonstrated elevated levels of glutathione synthase and glutathione S-transferase, suggesting an important function of thiols in the detoxifying of TiO_2 in tomatoes. Movahedi and Moieni (2024) analyzed that foliar application of 5 ml/L Farmax nano-fertilizer in an aeroponic system significantly enhanced the morphological traits and photosynthetic pigments of chicory (*Cichorium intybus* L.), yielding the highest plant height, root length, leaf number, biomass (fresh and dry weight of shoot and root), leaf area, chlorophyll a, chlorophyll b, and carotenoid content.

5.6.3 Nano-Elicitation for Secondary Metabolite Enhancement

Nano-elicitation is now recognized as an effective technique for boosting the synthesis of plant secondary metabolites, which are crucial for plant defense and environmental adaptability (Khan et al. 2021). These metabolites, including phenolics, flavonoids, and alkaloids, are highly sought after in the food, cosmetic, pharmaceutical, and nutraceutical industries due to their perceived safety compared to synthetic alternatives. The pharmaceutical attributes of many herbs are ascribed to the natural extracts or tinctures rather than to the specific components. Consequently, any modification in the metabolic pathway of specialized metabolites in plants could impact their therapeutic efficacy and economic worth. In hydroponics, nano-elicitation is an effective strategy to produce high-quality biomass, planting material, and enhanced secondary metabolites for pharmaceutical and medicinal significance

(Wang 2019; Jadoon et al. 2024). Mubeen et al. (2022) cultivated *Silybum marianum* hydroponically and documented improved phenolic and flavonoid content, nitric oxide (NO), and SOD content subjected to treatment with *Aspergillus niger* (0.2 g/L), MeJA (100 μM), and AgNPs NPs as elicitors. The combination of treatment with different elicitors substantially enhanced these metabolic responses. Similarly, biologically synthesized AgNPs and CuNPs increase the levels of phenolics, flavonoids, alkaloids, and saponins in *Bacopa monnieri* (Lala 2020). Hayat et al. (2021) revealed that AgNPs and CuNPs produced using *Moringa oleifera* extract notably enhanced the phenolic and flavonoid content of *Artemisia absinthium*, resulting in an antioxidant activity increase of up to 86%. Furthermore, the utilization of CuO and ZnO NPs enhanced both antioxidant levels and pest resistance in *Amaranthus hybridus* (Francis et al. 2022). Krishnaraj et al. (2012) observed that AgNP treatment in *B. monnieri* enhanced protein and carbohydrate levels while diminishing total phenolic content (TPC), CAT, and POX activity. Karimi et al. (2018) noted a substantial enhancement in flavonoid content in *Echinacea purpurea* subjected to green-synthesized ZnO NPs. Soltanabad et al. (2020) demonstrate that the foliar application of 200 ppm of AgNPs boosts the production of carnosic acid and antioxidant content in *Rosmarinus officinalis*. The utilization of aluminum and ZnO NPs resulted in a substantial enhancement in the synthesis of saponin, phenolic content, and iridoids in hydroponically cultivated *N. arvensis* (Chahardoli et al. 2020). Baharmoghadam et al. (2024) noted a significant enhancement in anthocyanin and TPC in *Carum copticum* shoots subjected to TiO_2 NPs (300 ppm) under white-blue light. The synergistic application of TiO_2 NPs with light spectrums has been shown to boost the synthesis of phytochemicals. ZnO NPs are more bioavailable than ionic zinc ($ZnSO_4$), making ZnO NPs effective for enhancing the nutritional quality and potential therapeutic properties of *Stevia rebaudiana*. Treatment with 50 mg/L of phyto-nanoparticles increases stevioside concentration in leaves compared to the control, whereas 100 mg/L reduces rebaudioside A content (Velázquez-Gamboa et al. 2021). These findings collectively emphasize the potential of nano-elicitation in hydro-aeroponic systems to address the constraints of phytochemical production, facilitating enhanced yield and bioactive metabolite synthesis across diverse plant species.

5.6.4 Biofortification Through Nanoparticles

Biofortification is an agricultural approach intended to enhance the amount of micronutrients in crops (Hotz 2009). Organic fertilizers employed in agricultural biofortification demonstrate sustained absorption of nutrients but possess reduced mineral content, whereas traditional inorganic fertilizers with bigger particles (>100 nm) are susceptible to volatilization and loss from leaching. Nanotechnology and hydro-aeroponics present considerable potential to address these difficulties in agriculture. NMs provide accurate nutrient management, improving fertilizer efficacy, reducing the negative environmental effects, and augmenting nutrient absorption in plants (Sun et al. 2022). Pourebrahimi et al. (2023) studied the effect of combined hydrogen sulfide (H_2S) and selenium (Se) treatment to achieve the highest free radical inhibition (84%), along with increased anthocyanin, vitamin C, and antioxidant enzyme activity (CAT, SOD, and POX). Sodium selenate treatment also enhances selenium accumulation in strawberries under both salt-stress and non-stress conditions, promoting selenium biofortification. The foliar treatment of ZnO NPs led to a little increase in Zn accumulation in the hydroponically grown lettuce leaves. This impact was most pronounced in treatments with elevated ZnO NPs concentrations, resulting in an augmentation of antioxidant substances, such as phenols and flavonoids, alongside enhanced POD enzyme activity. As a result, these alterations enhanced the nutritional value of the lettuce. Skiba et al. (2020) investigated the influence of five different kinds of ZnO NPs on the absorption of copper, manganese, and iron in *Pisum sativum* cultivated in Hoagland nutrient solutions. The findings indicated that Zn, in both molecular and nanoparticulate forms, affected the absorption and transport of Cu, Mn, and Fe in *P. sativum* plants, with the presence of ZnO NPs resulting in a significant decrease in Mn and Fe levels, while Cu accumulation increased. A new Se (II) molecule, 2-iminoselenazolidin-4-ones (ISeA), in nanoscale (1–5 nm) solutions was evaluated for the biofortification of Swiss chard and komatsuna plants through foliar (2 mg L^{-1}) and hydroponic (10 mg L^{-1}) applications. ISeA exhibited substantial uptake and accumulation, elevating Se levels, biomass, and quality characteristics, with biomass increasing by up to 68%, alongside enhancements in the dry-to-fresh weight ratio, leaf development, photosynthetic pigments, and anthocyanin concentration (Semenova et al. 2024).

5.6.5 Nanoparticles in Abiotic Stress Management

Abiotic stressors, including extreme temperatures, inadequate or surplus water, high salt, metalloid toxicity, and UV radiation, adversely affect plant growth and development, resulting in significant reductions in crop yields globally (He et al. 2018; Al-Khayri et al. 2023). The actual implementation of NPs and NMs can enhance the functionality and efficacy of physio-chemical and cellular processes in plants subjected to stress, resulting in improved stress tolerance (Raza et al. 2023). Selenium (Se) and nano-selenium (N-Se) NPs are utilized to assess their impact on the growth of tomato plants under varying temperature stress. Among the treatments, 2.5 μM Se greatly enhanced plant growth under both high- and low-temperature environments (Haghighi et al. 2014). Iqbal et al. (2021) examined the impact of conventional silicon (Si-C) and SiNPs on flooding-induced hypoxia on hydroponically grown southern highbush blueberry. Both Si-C and SiNPs effectively mitigated hypoxia-induced oxidative and osmotic cellular damage by augmenting the activities of enzymatic antioxidants (SOD, CAT, and POD), enhancing non-enzymatic antioxidants (ascorbic acid and glutathione levels), and promoting the synthesis of compatible solutes (proline and glycine betaine) in leaves and roots. Nonetheless, the SiNPs demonstrated superior efficacy compared to Si-C in enhancing antioxidant properties and osmolyte synthesis. The SiNPs modulated nutrient absorption by elevating the amounts of K, N, P, and Zn while reducing Fe and Mn levels to a less hazardous range. The interactions between cerium oxide (CeO_2) and zinc oxide (ZnO) NPs with inorganic arsenic (As) were examined under hydroponically cultivated rice seedlings. ZnO NPs significantly decreased As(III) accumulation in rice while providing needed Zn for plant growth. These findings demonstrate the possibility of ZnO NPs as an agricultural tool to promote plant development and reduce As accumulation (Iqbal et al. 2021). The simultaneous application of AgNPs with antimony Sb(V) leads to increased Ag accumulation in plant tissues relative to other treatments. This combination had more pronounced effects on pigment concentration, enzyme activity, and lipid peroxidation compared to individual treatments. Conversely, the combined effect of AgNPs with Sb(III) markedly elevated ROS levels in *glycine max* root (Cao et al. 2020). El-Shal et al. (2022) examined the impact of nano-fertilizers, comprising macronutrient (K_2SO_4) and micronutrients (ZnO, SiO_2), on two alfalfa genotypes: Bulldog 505 (susceptible) and Mesa-Sirsa (tolerant), under saline stress (6 and 10 dS $^{m-1}$) utilizing a split-split plot design. Findings suggested Nano-K_2SO_4 improved biomass and morphological metrics, including shoot dry weight, plant height, flower and tiller counts, root length, and overall biomass, under varying salt concentrations in hydroponic systems. Furthermore, it enhanced relative water content and lessened the leaching of electrolytes in both genotypes during saline stress. Recently, Singh et al. (2022) examined the effect of ZnO NPs (50 mg/L) on NaCl salt tolerance (0, 60, 80, and 100 mM) in two rice genotypes. The results indicate that salt-stress reduces plant height, photosynthetic pigments, Zn content, and K^+ uptake in both genotypes. Subsequently, the application of ZnO NPs alleviated these effects by promoting pigment accumulation, K^+ levels, stomatal and trichome structure, and antioxidant enzyme activity. Titanium dioxide (TiO_2) NPs were applied foliar to saffron plants exposed to UV-B stress to evaluate their effectiveness in reducing photo-oxidative damage. The findings revealed that TiO_2 NPs alleviated the adverse effects of UV-B radiation and potentially enhanced the nutritional value of saffron by boosting the antioxidant capacity of its stigmas (Rikabad et al. 2019). Ma et al. (2017) studied the combined effect of TiO_2 NPs and tetracycline (TC) in rice under hydroponics. Results highlighted that TiO_2 NPs markedly improved the growth of TC-treated rice, as evidenced by increased plant growth and total biomass. Moreover, TiO_2 NPs noticeably diminished the level of TC in the rice shoot and root part, along with oxidative stress enzymes. Rice seedlings were subjected to various combinations of 1 mg/L of As(III) or As(V) with 100 mg/L of CeO_2 NPs and ZnO NPs. ZnO NPs markedly decreased the accumulation of As(III) and As(V) in both rice roots and shoot parts (Wang et al. 2018). Ekim et al. (2024) investigated the effects of polyvinylpyrrolidone (PVP) coated Cu NPs (N1; 100 mg L^{-1}, N2: 400 L^{-1}) on *Triticum aestivum* subjected to salinity (S; 150 mM NaCl) and drought (D; 10% PEG-6000) stresses. Salinity and drought stress decreased seedling growth by 51% and 22%, respectively, whereas the combined stress (S + D) lowered osmotic potential by approximately threefold. Treatment with PVP-Cu NPs (100 mg L^{-1}) enhanced plant growth and water control in hydroponically cultivated stressed

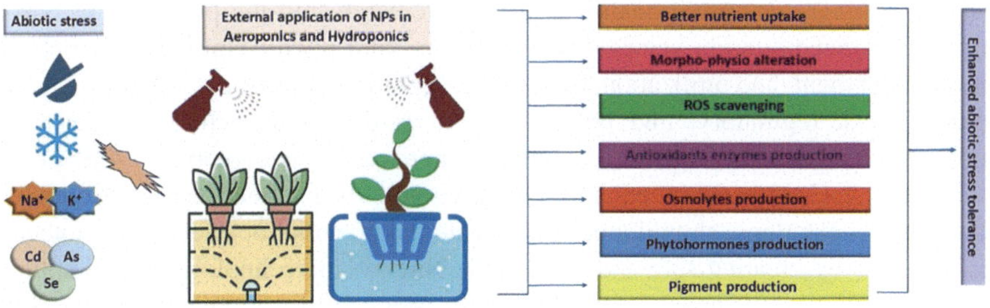

Fig. 5.3 Role of Nanoparticles (NPs) in abiotic stress tolerance in plants

plants. The findings indicate that NPs may serve as an innovative method to augment plant stress tolerance and promote crop yield, necessitating additional research to clarify the fundamental processes and investigate potential agricultural uses (Fig. 5.3).

5.7 Conclusion and Future Direction

This present book chapter discusses multiple functions of NPs in hydroponics and aeroponics. Based on previous research on the influence of NPs on using soilless cultivated crops, it indicates that integrating both technologies has huge possibilities to enhance crop yield and productivity (Lakhiar et al. 2018; Jadoon et al. 2024). NPs exhibited enhanced properties as compared to conventional fertilizers in terms of solubility, stability, thermal tolerance, the absorption of nutrients, and cytotoxicity abatement (Santos et al. 2024). Hydro-aeroponics systems can effectively address numerous issues inherent in traditional agricultural approaches, including land utilization, reliance on pesticides, shortages of water and nutrients, infection rates, and chances of soil-borne disease and contaminations. Hydro-aeroponics enhances agricultural efficiency and sustainability by optimizing resource utilization, increasing productivity, and mitigating environmental losses associated with conventional cultivation practices (Ghosh et al. 2023; Joshi et al. 2024). The application of NPs in soilless cultivation has emerged as a promising strategy to mitigate abiotic stress in plants. Various NPs, including Se, Si, ZnO, TiO_2, CeO_2, and Cu, have been shown to enhance plant resistance to environmental stressors, including temperature, salt, and UV-B radiation, by modulating physiological and biochemical processes, including antioxidant defense, osmolyte production, nutrient intake, and phytohormone signaling (Iqbal et al. 2021; Wu et al. 2023; Shang et al. 2024). Additionally, NP-based fertilizers, such as nano-K_2SO_4, demonstrate the capacity to enhance plant development and nutrient uptake under stress conditions (El-Shal et al. 2022).

The utilization of NMs in crop improvement requires additional investigation regarding developing and manufacturing safe NMs that do not adversely affect plant growth, environmental integrity, or human health (Mittal and Bhukal 2024; Santas-

Miguel et al. 2023). Additionally, as well as research into the processes of NMs intake and transportation across different tissues, and interdisciplinary research in the development and utilization of NPs in agricultural practices. While ethical, toxicological, regulatory, and policy issues regarding nanotechnology must be meticulously evaluated, they also facilitate a more complete evolution of the scientific research. Consequently, enhanced research initiatives, effectively sponsored by funding organizations, should aim to eliminate the potential obstacles to nanotechnology uses by creating safe and sustainable NMs. Nanotechnology in agriculture remains in its nascent phase, requiring further research to comprehend its possible health and environmental significance. Further work is essential to optimize many factors, including the type, dose, and duration of NPs across different plant species in soilless and soil-based systems.

References

Abbas Q, Yousaf B, Ali MU, Munir MAM, El-Naggar A, Rinklebe J, Naushad M (2020) Transformation pathways and fate of engineered nanoparticles (ENPs) in distinct interactive environmental compartments: review. Environ Int 138:105646. https://doi.org/10.1016/j.envint.2020.105646

Abu-Elsaad NI, Abdel hameed RE (2019) Copper ferrite nanoparticles as nutritive supplement for cucumber plants grown under hydroponic system. J Plant Nutr 42(14):1645–1659. https://doi.org/10.1080/01904167.2019.1630428

Adhikary T, Basak P (2024) The paradigm shift towards utilizing nanofertilizers paves a promising future in industrial agriculture and soilless cultivation. Hybrid Adv 6:100255

Al-Juthery HW, Hassan AH, Kareem FK, Musa RF, Khaeim HM (2019) The response of wheat to foliar application of nano-micro nutrients. Plant Arch 19(2):827–831. e-ISSN:2581-6063 (online), ISSN:0972-5210

Al-Khayri JM, Rashmi R, Surya Ulhas R, Sudheer WN, Banadka A, Nagella P, Aldaej MI, Rezk AAS, Shehata WF, Almaghasla MI (2023) The role of nanoparticles in response of plants to abiotic stress at physiological, biochemical, and molecular levels. Plants 12(2):292

Alsammarraie FK, Wang W, Zhou P, Mustapha A, Lin M (2018) Green synthesis of silver nanoparticles using turmeric extracts and investigation of their antibacterial activities. Colloids Surf B Biointerfaces 171:398–405. https://doi.org/10.1016/j.colsurfb.2018.07.059

Arora A, Nandal P, Singh J, Verma ML (2020) Nanobiotechnological advancements in lignocellulosic biomass pretreatment. Mater Sci Energy Technol 3:308–318. https://doi.org/10.1016/j.mset.2019.12.003

Baharmoghadam M, Keramat B, Nadernejad N, Mozafari H, Pourseyedi S (2024) Improving essential oil production and antioxidant properties of *Carum copticum* L. by titanium dioxide nanoparticles and light spectrum interaction under hydroponic conditions. Plant Cell Tissue Organ Cult 159(2):53. https://doi.org/10.1007/s11240-024-02897-0

Balusamy SR, Joshi AS, Perumalsamy H, Mijakovic I, Singh P (2023) Advancing sustainable agriculture: a critical review of smart and eco-friendly nanomaterial applications. J Nanobiotechnol 21(1):372. https://doi.org/10.1186/s12951-023-02135-3

Bathla M, Saini TC, Pal PK, Acharya A (2024) Applications of smart nanostructures in crop production and protection. In: Nanotechnology and nanomaterials in the agri-food industries. Elsevier, pp 125–150. https://doi.org/10.1016/B978-0-323-99682-2.00001-3

Benke K, Tomkins B (2017) Future food-production systems: vertical farming and controlled-environment agriculture. Sustain: Sci Pract Policy 13(1):13–26. https://doi.org/10.1080/15487733.2017.1394054

Cao W, Gong J, Zeng G, Song B, Zhang P, Li J, Fang S, Qin L, Ye J, Cai Z (2020) Mutual effects of silver nanoparticles and antimony (iii)/(v) co-exposed to *Glycine max* (L.) Merr. in hydroponic systems: uptake, translocation, physiochemical responses, and potential mechanisms. Environ Sci Nano 7(9):26912707. https://doi.org/10.1002/anie.202005577

Chahardoli A, Karimi N, Ma X, Qalekhani F (2020) Effects of engineered aluminum and nickel oxide nanoparticles on the growth and antioxidant defense systems of *Nigella arvensis* L. Sci Rep 10(1):3847. https://doi.org/10.1038/s41598-020-60841-6

Dağhan H (2018) Effects of TIO_2 nanoparticles on maize (*zea mays* l.) growth, chlorophyll content and nutrient uptake. Appl Ecol Environ Res 16:6873–6883. https://doi.org/10.15666/aeer/1605_68736883

Dağhan H, Gülmezoğlu N, Köleli N, Karakaya B (2020) Impact of titanium dioxide nanoparticles (TiO2-NPs) on growth and mineral nutrient uptake of wheat (*Triticum vulgare* L.). Biotech Stud 29(2):69–76. https://doi.org/10.38042/biost.2020.29.02.03

Ekim R, Arikan B, Alp-Turgut FN, Koyukan B, Ozfidan-Konakci C, Yildiztugay E (2024) Polyvinylpyrrolidone-coated copper nanoparticles dose-dependently conferred tolerance to wheat under salinity and/or drought stress by improving photochemical activity and antioxidant system. Environ Res 241:117681. https://doi.org/10.1016/j.envres.2023.117681

El-Shal RM, El-Naggar AH, El-Beshbeshy TR, Mahmoud EK, El-Kader NIA, Missaui AM, Du D, Ghoneim AM, El-Sharkawy MS (2022) Effect of nano-fertilizers on alfalfa plants grown under different salt stresses in hydroponic system. Agriculture 12(8):1113. https://doi.org/10.3390/agriculture12081113

Erekath S, Seidlitz H, Schreiner M, Dreyer C (2024) Food for future: exploring cutting-edge technology and practices in vertical farm. Sustain Cities Soc 106:105357. https://doi.org/10.1016/j.scs.2024.105357

Fatima RT, Dantas MV, Lima GS, Oliveira VKN, Soares LAA, Silva AAR, Gheyi HR, Guedes MA, Nóbrega JS, Fernandes PD (2023) Salicylic acid does not relieve salt stress on gas exchange, chlorophyll fluorescence, and hydroponic melon growth. Braz J Biol 83:e274595. https://doi.org/10.1590/1519-6984.274595

Francis DV, Sood N, Gokhale T (2022) Biogenic CuO and ZnO nanoparticles as nanofertilizers for sustainable growth of *Amaranthus hybridus*. Plants 11(20):2776. https://doi.org/10.3390/plants11202776

Garzón J, Montes L, Garzón J, Lampropoulos G (2023) Systematic review of technology in aeroponics: introducing the technology adoption and integration in sustainable agriculture model. Agronomy 13(10):2517. https://doi.org/10.3390/agronomy13102517

Ghidan AY, Al Antary TM (2019) Applications of nanotechnology in agriculture. In J Nanobiotechnol IntechOpen. https://doi.org/10.5772/intechopen.88390

Ghosh B, Roy S, Ahmed, De D (2023) Dew aeroponics: dew-enabled smart aeroponics system in agriculture 4.0. In: Dew computing: the sustainable IoT perspectives. Springer Nature Singapore, Singapore, pp 261–287. https://doi.org/10.1007/978-981-99-4590-0_13

Gomiero T (2016) Soil degradation, land scarcity and food security: reviewing a complex challenge. Sustainability 8(3):281. https://doi.org/10.3390/su8030281

Hadi Soltanabad M, Bagherieh-Najjar MB, Mianabadi M. (2020) Carnosic acid content increased by silver nanoparticle treatment in rosemary (Rosmarinus officinalis L.). Appl. Biochem. Biotechnol. 191:482–495. https://doi.org/10.1007/s12010-019-03193-w

Haghighi M, Abolghasemi R, da Silva JAT (2014) Low and high temperature stress affect the growth characteristics of tomato in hydroponic culture with Se and nano-Se amendment. Sci Hortic 178:231–240. https://doi.org/10.1016/j.scienta.2014.09.006

Hayat K, Ali S, Ullah S, Fu Y, Hussain M (2021) Green synthesized silver and copper nanoparticles induced changes in biomass parameters, secondary metabolites production, and antioxidant activity in callus cultures of *Artemisia absinthium* L. Green Process Synth 10(1):61–72. https://doi.org/10.1515/gps-2021-0010

He M, He CQ, Ding N (2018) Abiotic stresses: general defenses of land plants and chances for engineering multistress tolerance. Front Plant Sci 9:1771. https://doi.org/10.3389/fpls.2018.01771

Hoag GE, Collins JB, Holcomb JL, Hoag JR, Nadagouda MN, Varma RS (2009) Degradation of bromothymol blue by 'greener'nano-scale zero-valent iron synthesized using tea polyphenols. J Mater Chem 19(45):8671–8677. https://doi.org/10.1039/B909148C

Horwat D, Zakharov DI, Endrino JL, Soldera F, Anders A, Migot S, Karoum R, Vernoux P, Pierson JF (2011) Chemistry, phase formation, and catalytic activity of thin palladium-containing oxide films synthesized by plasma-assisted physical vapor deposition. Surf Coat Technol 205:S171–S177. https://doi.org/10.1016/j.surfcoat.2010.12.021

Hotz C (2009) The potential to improve zinc status through biofortification of staple food crops with zinc. Food Nutr Bull 30:S172–S178. https://doi.org/10.1177/15648265090301S109

Ikram S (2015) Synthesis of gold nanoparticles using plant extract: an overview. Nano Res 1(1):5. http://nanotechnology.imedpub.com/archive.php

Iqbal Z, Sarkhosh A, Balal RM, Rauf S, Khan N, Altaf MA, Camara-Zapata JM, Garcia-Sanchez F, Shahid MA (2021) Silicon nanoparticles mitigate hypoxia-induced oxidative damage by improving antioxidants activities and concentration of osmolytes in southern highbush blueberry plants. Agron J 11(11):2143. https://doi.org/10.3390/agronomy11112143

Isah T (2019) Stress and defense responses in plant secondary metabolites production. Biol Res 52: 39. https://doi.org/10.1186/s40659-019-0246-3

Jadoon L, Gul A, Fatima H, Babar MM (2024) Nano-elicitation and hydroponics: a synergism to enhance plant productivity and secondary metabolism. Planta 259(4):80. https://doi.org/10.1007/s00425-024-04353-x

Jaeger WK, Antle J, Gingerich SB, Bigelow D (2024) Advancing sustainable groundwater management with a hydro-economic system model: investigations in the Harney basin oregon. Water Resour Res 60(11). https://doi.org/10.1029/2023WR036972p.e2023WR036972

Jeyasubramanian K, Thoppey UUG, Hikku GS, Selvakumar N, Subramania A, Krishnamoorthy K (2016) Enhancement in growth rate and productivity of spinach grown in hydroponics with iron oxide nanoparticles. RSC Adv 6(19):15451–15459. https://doi.org/10.1039/C5RA23425E

Joshi N, Pathak A, Upadhyaya DC, Upadhyaya CP (2024) Impact of metal oxide nanoparticles on potato (*Solanum tuberosum* L.) tuber yield in hydroponics. Am J Potato Res 101(4):322–335. https://doi.org/10.1007/s12230-024-09961-7

Kang X, Zhao L, Liu X (2023) Calcium signaling and the response to heat shock in crop plants. Int J Mol Sci 25(1):32. https://doi.org/10.3390/ijms25010324

Karimi N, Behbahani M, Dini G, Razmjou A (2018) Enhancing the secondary metabolite and anticancer activity of Echinacea purpurea callus extracts by treatment with biosynthesized ZnO nanoparticles. Adv Nanosci Nanotechnol 9(4):045009. https://doi.org/10.1088/2043-6254/aaf1af

Khan AK, Kousar S, Tungmunnithum D, Hano C, Abbasi BH, Anjum S (2021) Nano-elicitation as an effective and emerging strategy for in vitro production of industrially important flavonoids. Appl Sci 11(4):1694. https://doi.org/10.3390/app11041694

Krishnaraj C, Jagan EG, Ramachandran R, Abirami SM, Mohan N, Kalaichelvan PT (2012) Effect of biologically synthesized silver nanoparticles on *Bacopa monnieri* (Linn.) Wettst plant growth metabolism. Process Biochem 47(4):651–658. https://doi.org/10.1016/j.procbio.2012.01.006

Kumar A, Partap M, Warghat AR (2024) Jasmonic acid: a versatile phytohormone regulating growth, physiology, and biochemical responses. J Plant Growth Regul 44:131–154. https://doi.org/10.1007/s00344-024-11376-x

Lakhiar IA, Gao J, Syed TN, Chandio FA, Buttar NA (2018) Modern plant cultivation technologies in agriculture under controlled environment: a review on aeroponics. J Plant Interact 13(1): 338–352. https://doi.org/10.1080/17429145.2018.1472308

Lala S (2020) Enhancement of secondary metabolites in *Bacopa monnieri* (L.) Pennell plants treated with copper-based nanoparticles in vivo. IET Nanobiotechnol 14(1):78–85. https://doi.org/10.1049/iet-nbt.2019.0124

Landa P, Vankova R, Andrlova J, Hodek J, Marsik P, Storchova H, White JC, Vanek T (2012) Nanoparticle-specific changes in *Arabidopsis thaliana* gene expression after exposure to ZnO, TiO2, and fullerene soot. J Hazard Mater 241:55–62. https://doi.org/10.1016/j.jhazmat.2012.08.059

Ma C, Liu H, Guo H, Musante C, Coskun SH, Nelson BC, White JC, Xing B, Dhankher OP (2016) Defense mechanisms and nutrient displacement in Arabidopsis thaliana upon exposure to CeO 2 and In2O3 nanoparticles. Environ Sci Nano 3(6):1369–1379. https://doi.org/10.1039/C6EN00189K

Ma C, Liu H, Chen G, Zhao Q, Eitzer B, Wang Z, Cai W, Newman LA, White JC, Dhankher OP, Xing B (2017) Effects of titanium oxide nanoparticles on tetracycline accumulation and toxicity in *Oryza sativa* (L.). Environ Sci Nano 4(9):1827–1839

Maluin FN, Katas H (2022) Chitosan functionalization of metal-and carbon-based nanomaterials as an approach toward sustainability tomorrow. Nanotoxicology 16(4):425–449. https://doi.org/10.1080/17435390.2022.2090025

Marslin G, Sheeba CJ, Franklin G (2017) Nanoparticles alter secondary metabolism in plants via ROS burst. Front Plant Sci 8:832. https://doi.org/10.3389/fpls.2017.00832

Mirzajani F, Askari H, Hamzelou S, Schober Y, Römpp A, Ghassempour A, Spengler B (2014) Proteomics study of silver nanoparticles toxicity on *Oryza sativa* L. Ecotoxicol Environ Saf 108:335–339. https://doi.org/10.1016/j.ecoenv.2014.07.013

Mittal R, Bhukal S (2024) Sustainable approach for agriculture and environmental remediation using hydroponics and their perspectives. In: Hydroponics and environmental bioremediation: wastewater treatment. Springer Nature Switzerland, Cham, pp 65–90. https://doi.org/10.1007/978-3-031-53258-0_3

Rabary PM, Movahedi Z, Ghabooli M, Rostami M (2022) Effects of foliar application of zinc oxide nanoparticles on traits of several medicinal plants under aeroponic system conditions. Int J Hort Sci Technol 9(4):445–452. https://doi.org/10.22059/ijhst.2021.316933.433

Movahedi Z, Moieni A (2024) Effects of Farmax nano fertilizer and amino acid on morological traits and photosynthetic pigments of chicory in aeroponic system. Iran J Soil Water Res 5(6):889–902. https://doi.org/10.22059/ijswr.2024.371442.669651

Mubeen B, Hasnain A, Mehboob R, Rasool R, Riaz A, Elaskary SA, Shah MM, Faridi TA, Ullah I (2022) Hydroponics and elicitation, a combined approach to enhance the production of designer secondary medicinal metabolites in *Silybum marianum*. Front Plant Sci 13:897795. https://doi.org/10.3389/fpls.2022.897795

Mubeen B, Hasnain A, Jie W, Zheng H, Peijnenburg WJ, Rozali SE, Rasool R, Naqvi SAH, Rao MJ, Sohail MA, Moustafa M (2023) Enhanced production of active photosynthetic and biochemical molecules in *Silybum marianum* L. using biotic and abiotic elicitors in hydroponic culture. Molecules 28(4):1716. https://doi.org/10.3390/molecules28041716

Nazarideljou MJ, Rahmani S, Kalanaky S, Hojati M, Nazaran MH (2024) Nanochelating-based fertilizers: innovations for yield enhancement in hydroponic cultivation. Sci Hortic 326:112776. https://doi.org/10.1016/j.scienta.2023.112776

Oraibi AG, Yahia HN, Alobaidi KH (2022) Green biosynthesis of silver nanoparticles using *Malva parviflora* extract for improving a new nutrition formula of a hydroponic system. Scientifica 2022(1):4894642. https://doi.org/10.1155/2022/4894642

Partap M, Kumar P, Kumar A, Joshi R, Kumar D, Warghat AR (2020) Effect of elicitors on morpho-physiological performance and metabolites enrichment in *Valeriana jatamansi* cultivated under aeroponic conditions. Front Plant Sci 11:01263. https://doi.org/10.3389/fpls.2020.01263

Partap M, Rattan S, Kanika A, Sood A, Kumar P, Warghat AR (2021) Hydroponic and aeroponic cultivation of economically important crops for production of quality biomass in agricultural biotechnology latest trends. Springer, pp 573–585. https://doi.org/10.1007/978-981-16-2339-4_25

Partap M, Kumar A, Kumar P, Kumar D, Warghat AR (2024) Elicitors mediated enhancement of picrosides and their pathway precursors in dedifferentiated cell suspension culture of *Picrorhiza kurroa* royle ex Benth. J Plant Growth Regul 44:247–265. https://doi.org/10.1007/s00344-024-11537-y

Pourebrahimi M, Eshghi S, Ramezanian A, Faghih S (2023) Selenium and hydrogen sulfide mitigate the adverse effects of salinity on growth, yield, and mineral nutrient status in strawberry. J Soil Sci Plant Nutr 23(3):4218–4232. https://doi.org/10.1007/s42729-023-01343-y

Rahmawati RS, Sukma D, Ardie SW, Sudarsono S 2021. Postharvest physiological deterioration in cassava: potential problems, possible inhibition, and resistant level identification. In IOP conference series: environ. Earth Sci 694(1) 012035). IOP Publishing. https://doi.org/10.1088/1755-1315/694/1/012035

Ran M, Wu J, Jiao Y, Li J (2024) Biosynthetic selenium nanoparticles (Bio-SeNPs) mitigate the toxicity of antimony (Sb) in rice (*Oryza sativa* L.) by limiting Sb uptake, improving antioxidant defense system and regulating stress-related gene expression. J Hazard Mater 470:134263. https://doi.org/10.1016/j.jhazmat.2024.134263

Rattan S, Partap M, Kumar S, Warghat AR (2022) Nutrient feeding approach enhances the vegetative growth biomass, volatile oil composition, and myristicin content in hydroponically cultivated *Petroselinum crispum* (Mill.). J Appl Res Med Aromat Plants 26:100359. https://doi.org/10.1016/j.jarmap.2021.100359

Raza A, Charagh S, Salehi H, Abbas S, Saeed F, Poinern GE, Siddique KH, Varshney RK (2023) Nano-enabled stress-smart agriculture: can nanotechnology deliver drought and salinity-smart crops. J Sustain Agric Environ 2(3):189–214. https://doi.org/10.1002/sae2.12061

Rikabad MM, Pourakbar L, Moghaddam SS, Popović-Djordjević J (2019) Agrobiological, chemical and antioxidant properties of saffron (*Crocus sativus* L.) exposed to TiO2 nanoparticles and ultraviolet-B stress. Ind Crop Prod 137:137–143. https://doi.org/10.1016/j.indcrop.2019.05.017

Roosta HR, Estaji A, Niknam F (2018) Effect of iron, zinc and manganese shortage-induced change on photosynthetic pigments, some osmoregulators and chlorophyll fluorescence parameters in lettuce. Photosynthetica 56:606–615. https://doi.org/10.1007/s11099-017-0696-1

Santás-Miguel V, Arias-Estévez M, Rodríguez-Seijo A, Arenas-Lago D (2023) Use of metal nanoparticles in agriculture. A review on the effects on plant germination. Environ Pollut 334:122222. https://doi.org/10.1016/j.envpol.2023.122222

Santos O, Vaz D, Sebastião F, Sousa H, Vieira J (2024) Wastewater as a nutrient source for hydroponic production of lettuce: summer and winter growth. Agric Water Manag 301:108966. https://doi.org/10.1016/j.agwat.2024.108966

Savvas D, Özteki GB, Tepecik M, Ropokis A, Tüzel Y, Ntatsi G, Schwarz D (2017) Impact of grafting and rootstock on nutrient-to-water uptake ratios during the first month after planting of hydroponically grown tomato. J Hort Sci Biotechnol 92(3):294–302. https://doi.org/10.1080/14620316.2016.1265903

Seleiman MF, Almutairi KF, Alotaibi M, Shami A, Alhammad BA, Battaglia ML (2020) Nano-fertilization as an emerging fertilization technique: why can modern agriculture benefit from its use. Plants 10(1):2. https://doi.org/10.3390/plants10010002

Selvakesavan RK, Kruszka D, Shakya P, Mondal D, Franklin G (2023) Impact of nanomaterials on plant secondary metabolism. Nanomaterial interactions with plant cellular mechanisms and macromolecules and agricultural implications. Springer, pp 133–170. https://doi.org/10.1007/978-3-031-20878-2_6

Semenova NA, Nikulina EA, Tsirulnikova NV, Godyaeva MM, Uyutova NI, Baimler IV, Simakin AV, Stepanova EV, Gudkov SV (2024) Application of 2-Iminoselenazolidin-4-Ones (ISeA) for *Beta vulgaris* L. and *Brassica rapa* L. Plants Se-Biofortification. Agron J 14(7):1407. https://doi.org/10.3390/agronomy14071407

Sena S, Ochatt SJ, Kumar V (2023) Application of green synthesized nanoparticles in medicinal plant research: revisiting an emerging eco-friendly approach. Plant Cell Tissue Organ Cult 155(2):345–384. https://doi.org/10.1007/s11240-023-02613-4

Shahid S, Imtiaz H, Rashid J, Xu M, Vithanage M, Ahmad M (2024) Uptake, translocation, and nutrient efficiency of nano-bonechar as a plant growth regulator in hydroponics and soil systems. Environ Res 251:118695. https://doi.org/10.1016/j.envres.2024.118695

Shang H, Li C, Cai Z, Hao Y, Cao Y, Jia W, Han L, White JC, Ma C, Xing B (2024) Biosynthesized selenium nanoparticles as an effective tool to combat soil metal stresses in rice (*Oryza sativa* L.). ACS Nano 18(30):19636–19648. https://doi.org/10.1021/acsnano.4c04215

Sharma D, Partap M, Warghat AR, Bhargava B (2024) Hydroponic cultivation enhances the morpho-physiological traits and quality flower production in three cultivars (marigold scarlet red, marigold orange, and marigold yellow) of French marigold (*Tagetes patula* L.). Sci Hort 327:112803. https://doi.org/10.1016/j.scienta.2023.112803

Shukla K, Mishra V, Singh J, Varshney V, Verma R, Srivastava S (2024) Nanotechnology in sustainable agriculture: a double-edged sword. J Sci Food Agric 104(10):5675–5688. https://doi.org/10.1002/jsfa.13342

Singh A, Sengar RS, Shahi UP, Rajput VD, Minkina T, Ghazaryan KA (2022) Prominent effects of zinc oxide nanoparticles on roots of rice (*Oryza sativa* L.) grown under salinity stress. Stress 3(1):33–46. https://doi.org/10.3390/stresses3010004

Singh A, Rajput VD, Lalotra S, Agrawal S, Ghazaryan K, Singh J, Minkina T, Rajput P, Mandzhieva S, Alexiou A (2024a) Zinc oxide nanoparticles influence on plant tolerance to salinity stress: insights into physiological, biochemical, and molecular responses. Environ Geochem Health 46(5):148. https://doi.org/10.12911/22998993/169142

Singh M, Goswami SP, Sachan P, Sahu DK, Beese S, Pandey SK (2024b) Nanotech for fertilizers and nutrients-improving nutrient use efficiency with nano-enabled fertilizers. J Exp Bot 46(5): 220–247. https://doi.org/10.9734/jeai/2024/v46i52372

Skiba E, Michlewska S, Pietrzak M, Wolf WM (2020) Additive interactions of nanoparticulate ZnO with copper, manganese and iron in *Pisum sativum* L., a hydroponic study. Sci Rep 10(1): 13574. https://doi.org/10.1038/s41598-020-70303-8

Soni RA, Rizwan MA, Singh S (2022) Opportunities and potential of green chemistry in nanotechnology. Nanotechnol Environ 7(3):661–673. https://doi.org/10.1007/s41204-022-00233-5

Sosan A, Svistunenko D, Straltsova D, Tsiurkina K, Smolich I, Lawson T, Subramaniam S, Golovko V, Anderson D, Sokolik A, Colbeck I (2016) Engineered silver nanoparticles are sensed at the plasma membrane and dramatically modify the physiology of *Arabidopsis thaliana* plants. Plant J 85(2):245–257. https://doi.org/10.1111/tpj.13105

Srivastava S, Bhargava A (2022) Green nanoparticles: the future of nanobiotechnology. Springer, pp 1–13. https://doi.org/10.1007/978-981-16-7106-7

Sun C, Sun N, Ou Y, Gong B, Jin C, Shi Q, Lin X (2022) Phytomelatonin and plant mineral nutrition. J Exp Bot 73(17):5903–5917. https://doi.org/10.1093/jxb/erac289

Tangley L (1987) Beyond the green revolution. Bioscience 37(3):176–180. https://doi.org/10.2307/1310514

Thakur K, Partap M, Kumar D, Warghat AR (2019) Enhancement of picrosides content in *Picrorhiza kurroa* Royle ex Benth. mediated through nutrient feeding approach under aeroponic and hydroponic system. Ind Crops Prod 133:160–167. https://doi.org/10.1016/j.indcrop.2019.03.021

Tiwari M, Sharma NC, Fleischmann P, Burbage J, Venkatachalam P, Sahi SV (2017) Nanotitania exposure causes alterations in physiological, nutritional and stress responses in tomato (*Solanum lycopersicum*). Front Plant Sci 8:633. https://doi.org/10.3389/fpls.2017.00633

Tofighi Alikhani T, Tabatabaei SJ, Mohammadi Torkashvand A, Khalighi A, Talei D (2021) Effects of silica nanoparticles and calcium chelate on the morphological, physiological and biochemical characteristics of gerbera (*Gerbera jamesonii* L.) under hydroponic condition. J Plant Nutr 44(7):1039–1053. https://doi.org/10.1080/01904167.2020.1867578

Tombuloglu H, Slimani Y, Tombuloglu G, Almessiere M, Baykal A, Ercan I, Sozeri H (2019) Tracking of NiFe2O4 nanoparticles in barley (*Hordeum vulgare* L.) and their impact on plant growth, biomass, pigmentation, catalase activity, and mineral uptake. Environ Nanotechnol Monit Manag 11:100223. https://doi.org/10.1016/j.chemosphere.2019.03.075

Van Aken O, Van Breusegem F (2015) Licensed to kill: mitochondria, chloroplasts, and cell death. Trends Plant Sci 20(11):754–766. https://doi.org/10.1016/j.tplants.2015.08.002

Velázquez-Gamboa MC, Rodríguez-Hernández L, Abud-Archila M, Gutiérrez-Miceli FA, González-Mendoza D, Valdez-Salas B, González-Terreros E, Luján-Hidalgo MC (2021) Agronomic biofortification of *Stevia rebaudiana* with zinc oxide (ZnO) phytonanoparticles and antioxidant compounds. Sugar Tech 23(2):453–460. https://doi.org/10.1007/s12355-020-00897-w

Verma A, Gautam SP, Bansal KK, Prabhakar N, Rosenholm JM (2019) Green nanotechnology: advancement in phytoformulation research. Medicines 6(1):39. https://doi.org/10.3390/medicines6010039

Wang Q, Ma X, Zhang W, Pei H, Chen Y (2012) The impact of cerium oxide nanoparticles on tomato (*Solanum lycopersicum* L.) and its implications for food safety. Metallomics 4(10): 1105–1112. https://doi.org/10.1039/c2mt20149f

Wang X, Sun W, Zhang S, Sharifan H, Ma X (2018) Elucidating the effects of cerium oxide nanoparticles and zinc oxide nanoparticles on arsenic uptake and speciation in rice (*Oryza sativa*) in a hydroponic system. Environ Sci Technol 52(17):10040–10047. https://doi.org/10.1016/j.cej.2019.123802

Wang Y, Wang L, Ma C, Wang K, Hao Y, Chen Q, Mo Y, Rui Y (2019) Effects of cerium oxide on rice seedlings as affected by co-exposure of cadmium and salt. Environ Pollut 252:1087–1096. https://doi.org/10.1016/j.envpol.2019.06.007

Wang Y, Zhang P, Li M, Guo Z, Ullah S, Rui Y, Lynch I (2020) Alleviation of nitrogen stress in rice (*Oryza sativa*) by ceria nanoparticles. Environ Sci Nano 7(10):2930–2940. https://doi.org/10.1039/d0en00757a

Wu YM, Liu SY, Shi BY, Peng JY, Kao ZW, Chen YY, Hsieh TY, Chung HY, Lin CY, Fang W, Chen LC (2023) IoT-interfaced solid-contact ion-selective electrodes for cyber-monitoring of element-specific nutrient information in hydroponics. Comput Electron Agric 214:108266. https://doi.org/10.1016/j.compag.2023.108266

Yadav R, Kumar M, Tomar RS (2024) Cyanobacteria based nanoformulation of biogenic CuO nanoparticles for plant growth promotion of rice under hydroponics conditions. Curr. Microbiol. 81(5):118. https://doi.org/10.1007/s00284-024-03619-7

Yang Y, Li R, Wang L, Bi H, Peng Z, Li J, Guo C, Bi Y, Lai Y, Guo D (2023) Enhanced germination and growth of alfalfa with seed presoaking and hydroponic culture in Fe2O3 magnetic nanoparticles. J Nanomater 1:9783977. https://doi.org/10.1155/2023/9783977

Zhang T, Zhang YH, Yang JX, Wang XZ, Yang QQ, Zhu XJ, Cao XY (2022) Transcriptome and targeted metabolome analysis revealed the effects of combined red and blue light on the growth and secondary metabolism of *Scutellaria baicalensis* Georgi. Ind Crop Prod 188:115598. https://doi.org/10.1016/j.indcrop.2022.115598

Zoufan P, Baroonian M, Zargar B (2020) ZnO nanoparticles-induced oxidative stress in *Chenopodium murale* L, Zn uptake, and accumulation under hydroponic culture. Environ Sci Pol 27:11066–11078. https://doi.org/10.1007/s11356-020-07735-2

Chapter 6
Nanotechnology in Agritech Revolution: Opportunities and Challenges

Ashutosh Bhaik, Andy Leader, and Courtney Gallup

Abstract Nanotechnology holds transformative potential for the agricultural sector, promising enhanced crop production, protection, and sustainability. This book chapter critically examines the opportunities and challenges of deploying nanotechnology in the agri-input sector of agriculture, emphasizing the need for robust research and regulatory frameworks. Nanotechnology offers a wide range of applications in agriculture, providing innovative solutions for enhancing crop nutrition, improving pest control and delivery systems, and promoting sustainable farming practices. The development and adoption of nano-fertilizers, nano-pesticides, nanosensors, and other nano-enabled products have the potential to revolutionize the agricultural industry. However, it is essential to consider the environmental and health impacts of these technologies and establish regulatory frameworks to ensure their safe use. Continued research and development in agricultural nanotechnology will contribute to more efficient, sustainable, and resilient food production systems.

Keywords Precision agriculture · Nano-based seed treatments · Nanosensors · Nano-enabled products · Regulatory frameworks

6.1 Introduction

The global agricultural sector faces unprecedented challenges that threaten food security and sustainability. Rising demand for food, driven by population growth, environmental stress from climate change, and resource depletion, are major concerns (FAO 2021). By 2050, the global population is expected to reach nearly 10 billion, significantly increasing the demand for agricultural products. Concurrently, climate change is exacerbating environmental stressors, leading to unpredictable weather patterns, more frequent extreme weather events, and shifting

A. Bhaik (✉)
Corteva Agriscience, Hyderabad, Telangana, India

A. Leader · C. Gallup
Corteva Agriscience, Indianapolis, USA

S. K. Yadav, A. Kumari (eds.), *Nanobiotechnology in Agri-food Sector*, Sustainability Sciences in Asia and Africa, https://doi.org/10.1007/978-981-96-7756-6_6

agricultural zones, which adversely affect crop yields and livestock productivity. Resource depletion, particularly of arable land, water, and soil nutrients, further compounds these challenges, necessitating more efficient and sustainable farming practices. Addressing these issues requires innovative approaches, including advancements in technology, improved resource management, and policies that promote sustainable agricultural practices to ensure a stable and resilient food supply for future generations.

The word nano comes from the Greek word for “dwarf.” A nanometer (nm) is a unit measuring one billionth of a meter in length or 10^{-9} M. To put this into context, consider that an average human hair is 80,000–100,000 nm wide, viruses fall between 30 and 50 nm, a strand of human DNA is 2.5 nm, and a water molecule is less than 1 nm across, so we are considering the very smallest of molecules (Nano.gov 2018). Nanotechnology offers innovative solutions through enhanced efficiency and precision in agricultural practices (Scott and Chen 2013). The manipulation of matter on an atomic, molecular, and supramolecular scale is emerging as a revolutionary field with promising applications in agriculture (Singh et al. 2020). In this chapter, we would like to explore the fundamental principles of nanotechnology and its potential to transform agricultural practices. By leveraging nanoscale materials and technologies, it is possible to enhance the efficiency of agricultural inputs, improve crop protection against pests, and increase overall productivity. Applications include nano-fertilizers that optimize nutrient delivery, nano-pesticides that target specific pests with minimal environmental impact, and nanosensors that provide real-time monitoring of crop health and soil conditions. Additionally, nanotechnology can contribute to food preservation and reduce post-harvest losses through advanced packaging and storage solutions. The integration of nanotechnology into agriculture holds the promise of addressing critical challenges such as resource efficiency, environmental sustainability, and food security, paving the way for a more resilient and productive agricultural sector.

6.1.1 The Origins of Nanotechnology

The concept of nanotechnology was first introduced by physicist Richard Feynman in his famous 1959 lecture titled “There’s Plenty of Room at the Bottom,” where he discussed the possibility of manipulating materials at the atomic and molecular levels (Feynman 1960). However, the term “nanotechnology” was popularized by K. Eric Drexler in the 1980s when he talked about the potential of building machines on the scale of molecules through a hypothetical process called molecular nanotechnology (Drexler 1986).

The development of the scanning tunneling microscope (STM) in 1981 by Gerd Binnig and Heinrich Rohrer, which could view individual atoms, significantly advanced the field, allowing scientists to observe and manipulate nanoscale materials (Binnig and Rohrer 1982).

6.2 Applications of Nanotechnology in Agriculture

The integration of nanotechnology into agriculture is a more recent phenomenon, primarily gaining momentum in the early 2000s. Early examples of nanotechnology in agriculture involved the use of nanoparticles (NPs) to improve the delivery systems for pesticides, herbicides, and fertilizers, enhancing their effectiveness and reducing environmental impacts.

Early Examples of Nanotechnology in Agriculture

Nano-encapsulated Agrochemicals	In the early 2000s, researchers began experimenting with nanoencapsulation techniques to improve the delivery of agrochemicals. Nair et al. (2010) discuss how NPs can enhance the delivery of active ingredients in plant systems, improving uptake and reducing runoff.
Nanosensors	Around the same time, the development of nanosensors designed to detect soil moisture and nutrient levels provided farmers with tools to make more informed decisions regarding irrigation and fertilization. Pérez-de-Luque (2017) highlights how nanosensors can detect changes at the nanoscale, providing precise data for optimizing farm management practices.
Nanoparticles for Disease Detection and Protection:	Another early application was the use of NPs to protect plants from pests and diseases. Sharon et al. (2010) discuss the antimicrobial properties of silver NPs used to protect plants against pathogens without the environmental and health risks associated with traditional chemical treatments.

These applications were among the pioneers in leveraging nanotechnology for agricultural improvement, setting the stage for more sophisticated developments in the years that followed. As research has progressed, applications have expanded dramatically, encompassing areas like nano-formulated growth enhancers, biodegradable NMs for reduced ecological impact, and even nano-biotechnology for crop improvement at the genetic level.

Nanotechnology is revolutionizing the agriculture sector by offering innovative solutions to age-old problems. By manipulating materials at the nanoscale, researchers have developed products that enhance crop nutrition, improve pest control, and promote sustainable agricultural practices. The following sections explore key applications of nanotechnology in agriculture, including nano-fertilizers, nano-pesticides, nanosensors, nano-based seed treatments, smart packaging, water management, soil improvement, and enhanced genetic engineering tools.

6.2.1 *Nano-Fertilizers: Enhanced Crop Nutrition*

Types and Benefits of Nano-Fertilizers Nano-fertilizers are engineered particles that provide nutrients to plants more efficiently than conventional fertilizers. They come in various forms, such as nano-encapsulated nutrients, nano-emulsions, and nanoparticle-coated nutrients. These fertilizers offer several advantages, including enhanced nutrient delivery, reduced leaching, and minimized environmental impact.

Impact on Nutrient Uptake and Crop Yields Nano-fertilizers improve nutrient uptake by increasing the surface area available for absorption and enhancing the solubility and mobility of nutrients in the soil. This leads to better root penetration and nutrient assimilation, resulting in higher crop yields. Studies have shown that nano-fertilizers can increase crop productivity by 20–30% compared to traditional fertilizers (Solanki et al. 2015).

Case Studies and Commercial Products One example of a successful nano-fertilizer is the Nano Urea developed by the Indian Farmers Fertilizer Cooperative (IFFCO). Field trials have demonstrated its effectiveness in enhancing nitrogen use efficiency and crop yields. Another example is the use of zinc oxide NPs to improve rice growth and yield in various regions of Asia (Raliya et al. 2016).

6.2.2 *Nano-Pesticides: Targeted Pest Control*

Development of Nano-Pesticides and Their Targeted Delivery Mechanisms Nano-pesticides, including fungicides, insecticides, and virucides, are formulated to deliver active ingredients more precisely and effectively than conventional pesticides. These products utilize nanocarriers, such as liposomes and polymers, to encapsulate active ingredients and release them in a controlled manner. This targeted delivery reduces the amount of pesticide needed and minimizes off-target effects. Much can be learned from nanomedicine research. Conventional nanomaterials, also known as nanocarriers or nanomedicines, have been extensively used to improve drug delivery efficiency and specificity. Recently, a novel concept, medical nanorobots, has been introduced, showing great potential in cancer treatments and other biomedical applications (Kong et al. 2023).

Advantages Over Conventional Pesticides Nano-pesticides offer several benefits over traditional pesticides, including increased solubility of hydrophobic ingredients, prolonged shelf life, and reduced environmental impact. They are designed to adhere more effectively to plant surfaces, enhancing the efficacy of pest control while reducing the overall chemical load on the environment (Kah et al. 2013).

Environmental and Health Impacts While nano-pesticides have the potential to reduce the negative environmental impact of conventional pesticides, concerns remain about their long-term effects on ecosystems and human health. It is essential to conduct thorough risk assessments and establish regulatory frameworks to ensure their safe use (Carmen and Ignacio 2015).

6.2.3 Nanosensors: Precision Agriculture

Role of Nanosensors in Precision Agriculture Nanosensors are tiny devices capable of detecting and transmitting information about environmental conditions and plant health. They play a crucial role in precision agriculture by providing real-time data on soil moisture, nutrient levels, and pest infestations. This information enables farmers to make informed decisions about crop management, optimize resource use, and improve yields.

Applications in Soil Health, Crop Monitoring, and Disease Detection Nanosensors are used to monitor soil health by detecting nutrient levels, pH, and microbial activity. They also aid in crop monitoring by identifying early signs of disease or pest attacks. For example, carbon nanotube-based sensors can detect volatile organic compounds released by plants under stress, allowing for timely intervention (Rai et al. 2015).

Real-Time Data Collection and Analysis The integration of nanosensors with wireless networks enables real-time data collection and analysis, facilitating timely decision-making and efficient farm management. This technology supports precision agriculture practices, reducing input costs and environmental impact (Parisi et al. 2015).

6.2.4 Nano-Based Seed Treatments: Enhanced Seed Protection

Enhancements in Seed Protection and Germination Nano-based seed treatments involve coating seeds with NPs that provide protection against pathogens and enhance germination. These treatments improve seedling vigor, increase resistance to diseases, and promote early growth. NPs can also deliver growth-promoting substances directly to the seed, enhancing overall plant health (Liu and Lal 2015).

Examples of Commercially Available Products Several companies have developed nano-based seed treatment products. For instance, Syngenta's "Avicta®" uses a combination of NPs and traditional nematicides to protect seeds from nematode infestations. Similarly, Bayer's "Poncho®" employs a nano-coating to deliver insecticides more effectively (Nair et al. 2010).

6.2.5 Nano-Enabled Smart Packaging: Reducing Post-Harvest Losses

Innovations in Food Packaging to Reduce Post-Harvest Losses Nano-enabled smart packaging incorporates NPs that enhance the protective qualities of packaging materials, reducing post-harvest losses. These materials can include antimicrobial agents, oxygen scavengers, and moisture absorbers, which extend the shelf life of agricultural products and reduce spoilage (Silvestre et al. 2011).

Impact on Food Safety and Shelf Life Nano-enabled packaging not only improves food safety by preventing contamination but also enhances shelf life by maintaining optimal storage conditions. For example, silver NPs have been used for their antimicrobial properties to prevent bacterial growth in packaged food products (Emamifar et al. 2011).

6.2.6 Water Management

Nanotechnology plays a vital role in water management by offering solutions for filtration, purification, and efficient use of water resources in agriculture. NMs such as carbon nanotubes and graphene oxide are used in water filters to remove contaminants, including heavy metals and pathogens, from irrigation water. Additionally, nano-enabled water retention products improve soil moisture retention, reducing the need for frequent irrigation and conserving water resources (Prasad et al. 2014).

6.2.7 Soil Improvement

Nanotechnology contributes to soil improvement by enhancing soil structure, fertility, and nutrient availability. Nano-clays and nano-composites are used to improve soil aeration and water-holding capacity, promoting healthier root growth. Furthermore, NPs can be used to deliver essential nutrients and soil amendments directly to the rhizosphere, increasing nutrient uptake efficiency and promoting plant growth (Raliya et al. 2016).

6.2.8 Enhanced Genetic Engineering Tools

Nanotechnology provides advanced tools for genetic engineering, facilitating the delivery of genes and biomolecules into plant cells. Nanocarriers such as liposomes, dendrimers, and NPs are used to transport DNA, RNA, and proteins across cell membranes, enhancing the efficiency of genetic transformation. This technology enables the development of genetically modified crops with improved traits, such as increased resistance to diseases, pests, and environmental stress (Cao et al. 2018).

6.3 Impact of Nanotechnology on the Agricultural Input Industry

Nanotechnology, the science of manipulating materials at the nanoscale, is revolutionizing various industries, and agriculture is no exception. The application of nanotechnology in the agricultural input industry has the potential to address challenges such as increasing food demand, resource depletion, and environmental sustainability. This article explores the market dynamics, economic potential, corporate adoption, and recent innovations driven by nanotechnology in agriculture.

6.3.1 Market Dynamics and Economic Potential

6.3.1.1 Current Market Trends and Future Projections

The agricultural input industry, comprising fertilizers, pesticides, seeds, and other inputs, is witnessing a transformative shift with the integration of nanotechnology. Recent market trends indicate a growing adoption of nano-enabled products, driven by the need for sustainable farming practices and improved crop yields.

The global market for agricultural nanotechnology was valued at approximately $4 billion in 2020 and is projected to reach $8.1 billion by 2025, growing at a compound annual growth rate (CAGR) of 11.5%. This growth is attributed to several factors:

Enhanced Efficacy Nanotechnology improves the efficacy of agricultural inputs by increasing the bioavailability of nutrients and active ingredients. For instance, nano-pesticides offer targeted delivery, reducing the amount of chemicals needed and minimizing environmental impact.

Sustainability As environmental concerns grow, the demand for eco-friendly agricultural practices is increasing. Nano-formulations of fertilizers and pesticides promote sustainable agriculture by reducing leaching and volatilization, thereby enhancing environmental safety.

6.3.1.2 Economic Benefits for Farmers and Agricultural Businesses

The economic potential of nanotechnology in agriculture is significant, offering various benefits to farmers and agricultural businesses:

Cost Savings Nano-enabled products often require lower application rates due to their increased efficiency, leading to cost savings for farmers. For example, using nano-fertilizers can reduce the amount of fertilizer needed by up to 30% while maintaining or enhancing crop yields.

Increased Crop Yields By enhancing nutrient uptake and protecting plants from pests and diseases, nanotechnology can lead to higher crop yields, translating into increased profitability for farmers.

Market Competitiveness For agricultural businesses, adopting nanotechnology can be a competitive advantage, allowing them to offer innovative products that meet the evolving demands of modern agriculture.

6.3.2 Corporate Adoption and Innovations

Examples of Multinational Companies Adopting Nanotechnology
Several multinational companies have recognized the potential of nanotechnology in agriculture and are actively investing in research and development to incorporate it into their products. Some notable examples are included in Table 6.1.

These companies reflect just a portion of the ongoing efforts within the agricultural sector to harness nanotechnology for improved product performance. The trend is toward creating smarter, more efficient agricultural inputs that reduce environmental impact and enhance crop productivity. As nanotechnology continues to evolve, it is expected to play an increasingly prominent role in the agricultural input industry.

6.3.3 Innovations Driving Corporate Growth

The continuous innovation in agricultural nanotechnology is driving growth and offering new opportunities for businesses in the sector:

R&D Investments Companies are investing heavily in research and development to explore new applications of nanotechnology, leading to innovative products that cater to the diverse needs of modern agriculture.

Collaborative Efforts Collaborations between academic institutions, research organizations, and corporations are accelerating the development of nano-enabled agricultural technologies. These partnerships foster knowledge exchange and facilitate the commercialization of cutting-edge solutions.

Regulatory Support As governments recognize the benefits of nanotechnology in agriculture, supportive regulations are being implemented to encourage its adoption and ensure safety standards are met.

Table 6.1 Products containing nanoparticles in the market

Company	Product	Description
BASF	Maglis NanoStrengthen®	*Description*: Maglis is an integrated platform that uses advanced NMs for soil and crop health management. It includes nano-enhanced fertilizers and pesticides that ensure precise delivery of nutrients and active ingredients to plants. The platform provides real-time data and analytics, helping farmers make informed decisions about crop management. BASF has developed various nanotechnology applications, including NanoStrengthen®, which enhances the delivery and performance of crop protection products. The company has also explored nanoscale nutrient delivery systems to improve the uptake and efficiency of fertilizers. *Benefits*: Improved nutrient uptake, reduced environmental impact, and increased crop yields.
Bayer (formerly Monsanto)	BioDirect™ technology	*Description*: Before its acquisition by Bayer, Monsanto was working on BioDirect™ agricultural biologicals, which are aimed at pest, virus, and weed control through naturally occurring processes that could potentially employ nanotechnology to enhance delivery and efficacy. *Benefits*: Enhanced efficacy of weed control, reduced herbicide resistance, and lower environmental toxicity.
Syngenta	Avicta® Complete Beans (Nano-Based Seed Treatments)	*Description*: Syngenta offers nano-based seed treatments that protect seeds from pests and diseases. These treatments use nanomaterials to provide a protective coating that delivers active ingredients directly to the seed, ensuring efficient uptake and long-lasting protection. Syngenta has been exploring nanotechnology for pesticide formulation to improve the solubility and efficacy of its products. Although specific product names are not always disclosed due to proprietary formulations, the company has invested in developing nanoencapsulation technologies that allow pesticides to be delivered more effectively and safely. *Benefits*: Improved seedling vigor, enhanced resistance to pests and diseases, and increased crop productivity.
Thermo Fisher Scientific Inc.	Nano-sensor	Description: Develops nanoscale sensors that help monitor soil health, nutrient levels, and crop conditions in real-time.

(continued)

Table 6.1 (continued)

Company	Product	Description
ICL Group	Agroblen® Controlled Release Fertilizers	*Description*: ICL has developed a range of nano-fertilizers under the Agroblen® brand that uses nanotechnology for the controlled release of nutrients. These fertilizers provide a steady supply of nutrients to crops, improving efficiency and reducing nutrient runoff. *Benefits*: Enhanced nutrient use efficiency, reduced fertilizer application rates, and minimized environmental impact.
Tandem Nano	Solid Drug Nanoparticles (SDN) for agrochemical delivery	Description: Specializes in nanoformulations that improve the delivery of active compounds in agrochemicals, enhancing effectiveness and reducing usage.

6.4 Potential Risk of the Technology

To strengthen the understanding and validation of the potential risks associated with the use of nanotechnology in the agriculture industry, it is crucial to reference current research and authoritative sources. Below, I'll integrate appropriate references that you can consult to explore these issues further:

6.4.1 Environmental Risks

The use of nanotechnology in agriculture presents significant opportunities for improving productivity and sustainability. However, it also introduces potential environmental risks that must be carefully considered. These risks primarily revolve around biodiversity impact, pollution and accumulation, and the persistence of NPs in the environment.

Biodiversity Impact NPs have the potential to affect non-target species and ecosystems, leading to unintended consequences on biodiversity. According to Handy et al. (2008), NPs can interact with biological systems in ways that are not yet fully understood, posing ecotoxicological risks to aquatic and terrestrial organisms. These interactions may disrupt physiological and biochemical processes in non-target species, affecting growth, reproduction, and survival rates. The study emphasizes the need for comprehensive research to fill the knowledge gaps in understanding the mechanisms of NP toxicity and their broader ecological impacts (Handy et al. 2008).

Pollution and Accumulation NPs can enter various environmental compartments, including soil, water, and air, leading to pollution and accumulation. Nowack and Bucheli (2007) highlight the environmental fate and behavior of NPs, noting that their small size and unique properties can lead to widespread dispersion and persistence in the environment. Once released, NPs can accumulate in soil and water bodies, potentially affecting microbial communities and food chains. This accumulation can lead to the bioaccumulation of NPs at higher trophic levels, raising concerns about the long-term impacts on wildlife and human health (Nowack and Bucheli 2007).

Persistence in the Environment One of the critical concerns regarding NPs is their persistence in the environment. Keller et al. (2013) examine the lifecycle and environmental persistence of engineered NMs, noting that they can remain in ecosystems for extended periods due to their stability and resistance to degradation. This persistence poses challenges for environmental management, as NPs can continuously interact with living organisms and abiotic components, potentially leading to chronic exposure and effects. The study calls for further research to understand the degradation pathways and long-term environmental impacts of NPs (Keller et al. 2013).

Addressing Environmental Risks To mitigate the environmental risks associated with nanotechnology in agriculture, it is crucial to adopt a precautionary approach and develop comprehensive risk assessment frameworks. This includes conducting thorough ecotoxicological studies, monitoring the environmental fate and behavior of NPs, and implementing regulatory measures to control their release into the environment. Additionally, promoting the development of biodegradable and eco-friendly NPs can help minimize their environmental impact.

6.4.2 Health Risks

Nanotechnology presents promising advances in agriculture, yet it also poses potential health risks, particularly concerning human exposure and bioaccumulation. Understanding these risks is essential for ensuring the safe use of nanotechnology in agricultural practices.

Human Exposure As nanotechnology becomes more prevalent in agriculture, the potential for human exposure to engineered NMs increases. Savolainen et al. (2010) highlight the need to address occupational health and safety concerns for workers who handle these materials. Due to their small size, NPs can enter the human body through inhalation, skin contact, or ingestion. Once inside, they may penetrate cells and accumulate in organs, posing potential long-term health risks. To minimize these risks, it is crucial to implement protective measures such as safety equipment, improved ventilation, and guidelines for handling and disposal. Continuous monitoring of exposure levels is essential to safeguard the health of agricultural workers.

Bioaccumulation NPs also pose risks through bioaccumulation, where they can enter ecosystems and the food chain. Robichaud et al. (2009) discuss the potential for NPs like nano-TiO2 to accumulate in plants and animals, eventually reaching humans. This bioaccumulation could result in chronic exposure and unknown health impacts. Addressing bioaccumulation requires further research to understand its pathways and impacts on human health. Developing regulations and guidelines can help ensure the responsible use of nanomaterials in agriculture, preventing them from entering the food chain.

6.4.3 Ethical and Social Risks

Nanotechnology holds significant potential for transforming agriculture, but it also presents ethical and social risks, particularly concerning access and equity and public perception and trust.

Access and Equity Nanotechnology's impact on agriculture may exacerbate existing inequalities. The ETC Group (2004) highlights that while advanced technologies could benefit large-scale farms and well-resourced regions, smaller and resource-poor farms may lack access to these innovations. This disparity can widen the gap between affluent and marginalized agricultural sectors, potentially reinforcing global inequities.

Public Perception and Trust Public acceptance of nanotechnology is critical for its successful integration into food systems. Siegrist et al. (2007) found that consumer trust and perceptions significantly influence the acceptance of nanotechnology in food and packaging. Negative perceptions or lack of trust can hinder the adoption of nanotechnology, affecting its potential benefits and market acceptance. Addressing these issues through transparent communication and equitable access strategies is essential for the responsible development and acceptance of nanotechnology in agriculture.

6.4.4 Regulatory and Legal Risks

Nanotechnology faces significant regulatory and legal challenges. Bowman and Hodge (2007) review the global landscape of nanotechnology regulation, highlighting a lack of adequate, unified regulatory frameworks to manage its risks effectively. This regulatory gap can lead to inconsistent safety standards and oversight. Additionally, Bawa (2007) addresses intellectual property issues in nanotechnology, noting that complex patent landscapes and disputes can hinder innovation and commercialization. Clear regulations and streamlined intellectual property policies are crucial for addressing these challenges and ensuring the safe and equitable development of nanotechnology in agriculture.

6.4.5 Economic Risks

Nanotechnology faces significant economic challenges, particularly high development and implementation costs. According to Lux Research (2014), the substantial investment required for nanotechnology research and commercialization can be a barrier for widespread adoption. Additionally, market volatility is a concern, as highlighted by Roco (2011), who discusses how fluctuations in the nanotechnology market can impact long-term investment and development. The economic uncertainty and high costs associated with nanotechnology necessitate careful planning and risk management to ensure sustainable progress and market stability.

6.5 Future Potential and Directions

As the agricultural sector grapples with significant challenges like food security, resource depletion, and climate change, nanotechnology offers innovative solutions poised to transform farming practices. This section explores the future potential of nanotechnology in agriculture, highlighting emerging technologies, integration with other technologies, and contributions to sustainable farming practices.

6.5.1 Innovations and Emerging Technologies

6.5.1.1 Emerging Trends and Future Directions in Agricultural Nanotechnology

Nanotechnology in agriculture is evolving rapidly, with key trends that promise to address current agricultural challenges:

Precision Agriculture Nanotechnology enhances precision agriculture through nanosensors that monitor soil health, nutrient levels, and plant growth in real time. This allows for precise application of resources, reducing waste and improving efficiency (Kah et al. 2018).

Smart Delivery Systems Nanocarriers are being developed for targeted delivery of agrochemicals, such as pesticides and fertilizers. These systems improve the efficacy of active ingredients while minimizing environmental impact and reducing chemical usage (Ghormade et al. 2011).

Nanobiosensors Emerging nanobiosensors provide early detection of plant diseases and pests, offering farmers data for preventative measures. These sensors are highly sensitive, detecting changes at the molecular level (Ray et al. 2022).

6.5.1.2 Potential Breakthroughs and Research Areas

Ongoing research continues to uncover new possibilities for nanotechnology in agriculture:

Nanophotonics for Enhanced Photosynthesis Advances in nanophotonics aim to improve photosynthesis efficiency, potentially leading to higher crop yields by engineering nanomaterials that more effectively capture and convert light (Zhang et al. 2018).

Gene Delivery Systems Nanotechnology facilitates precise gene delivery for plant breeding and genetic engineering, potentially creating crops with improved traits like drought resistance and pest tolerance (Cunningham et al. 2018).

Biodegradable Nanoparticles Developing biodegradable NPs addresses environmental concerns, aiming to deliver the benefits of nanotechnology while ensuring they break down harmlessly in the environment (Silva et al. 2019a, b).

6.5.2 Integration with Other Technologies

6.5.2.1 Synergies with Biotechnology, Internet of Things (IoT), and Artificial Intelligence (AI) in Agriculture

The integration of nanotechnology with biotechnology, the Internet of Things (IoT), and AI enhances agricultural capabilities:

Biotechnology and Nanotechnology Combining these fields can improve crop traits and protection, leading to enhanced nutritional profiles and resistance to stresses (Chaudhry et al. 2017).

IoT and Nanosensors IoT integration with nanosensors enables remote monitoring and management, facilitating precision agriculture and optimizing decision-making (Amin et al. 2022).

AI and Predictive Analytics AI analyzes data from nanosensors, providing predictive insights to optimize farming practices, improve resource allocation, and enhance productivity (Sharma et al. 2023).

6.5.2.2 Holistic Approaches to Smart Farming

The convergence of nanotechnology with IoT and AI is leading to holistic smart farming approaches:

Automated Farming Systems Automation technologies, such as drones and robotics, integrated with nanosensors and AI, perform tasks like planting and harvesting, improving efficiency and reducing labor costs (Walter et al. 2017).

Sustainable Resource Management These holistic approaches focus on sustainable resource management, optimizing the use of water, energy, and agrochemicals, with nanotechnology playing a critical role (Rai et al. 2018).

6.6 Sustainable Agriculture and Nanotechnology

Nanotechnology can significantly promote sustainable agriculture by enhancing resource efficiency and reducing environmental impact:

Water Conservation Nanotechnology contributes to water conservation through nano-filtration systems that purify and recycle agricultural water, managing resources effectively, especially in water-scarce regions (Homaeigohar et al. 2021a, b).

Soil Health Management Nano-formulations improve nutrient delivery and uptake, maintaining soil health and fertility, and supporting sustainable farming practices for future generations (Siddiqui et al. 2015).

Carbon Sequestration NMs for carbon capture and sequestration offer promising avenues for reducing greenhouse gas emissions, enhancing carbon storage in soils and plants, and mitigating climate change impacts (Zhao et al. 2020).

6.6.1 Long-Term Benefits and Sustainability Goals

Nanotechnology is essential for achieving long-term sustainability goals in agriculture:

Enhanced Food Security By improving crop yields, reducing losses, and optimizing resource use, nanotechnology contributes to food security, ensuring a stable supply for the growing global population (Khan et al. 2019).

Reduction of Environmental Impact The adoption of nanotechnology in agriculture can significantly reduce environmental impact, minimize harmful chemical usage, and promote eco-friendly practices (Servin et al. 2015).

Economic Viability and Social Impact As nanotechnology becomes more accessible, it enhances the economic viability of farming operations and improves livelihoods worldwide, particularly in developing regions (Singh et al. 2020).

6.7 Future Outlook

The future of nanotechnology in agriculture is promising, particularly with its integration with other technologies such as IoT and AI, promoting smart farming solutions (Sharma et al. 2024). Its focus on sustainability aligns with increasing environmental concerns, which will drive further adoption (Homaeigohar et al. 2021a, b).

6.8 Overcoming Challenges

Addressing safety, cost, and accessibility issues is crucial. Ensuring environmental safety through rigorous testing and making technology affordable can overcome barriers (Khan et al. 2020a, b; Silva et al. 2019a, b). Collaborative efforts among stakeholders can accelerate innovation and adoption (Chaudhry et al. 2022).

6.9 Conclusion

Nanotechnology is rapidly emerging as a transformative force in the agricultural input industry, offering innovative solutions to some of the most pressing challenges facing global agriculture. Throughout this chapter, we have explored the various applications of nanotechnology in agriculture, examining its impact on market dynamics, economic potential, and corporate adoption.

In conclusion, nanotechnology presents significant opportunities to revolutionize agriculture. However, there will be many scientific, environmental, and political challenges, and strategic efforts will be needed to harness its full potential for a sustainable future.

References

Amin R, Sharma S, Vardhan D (2022) IoT and nanosensors in agriculture: enhancing precision farming. J Precis Agric 12(3):145–160. https://doi.org/10.1007/s11269-022-03456-2

Bawa R (2007) Nanotechnology patenting in the United States. Nanotechnol Law Bus J 4(1):31–49. https://doi.org/10.1016/j.jbusres.2006.10.012

Binnig G, Rohrer H (1982) Scanning tunneling microscopy. Helv Phys Acta 55:726–735. https://doi.org/10.1103/PhysRevLett.49.57

Bowman DM, Hodge GA (2007) A small matter of regulation: an international review of nanotechnology regulation. Colum Sci Tech Law Rev 8:1–32. https://doi.org/10.2139/ssrn.995160

Cao Y, Wu W, Wang Q, Zeng X et al (2018) Nanocarriers: a versatile tool for crop breeding. Mol Plant 11(2):261–263. https://doi.org/10.1016/j.molp.2017.10.010

Carmen L, Ignacio J (2015) Nanotechnology and its risk perception: factors influencing public acceptance. Sci Eng Ethics 21(4):903–929. https://doi.org/10.1007/s11948-014-9561-8

Chaudhry Q, Sloth J, Stone D (2017) Biotechnology and nanotechnology: synergies in agriculture. Biotechnol Adv 35(2):132–143. https://doi.org/10.1016/j.biotechadv.2016.11.007

Chaudhry Q, Hadi H, Zhang Y (2022) Collaborative efforts in advancing nanotechnology in agriculture. J Agric Sci 160(7):897–908. https://doi.org/10.1017/S0021859622000452

Cunningham FJ, Goh NS, Demirer GS, Matos JL et al (2018) Nanoparticle-mediated delivery towards advancing plant genetic engineering. Trends Biotechnol 36(9):882–897. https://doi.org/10.1016/j.tibtech.2018.04.003

Drexler KE (1986) Engines of creation: the coming era of nanotechnology. Anchor Books, New York

Emamifar A, Kadivar M, Shahedi M, Soleimanian-Zad S (2011) Evaluation of nanocomposite packaging containing Ag and ZnO on shelf life of fresh oranges. Int J Food Sci Technol 46(9): 1811–1818. https://doi.org/10.1111/j.1365-2621.2011.02666.x

ETC Group (2004) Down on the farm: The impact of nano-scale technologies on food and agriculture. Retrieved from http://www.etcgroup.org/content/down-farm-impact-nano-scale-technologies-food-and-agriculture

FAO (2021) The state of food and agriculture 2021. Food and Agriculture Organization of the United Nations. http://www.fao.org/publications/sofa/2021/en/

Feynman RP (1960) There's plenty of room at the bottom. Eng Sci 23(5):22–36. https://doi.org/10.1103/PhysRev.118.498

Ghormade V, Deshpande MV, Paknikar KM (2011) Perspectives for nano-biotechnology enabled protection and nutrition of plants. Biotechnol Adv 29(6):792–803. https://doi.org/10.1016/j.biotechadv.2011.06.007

Handy RD, von der Kammer F, Lead JR, Hassellöv M, Owen R, Crane M (2008) The ecotoxicology and chemistry of manufactured nanoparticles. Ecotoxicology 17(4):287–314. https://doi.org/10.1007/s10646-008-0199-8

Homaeigohar S, Elbahri M, Abdelaziz R (2021a) Nanomaterials for water purification: opportunities and challenges. Chem Soc Rev 50(4):1042–1079. https://doi.org/10.1039/D0CS00398K

Homaeigohar S, Noroozi A, Li J (2021b) Water conservation through nanotechnology: advances and challenges. J Clean Prod 280:124362. https://doi.org/10.1016/j.jclepro.2020.124362

Kah M, Beulke S, Tiede K, Hofmann T (2013) Nanopesticides: state of knowledge, environmental fate, and exposure modeling. Crit Rev Environ Sci Technol 43(16):1823–1867. https://doi.org/10.1080/10643389.2012.671750

Kah M, Tufenkji N, White JC (2018) Nano-enabled strategies to enhance crop nutrition and protection. Nat Nanotechnol 13(8):733–743. https://doi.org/10.1038/s41565-018-0201-9

Keller AA, McFerran S, Lazareva A, Suh S (2013) Global life cycle releases of engineered nanomaterials. J Nanopart Res 15(6):1692. https://doi.org/10.1007/s11051-013-1692-4

Khan Y, Rizwan M, Shahid M (2019) Nanotechnology and food security: enhancing agricultural sustainability. J Agric Food Chem 67(15):4076–4090. https://doi.org/10.1021/acs.jafc.8b06255

Khan MA, Asiri AM, Khan SB (2020a) Nanomaterials: applications in biofuels and bioenergy production. Mater Today: Proc 36(2):260–265. https://doi.org/10.1016/j.matpr.2020.04.154

Khan Y, Sajjid M, Shahid M (2020b) Safety and environmental impact of nanotechnology in agriculture. Sci Total Environ 724:138120. https://doi.org/10.1016/j.scitotenv.2020.138120

Kong Y, Chen G, Liu X, Wang Y et al (2023) Recent advances in nanomedicine and nanorobots: a new horizon in cancer treatment. Adv Drug Deliv Rev 197:114131. https://doi.org/10.1016/j.addr.2023.114131

Liu R, Lal R (2015) Nano-enhanced materials for soil remediation. In: Gerhardt C (ed) Environmental soil remediation and management. Springer, New York, pp 213–230

Lux Research (2014) Sizing up the $3.8 billion nanotechnology opportunity in food and agriculture Lux Research. Retrieved from https://www.luxresearchinc.com/

Nair R, Varghese SH, Nair BG, Maekawa T et al (2010) Nanoparticulate material delivery to plants. Plant Sci 179(3):154–163. https://doi.org/10.1016/j.plantsci.2010.04.012

Nano.gov (2018) What is nanotechnology? National Nanotechnology Initiative. Retrieved from https://www.nano.gov/nanotech-101/what/definition

Nowack B, Bucheli TD (2007) Occurrence, behavior, and effects of nanoparticles in the environment. Environ Pollut 150(1):5–22. https://doi.org/10.1016/j.envpol.2007.06.006

Parisi C, Vigani M, Rodríguez-Cerezo E (2015) Agricultural nanotechnologies: what are the current possibilities? Nano Today 10(2):124–127. https://doi.org/10.1016/j.nantod.2014.09.009

Pérez-de-Luque A (2017) Interaction of nanomaterials with plants: what do we need for real applications in agriculture? Front Environ Sci 5:12. https://doi.org/10.3389/fenvs.2017.00012

Prasad R, Bhattacharyya A, Nguyen QD (2014) Nanotechnology in sustainable agriculture: recent developments, challenges, and perspectives. Front Microbiol 5:571. https://doi.org/10.3389/fmicb.2014.00571

Rai M, Ingle AP, Pandit R, Paralikar P et al (2015) Nanotechnology in crop protection: status and future prospects. In: Husaini CM (ed) Translational genomics for crop breeding: biotic stress. Wiley, Hoboken, pp 129–150. https://doi.org/10.1002/9781118936072.ch7

Rai M, Ingle AP, Paralikar P, Gupta I et al (2018) Nanotechnology-based approaches for pest control and crop improvement: innovations and challenges. Biotechnol Lett 40(9):1237–1254. https://doi.org/10.1007/s10529-018-2579-5

Raliya R, Tarafdar JC, Biswas P (2016) Enhancing the mobilization of native phosphorus in Mung bean rhizosphere using ZnO nanoparticles synthesized by Aspergillus fumigatus TFR-8. J Agric Food Chem 64(16):3111–3118. https://doi.org/10.1021/acs.jafc.6b00657

Ray PC, Yu H, Fu PP (2022) Nanoscience for nanotoxicology and nanomedicine. J Food Drug Anal 30(2):202–210. https://doi.org/10.38212/2224-6614.1055

Robichaud CO, Uyar AE, Darby MR, Zucker LG et al (2009) Estimates of upper bounds and trends in nano-TiO2 production as a basis for exposure assessment. Environ Sci Technol 43(12): 4227–4233. https://doi.org/10.1021/es8032549

Roco MC (2011) The long view of nanotechnology development: the National Nanotechnology Initiative at 10 years. J Nanopart Res 13:427–445. https://doi.org/10.1007/s11051-010-0177-2

Savolainen K, Alenius H, Norppa H, Pylkkänen L et al (2010) Nanotechnologies, engineered nanomaterials and occupational health and safety—a review. Saf Sci 48(8):957–963. https://doi.org/10.1016/j.ssci.2010.03.005

Scott N, Chen H (2013) Nanotechnology applications in agriculture: a sustainable approach. Int J Agric Biol 15(6):975–982. https://doi.org/10.17957/IJAB/15.6.13

Servin A, Elmer W, Mukherjee A, De la Torre-Roche R et al (2015) A review of the use of engineered nanomaterials to suppress plant disease and enhance crop yield. J Nanopart Res 17(2):92. https://doi.org/10.1007/s11051-015-2896-0

Sharma S, Agrawal A, Chaurasia V (2023) AI and predictive analytics in smart farming: the future of agriculture. Comput Electron Agric 194:106865. https://doi.org/10.1016/j.compag.2022.106865

Sharma S, Gupta R, Tiwari V (2024) Future outlook on nanotechnology integration with IoT in agriculture. J Agric Informatics 13(1):25–40. https://doi.org/10.13031/jaa.2024.1.12345

Sharon M, Choudhary AK, Kumar R (2010) Nanotechnology in agricultural diseases and food safety. J Phytology 2(4):83–92

Siddiqui MH, Al-Whaibi MH, Mohammad F (2015) Nanotechnology and plant sciences: nanoparticles and their impact on plants. Springer International Publishing. https://doi.org/10.1007/978-3-319-14502-0

Siegrist M, Cousin ME, Kastenholz H, Wiek A (2007) Public acceptance of nanotechnology foods and food packaging: the influence of affect and trust. Appetite 49(2):459–466. https://doi.org/10.1016/j.appet.2007.03.002

Silva BN, Khan M, Sulaiman S (2019a) Biodegradable nanoparticles for agricultural applications: progress and challenges. J Nanosci Nanotechnol 19(4):2141–2152. https://doi.org/10.1166/jnn.2019.15943

Silva GA, Lima MD, Andrade RG, Leite FL (2019b) Environmental impact and toxicity of nanomaterials in agriculture. Environ Nanotechnol Monit Manag 11:100211. https://doi.org/10.1016/j.enmm.2019.100211

Silvestre C, Duraccio D, Cimmino S (2011) Food packaging based on polymer nanomaterials. Prog Polym Sci 36(12):1766–1782. https://doi.org/10.1016/j.progpolymsci.2011.02.003

Singh R, Gautam RK, Chattopadhyay A (2020) Nanotechnology in modern agriculture: challenges and opportunities. J Appl Biotechnol Rep 4(2):115–125. https://doi.org/10.29252/jabr.4.2.115

Solanki P, Bhargava A, Chhipa H, Jain N et al (2015) Nano-fertilizers and their smart delivery system. In: Advances in nano-fertilizers for sustainable agriculture. Springer, pp 1–23. https://doi.org/10.1007/978-81-322-2023-0_1

Walter A, Finger R, Huber R, Buchmann N (2017) Opinion: smart farming is key to developing sustainable agriculture. Proc Natl Acad Sci 114(24):6148–6150. https://doi.org/10.1073/pnas.1707462114

Zhang C, Zhao Y, Du J, Guo X et al (2018) Nanophotonics for energy-efficient artificial photosynthesis. Nat Rev Mater 3(5):170–183. https://doi.org/10.1038/s41578-018-0024-8

Zhao L, Sun Y, Hernandez-Viezcas JA, Hong J et al (2020) Nanomaterials for carbon capture: opportunities and challenges. Sci Total Environ 703:135479. https://doi.org/10.1016/j.scitotenv.2019.135479

Chapter 7
Harnessing Nanotechnology and Artificial Intelligence to Revolutionize Sustainable Smart Agriculture

Anurag Yadav and Kusum Yadav

Abstract This chapter explores the transformative potential of integrating nanotechnology and artificial intelligence (AI) in precision agriculture, offering a paradigm shift in farming practices through data-driven, targeted interventions. Precision agriculture focuses on optimizing resource use and enhancing yield by tailoring farming practices to the specific needs of crops, utilizing high-resolution data. Nanotechnology plays a crucial role by providing advanced tools such as nanosensors and nanocarriers, which enable real-time monitoring and precise delivery of inputs like fertilizers and pesticides. These technologies significantly improve the efficiency and sustainability of agricultural practices by reducing waste and environmental impact. AI complements these advancements by analyzing complex datasets from various sources, including nanosensors, to provide actionable insights. Machine learning models and deep learning techniques are employed to predict crop health, weather patterns, and optimal resource allocation, enhancing farmers' decision-making processes. The chapter demonstrates the potential for developing sophisticated, responsive agricultural systems that boost productivity and contribute to environmental conservation by integrating AI and nanotechnology. This interdisciplinary approach to farming represents a significant leap forward in addressing global food security challenges while promoting sustainable agricultural practices.

Keywords Machine learning · IoT · Deep learning · agriculture · AI · Pest detection

---

A. Yadav (✉)
Department of Microbiology, C. P. College of Agriculture, Sardarkrushinagar Dantiwada Agricultural University, Sardarkrushinagar, Gujarat, India
e-mail: anuragyadav123@gmail.com

K. Yadav
Department of Biochemistry, University of Lucknow, Lucknow, Uttar Pradesh, India

S. K. Yadav, A. Kumari (eds.), *Nanobiotechnology in Agri-food Sector*, Sustainability Sciences in Asia and Africa, https://doi.org/10.1007/978-981-96-7756-6_7

7.1 Introduction

The rapid growth of the global population and the escalating demand for food have placed immense pressure on traditional agricultural systems. These systems are increasingly challenged by inefficient resource use, environmental degradation, and heightened vulnerability to climate change. To meet these challenges and ensure a sustainable food supply, a transformative shift in farming practices is essential.

This shift is manifesting in the adoption of precision agriculture (PA), an innovative approach that leverages advanced technologies to optimize resource utilization, boost productivity, and enhance sustainability (Wolfert et al. 2017). Precision agriculture diverges from the traditional "one-size-fits-all" methods by focusing on the observation, measurement, and responsive management of field variability. This approach allows for tailored interventions that address the specific needs of crops and field conditions, enabling farmers to make informed, data-driven decisions.

The benefits of precision agriculture are manifold. It improves resource efficiency, increases crop yields, reduces environmental impacts, and enhances the quality of agricultural produce. However, achieving the high level of precision required by PA necessitates more granular and accurate monitoring tools than traditional methods can provide. This is where the integration of nanotechnology and artificial intelligence (AI) becomes pivotal.

Nanotechnology introduces highly sensitive nanosensors and targeted delivery systems that facilitate real-time monitoring of crop and soil health at the nanoscale. These innovations enable precise interventions, significantly improving the effectiveness of agricultural practices. Complementing this, AI brings powerful machine learning and data analysis capabilities that can process the vast amounts of data generated by nanosensors. AI not only identifies patterns and predicts outcomes but also optimizes decision-making processes, enhancing the overall effectiveness of PA.

Together, nanotechnology and AI form a synergistic partnership that elevates precision agriculture to new levels of efficiency and sustainability. By combining detailed data collection with actionable insights through advanced machine learning models, these technologies offer a comprehensive solution to the challenges faced by modern agriculture.

This chapter delves into the integrated applications of nanotechnology and AI in promoting sustainable and precision agriculture. It examines how these cutting-edge technologies can address critical challenges, revolutionize agricultural practices, and foster a more productive and environmentally responsible agrarian sector.

7.2 Minimizing Resource Use and Waste Generation

7.2.1 Targeted and Need-Based Application of Inputs

The integration of nanotechnology and AI offers a transformative solution to the problem of agricultural input overuse. Nanoscale formulations of fertilizers and pesticides can vastly increase nutrient uptake efficiency and target pest control while reducing the needed product (Chhipa 2016; Kah et al. 2018). Controlled release mechanisms enabled by nanotechnology minimize losses through leaching and runoff, protecting water resources and preventing environmental pollution from excess agrochemicals. AI augments the precision by analyzing comprehensive datasets, including soil nutrient levels (collected by nanosensors), weather patterns, crop models, and historical data. This analysis enables AI models to generate optimized fertilization strategies tailored to the specific conditions, ensuring inputs are applied only when and where needed. AI-enabled, real-time soil nutrient mapping combined with slow-release, nano-enabled fertilizers has remarkably reduced fertilizer use without compromising crop yields.

7.2.2 Precision Irrigation Optimization

Intelligent irrigation plays a crucial role in conserving water, a precious and often limited resource in agriculture. Nanosensors embedded in the soil provide highly accurate, granular data on soil moisture levels at different depths, surpassing the capabilities of traditional moisture meters (Lekshmi et al. 2014). AI algorithms integrate this nanosensor data with weather forecasts, crop-specific water requirements, and evapotranspiration models to create dynamic irrigation schedules. These schedules match crop needs and field conditions, leading to significant water savings. The nanosensor-based irrigation systems coupled with AI models can reduce water use drastically without impacting crop health or productivity.

7.2.3 Nanoscale Remediation for Soil and Water Health

Beyond optimizing resource use, the integration of AI and nanotechnology extends to environmental protection. Nanoscale remediation, supported by AI-driven monitoring systems, is now a crucial strategy for maintaining soil and water health in agriculture. Leveraging nanomaterials' high reactivity, these techniques efficiently address contaminants like heavy metals, pesticide residues, and other pollutants. These nanomaterials can significantly enhance the remediation process, making it more effective in restoring the quality of farming ecosystems (Ali et al. 2023).

In addition to the direct application of nanomaterials, the integration of AI has further advanced the field of environmental remediation. AI-assisted monitoring systems, which utilize data from water quality nanosensors, provide critical insights that inform targeted remediation efforts. By analyzing data in real time, AI helps optimize remediation strategies and continuously tracks their effectiveness. For example, field trials have demonstrated the successful removal of heavy metals like cadmium and lead from contaminated soils using iron oxide nanoparticles, which not only removes the pollutants but also restores soil health and productivity (Gil-Díaz et al. 2014).

Furthermore, the application of AI in agriculture extends to data-driven optimization for long-term sustainability. Through analyzing trends and patterns, AI can assist in promoting sustainable farm management practices. By continuously refining and adjusting remediation efforts based on real-time data, AI ensures that these strategies contribute to the overall sustainability of agricultural ecosystems, helping to maintain soil and water health over the long term.

7.3 Nanotechnology in precision agriculture

Artificial nanotechnology has emerged as a transformative tool in precision agriculture, offering innovative solutions across various domains. One such innovation is the development of nanosensors, which have shown great promise in soil analysis, crop health monitoring, and environmental sensing. These nanosensors, categorized into chemical and biological types, provide a significant advantage due to their superior sensitivity, real-time detection capabilities, and minimal sample requirements. By precisely monitoring agricultural conditions, nanosensors can help farmers make informed decisions, leading to optimized resource use and improved crop yields (Prasad et al. 2017).

In addition to nanosensors, nanocarriers represent another crucial application of nanotechnology in agriculture. These carriers are employed in the targeted delivery of fertilizers, pesticides, and growth regulators, allowing more efficient use of these inputs. Nanocarriers' precision enhances agricultural chemicals' effectiveness and reduces their environmental impact. Furthermore, nanocarriers facilitate controlled release mechanisms, ensuring that nutrients and chemicals are delivered at the right time and in the right quantities, thereby minimizing waste and improving overall agricultural sustainability (Chen and Yada 2011).

Another promising application of nanotechnology in agriculture is nano-based remediation. Such an approach holds significant potential for addressing the contamination of soil and water resources. Utilizing nanomaterials makes it possible to remediate pollutants more effectively than traditional methods, thereby restoring the health of agricultural ecosystems. The versatility and efficiency of nano-remediation techniques could be crucial in mitigating pollution's adverse effects and enhancing agricultural practices' sustainability (Kah et al. 2013).

7.4 Artificial Intelligence in Precision Agriculture

7.4.1 Big Data and Analytics

In agriculture, AI techniques for data analysis play a pivotal role in harnessing the potential of massive datasets collected through various sensors. These techniques encompass machine learning, deep learning, and data mining, each contributing uniquely to the advancement of agricultural practices.

7.4.1.1 Machine Learning

A core subset of AI includes both supervised and unsupervised learning algorithms. In supervised learning, algorithms such as decision trees, random forests, and support vector machines (SVM) are employed to predict outcomes based on historical, labeled data. For instance, these models can forecast crop yields, detect disease outbreaks, and assess soil health by learning from past data patterns (Liakos et al. 2018). Supervised learning's ability to provide predictive insights makes it invaluable for planning and decision-making in agriculture. On the other hand, unsupervised learning algorithms, including clustering techniques like K-means, analyze data without predefined labels. This approach is essential for identifying inherent patterns and groupings within the data, such as zones within a field that share similar soil properties or crop growth stages (Jha et al. 2019). By uncovering these natural groupings, unsupervised learning facilitates more targeted and efficient management practices (Fig. 7.1).

7.4.1.2 Deep Learning

Deep learning (DL) techniques have significantly advanced precision agriculture by enabling complex and high-dimensional data analysis, particularly images and temporal sequences. Convolutional Neural Networks (CNNs), designed specifically for image analysis, have proven highly effective in identifying crop diseases, nutrient deficiencies, and weed infestations through aerial and ground-based imagery. By automatically extracting critical features from images, CNNs provide precise insights into crop health, facilitating early intervention and mitigation (Pantazi et al. 2022). For example, CNNs can detect early signs of diseases such as powdery mildew or rust on leaves, allowing farmers to take targeted actions to protect their crops. They can also identify nitrogen deficiencies in crops by analyzing the color and texture of leaves, enabling optimized fertilizer application and reducing environmental impact (Kumar and Kukreja 2021). Additionally, CNNs are instrumental in distinguishing between crops and weeds, thus improving automated weed removal systems and enhancing crop yields while minimizing herbicide use (Anagnostis et al. 2020) (Fig. 7.2).

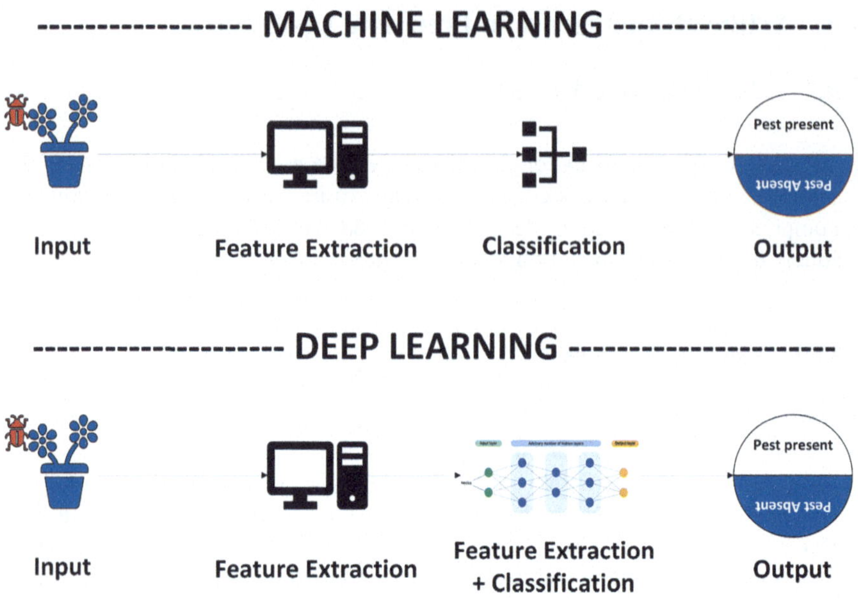

Fig. 7.1 Comparison of machine learning (ML) and deep learning (DL) approaches for pest detection in smart agriculture. ML involves separate feature extraction and classification steps, while DL integrates both processes within a neural network for improved accuracy and efficiency

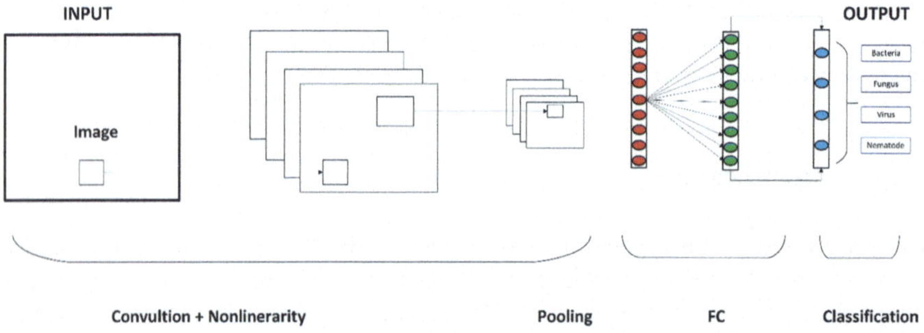

Fig. 7.2 CNN architecture for pathogen detection in agriculture. The network processes input images through convolution, pooling, and fully connected layers to classify pathogens like bacteria, fungi, viruses, and nematodes

On the other hand, Recurrent Neural Networks (RNNs) excel at processing sequential data and are widely used to predict weather patterns and their agricultural impacts. By analyzing historical weather data, RNNs can forecast future conditions, helping farmers make informed decisions about planting, irrigation, and harvesting. For instance, RNNs can predict rainfall patterns, enabling more efficient irrigation scheduling and ensuring crops receive adequate water without waste (Jha et al. 2019). Moreover, RNNs are crucial in forecasting adverse weather conditions, such as droughts or floods, allowing farmers to implement preemptive measures to

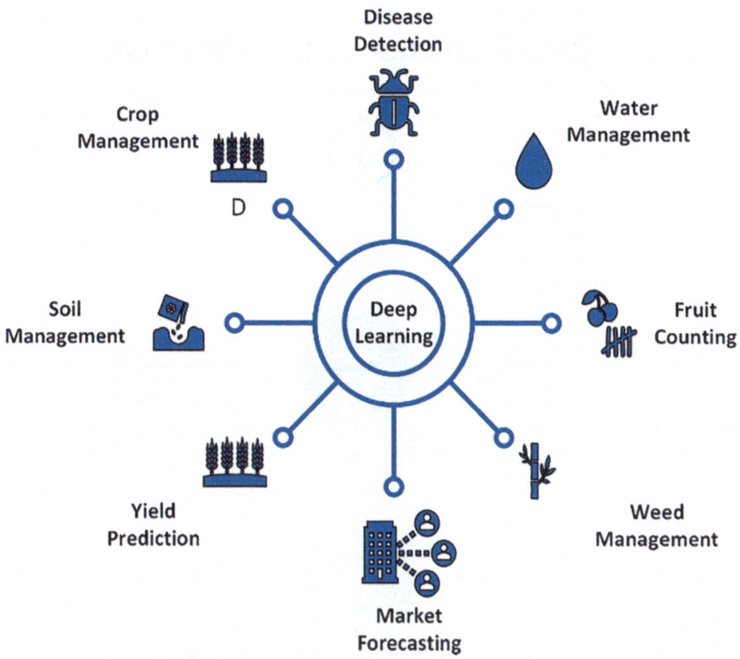

Fig. 7.3 Key applications of deep learning in agriculture

mitigate the impact on crop production. Such capability is precious in regions with unpredictable weather, where accurate forecasts can significantly enhance agricultural resilience and productivity (Jha et al. 2019).

The integration of CNNs and RNNs in agriculture has developed advanced tools and applications that further refine precision farming practices. Drone-mounted cameras with CNNs can swiftly survey large fields, identifying areas that need attention and providing actionable insights to farmers. Similarly, mobile applications powered by deep learning algorithms allow farmers to upload crop images and receive instant diagnoses and recommendations, democratizing access to advanced agricultural technologies (Pantazi et al. 2022; Jha et al. 2019) (Fig. 7.3).

7.4.1.3 Data Mining Complements

These AI techniques function by extracting hidden patterns and correlations from vast datasets. This process involves exploring data to uncover relationships that are not immediately apparent, thereby generating valuable insights for agricultural practices. For example, data mining can reveal how specific combinations of soil properties, weather conditions, and crop management practices influence yield outcomes (Liakos et al. 2018). These insights enable farmers to optimize their practices, improving productivity and sustainability (Fig. 7.4).

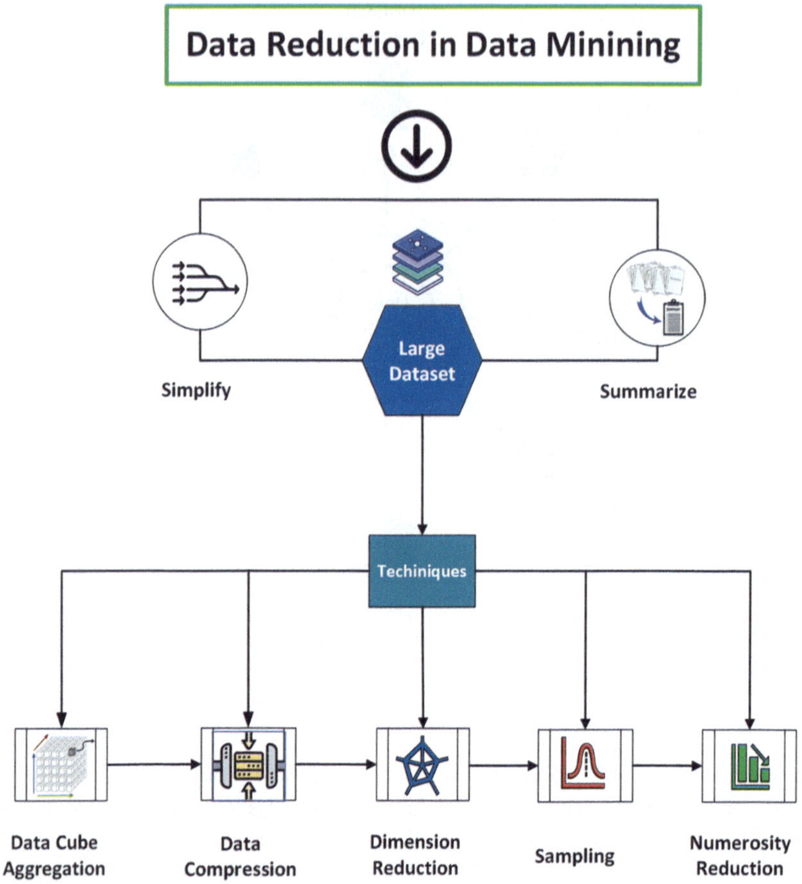

Fig. 7.4 Data reduction techniques in data mining, including data cube aggregation, compression, dimension reduction, sampling, and numerosity reduction, simplify and summarize large datasets for efficient analysis

Collectively, these AI techniques transform raw sensor data into actionable knowledge, driving the shift toward precision agriculture (Table 7.1). Machine learning (ML) models provide predictive analytics, deep learning algorithms enhance image and temporal data processing, and data mining uncovers underlying patterns. They empower farmers with the information to make data-driven decisions, optimize resource use, and enhance agricultural productivity. As these technologies evolve, their integration into farming systems promises to revolutionize the industry further, addressing global challenges such as food security and environmental sustainability.

Table 7.1 Summary of different AI algorithms and their specific uses in agriculture

AI Algorithm	Specific use in agriculture	Description	References
Machine learning	Yield prediction	Uses historical data to predict crop yields.	Kamilaris and Prenafeta-boldú (2018)
	Disease detection	Identifies plant diseases based on patterns in the data.	Bharadiya et al. (2023)
	Pest detection	Detects the presence of pests in crops through analysis of sensor data.	Mittal et al. (2024)
	Soil quality monitoring	Assesses soil quality and fertility.	Diaz-Gonzalez et al. (2022)
	Weather forecasting	Predicts weather patterns affecting agricultural activities.	Shin et al. (2022)
Deep learning	Image recognition	Analyzes images of crops for health assessment and pest detection.	Tetila et al. (2020)
	Disease classification	Classifies types of plant diseases from images.	Kumar (2020)
	Weed detection	Identifies and classifies weeds in crop fields.	Hasan et al. (2021)
	Plant phenotyping	Assesses plant traits for breeding and research.	Arya et al. (2022)
Data mining	Data analysis and pattern recognition	Extracts valuable patterns from large datasets.	Bauckhage and Kersting (2013)
	Market trend analysis	Analyzes market trends for better decision-making in crop sales.	Tian et al. (2022)
Predictive analytics	Crop yield forecasting	Uses statistical models to predict future crop yields based on historical data.	Surya and Aroquiaraj (2018)
	Resource allocation optimization	Optimizes the allocation of resources like water, fertilizer, and labor.	Amaruchkul (2023)
Neural networks	Irrigation scheduling	Determines optimal irrigation schedules to maximize water efficiency.	Farooque et al. (2022)
	Fertilization management	Recommends fertilization schedules based on crop needs and soil conditions.	Zhu et al. (2023)
Support vector machines (SVM)	Crop classification	Classifies different types of crops using spectral data from remote sensors.	Surase and Kale (2015)
	Pest and disease identification	Identifies pests and diseases based on sensor data and images.	Rajan et al. (2016)
Genetic algorithms	Breeding program optimization	Optimizes breeding programs to enhance crop traits.	Akdemir et al. (2015)

(continued)

Table 7.1 (continued)

AI Algorithm	Specific use in agriculture	Description	References
Natural language processing (NLP)	Text analysis of research papers	Analyzes scientific literature to extract relevant information for agriculture.	Domínguez et al. (2024)
Reinforcement learning	Autonomous farming equipment	Trains autonomous vehicles and robots for farming operations.	Bini et al. (2020)
Generative models	Generative adversarial networks (GAN), Variational autoencoders (VAE)	Synthetic data generation for training models, simulating various agricultural scenarios (e.g., crop growth under different conditions)	Akkem et al. (2024)
Computer vision	Image segmentation, object detection	Fruit counting and quality assessment, weed mapping, crop growth monitoring	Tian et al. (2020)

7.4.2 Image Recognition for Plant Health, Pest Identification, and Weed Detection

7.4.2.1 Image Recognition

Image recognition, powered by machine learning and artificial intelligence, is revolutionizing agriculture by providing rapid, accurate, and non-invasive methods for assessing plant health, identifying pests, and detecting weeds. Such technological advancement transforms traditional farming practices, enabling farmers to make data-driven decisions that enhance productivity, reduce costs, and promote sustainable agriculture.

Plant health assessment is one of the primary applications of image recognition in agriculture. Using sophisticated algorithms, image recognition systems can analyze visual data captured by cameras mounted on drones, satellites, or handheld devices to detect signs of plant stress, disease, or nutrient deficiencies. For instance, CNNs' algorithms have been trained to recognize specific patterns and features associated with various plant diseases. These networks can differentiate between healthy and diseased plants by analyzing leaf color, texture, and morphology. Early detection of plant diseases through image recognition allows farmers to implement timely interventions, such as applying targeted treatments or adjusting cultivation practices, thereby preventing the spread of disease and minimizing crop losses (Kamilaris and Prenafeta-boldú 2018).

In addition to plant health monitoring, image recognition is instrumental in pest identification. Traditional methods of pest detection often involve manual inspection, which can be time-consuming and prone to human error. Image recognition technologies, however, automate this process by analyzing images of crops to identify the presence and type of pests. Advanced image recognition systems can distinguish between different species of pests, providing precise information on the

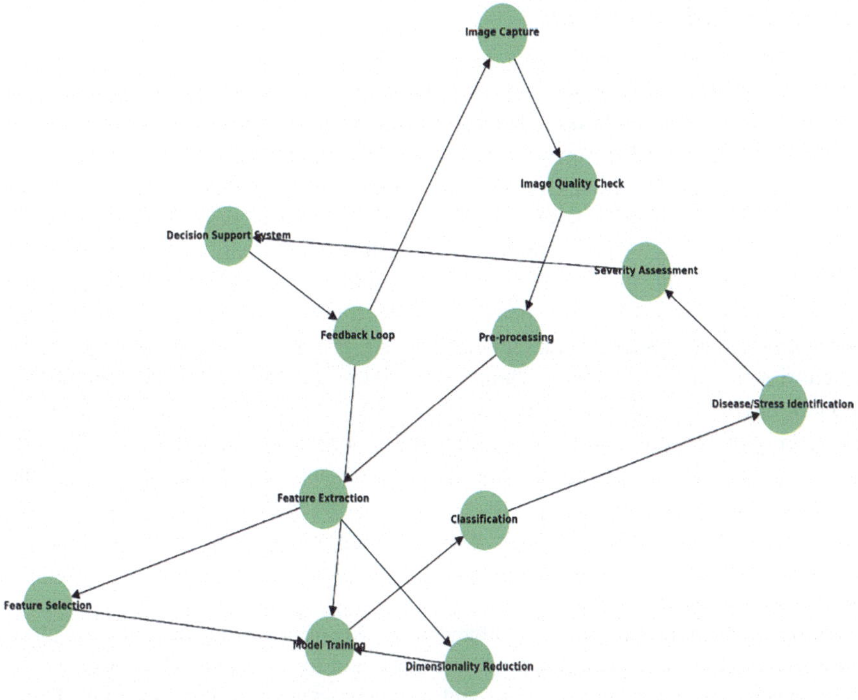

Fig. 7.5 AI-driven workflow for plant disease identification

infestation. Such capability is precious for integrated pest management (IPM) programs, which rely on accurate pest identification to determine the appropriate control measures (Fig. 7.5). For example, image recognition can help identify specific insects, such as aphids, caterpillars, or beetles, enabling farmers to choose the most effective and environmentally friendly pest control strategies (Batz et al. 2023).

7.4.2.2 Plant Health Assessment

Early and accurate detection of plant diseases is crucial for effective disease management and yield preservation. Image recognition models can analyze visual cues like leaf discoloration, wilting, and lesions to diagnose powdery mildew, rust, and bacterial spots. For instance, a study by Mohanty et al. (2016) demonstrated the successful use of deep CNNs to accurately classify crop species and diseases.

Moreover, image recognition can go beyond disease detection and assess overall plant health by analyzing growth patterns, nutrient deficiencies, and stress responses. Hyperspectral imaging, which captures data beyond the visible spectrum, can provide valuable insights into plant physiology and detect subtle changes before they become visible to the naked eye.

7.4.3 Pest Identification

Timely identification of pests is essential for implementing targeted pest control measures. Image recognition models can be trained to recognize various pests based on their morphological features, such as color, shape, and size.

Unmanned aerial vehicles (UAVs) equipped with cameras can capture high-resolution images of large fields, enabling the detection of pests at an early stage and over vast areas. The technology is beneficial for identifying pests that are difficult to spot with the naked eye, such as aphids and spider mites. Integrating nanotechnology and AI in pest detection offers a transformative approach to agricultural practices. Nanosensors, enhanced by AI, provide significant advancements in real-time monitoring systems. Deployed in fields, these nanosensors continuously monitor environmental conditions and plant health. AI algorithms process the data, identifying patterns and anomalies indicating pest presence, enabling timely alerts and early interventions. Incorporating nanosensors into smart traps allows for the detection and identification of specific pest species based on biochemical signatures. AI systems analyze data from these traps, offering actionable insights for targeted pest control measures, reducing the need for broad-spectrum pesticides, and minimizing environmental impact.

Nano-encapsulated pesticides, enhanced by AI, offer further benefits. AI algorithms can determine optimal application times and locations based on nanosensor data, ensuring precise pesticide use and reducing environmental footprint. These pesticides can also be designed to release their active ingredients in response to specific triggers, such as pest-related enzymes. AI predicts pest behavior and environmental conditions for effective release.

7.4.3.1 Weed Detection

Weeds compete with crops for resources, yielding losses and increased production costs. Image recognition can differentiate between crops and weeds based on their visual characteristics. For instance, a study (Lottes et al. 2017) used a CNN-based approach to classify eight weed species with an accuracy of 95%.

Image recognition-based weed detection systems can be integrated with robotic weeders to automate weed removal, reducing the reliance on herbicides and manual labor. This process not only improves efficiency but also minimizes environmental impact.

Plantix is a mobile app that uses image recognition to diagnose plant diseases, pests, and nutrient deficiencies. It has been downloaded over ten million times and is used by farmers in over 180 countries (Ahmad et al. 2020). Another company named Blue River Technology developed a robotic weeder called See & Spray, which uses image recognition to identify and spray weeds with herbicides, reducing herbicide use by up to 90% (Vijayakumar et al. 2023). Yet another company, Prospera, offers a crop monitoring platform that uses image recognition to detect crop stress, pests, and diseases, enabling growers to make data-driven decisions for optimizing crop management (Huchchannanavar and Divate Anuja 2023).

7.4.4 Decision Support Systems

AI-powered tools provide real-time recommendations to farmers on crop management, irrigation scheduling, and disease control. AI-powered tools transform the agricultural landscape by providing real-time recommendations to farmers on various aspects of crop management, irrigation scheduling, and disease control. These tools act as decision support systems (DSS), empowering farmers with data-driven insights to optimize their practices and maximize yields.

7.4.5 Crop Management

AI-powered DSS can analyze many data points, such as weather patterns, soil conditions, crop growth stages, and historical yield data, to generate tailored recommendations for crop management practices. For instance, the Xarvio™ FIELD MANAGER by BASF utilizes AI algorithms to provide farmers with field-specific recommendations on seed selection, fertilizer application, and pest control measures, resulting in improved crop yields and reduced input costs (Nolte et al. 2023). Moreover, AI-powered tools can leverage remote sensing data from satellites and drones to monitor crop health and identify areas that require attention. This process enables farmers to implement targeted interventions, such as adjusting irrigation schedules or applying fertilizers only where needed, leading to more efficient resource utilization.

7.4.5.1 Irrigation Scheduling

Precision irrigation maximizes water use efficiency and ensures optimal crop growth. AI-powered DSS can analyze weather forecasts, soil moisture levels, crop water requirements, and evapotranspiration rates to develop irrigation schedules that match crop needs (Asolo et al. 2024).

Furthermore, AI-powered tools can integrate with smart irrigation systems to automate irrigation based on real-time data and weather conditions, optimizing water use and minimizing the risk of over- or under-watering.

Integrating nanotechnology within irrigation systems has unlocked unprecedented levels of precision in water management. Nanosensors, for instance, have emerged as pivotal tools, enabling real-time monitoring of soil moisture, temperature, nutrient levels, and salinity (Dasgupta et al. 2017; Neupane and Guo 2019). These minute devices, often smaller than a grain of sand, can be dispersed

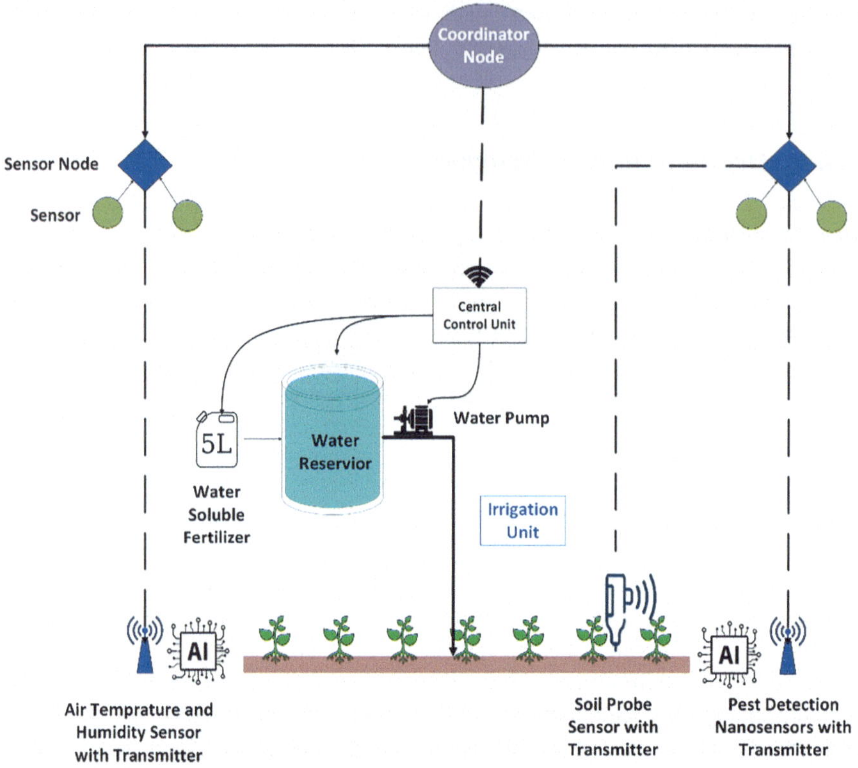

Fig. 7.6 Smart irrigation system integrating AI and IoT, with sensors for air temperature, soil moisture, and pest detection, connected to a central control unit for precise water and fertilizer management

throughout the soil profile, providing a continuous stream of data that accurately informs irrigation decisions. The real-time feedback loop ensures crops receive water only when and where needed, minimizing wastage and maximizing resource utilization.

Nanomaterials, another facet of nanotechnology, are being explored for their potential to enhance soil properties and water retention. Nanoparticles, such as zeolites and hydrogels, can be incorporated into the soil matrix to improve its water-holding capacity, thereby reducing irrigation frequency and mitigating the effects of drought (Dasgupta et al. 2017) . Additionally, nanofiltration membranes are being developed to remove impurities from irrigation water, ensuring the delivery of clean and safe water to crops (Dasgupta et al. 2017).

While nanotechnology provides the eyes and ears of smart irrigation systems, AI serves as the brain, analyzing the vast amounts of data collected by nanosensors and making informed decisions about irrigation scheduling (Fig. 7.6). Machine learning algorithms, a subset of AI, are particularly adept at identifying patterns and correlations within this data, allowing them to predict future water demand and optimize irrigation schedules accordingly. AI-powered irrigation systems can also leverage weather forecasts and historical data to anticipate irrigation needs, enhancing efficiency (Abioye et al. 2022). For example, a study demonstrated that an AI-based irrigation system drastically reduces water consumption compared to traditional methods while maintaining or even improving crop yields (Preite and Vignali 2024).

Several case studies illustrate the successful integration of nanotechnology and AI in irrigation management. Netafim, an Israeli company, has developed a suite of precision irrigation solutions that leverage nanotechnology and AI to optimize water use in agriculture. Their drip irrigation systems, equipped with nanosensors and AI-powered controllers, have been deployed in numerous countries, resulting in significant water savings and yield improvements (Eisenstadt and Pollock 2021). Jain Irrigation Systems, an Indian multinational, has pioneered the use of nanotechnology in irrigation, particularly in the development of nano-fertilizers and nano-pesticides, which offer enhanced nutrient delivery and pest control, contributing to sustainable agriculture practices (Bhadauria and Singh 2021).

7.4.5.2 Disease Control

Early detection and timely intervention are critical for effective disease management in crops. AI-powered DSS can analyze images of crops to identify early signs of diseases, such as leaf lesions, discoloration, and wilting. For instance, the PlantVillage Nuru app uses AI algorithms to diagnose several crop diseases accurately, enabling farmers to take prompt action to protect their crops (Mrisho et al. 2020).

Additionally, AI-powered tools can predict disease outbreaks based on weather patterns, crop susceptibility, and historical disease data, allowing farmers to implement preventive measures such as applying fungicides before the disease spreads.

CropX provides an AI-powered soil sensing and irrigation management system that helps farmers optimize irrigation scheduling and reduce water consumption (Tetala 2023). Yet another company, *PEAT,* developed the Plantix app, which utilizes AI to diagnose plant diseases, pests, and nutrient deficiencies, offering farmers personalized treatment recommendations (Kaur and Prashar 2019).

7.4.6 Synergies of Nanotechnology and AI

The intersection of nanotechnology and artificial intelligence (AI) presents a transformative approach to agriculture, demonstrating great potential to increase nutrient and pesticide efficacy, reduce waste, improve yields, optimize resource efficiency, and minimize environmental impact. One example of such synergy is developing zeolite-based nano-fertilizers that synchronize urea release with maize growth needs (Manikandan and Subramanian 2016). This optimization increases yields and minimizes nitrogen leaching, highlighting economic and environmental benefits.

Beyond delivery, AI empowers intelligent decision-making. Researchers can predict pathogen behavior based on environmental data by integrating AI-driven models with chitosan nanoparticles carrying specific biocontrol agents (Elmer and White 2016). These nanoparticles release their payload only when the model indicates a high risk of infection, minimizing unnecessary applications. Similarly, AI optimizes pest control through smart nanocarriers containing RNA molecules (Christiaens et al. 2020). AI algorithms guide carrier design and dosage calculations for maximum effect and minimal environmental impact.

AI also revolutionizes precision management practices through large-scale sensor networks (García et al. 2020). AI algorithms optimize sensor placement and analyze data, providing farmers with real-time information on soil moisture, nutrient levels, and crop health. This knowledge empowers precision farming techniques tailored to specific field conditions. Even more futuristic applications combine nanotechnology with drone technology (Parisi et al. 2015). In this scenario, AI-equipped drones target hotspots with nanoformulations of pesticides or fertilizers, reducing overall application quantities and maximizing treatment efficacy.

While challenges like data acquisition, integration, cost, and accessibility of these technologies remain, the convergence of nanotechnology and AI holds immense promise for sustainable agriculture. It's a field where resources are delivered precisely and efficiently, promoting increased crop yields and environmental protection.

7.4.7 Smart Delivery Systems

AI algorithms are revolutionizing the agricultural sector by optimizing the release and behavior of nanocarriers, enhancing their efficacy, and reducing environmental impact. Due to their unique properties, such as high surface-area-to-volume ratio and controllable release mechanisms, nanocarriers are gaining traction in various agricultural applications, including pesticide delivery, nutrient supply, and plant disease diagnostics.

AI algorithms can analyze vast datasets from multiple sources, including environmental sensors, plant physiological data, and nanocarrier properties, to model and predict the behavior of nanocarriers in complex agricultural settings. This process enables the development of intelligent nanocarrier systems that can adapt to changing environmental conditions and release their payloads at optimal rates for maximum effectiveness. One of the critical challenges in using nanocarriers is achieving controlled release kinetics. AI algorithms, particularly machine learning models, can be trained on experimental data to identify the key parameters influencing release rates, such as temperature, pH, and enzyme activity (Alshawwa et al. 2022). Such knowledge can be used to design nanocarriers with tailored release profiles that match the specific needs of crops and agricultural applications. For instance, AI algorithms can be used to develop "smart" nanocarriers that respond to specific environmental triggers, such as plant signals or pest infestations. This process allows for the targeted delivery of pesticides or nutrients only when and where needed, minimizing waste and reducing the risk of environmental contamination (Sagdic et al. 2022).

Another promising application of AI in nanocarrier design is developing "self-destructing" nanocarriers. These nanocarriers are designed to degrade after their payload, minimizing their persistence in the environment and reducing the risk of unintended consequences (Agrahari et al. 2020). AI algorithms can identify biodegradable materials and design nanocarriers that degrade at specific rates, ensuring they are effective without posing long-term environmental risks (Motadayen et al. 2024).

AI algorithms can be used to optimize the design of nanocarriers, improving their ability to adhere to and be absorbed by plants (Hsueh et al. 2023). This targeted approach could lead to more efficient delivery of pesticides, nutrients, and other beneficial agents, ultimately improving crop health and productivity. Furthermore, AI could be used to fine-tune the release kinetics of these substances, ensuring they are delivered at the optimal time and rate for maximum effectiveness (de Oliveira et al. 2019).

7.4.8 Sensor Network Optimization

Large-scale sensor networks (LSNs) are instrumental in various fields, including agriculture, environmental monitoring, and industrial processes (Shen 2021). However, sensor placement, data interpretation, and self-calibration challenges often hinder their efficiency and accuracy. AI offers promising solutions to these challenges, revolutionizing LSNs' capabilities.

7.4.8.1 Sensor Placement Optimization

Optimal sensor placement is crucial for maximizing the information gathered by LSNs while minimizing costs and resource consumption. Traditional methods often rely on heuristic approaches or expert knowledge, which can be time-consuming and suboptimal. AI algorithms, such as reinforcement learning and genetic algorithms, can learn optimal sensor placements through iterative simulations and evaluations.

In a study (Krause et al. 2008), a decentralized algorithm based on reinforcement learning was used to determine the optimal placement of sensors for environmental monitoring. The algorithm significantly outperformed traditional methods in terms of coverage and energy efficiency.

Coverage-based sensor placement focuses on ensuring that all areas of the field are adequately monitored. Algorithms such as the Genetic Algorithm (GA) or Particle Swarm Optimization (PSO) are often used to determine the optimal locations for sensors, considering field geometry, the range of sensors, and obstacles that may interfere with signal transmission. This method ensures comprehensive data collection, leading to better-informed decision-making (Fan and Liang 2022).

In contrast, data-driven sensor placement utilizes historical and real-time data to identify critical field areas requiring more intensive monitoring. Machine learning models analyze soil moisture, nutrient levels, and crop health patterns to pinpoint zones with the highest variability. Sensors are strategically placed in these zones to capture the most significant data variations, enabling targeted interventions. This approach is particularly beneficial in heterogeneous fields where conditions vary widely over short distances (Folorunso et al. 2023; Shukla et al. 2021).

Cost-effective sensor placement aims to balance the trade-off between data quality and installation costs. Techniques such as the Simulated Annealing (SA) algorithm help identify the minimum number of sensors needed to achieve acceptable data accuracy. By considering both the spatial distribution of sensors and the economic constraints of the farming operation, farmers can achieve significant savings while still benefiting from advanced monitoring capabilities.

7.4.8.2 Data Interpretation Enhancement

LSNs generate massive amounts of data, often containing noise and outliers. Traditional data analysis methods may struggle to extract meaningful insights from such data. AI-powered techniques, such as deep learning and anomaly detection, can identify patterns, anomalies, and correlations that may be difficult to discern using traditional methods. A study (Lei et al. 2024) employed deep learning to analyze sensor data from a large-scale agricultural network. The model accurately predicted crop yields and identified areas of stress, enabling farmers to take targeted actions for improved crop management.

7.4.8.3 Self-Calibration Facilitation

Sensor calibration is essential for ensuring the accuracy and reliability of LSNs. However, manual calibration can be expensive and time-consuming, especially in large-scale networks. AI algorithms can automate calibration by learning the relationship between sensor readings and ground truth data. Such capability enables sensors to self-calibrate in real time, reducing the need for manual intervention and improving the overall accuracy of LSNs.

Self-calibrating sensor networks for environmental monitoring have been developed that use machine learning algorithms to calibrate sensors based on data from neighboring nodes, resulting in improved accuracy and reduced drift (Ahmad et al. 2023) (Fig. 7.7.).

IBM's Deep Thunder is an AI-powered weather forecasting system that utilizes LSN data to provide highly accurate hyperlocal weather predictions. The system's AI algorithms analyze sensor data from various sources, including weather stations, radars, and satellites, to generate real-time forecasts tailored to specific locations (Treinish and Praino 2004).

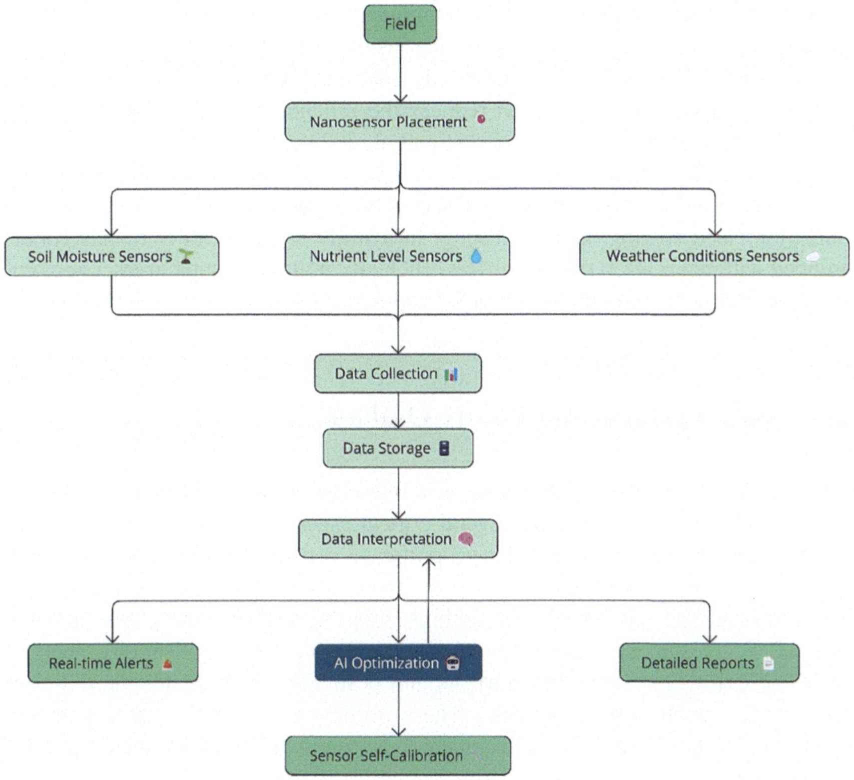

Fig. 7.7 Smart agriculture system using nanosensors for monitoring soil moisture, nutrients, and weather, with AI optimization enabling real-time alerts, detailed reports, and sensor self-calibration

7.4.9 Precision Farming Robots

AI-driven autonomous robots with nanosensors and nanoactuators can navigate complex field environments, identify individual plants, and perform targeted interventions with unparalleled accuracy. This level of precision minimizes collateral damage to non-target organisms and reduces the overall environmental impact of agricultural practices.

Moreover, nanosensors integrated into autonomous robots can monitor plant health in real time, detecting early signs of stress, disease, or nutrient deficiencies. Such integration allows for timely interventions, preventing yield losses and reducing the need for broad-spectrum chemical treatments.

A study used an AI-powered robotic platform with nanosensors to detect and control grapevine powdery mildew (Oberti et al. 2016). The robot autonomously navigated the vineyard, identified infected leaves, and applied a high-precision nano-formulated fungicide, significantly reducing disease incidence and improving grape quality.

Another example of targeted field operations is the use of autonomous robots for selective weeding. Equipped with machine vision and AI algorithms, these robots can accurately differentiate crops and weeds, enabling precise weed removal without harming the crops. Such a method reduces the reliance on herbicides, promoting sustainable agricultural practices (Visentin et al. 2023).

The potential of AI-driven autonomous robots integrated with nanotechnology extends beyond crop management. These robots can also be used for targeted pollination, seed sowing, and harvesting. In a study, a swarm of miniature robots equipped with nano-sized pollen carriers efficiently pollinated apple trees, overcoming the challenges of declining pollinator populations (Pinheiro et al. 2023).

7.5 Considerations and Future Outlook

The future outlook for integrating image recognition and other AI-driven technologies in agriculture must consider several critical aspects to ensure their successful adoption and sustained impact. One significant consideration is the safety and regulation of the technologies involved. These technologies often incorporate advanced materials and complex algorithms, so a thorough biosafety assessment is essential. For instance, using nanomaterials in agricultural applications, such as sensors or delivery systems for nutrients and pesticides, necessitates the development of responsible use frameworks. These frameworks should address potential risks to human health and the environment, ensuring that deploying such technologies does not inadvertently cause harm (Kah et al. 2013).

Cost and accessibility are also pivotal in adopting AI and image recognition technologies in agriculture. While these technologies offer substantial benefits, they must be made affordable and accessible to farmers of all scales, including smallholder farmers who constitute a significant portion of the global agricultural community. Developing and disseminating cost-effective solutions are essential to avoid creating a technological divide that could exacerbate agricultural productivity and income inequalities. Ensuring broad accessibility may involve public and private sector collaborations to subsidize costs, provide training, and support infrastructure development (Wolfert et al. 2017).

Cross-disciplinary collaboration is another crucial factor in advancing the application of AI and nanotechnology in agriculture. The convergence of expertise from fields such as nanotechnology, artificial intelligence, and agricultural sciences can drive innovative solutions tailored to the specific needs and challenges of modern farming. For example, integrating nanotechnology with AI can enhance the precision and effectiveness of image recognition systems. Nanomaterials can be used to create more sensitive and accurate sensors, which, when combined with advanced AI algorithms, can provide even more detailed and actionable insights into crop health and field conditions. Such an interdisciplinary approach ensures that technological developments are grounded in practical agricultural knowledge and can address real-world issues effectively (Mora et al. 2017).

Looking ahead, the future of AI in agriculture is promising, with continuous advancements expected to revolutionize the industry further. Innovations in image recognition will likely lead to even more sophisticated systems capable of diagnosing a wider range of plant diseases, pests, and nutrient deficiencies with greater accuracy and speed. Moreover, integrating AI with other emerging technologies, such as blockchain for supply chain transparency and IoT for real-time data collection, will create a more interconnected and responsive agricultural ecosystem (Kamilaris and Prenafeta-boldú 2018).

7.6 Conclusion

The convergence of nanotechnology and AI marks a transformative epoch in agriculture, creating unprecedented opportunities for sustainable, efficient, and data-driven farming practices. By integrating nanotechnology's precision with AI's predictive power, we stand on the brink of a revolution that can address some of the most pressing challenges facing global food systems. This fusion offers the tools to enhance crop yield, manage resources more effectively, and mitigate environmental impact and the potential to reshape agricultural paradigms fundamentally.

Future research must focus on refining these technologies to address global food security challenges. Nanotechnology can enhance the delivery of nutrients and pesticides, reduce waste, and improve the quality of agricultural produce. Meanwhile, AI's ability to process vast amounts of real-time data enables more accurate predictions and decision-making, optimizing farm management. Integrating these technologies could be pivotal in meeting the rising food demands of a growing global population while minimizing environmental degradation.

Moreover, the critical role of nanotechnology and AI in building sustainable and resilient food production systems cannot be overstated. These technologies have the power to enhance the efficiency and sustainability of agricultural practices, making them essential components of future food systems. However, the journey ahead requires continuous innovation, responsible implementation, and a commitment to inclusivity. These advancements must be accessible to farmers worldwide, particularly those in developing regions most vulnerable to climate change and food insecurity.

In conclusion, the combined power of nanotechnology and AI can revolutionize precision agriculture, offering a path forward that is more productive, sustainable, and equitable. By embracing these technologies, we can cultivate a future where agriculture feeds the world and preserves the planet for generations.

References

Abioye EA, Hensel OE, Elijah TJ, Abidin O et al (2022) Precision irrigation management using machine learning and digital farming solutions. Agri 4:70–103

Agrahari V, Agrahari V, Chou ML, Chew CH et al (2020) Intelligent micro−/nanorobots as drug and cell carrier devices for biomedical therapeutic advancement: promising development opportunities and translational challenges. Biomaterials 260:120163

Ahmad J, Jan B, Farman H, Ahmad W et al (2020) Disease detection in plum using convolutional neural network under true field conditions. Sensors 20:5569

Ahmad R, Rinner B, Wazirali R, Abujayyab SK et al (2023) Two-level sensor self-calibration based on interpolation and autoregression for low-cost wireless sensor networks. IEEE Sensors J 23: 25242

Akdemir D, Sanchez JI, Jannink JL (2015) Optimization of genomic selection training populations with a genetic algorithm. Genet Sel Evol 47:1–10

Akkem Y, Biswas SK, Varanasi A (2024) A comprehensive review of synthetic data generation in smart farming by using variational autoencoder and generative adversarial network. Eng Appl Artif Intell 131:107881

Ali Q, Zia MA, Kamran M, Shabaan M et al (2023) Nanoremediation for heavy metal contamination: a review. Hybrid Adv 4:100091

Alshawwa SZ, Kassem AA, Farid RM, Mostafa SK et al (2022) Nanocarrier drug delivery systems: characterization, limitations, future perspectives and implementation of artificial intelligence. Pharmaceutics 14

Amaruchkul K (2023) Multiobjective land-water allocation model for sustainable agriculture with predictive stochastic yield response. Int Trans Oper Res 30:1647–1672

Anagnostis A, Asiminari G, Papageorgiou E, Bochtis D (2020) A convolutional neural networks based method for anthracnose infected walnut tree leaves identification. Appl Sci 10:469

Arya S, Sandhu KS, Singh J, Kumar S (2022) Deep learning: as the new frontier in high-throughput plant phenotyping. Euphytica 218:47

Asolo E, Gil-ozoudeh I, Ejimuda C (2024) AI-powered decision support systems for sustainable agriculture using ai-chatbot solution. J Digital Food Energy Water Syst 5

Batz P, Will T, Thiel S, Ziesche TM et al (2023) From identification to forecasting: the potential of image recognition and artificial intelligence for aphid pest monitoring. Front Plant Sci 14

Bauckhage C, Kersting K (2013) Data mining and pattern recognition in agriculture. KI-Künstl Intell 27:313–324

Bhadauria S, Singh V (2021) Organization culture and open innovation in sustainable agri-business: a case study of jain irrigation systems limited. J Contemp Issues Bus Gov 27:1570–1586

Bharadiya JP, Tzenios NT, Reddy M (2023) Forecasting of crop yield using remote sensing data, agrarian factors and machine learning approaches. J Eng Res Rep 24:29–44

Bini D, Pamela D, Prince S (2020) Machine vision and machine learning for intelligent agrobots: a review. In: 5th International Conference on Devices, Circuits and Systems (ICDCS). IEEE, pp 12–16

Chen H, Yada R (2011) Nanotechnologies in agriculture: new tools for sustainable development. Trends Food Sci Technol 22:585–594

Chhipa H (2016) Nanofertilizers and nanopesticides for agriculture. Environ Chem Lett 15:15–22

Christiaens O, Whyard S, Vélez AM, Smagghe G (2020) Double-stranded rna technology to control insect pests: current status and challenges. Front Plant Sci 11:451

Dasgupta N, Ranjan S, Ramalingam C (2017) Applications of nanotechnology in agriculture and water quality management. Environ Chem Lett 15:591–605

de Oliveira JL, EVR C, de Carvalho LB et al (2019) Nanotechnology-based delivery systems: highlights in agricultural applications. J Sib Fed Univ 12:311–328

Diaz-gonzalez FA, Vuelvas J, Correa CA, Vallejo VE (2022) Machine learning and remote sensing techniques applied to estimate soil indicators-review. Ecol Indic 135:108517

Domínguez AG, Roig-tierno N, Chaparro-banegas N, García-álvarez-coque JM (2024) Natural language processing of social network data for the evaluation of agricultural and rural policies. J Rural Stud 109:103341

Eisenstadt M, Pollock D (2021) Asset test 2021: how the us can keep benefiting from its alliance with Israel. In: The washington institute for near east policy

Elmer WH, White JC (2016) The use of metallic oxide nanoparticles to enhance growth of tomatoes and eggplants in disease infested soil or soilless medium. Environ Sci Nano 3:1072–1079

Fan YA, Liang CK (2022) Hybrid discrete particle swarm optimization algorithm with genetic operators for target coverage problem in directional wireless sensor networks. Appl Sci 12:8503

Farooque AA, Afzaal H, Abbas F, Bos M et al (2022) Forecasting daily evapotranspiration using artificial neural networks for sustainable irrigation scheduling. Irrig Sci 40:1–15

Folorunso O, Ojo O, Busari M, Adebayo M et al (2023) Exploring machine learning models for soil nutrient properties prediction: a systematic review. Big Data Cogn Comput 7:113

García L, Parra L, Jimenez JM (2020) Iot-based smart irrigation systems: an overview on the recent trends on sensors and Iot systems for irrigation in precision agriculture. Sensors 20(4):1042. https://doi.org/10.3390/s20041042

Gil-díaz M, Alonso J, Rodríguez-valdés E, Pinilla P et al (2014) Reducing the mobility of arsenic in brownfield soil using stabilised zero-valent iron nanoparticles. J Environ Sci Health A Tox Hazard Subst Environ Eng 49:1361–1369

Hasan AM, Sohel F, Diepeveen D, Laga H et al (2021) A survey of deep learning techniques for weed detection from images. Comput Electron Agric 184:106067

Hsueh HT, Chou RT, Rai U, Liyanage W et al (2023) Machine learning-driven multifunctional peptide engineering for sustained ocular drug delivery. Nat Commun 14:2509

Huchchannanavar S, Divate Anuja D (2023) Artificial intelligence (ai) and iots in agriculture: a concept and reality. In: Emerging trends in Agronomy Publisher. Biotech Books

Jha K, Doshi A, Patel P, Shah m. (2019) A comprehensive review on automation in agriculture using artificial intelligence. Artif Intell Agric 2:1–12

Kah M, Beulke S, Tiede K, Hofmann T (2013) Nanopesticides: state of knowledge, environmental fate, and exposure modeling. Crit Rev Environ Sci Technol 43:1823–1867

Kah M, Kookana RS, Gogos A, Bucheli TD (2018) A critical evaluation of nanopesticides and nanofertilizers against their conventional analogues. Nat Nanotechnol 13:677–684

Kamilaris A, Prenafeta-boldú FX (2018) Deep learning in agriculture: a survey. Comput Electron Agric 147:70–90

Kaur H, Prashar D (2019) Applications of machine learning in plant disease detection. Think India J 22:3100–3106

Krause A, Singh A, Guestrin C (2008) Near-optimal sensor placements in gaussian processes: theory, efficient algorithms and empirical studies. J Mach Learn Res 9

Kumar D, Kukreja V (2021) N-cnn based transfer learning method for classification of powdery mildew wheat disease. In: International Conference on Emerging Smart Computing and Informatics (ESCI). IEEE, pp 707–710. https://doi.org/10.1109/ESCI50559.2021.9396972

Kumar S (2020) Deep learning for plant disease classification. Curr Appl Sci Tech 24(4):1–24

Lei L, Yang Q, Yang L, Shen T et al (2024) Deep learning implementation of image segmentation in agricultural applications: a comprehensive review. Artif Intell Rev 57:149

Lekshmi S, Singh D, Shojaei baghini M. (2014) A critical review of soil moisture measurement. Measurement 54:92–105

Liakos KG, Busato P, Moshou D, Pearson S et al (2018) Machine learning in agriculture: a review. Sensors 18:2674

Lottes P, Khanna R, Pfeifer J, Siegwart R et al (2017) Uav-based crop and weed classification for smart farming. IEEE. https://doi.org/10.1109/ICRA.2017.7989347

Manikandan A, Subramanian K (2016) Evaluation of zeolite based nitrogen nano-fertilizers on maize growth, yield and quality on inceptisols and alfisols. Int J Plant Soil Sci 9:1–9

Mittal M, Gupta V, Aamash M, Upadhyay T (2024) Machine learning for pest detection and infestation prediction: a comprehensive review. Wiley Interdiscip Rev Data Min Knowl Discov 14:e1551

Mohanty SP, Hughes DP, Salathé M (2016) Using deep learning for image-based plant disease detection. Front Plant Sci 7:1419

Mora H, Gil D, Munoz Terol R, Azorín J et al (2017) An iot-based computational framework for healthcare monitoring in mobile environments. Sensors 17:2302

Motadayen M, Nehru D, Agarwala S (2024) Advancing sustainability: biodegradable electronics and materials discovery through artificial intelligence. Int J AI Mater Des 1(2):1–20

Mrisho L, Mbilinyi N, Ndalahwa M, Ramcharan A et al (2020) Evaluation of the accuracy of a smartphone-based artificial intelligence system, plantvillage nuru, in diagnosing of the viral diseases of cassava. Biorxiv. https://doi.org/10.1101/2020.01.26.919449

Neupane J, Guo W (2019) Agronomic basis and strategies for precision water management: a review. Agronomy 9:87

Nolte M, Tewes A, Hoffmann H (2023) Xarvio® digital farming solutions. Precision agriculture: modelling. Springer

Oberti R, Marchi M, Tirelli P, Calcante A et al (2016) Selective spraying of grapevines for disease control using a modular agricultural robot. Biosyst Eng 146:203–215

Pantazi XE, Lagopodi AL, Tamouridou A, Kamou NN et al (2022) Diagnosis of induced resistance state in tomato using artificial neural network models based on supervised self-organizing maps and fluorescence kinetics. Sensors 22:5970

Parisi C, Vigani M, Rodríguez-cerezo E (2015) Agricultural nanotechnologies: what are the current possibilities? Nano Today 10:124–127

Pinheiro I, Aguiar A, Figueiredo A, Pinho T et al (2023) Nano aerial vehicles for tree pollination. Appl Sci 13:4265

Prasad R, Bhattacharyya A, Nguyen QD (2017) Nanotechnology in sustainable agriculture: recent developments, challenges, and perspectives. Front Microbiol 8:1014. https://doi.org/10.3389/fmicb.2017.01014

Preite L, Vignali G (2024) Artificial intelligence to optimize water consumption in agriculture: a predictive algorithm-based irrigation management system. Comput Electron Agric 223:109126

Rajan P, Radhakrishnan B, Suresh LP (2016) Detection and classification of pests from crop images using support vector machine. In: International Conference on Emerging Technological Trends (ICETT). IEEE, pp 1–6

Sagdic K, Eş I, Sitti M, Inci F (2022) Smart materials: rational design in biosystems via artificial intelligence. Trends Biotechnol 40:987–1003

Shen Y (2021) Construction of a wireless sensing network system for leisure agriculture for cloud-based agricultural internet of things. J Sens 2021:3021771

Shin JY, Min B, Kim KR (2022) High-resolution wind speed forecast system coupling numerical weather prediction and machine learning for agricultural studies—a case study from South Korea. Int J Biometeorol 66:1429–1443

Shukla R, Dubey G, Malik P, Sindhwani N et al (2021) Detecting crop health using machine learning techniques in smart agriculture system. J Sci Ind Res 80:08. https://doi.org/10.56042/jsir.v80i08.44034

Surase RR, Kale KV (2015) Multiple crop classification using various support vector machine kernel functions. J Eng Res Appl 5:103–109

Surya P, Aroquiaraj IL (2018) Crop yield prediction in agriculture using data mining predictive analytic techniques. Int J Res Anal Rev 5:783–787

Tetala SSDR (2023) Artificial intelligence powered personalized agriculture. Colorado State University

Tetila EC, Machado BB, Astolfi G, De Souza Belete NA et al (2020) Detection and classification of soybean pests using deep learning with uav images. Comput Electron Agric 179:105836

Tian H, Wang T, Liu Y, Qiao X et al (2020) Computer vision technology in agricultural automation-a review. Inf Process Agric 7:1–19

Tian T, Zhang Y, Mei Y (2022) Intelligent analysis of precision marketing of green agricultural products based on big data and gis. Earth Sci Inf 15:1395–1406
Treinish LA, Praino AP (2004) Applications and implementation of a mesoscale numerical weather prediction and visualization system. In: Proceedings of the 20th conference on weather analysis and forecasting/16th conference on numerical weather prediction
Vijayakumar V, Ampatzidis Y, Schueller JK, Burks T (2023) Smart spraying technologies for precision weed management: a review. Smart Agric Tech 100337:100337
Visentin F, Cremasco S, Sozzi M, Signorini L et al (2023) A mixed-autonomous robotic platform for intra-row and inter-row weed removal for precision agriculture. Comput Electron Agric 214: 108270
Wolfert S, Ge L, Verdouw C, Bogaardt MJ (2017) Big data in smart farming–a review. Agric Syst 153:69–80
Zhu F, Zhang L, Hu X, Zhao J, Meng Z, Zheng Y (2023) Research and design of hybrid optimized backpropagation (bp) neural network pid algorithm for integrated water and fertilizer precision fertilization control system for field crops. Agronomy 13:1423

Chapter 8
Role of Nanotechnology in Enhancing Plant Genetic Engineering

Amit, Dimpal Mehla, and Manoj K. Sharma

Abstract Nanotechnology has transformed several scientific domains, including plant genetic engineering. Here we explore the role of nanotechnology in advancing plant genetic engineering, specifically focusing on the use of nanoparticles as a gene delivery vehicle. We examine several nanoparticle forms, such as organic, inorganic, and carbon-based nanoparticles, and the special qualities that render them appropriate for gene delivery applications.

Conventional direct and indirect gene delivery techniques have drawbacks, such as low transformation efficiency, insertion of multiple gene copies, tissue damage, and limited target plant species. On the other hand, nanoparticles are flexible and effective substitutes for gene delivery in plants. We address the benefits of gene transfer using nanoparticles over traditional approaches, which emphasizes how different types of nanoparticles can overcome the limitations of conventional methods. We present a thorough review of nanoparticles' function in gene transport into plants. The chapter highlights the significance of target-specific gene transport, particularly delivery into mitochondria and chloroplasts, to achieve precise genetic alterations.

We also discussed the synthesis, surface modification, and mechanisms of action of the different kinds of nanoparticles, like mesoporous, magnetic, and gold nanoparticles, utilized in gene delivery. The benefits and drawbacks of each kind of nanoparticle are covered in detail in this chapter. The chapter emphasizes how nanoparticles can improve gene delivery's effectiveness and specificity, hastening the creation of genetically modified plants with desired characteristics. This chapter offers a thorough and perceptive examination of how nanotechnology may enhance plant genetic engineering.

Keywords Nanoparticles · Gene transfer · Targeted gene delivery · Gene editing · Plant transformation

---

Amit · D. Mehla · M. K. Sharma (✉)
Crop Genetics & Informatics Group, School of Biotechnology, Jawaharlal Nehru University, New Delhi, Delhi, India
e-mail: mksharma@jnu.ac.in

S. K. Yadav, A. Kumari (eds.), *Nanobiotechnology in Agri-food Sector*, Sustainability Sciences in Asia and Africa, https://doi.org/10.1007/978-981-96-7756-6_8

8.1 Introduction

In changing climatic conditions, biotic stress like infections of bacteria, fungi, or insects and abiotic stress conditions like drought, salinity, or high temperature are the key factors that result in heavy losses in crop productivity. The challenge is not only to improve the crop yield, but changing environmental conditions would make it difficult to maintain the existing yield too. Over time, conventional plant breeding has played a significant role in developing environmentally resilient crops and improving overall productivity. However, this is a slow and time-consuming process and would not be enough to meet the requirements of a growing population. Furthermore, it does not provide an opportunity to introduce new traits that do not exist within a species (Ahmar et al. 2020). On the other hand, genetic engineering (GE) is a powerful tool that has been successfully used in the past for crop improvement and has resulted in the development of new crops that are resistant to pathogen infections, tolerant to herbicides, have better nutrition profiles, or have superior performance under environmental (biotic/abiotic) stress conditions (Sircaik et al. 2021). It involves the manipulation of the genetic material of a species or the addition of new genetic material of interest to the genome of the target species, followed by the regeneration of plants with an altered genotype or phenotype. It has not only complemented the existing conventional breeding tools to speed up the process but also opened up the possibility of transferring novel traits to a species that did not exist in that species or helping in using plants as bioreactors for the production of pharmaceutically/industrially important molecules (Madhu et al., 2023). For example, the glucocerebrosidase enzyme, which is used for the treatment of Gaucher's disease, is being produced in carrot cultures and has been approved for human use (Naphatsamon et al. 2018). Similarly, rice has been engineered to produce vitamin A, which can play a significant role in reducing malnutrition in rice-dependent Asian countries in a sustainable manner (Paine et al. 2005), or the transgenic virus-resistant papaya "Rainbow," which helped the Hawaiian papaya industry fight the papaya ringspot virus and has been stably integrated into the consumer markets (Jia et al. 2017). New DNA segments containing one or many genes that are transferred to the genome of the host plant may belong to the same species or may be from a totally unrelated organism, thereby helping in breaking the sexual incompatibility barrier.

A diversity of tools have been developed for the delivery and integration of DNA/genes into the host plant genome. Some of the popular plant transformation methods are particle bombardment, microinjection, and *Agrobacterium*- or viral-vector-mediated plant transformation. Though significant progress has been made over the decades and transformation efficiency has been improved, there are several limitations that are limiting the further improvement of these tools. The presence of a rigid cell wall is a major barrier to successful transformation. Most of these genetic manipulations are done at the cellular level, and therefore, optimized regeneration protocols are required for developing transformed plants.

Nanomaterials (NMs) have emerged as promising tools for improving agricultural productivity and are being used extensively in agriculture as well as in the food industry. They regulate plant growth and development and find applications in various forms like nano-fertilizers, nano-herbicides, nano-pesticides, or nutrients and have been reviewed elsewhere (Gade et al. 2023). They are unique as their physicochemical properties, like shape, size, surface, and chemical nature, can be tuned as per the requirement.

The use of nanomaterials (NMs) as a delivery vehicle for biomolecules is well established in animal or microbial cells; their role as a delivery vehicle for biomolecules in plants is in its infancy (Jat et al. 2020). The plant cell wall is a major barrier for delivery vehicles, but nanoparticles (NPs), being small in size, have been shown to cross the cell wall barrier and hence can deliver biomolecules (protein, DNA, and RNA) into the plant cells (Rai et al. 2012). Several reports of nanoparticle-mediated successful delivery of nucleic acids or other biomolecules in plants have been published in the recent past as proof of concept, which highlights the potential of NMs in plant genetic engineering. This chapter highlights the progress made in NP-mediated DNA delivery in plant cells along with the opportunities and challenges ahead.

8.2 Gene Delivery Methods in Plants

Gene delivery tools are used to deliver DNA segments to plant genomes to alter their genotype or phenotype. This process includes the delivery of the DNA segments in the plant cell followed by its integration into the plant genome, which may result in genotypic and/or phenotypic changes. Several other components are also important for the regulation of foreign genes, like a suitable and specific promoter and transcriptional termination signal. For the expression of a gene, several regulatory components are required in addition to the gene sequence, which is used to construct the gene expression cassette. Some of the essential regulatory elements required for gene expression include a suitable promoter and a transcription termination signal. In addition to genes of interest, antibiotic marker genes and/or reporter protein gene expression cassettes are also constructed. Selection marker gene expression helps in the differential multiplication of transformed cells, and reporter protein is used for validation of the expression of transgenes (Low et al. 2018).

To target the accumulation of desired protein in specific locations/organs in the cell, respective signal sequences can be used. In a plant cell, foreign genes can be inserted either in the nuclear genome or organellar genomes like chloroplasts or mitochondria. For genetic engineering-mediated crop improvements, stable integration of the foreign in the nuclear genome is preferred as nuclear-expressed genes make eukaryotic-type proteins. Whereas organelles are prokaryotic in origin, chloroplast-expressed genes make prokaryotic-type proteins that lack post-translational modifications, which may be required for their function. However, the chloroplast is a suitable destination for achieving higher expression levels of transgenes. A plant cell can have a large number of chloroplasts, and each chloroplast has several genomes. Therefore, the integration of foreign genes into the chloroplast genome allows the expression of a large number of copies of candidate genes, thereby resulting in high expression of foreign protein (Meyers et al. 2010). Broadly, gene transfer methods can be classified into direct methods, where physical or chemical tools are used for gene delivery, or indirect methods, where biological systems, like *Agrobacterium*, are employed for the transfer of DNA segments to the plant genome. The selection of the appropriate method depends on the requirements and experimental demand.

8.2.1 Direct Methods

Several tools have been developed to deliver DNA into the plant cells using either physical forces or chemical methods. These tools are categorized as direct methods. Biolistic, microinjection, electroporation, or laser beam-induced gene transfer are all examples of direct methods.

Biolistic or particle gun-mediated gene delivery is commonly used as the most successful direct method for gene delivery in plants. Gold or tungsten particles of micrometer size are used as vehicles for gene transfer and are also known as microprojectiles. Plasmids containing genes of interest are coated on these microprojectiles and are bombarded on the plant tissues using high-pressure gas (Begum and Jayawardana 2021). Bombardment of high-velocity microcarriers causes cell leakage due to turgor pressure imbalance; therefore, plant cells are maintained on osmotic media during the bombardment. These DNA-coated particles enter plant cells and release DNA in the cell, which results in the integration of the DNA in the host genome. The process of DNA being released from the microparticle is still poorly understood. One of the major advantages of this method is that it can transfer DNA to all the plant cells independent of species or tissue type. Therefore, if the regeneration protocol is available, it can be used to transform any tissue type and hence is most popular among direct methods.

Use of injection pipette is another physical method used for DNA delivery. Injection pipette having fine needles of the size ranging from 0.5 to 10 μm can be used, depending upon the size of the target cell. Though it can be used to deliver DNA inside intact plant cells, protoplasts and other preferred explants. A holding capillary or several other methods of immobilization are developed to hold the protoplast/cell in the correct orientation for intranuclear injection. Generally, poly-L-lysine (PLL) and agarose are considered for the attachment of protoplasts on the coverslip (Nomura and Komamine 1986; Toyoda et al. 1988). Electroporation uses electric current to create microscopic transient pores in the plant cell membrane, through which nucleic acids or other molecules can enter the cell. Cell suspension and candidate DNA are placed in the cuvette, and an electric pulse is applied through the electrodes in the electroporation device. The electric field of optimal strength results in the electro-permeabilization of the cell membrane. Thermodynamics regulates the size of the transient pores and the capability of the membrane to regain its original form; therefore, the required electric threshold needs to be optimized. These transient aqueous pores up to a certain size provide temporary passage for biomolecule transport and can be easily sealed. Therefore, cells can be recovered with high efficiency after electroporation. *Arabidopsis thaliana* intact cells were used to demonstrate electroporation for Cre-mediated site-specific recombination (Furuhata et al. 2019).

While particle bombardment has been used successfully to transform several plants, other physical methods are not very successful. Particle bombardment often damages the target tissue and may compromise the regeneration capacity of the explant tissue. It often results in the integration of multiple transgenic copies, which is not desirable because of unintended segregation patterns and gene silencing (Altpeter et al. 2005). Similarly, during electroporation, a high-voltage electric pulse is applied, which can result in cell death too. All of these methods are laborious and need expert handling as well as costly instrumentation, which is not within the reach of every lab and hence restricts its widespread usage for plant transformation (Ozyigit 2020).

8.2.2 Indirect or Vector-Mediated Gene Transfer Methods

Agrobacterium- or virus vector-mediated DNA delivery methods are the indirect DNA delivery tools or biological methods that use living organisms for foreign DNA delivery in plant cells. Among them, *Agrobacterium*-mediated plant transformation is the most popular and is frequently used. *Agrobacterium* infects plants and transfers genes encoding for hormones and opines to the plant genome. Production of hormones results in uncontrolled cell division, thereby resulting in tumor formation, and opines are metabolized by the bacterium for nutrition (Tiwari et al. 2022). The DNA to be transferred, along with genes encoding regulatory proteins required for the infection, is present on a plasmid known as tumor-inducing (Ti) plasmids. Because the DNA transferred by *Agrobacterium* into the plant genome does not play any role in the transfer process, it can be replaced with a DNA segment carrying the gene of interest on the plasmid. These engineered plasmids are transformed into *Agrobacterium,* which, upon infecting plant tissue, transfers the DNA having a gene of interest into the plant genome. Transgenic plants can then be raised from this transformed tissue. As it does not require specialized and costly equipment, it can be performed routinely in a research lab with basic facilities. Furthermore, it produced a high frequency of single-copy transgenic plants, and therefore it is the method of choice. However, a major challenge is the narrow host range. *Agrobacterium* does not infect all the plants naturally, like monocot plants, and the efficiency of infection can also vary among different plant species; therefore, it cannot be used to transform these plants (Cheng et al. 2004; Wu et al. 2023a).

Viruses infect plants and use the host cell's machinery for their multiplication (Mahmood et al. 2023). During infection, viruses inject their genetic material into the plant cell, where it replicates and synthesizes new viral particles using host machinery. Several viral vectors, like retroviruses, adenoviruses, or adeno-associated viruses, have been constructed and have been used successfully for delivering nucleic acids. The gene of interest is inserted in the viral vector, and when these infect plant cells, they replicate rapidly in the plant cells, producing several copies of them. Though this system does not result in stable integration of the gene of interest, these vectors are being used for the commercial production of several useful proteins. The major disadvantage of the system is the lack of stable integration of the transgene in the host genome, and therefore stable transgenic plants cannot be generated using this system. Furthermore, as it is dependent on the capability of the virus to infect the host, it is also species-dependent and limits its applications.

Despite significant progress in the optimization of plant transformation tools, several challenges like species dependency, low transformation efficiency, and costly instrumentation exist for each method, and none of them can be applied universally to all plant species with high transformation efficiency. Nanotechnology has the potential to overcome the limitations of conventional methods. Tunable size and surface are a major advantage in the case of NPs, which allow us to explore a wide range of plant species for genetic manipulation.

Nanotechnology has revolutionized the biomolecule delivery tools in microbial and animal systems and is being used extensively for delivering biomolecules like nucleic acids, proteins, and peptides successfully. However, these tools cannot be directly translated for nucleic acid delivery in plant cells because of fundamental differences between plant cells and animal cells. The presence of the cell wall is the primary barrier to the delivery of biomolecules in plant cells (Yadav et al. 2023). After removal of the cell wall through enzymatic degradation, delivery tools used for animal cells can be extended for protoplast transformation successfully to plants too, but regeneration of transgenic plants from protoplasts is a major challenge. Furthermore, physical barriers like size exclusion, as well as the localization barriers that work at the cellular level in organs or subcellular level in cell organelles, may limit the delivery of biomolecules through NPs. Though NPs are being used as nano-fertilizers, nano-pesticides, nano-herbicides, or nano-insecticides and have been shown to internalize in plant cells, their mechanism of action is not well understood.

8.3 Nanoparticles and Their Role in Plants

NPs are very small particles, which range in size from 1 to 100 nm in size. Compared to bulk materials used to synthesize NPs, these have unique thermal, optical, catalytic, magnetic, or mechanical properties, which differ from the bulk materials because of surface and quantum effects (Guisbiers et al. 2012). NPs are being used for diverse applications in agriculture, which include protection from pathogens, improving growth and development, controlled release of agrochemicals, and making a significant impact on agricultural productivity in an environmentally friendly manner (dos Santos et al. 2024). It has been reported that NPs are required in very small amounts and may stimulate various physiological processes like nitrogen metabolism, photosynthetic efficiency, protein synthesis, sugar metabolism, and stress tolerance. Chemical fertilizers, insecticides, and pesticides have been used extensively to enhance crop productivity in the last few decades. However, these have deleterious effects on the soil microbiome and fertility, and their excessive use has led to adverse effects on the environment and nutritional quality as well as on human health. Use of nano-fertilizers where essential minerals and nutrients are coated over NPs along with nano-dimensional adsorbents, and controlled release of these nutrients has provided an environmentally friendly alternative of conventional fertilizers (Haydar et al. 2024).

Uptake of the NPs and their transport within the plant body are directly affected by their size and shape, concentration, surface charge, and exposure time. Furthermore, their surface properties are influenced by the pH of the solution or the presence of other ionic components in the solution and hence its transport too (Báez-Cruz et al. 2021). NPs may enter the plant systems through various openings like stomata, cracks, or root hairs and may move through the plant system either by bulk flow or diffusion (Khan et al. 2022). Furthermore, NPs have been shown to enter the plant cell cytoplasm and may travel from one cell to another cell through plasmodesmata. Based on the existing evidence, it can be concluded that NPs can move within the plant through apoplastic transport as well as through symplastic transport (Wang et al. 2023). However, how they regulate physiological and molecular functions in plant cells is not well understood.

NPs differ in their size, shape, and structure and can be categorized in several ways (Khan et al. 2019). Broadly, these are classified into carbon-based, organic, and inorganic NPs. The unique and versatile properties of NPs exhibit enormous opportunities for plant engineering applications. NPs are of different types, such as gold, silver, and carbon-based, and each comes with specific properties. For instance, gold NPs (AuNPs) are known for their biocompatible nature and their ability to facilitate the delivery of genetic material. Silver NPs exhibit strong antimicrobial properties, making them useful for protecting plants against pathogens while also aiding in gene transfer (Loo et al. 2018). Carbon NPs, single-walled and multi-walled carbon nanotubes, offer excellent mechanical strength and electric conductivity, which can be leveraged for efficient gene delivery. In addition to these NPs, specialized NPs like mesoporous NPs and magnetic nanoparticles (MNPs) can play an important role in gene delivery. MNPs can be guided by an external magnetic field and also ensure precise genetic material delivery. The highly porous structure of mesoporous NPs allows them to load a large amount of cargo and help in controlled release, thereby ensuring stable and sustainable delivery. Given these diverse capabilities, NPs not only enhance nutrient delivery and plant protection but also offer a groundbreaking approach to gene delivery in plants. By enabling targeted genetic modifications with minimal environmental impact, they could significantly advance plant biotechnology, leading to crops with improved resilience, productivity, and sustainability. The integration of various types of nanoparticles (Fig. 8.1) into gene delivery systems promises to transform

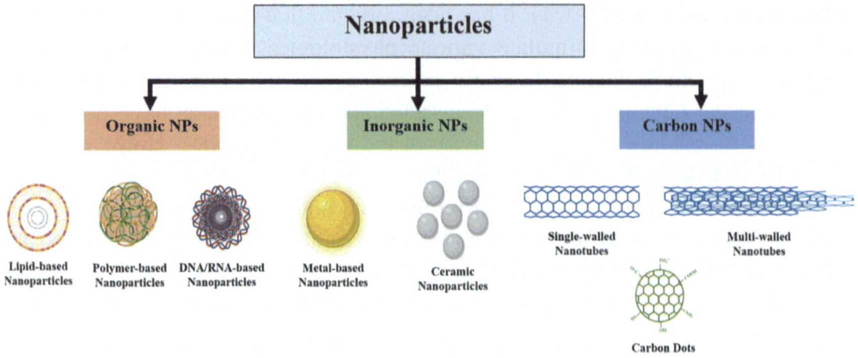

Fig. 8.1 Nanoparticle classification based on their synthesis material, i.e., organic nanoparticles, inorganic nanoparticles, and carbon nanoparticles

agricultural practices in an eco-friendly and efficient manner, paving the way for a new era of precision agriculture.

8.3.1 Inorganic NPs

Synthetic vectors like inorganic NPs are emerging as promising delivery vehicles due to their advanced features over traditional vectors like tenable size and surface properties. Metallic NPs like gold, silver, or iron oxides are among the most reliable for agricultural applications and can play a crucial role in developing advanced gene delivery tools and plant genetic engineering. Metal NPs are purely made from metal precursors, and depending upon the precursor material, these can be categorized into mono-, bi-, or polymetallic NPs. For example, bimetallic NPs are formed when the starting material used to synthesize them is an alloy. They have localized surface plasmon resonance (SPR) characteristics, because of which they have unique optical and electrical properties, making them useful for various biological applications (Mohanpuria et al. 2008; Kumar et al. 2013).

Gold NPs (AuNPs) are one of the potential gene delivery tools due to their properties like cytotoxicity, biodistribution, and in vivo excretion that can be regulated by particle size and surface functionality (Ding et al. 2014). Their ability to conjugate with DNA and other biomolecules has made them a preferred vehicle for delivering genetic materials to plants. The small size of AuNPs also allows them to penetrate cell walls that facilitate direct gene delivery into plant cells. Versatility in surface functionalization may give advantages in gene editing and silencing or other genetic illnesses (Kanu et al. 2022). Vijaykumar and coworkers used nanogold-loaded carbon bullets for gene delivery into tobacco and rice plants. They coated binary vectors on the nanogold particles and delivered them to the calli of rice and tobacco using a biolistic gun, and successful transformation was confirmed through the GUS expression. In this carbon-gold nanocomposite, carbon is used as an inert material, and gold provides the best support to DNA (Vijayakumar et al. 2010). Hao et al. (2013) used polyethylene glycol (PEG)-coated magnetic gold NPs for canola protoplast transformation. Similarly, gold nanoparticle-capped gene-loaded mesoporous silica nanoparticles (MSNs) were used for gene and chemical delivery in *Nicotiana tabacum* (Torney et al. 2007).

Silver NPs, along with antimicrobial properties (Loo et al. 2018), can also contribute to gene delivery by protecting (Gur 2022) them against mechanical damage during the transfer process, thereby ensuring their successful integration. Metal NPs may possess specialized magnetic or thermal properties that further broaden their applications in biological systems. Magnetic NPs, generally made up of iron oxides, can be manipulated by external magnetic fields due to their magnetic properties, enhancing their ability to control and target gene delivery. This makes them a system of choice for enhancing the efficiency of biomolecule delivery in plants. In a study, gold microparticles and silver NPs were compared for their plant transformation efficiency in a biolistic method, and it was reported that silver NPs showed several-fold higher transformation efficiency compared to gold microparticles (Rajkumari et al. 2021). Ceramic nanoparticles (CeNPs) are also a good delivery vehicle due to their unique set of characteristics, which protect biomolecules like enzymes, DNA, and RNA against pH and temperature fluctuations. CeNPs are primarily made up of inorganic NMs, including carbonates, carbides, phosphates, and oxides. They are synthesized at high temperatures followed by rapid cooling (Bhardwaj et al. 2021). Calcium triphosphate and nanohydroxyapatite [$Ca_{10}(PO4)_6(OH)_2$] are two examples of materials with great promise for tissue engineering and medication delivery (Ghaffari et al. 2023), which may make it easier to transfer genetic material into plant cells. These characteristics aid in the effective and focused transfer of genes and may add to or enhance the processes involved in genetic modification and plant transformation.

Mesoporous materials, large pore volumes, adjustable pore diameters, and controllable morphologies are important components in the field of nanomaterials. Mesoporous NPs are small, solid, porous structures with large surface areas. This large surface area and pores allow different functional groups to attach for targeted delivery. Mesoporous silica NPs have a honeycomb-like structure and an active surface. Low toxicity and high loading capacity (Bharti et al. 2015) make MSNs a potential carrier system for DNA delivery. Their application has been shown to improve growth and development in wheat and jupin (Sun et al. 2016). Furthermore, it is demonstrated that MSNs can enter the endodermis and vascular tissue by penetrating plant cell walls, thereby suggesting that these materials can be used to deliver biomolecules in the plant cells across plant cell walls. Martin-Ortigosa et al. (2012) functionalized mesoporous silica NPs with gold and used them for the codelivery of protein and nucleic acids using particle guns in plant cells.

In recent years, magnetic nanoparticles (MNPs) have gained enormous attention due to their applications in medicine, biosensing, agriculture, and environmental science (Jat et al. 2021). Magnetic nanoparticles are generally synthesized by metals and metal oxides (iron chloride, iron sulfates). MNPs show excellent biocompatibility and display superparamagnetic (SPM) behavior, i.e., these particles exhibit magnetism only when an external magnetic field is applied. Superparamagnetic behavior has several advantages: better nucleic acid loading capability, improved penetration, reduced production costs, and the capability to transport biomolecules when subjected to external magnetic fields (Berry 2009). Fe_3O_4 is the most popular superparamagnetic magnetite with excellent biocompatibility. A magnetic NP-mediated gene delivery system has been developed that employs an external magnetic force and is called "magnetofection." It does not require complicated equipment and has been used for gene delivery in plants successfully (Wu et al. 2015). Magnetic nanoparticles synthesized by magnetite of different shapes and sizes have been shown to alter root development with no phytotoxicity (Sun et al. 2023). MNPs have also been shown to penetrate across plant cell walls, and it has opened up several possibilities for plant transformation (Mohamed et al. 2019), though molecular understanding of their penetration through cells would be required to optimize the plant transformation tools.

8.3.2 Organic NPs

Organic nanoparticles (ONPs) are prepared from organic materials like proteins, carbohydrates, polymers, or other compounds and can be of different types, like liposomes, nanogels, micelles, dendrimers, and polymer-based NPs (Alshammari et al. 2023). As their precursors are biological in origin, typically these are biodegradable and non-toxic, which gives an extra advantage in gene delivery and genetic engineering applications. Additionally, lipid-based nanoparticles (LNPs), such as liposomes, have also been employed in delivering plasmid DNA and RNA into cells, offering a non-toxic and efficient means of genetic modification. LNPs mainly consist of four components, each contributing to their synthesis and structural properties. Phospholipids and cholesterol are essential for particle formation and stability of NPs; cationic lipids provide a positive charge to enable DNA binding, and PEGylated lipids enhance circulation time in the body. These lipids are similar to those found in cell membranes, giving LNPs a biomimetic nature that helps their passage through cell membranes to deliver different types of cargo (Mehta et al. 2023). Polymer-based ONPs are used as an alternative to inorganic NPs. Polymeric NPs allow surface modifications that reduce off-target effects through conjugation with target-specific receptors and antibodies (Amreddy et al. 2017).

ONPs are usually formed by non-covalent intermolecular interactions, and their clearance from the host system is easy. This growing body of research underscores the potential of organic nanoparticles as effective carriers for genetic material, opening new avenues for the development of resilient, high-yield crops through advanced genetic engineering techniques.

8.3.3 Carbon-Based NPs

These are purely made of carbon-based materials and consist solely of carbon atoms. They may exist in various forms: crystalline form such as in diamonds or graphite and amorphous form such as in coal or charcoal. The nanoparticles that are derived from carbon-based pure nanomaterials like fullerenes, quantum dots, graphenes, or their derivatives like graphene oxides have excellent electrical and thermal conductivity, high stability, and favorable mechanical properties like extreme stiffness or toughness (Maiti et al. 2019). Furthermore, these are highly biocompatible and have low toxicity in biological systems and hence can be used for various applications in an environmentally friendly manner. In plant sciences, the application of carbon nanotubes (CNTs) has the potential role of delivering nucleic acid, protein, or other biomolecules to cells and organs at targeted locations (Majeed et al. 2020). Small tubular-like cylindrical structures with a diameter in the nanometer range created by the folding of graphene sheets are called carbon nanotubes (CNTs). Depending upon the number of layers of graphene sheets, these can be categorized as single-walled CNTs (SWCNTs) or multi-walled CNTs (MWCNTs). These have been shown to have a significant effect on plant biomass enhancement (Khodakovskaya et al. 2011). Demirer and colleagues used single-walled CNTs (SWCNTs) for delivering siRNA molecules in tobacco leaves successfully and could silence the target gene (Demirer et al. 2020). Carbon-based NMs have also been used for the delivery of biomolecules and chemicals into cell organelles like chloroplasts (Santana et al. 2022).

Chitosan-complexed SWCNTs are involved in chloroplast-targeted gene delivery in *Nicotiana tabacum, Spinacia oleracea*, and *Eruca sativa* plant species (Kwak et al. 2019). Engineered peptides for targeted delivery to DNA-coated carbon nanotubes can successfully pass the plant cell wall barrier (Golestanipour et al. 2018). The nanocomposite of carbon [carbon dot (CD) with polyethylenimine (PEI)] for gene transfection in plants and animals was confirmed as a good vector. In rice and wheat, PEI-modified CDs efficiently carried DNA into leaves and roots for transient expression (Wang et al. 2020a). Recently, Dunbar and colleagues loaded plasmid DNA with reporter markers such as YFP, green fluorescent protein (GFP), and GUS on carbon NPs to demonstrate diffusion through plant cell walls. They show transient expressions of reporter genes in rice leaves through infiltration, as well as in embryo tissues. Quantitative and qualitative data indicate GFP, YEP, and GUS expressions (Dunbar et al. 2022).

Subcellular organelles are also important for their agronomic trial. So, Law and colleagues explored CNT's potential as a gene delivery tool for the subcellular level. The CNT-polymer hybrid model with a functional peptide delivers DNA into mitochondria to enhance efficiency (Law et al. 2022). The flexibility of polymer-coated CNTs for the conjugation with other biomolecules will allow targeting other organelles, which also creates more opportunities for plant genetic engineering. Studies of CNTs as an efficient tool for gene silencing are also explored by Demirer et al. (2020) and carbon tubes as a direct delivery tool for siRNA to plant cells.

8.4 Properties of Nanoparticles

8.4.1 Magnetic

The magnetic property of NPs depends upon the synthesis method and substrates (Saleh 2022). In general, compounds having nickel, iron, or cobalt have magnetic properties. However, at the nanoscale, many materials show magnetic properties due to irregular electron distribution. Usually, material composition, its structure, or magnetic anisotropy are the major parameters that determine the magnetic behavior; however, at the nanoscale, the size and shape of the NPs play a key role in the magnetic properties (Caizer 2016). The energy required for maintaining the magnetic moment in one direction is called magnetic anisotropy energy. One of the most important characteristics of NPs is superparamagnetism, which occurs when thermal energy surpasses the energy barriers associated with magnetic anisotropy and causes NPs to get magnetic character when they are exposed to an external magnetic field (Mohapatra et al. 2019).

Superparamagnetic (SPM) particles display magnetic behavior only in the presence of a magnetic field. They lose their magnetization properties when the magnetic field is removed. SPM NPs are useful for biological applications as they are capable of loading higher amounts of nucleic acids, improved protection, and better penetration capabilities (Berry 2009). Furthermore, MNPs are capable of transporting diverse types of biomolecules and have excellent biocompatibility properties. Surface interactions with surroundings at the nanoscale drive the distinct behaviors of NPs in comparison to the bulk of their counterparts (Issa et al. 2013); these magnetic characteristics of NPs have attracted a great deal of research. Their magnetic response is the main characteristic that defines these qualities, and it can be affected by various aspects like size, shape, composition, surface coating, and synthesis techniques. Coating with different polymers and functional groups for specific use offers a variety of applications (Andrade et al. 2011). M. Ferrite ($M.Fe_2O_3$) NPs, here M represents any metal that forms divalent bonds, are the most studied MNPs (Salih and Mahmood 2023). Their surface is often altered by silica, surfactants, and derivatives of phosphoric acids to enhance the stability of these NPs in solutions (Zeleňáková et al. 2024). The alignment of magnetic moments inside the material is responsible for the ferromagnetism and ferrimagnetism found in NPs. This alignment can be altered by varying the particle composition or size. Ferromagnetic NPs show a homogeneous arrangement of magnetic moments, resulting in strong and persistent magnetization, whereas ferrimagnetic materials have a net magnetic moment due to the opposing alignment of magnetic moments in various sublattices.

The size-dependent magnetic behavior of NPs is important because it enables precisely designed tunable magnetic characteristics for particular applications. Additionally, compared to bulk materials, NPs have a substantially higher surface-to-volume ratio, which may result in improved surface effects and interactions with the surrounding medium. The magnetic behavior of these NPs can be further modified, their biocompatibility can be enhanced, and their integration into intricate biological systems is made easier by these alterations and functionalization with various ligands or coatings.

8.4.2 Thermal

The NPs' ability to precisely control the release of biomolecules or drugs and target a desired location is made possible by their thermal properties, which also increase their potential as delivery vehicles (Colilla et al. 2015). NP size directly regulates their thermal conductivity capabilities. As the size of the NP decreases, a higher surface-to-volume ratio provides a large number of electrons; these electrons are the source of heat conductivity, and the capability to transfer heat increases with a reduction in the size of the particles (Ambreen and Kim 2020). Furthermore, the melting temperatures of the NPs are significantly lower than their bulk material (Gao and Gu 2015). The composition of the NPs is also important for their thermal stability. Generally, NPs prepared from bimetallic alloys have been shown to have higher thermal stabilities. Engineered NPs react to specific temperatures for the release of biomolecules, a property that can be used for the controlled release of biomolecules in biological systems. External stimuli are used to enhance the temperature of the target regions, resulting in the release of the drug specifically at the target tissues (Li et al. 2021).

Targeted medication delivery and therapeutic applications benefit greatly from exceptional versatility and efficacy in fine-tuning the thermal properties of NPs and may find applications in delivering nucleic acids or other biomolecules in plants in a precise manner, thereby enhancing their delivery efficiency.

8.4.3 Mechanical

Mechanical characteristics of material include strength, hardness, brittleness, elasticity, plasticity, and rigidity. Strength, flexibility, and stability are the main mechanical characteristics of nanoparticles (Wu et al. 2020), which are crucial for the efficacy of NPs as a delivery tool (Hui et al. 2019). NPs can be synthesized or developed with such robust mechanical properties (Wu et al. 2020) to endure the severe circumstances of biological environments, including enzyme degradation and mechanical stress; it also ensures the integrity of their cargo throughout transportation. NPs' mechanical flexibility and their high surface-to-charge or surface-to-volume ratio facilitate their deformation and adaptation to the microenvironment of the target tissue, as well as effective functionalization with targeting ligands (Ling et al. 2014). This flexibility is especially useful for getting past intricate biological barriers and through cellular membranes. Furthermore, tailoring an nanoparticle's mechanical characteristics, including its stiffness or elasticity, might improve its effectiveness in controlled release applications and enable the accurate and continuous delivery of biomolecules. Overall, NPs' mechanical toughness and versatility increase their dependability and effectiveness as delivery systems in a range of medical settings and can provide opportunities to improve biomolecule delivery systems in plants (Table 8.1).

8.5 Applications

8.5.1 Nanoparticle-Based Plant Transformation

NP-mediated targeted gene delivery in plants is an innovative approach that overcomes the limitations of traditional methods by enabling the precise transfer of DNA or RNA, either for transient expression or stable integration into the host genome, thereby producing stable transgenic plants. This technique provides opportunities through various types of NPs to efficiently deliver genetic material into the target-specific plant cells, thereby addressing major challenges such as the rigid plant cell wall and species-specific limitations. Some of the common groups of NPs used for this purpose include mesoporous NPs, metal-based NPs, magnetic NPs, and CNTs. In the following section, we categorize the role of NPs in delivering nucleic acids to plant cells, in general, or in organelles such as chloroplasts or mitochondria, and their expression within the respective compartments.

8.5.1.1 Nuclear Transformation

The developments in the field of nanobiotechnology have introduced innovative approaches for the development of transgenic plants by using NPs as gene carriers, thereby contributing to sustainable agricultural practices. NPs of diverse NMs, shapes, and sizes have been synthesized and have been optimized for the delivery of biomolecules in plant cells.

Mesoporous silica nanoparticles (MSNs) have attracted much attention because of their unique structural features, such as preserving their thermal and chemical stability, solid framework with a large surface area, and porous structure. MSNs have been extensively investigated as a novel approach for gene delivery as they have very low toxicity. Silica particles are widely recognized for their inert properties and are highly biocompatible. The Food and Drug Administration (FDA), United States, has classified silica in the "Generally Recognized as Safe (GRAS)" substance category (Food Additives Status List; https://www.fda.gov/food/food-additives-petitions/food-additive-status-list). Moreover, another distinctive advantage is their ability to easily functionalize the MSN NPs, due to silanol groups on the surface of silica (Si-OH), making them effective and stable tools for diverse applications (Bharti et al. 2015). These kinds of surface modifications onto the MSNs help in loading biomolecules/drugs and enable site-specific targeting. The utilization of silica NPs can play an important role in the delivery of biomolecules to plants.

Table 8.1 List of various nanoparticles with different functional properties used for gene delivery in plants

Sr. No.	Nanoparticle	Surface modification	Size	Delivery method	Cargo	Target tissue/host plant	References
1.	CNTs	Arginine-functionalized SWCNTs	20 μg Arg-SWCNT complex	Co-culture	Plasmid DNA	Tobacco root cells (*Nicotiana tabacum var. Virginia*)	Golestanipour et al. (2018)
2.	DNA Nanostructures	3D tetrahedron, 1D hairpin-tile (HT) monomer, and a high aspect ratio 1D nano strings	2–10 nm	Energy-dependent or independent internalization by infiltration	siRNA	*Nicotiana benthamiana, Nicotiana tabacum, Eruca sativa*, and *Nasturtium officinale* plants	Zhang et al. (2019)
3.	SWCNT	Chitosan-SWCNT, PEGylated CSSWNT, carboxylated SWCNTs	127–275 nm	Injection	Plasmid DNA	Chloroplast of arugula (*Eruca sativa*), watercress (*Nasturtium officinale*), tobacco, spinach (*Spinacia oleracea*)	Kwak et al. (2019)
4.	CNTs	Polyethylenimine (PEI)-functionalized SWCNTs	20 nm	Injection	Plasmid DNA	Tobacco leaves (*Nicotiana benthamiana*), Arugula (*Eruca sativa*) Wheat (*Triticum aestivum*), and Cotton (*Gossypium hirsutum*) leaves	Demirer et al. (2019)
5.	Carbon dots	PEI-coated	4 nm	Infiltration	Plasmid DNA	Rice (*Oryza Sativa japonica*), wheat (*Triticum aestivum*), and mung bean (*Phaseolus radiatus*)	Wang et al. (2020a)
6.	MSNs	PEI-coated MSNs	100 ± 8.7 nm	Ultrasound treatment	Plasmid DNA	*Nicotiana tabacum*	Zolghadrnasab et al. (2021)
7.	Silver and gold microcarriers	No modification	100 and 600 nm, respectively	Biolistic method	Plasmid DNA	*Nicotiana tabacum*	Rajkumari et al. (2021)

(continued)

Table 8.1 (continued)

Sr. No.	Nanoparticle	Surface modification	Size	Delivery method	Cargo	Target tissue/host plant	References
8.	AuNPs	PEI and SiO_2 coated	4–5 nm	Infiltration	DNA	*Cannabis sativa*	Ahmed et al. (2021)
9.	Layered double hydroxide (LDH), clay NPs	MgAl layered	50 nm	Co-incubation/co-culture	dsRNA	Tomato (*Solanum lycopersicum*) pollen	Yong et al. (2021)
10.	AuNPs	PEI-coated	1–2 nm	Infiltration	siRNA	*Nicotiana benthamiana*	Zhang et al. (2021)
11.	Lipid nanoparticles	Cationic lipid	100 nm	Polyethylene glycol and lipofectamine-mediated CRISPR/Cas9 plasmid delivery	CRISPR/Cas9 plasmid	Sweet orange (*Citrus sinensis* (L.) Osbeck) protoplast	Mahmoud et al. (2022)
12.	Carbon (SWCNTs) NPs	PEI-coated and target peptide-functionalized	~220 nm	Biorecognition approach/co-incubation	pATV1 plasmid	*Arabidopsis thaliana*	Santana et al. (2022)
13.	AuNPs	PEI-coated	33–38 nm	Infiltration	AmiRNA (Artificial miRNA)	*Arabidopsis thaliana*	Zhang et al. (2023)
15.	Mesoporous NPs	Cetyltrimethylammonium bromide template and NH_4OH	31 nm	Infiltration	siRNA	*Nicotiana benthamiana*	Cai et al. (2024)

Torney and colleagues used MSNs for the first time to deliver biomolecules in plant cells. They observed transient expression of GFP in tobacco protoplasts 36 h after delivering a plasmid containing the green fluorescent protein (GFP) gene controlled by a constitutive promoter, using MSN NPs. For delivery into the plant tissues directly, MSN NPs were capped with the surface-functionalized gold NPs, and plasmid DNA was loaded onto it. Nucleic acid-loaded NPs were delivered to tobacco cotyledon tissue using a particle gun. Capping with AuNPs helped to increase the density of MSN NPs, which could help in their transport across the cell wall and also protect the nucleic acids loaded onto them (Torney et al. 2007). Furthermore, they also bombarded these DNA-loaded NPs on the embryos isolated from immature maize seeds and cultured them on the in vitro media. GFP-expressing cell clusters were observed within 10 days of transformation.

In various plant functional genomic studies, functionalized MSNs have been used as a facile delivery method, such as Chang and collaborators (2013), who have developed a transient gene expression system. They have used MSNs (TMAPS/F-MSNs) as carriers to deliver foreign DNA without the use of any mechanical force into *Arabidopsis thaliana* roots. In the epidermal root tissues, epidermal layer, and more inner cortical tissues, fluorescence and antibody labeling have been observed (Chang et al. 2013). Noteworthy, Hajiahmadi et al. demonstrated various routes through which tomato plants were transformed with MSN-pDNA (35S-cryIAb-GUS) containing GUS. These include infiltrating the shoot or abaxial surface of the leaves and spraying the solution on the abaxial surface of the leaves. Treatment of the lower surface of the leaves was observed as the significant approach. The study aimed to provide resistance against severe damage caused by *Tuta absoluta*. Accordingly, this was the first time when in vivo conditions MSN-mediated gene transient transformation was successfully performed (Hajiahmadi et al. 2019). Interestingly, the same research team has developed a stable and novel gene transformation system using MSNs under the control of maize proteinase inhibitors (MPI) as an inducible promoter, and the putative transgenic tomato plants have been investigated against *Tuta absoluta.* Based on their previous studies, they used aminopropyl triethoxysilane (APTES) to functionalize the MSNs and the best possible binding ratio of the pDNA (plasmid) to MSNs, i.e., 1:100 (μg/μL). Plant transformation with the pDNA-MSN complex has been done through injection/infiltration into the early ripening stage of tomato red fruit (600 μL per fruit). Furthermore, putative transgenic tomato plants have been confirmed in selected seedlings through PCR, RT-PCR, and western blot analysis up to T2 generation. DNA-loaded MSNs have also been used for plant transformation with the assistance of ultrasound treatment (Zolghadrnasab et al. 2021). It is suggested that ultrasound treatment can puncture the cell wall, through which biomolecule cargo can enter the cell; however, this method can cause severe damage to the cell walls and can compromise the cell viability too. In conclusion, the MSN-mediated stable transformation system presents a promising and viable alternative approach to conventional plant genetic transformation systems, combining efficiency, safety, and effectiveness in enhancing plant traits.

Metal NPs have garnered much attention by offering significant advances because of their unique optical properties, such as surface plasmon resonance (SPR), which enables them to control optical fields. Metal NPs have been derived from several materials, including lead, aluminum, silver, gold, iron, cobalt, zinc, cadmium, and copper (Chandrakala et al. 2022). However, gold and silver NPs are prominent due to their tunable size and shape, ease of synthesis, flexible surface modification, bio-conjugation capabilities, and tunable optical properties (Jat et al. 2020). Furthermore, their biocompatibility with biological molecules such as nucleic acids is superior, making them a potential gene delivery system. Though these are used frequently for biomolecule delivery in animal cells or micronutrients in plants (Singh et al. 2018), their use for DNA delivery in plants is not well established. Vijaykumar and colleagues used nanogold-loaded carbon bullets for DNA delivery in plant cells via a biolistic particle delivery system. Gold NPs provide a high surface area, which provides a remarkable loading capacity of up to 2.4 μg of plasmid DNA on carbon bullets embedded with AuNPs. Interestingly, in the absence of gold NPs, the DNA could not bind effectively to the pure carbon material, highlighting the crucial role of AuNPs in enhancing binding and delivery. The HTC 600-Au composite, loaded with plasmid DNA containing the GUS reporter expression cassette, was successfully delivered to tobacco and rice cells. AuNPs enhanced the usable surface area and also protected DNA from enzymatic degradation. Furthermore, sharp edges of the carbon matrix were suggested to facilitate the delivery of DNA into plant cells (Vijayakumar et al. 2010). Similarly, zinc sulfide nanoparticles (ZnS NPs) in the size range of 3–5 nm, functionalized with cationic poly-L-lysine (PLL), have also been explored as carriers for gene delivery into plant cells. The positive charge of PLL enhances the binding affinity of ZnS NPs to negatively charged nucleic acids, facilitating efficient gene transfer. Plasmid DNA-ZnS NP Complex was successfully transferred to tobacco cells with the help of ultrasound. Authors reported stable integration of the exogenous DNA in the host genome, and transgenic tobacco plants were raised successfully (Li et al. 2012).

It has been reported that Au-MNPs in combination with PEI (polyethyleneimine) show higher transfection efficiency. PEI is widely used as a gene transfection vector for animal cells. PEI carries plasmids into the cells and then releases them in accordance with the proton sponge effect (Thomas and Klibanov 2003). Recently, Zhang and colleagues used the PEI reduction method to synthesize AuNPs-PEI for delivering different foreign DNA segments, i.e., NLS-mCherry, GFP, and NPR1-GFP, into the leaves of *Arabidopsis* of periwinkle (*Catharanthus roseus*) by infiltration. Delivery of DNA in the plant cells and their expression was validated. While mCherry fluorescence was observed in the nucleus within 6 hours of infiltration, GFP expression was observed at 12 h of infiltration, confirming the successful AuNPs-mediated delivery of DNA into plant cells (Zhang et al. 2022). Rajkumari et al. (2021) compared the DNA delivery efficiency of gold and silver NPs in *Nicotiana* leaves via particle bombardment. Gold microcarriers (size 0.6 micron) and silver NPs (size 100 nm), loaded with plasmid DNA having the GUS gene, were bombarded under varying helium pressures (450 psi, 650 psi, 900 psi, 1100 psi). Delivery of the transgene was confirmed through molecular analysis and GUS histochemical assay. Silver NPs at a concentration of 10 mg/L showed significantly higher transformed efficiency compared to gold particles (Rajkumari et al. 2021).

Recently, AuNPs were used to deliver artificial microRNA (AmiRNA) in *Arabidopsis*. AmiRNA-ATG6 (autophagy-related proteins) were loaded onto the AuNPs and were infiltrated in the *Arabidopsis* and *Nicotiana benthamiana* leaves. Fluorescence was observed after 3 or 5 hours of filtration in *Arabidopsis* and *Nicotiana*, respectively. Furthermore, it was observed that AuNPs-AmiRNA-ATG6 effectively silenced the expression of ATG6 both in *Arabidopsis* and in *Nicotiana* and reduced the plant's resistance toward *Pseudomonas syringae* infection (Zhang et al. 2023).

MNP-mediated gene delivery, or magnetofection, is an ideal cutting-edge approach to transfer genetic material that involves the uptake of MNPs loaded with DNA into the target cells under magnetic forces. Heretofore, various kinds of MNPs have been synthesized, including pure metal NPs like Fe, Co, or Ni; metal oxides like Fe_3O_4, γ-Fe_2O_3; ferrites like $CuFe_2O_4$, $NiFe_2O_4$, and $MnFe_2O_4$; and metal alloys such as FePt and CoPt (Ansari et al. 2022). MNPs, especially iron oxides, are very specific in biocompatibility and paramagnetism that enable their magnetic properties only in the presence of external magnetic fields. MNPs having a positive charge on them can transport the genetic material without any external assistance (Jat et al. 2020). Hao and coworkers synthesized mGNPs of about 20–30 nm in diameter and functionalized them with PEG, which resulted in free amino groups on the surface of the NPs. Fluorescein isothiocyanate (FITC) DNA was attached to the surface of mGNPs through a covalent amide bond and was delivered in the canola protoplasts and intact cells too using an external magnetic field. Authors reported successful transfer of mGNPs across plant cell walls (Hao et al. 2013). Later, Zhao et al. (2017) highlighted the potential of "pollen magnetofection" for transforming various plant species via direct production of transgenic seeds without in vitro tissue culture regeneration. The diameter of the surface aperture of most crop pollen is in the range of 5–10 μm, where the wall of pollen is very thin. Pollen collected from cotton plants was magnetofected with DNA-loaded magnetic fields using an external magnetic field, and these magnetofected pollens were used for pollination. It could transfer the external DNA to the host genome, which resulted in stable integration, and transgenic plants were regenerated successfully. Seven different DNA expression vectors have been used to validate this study, including those genes with pest and herbicide-resistant genes. Integration of transgenes and expression has been observed, and subsequently, these were successfully inherited by the next generations (Zhao et al. 2017).

Later, Fe_3O_4 MNPs bound with enhanced green fluorescent protein gene (EGFP) or bialaphos resistance gene (Bar) (MNP-DNA complex) were delivered to maize pollens and florets across five different maize varieties. Though in both cases transformation was observed, transfection efficiency through pollen was much higher compared to immature embryos or florets. Both eGFP and Bar genes were stably expressed and inherited by progenies through genotype-independent pollen transfection (Wang et al. 2022; Wang et al. 2020b). Similarly, magnetofection-mediated DNA delivery in pollen and rising transgenic plants has been extended to other plants successfully. While Farooq and colleagues transformed Okra via magnetofection using green iron NPs, Park and colleagues have reported the

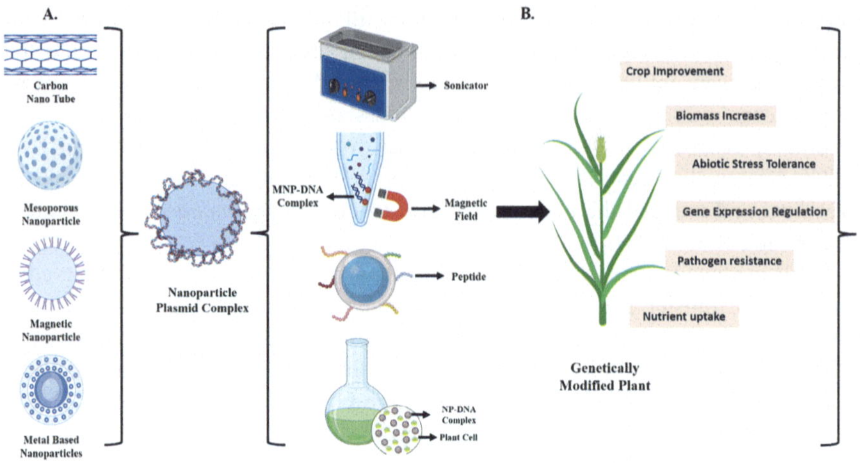

Fig. 8.2 The diagram represents (**a**) different types of nanoparticles. (**b**) Delivery methods of nanoparticle-DNA complex into the plants and their applications. MNP Magnetic nanoparticles, NP Nanoparticles

generation of genetically modified *Cucumis sativus* plants through magnetofection (Farooq et al. 2022; Park et al. 2024). This is specifically important as it bypasses the in vitro tissue culture and regeneration and does not depend on the genotype of the host plant. Therefore, it can address the requirement of optimized regeneration protocols for genetic engineering-mediated genome engineering.

Carbon NP-mediated biomolecule delivery has emerged as a promising tool for the delivery of nucleic acids or other biomolecules in plants. Carbon-based NPs (CNPs) have versatile features such as surface morphology, stable molecular architecture, and increased chemical reactivity that make them valuable materials in engineering and chemical sciences. CNPs are classified based on their geometrical structure (tube-shaped, horn-shaped, spherical, or ellipsoidal) (Chandrasekaran et al. 2020).

CNPs like carbon dots can be coated or modified with polyethyleneimine (PEI), which is a cationic polymer used to bind negatively charged DNA. These modified NPs are called amine-functionalized carbon dots (A-CDs), which are positively charged at pH 5.7. Because of their small size, i.e., <10 nm, these NPs are capable of crossing plant cell walls, and in the cell, an amine group releases protons, which help in DNA release in a weakly alkaline environment. Biomimicking amine-functionalized carbon NPs-GFP complex was used to transfect model plant tobacco leaves by injection, and expression of genes was observed under a confocal microscope; green fluorescence was observed, which indicates successful delivery of genes into plants (Liu et al. 2023). In the recent past, several reports have been published on carbon nanotube (CNT)-mediated DNA delivery in plants (Fig. 8.2). CNTs are rolled-up sheets of single-layer carbon atoms, and depending upon the number of layers, these are classified as single walled CNT (SWCNT) or MWCNTs. CNTs have been shown to infiltrate through plant cell walls and deliver their cargo into the cell successfully (Liu and Xing 2009). Furthermore, it is suggested that cellular uptake of the CNTs may occur through endocytosis. Incubation time and temperature at which cells were exposed to cargo-loaded CNTs had a direct impact on the CNT uptake efficiency of cells. Longer incubation and room temperature were found ideal for achieving maximum transfection efficiency. In order to enhance the CNT transport efficiency across the plant cell wallGiraldo et al. (2014) have examined the impact of CNT surface patterns as well as the charge on cell membrane penetration. This study concluded that chitosan coating helped in the passive passage of NPs across the chloroplast membrane, whereas CNTs coated with lipids or polyvinyl alcohol were unable to penetrate the chloroplast membranes. The authors suggested that glycerolipids on the chloroplast membrane wrap around the CNTs at the time of their interaction with the membrane, helping their passage through the membrane (Giraldo et al. 2014).

Furthermore, to strengthen pDNA loading on CNTs and improve their delivery in the chloroplast, Kwak and coworkers used four types of chitosan-complexed SWCNTs, using both non-covalent covering and covalent refinement of the nanotube sidewalls. These SWCNTs were used to deliver DNA in the chloroplasts in *Eruca sativa, Nasturtium officinale, Nicotiana tabacum*, and *Spinacia oleracea* plants. Delivery of the DNA in the cells was confirmed via the expression of the marker genes (Kwak et al. 2019). Burlaka and colleagues loaded cellulase enzymes on the CNTs along with the cargo to be transported. When these cellulase-loaded CNTs were incubated with plant cells, they digested the cell wall and created nanoholes on the wall. Delivery of the cargo was monitored using GFP protein, confirming successful delivery in the plant cells. While both SWCNTs and MWCNTs could successfully deliver DNA in the protoplasts, only SWCNTs could efficiently pass through the plant cell wall and deliver DNA in the intact plant cells. The plant cell wall-penetrating capacity of MWCNTs was very low (Burlaka et al. 2015).

Loading of DNA cargo on the CNT is an important factor and can directly impact the transformation efficiency. Demirer and colleagues compared two different approaches to load plasmid DNA on CNT and study their carrying capacity. They coated DNA on the CNTs either through the dialysis method or through electrostatic interactions between the positive charge on the CNT and the negative charge on the DNA. It was observed that DNA loading capacity on CNTs through electrostatic interactions was significantly higher than the dialysis method. However, both of them could deliver DNA to the plant cells. The capability of CNT to deliver DNA was tested in several plants, and successful delivery was reported in the leaves of plants belonging to monocots as well as dicots. Confirmation of the delivery of the DNA in the plant cells was confirmed through RT-PCR of the reporter gene expression. Authors also reported that CNTs up to 20 nm in size could deliver DNA in the plant cells (Demirer et al. 2019). In another study, PEI bounded CNTs to transfer ssDNA-FITC to *German chamomile* cells, and the ability of NPs to protect DNA against DNA-digesting enzymes was also analyzed. Scanning electron microscopy verified the shape and actual size of pUC-18-bound CNTs. A bath sonicator or ultrasonic sound was used to assist the CNT delivery in the cells. Expression of the transgene was confirmed by fluorescence microscopy. Furthermore, it was concluded that CNTs protected DNA cargo from nuclease degradation as well as damage from ultrasonic waves, and assistance from ultrasonic waves enhanced the gene delivery by creating a cavity in the cell wall to enter nanoparticles (Ghaghelestany et al. 2020). Overall use of CNTs as DNA delivery vehicles has great potential and can have a significant impact on improving transformation efficiency in plants.

8.5.1.2 Nanoparticle-Mediated Organelle-Targeted Gene Delivery

Chloroplasts and mitochondria have their own genomes that regulate several essential functions in plants. In addition, chloroplast is a good target for plant molecular farming because several copies of that same gene can be expressed in the same cell, as each cell contains many chloroplasts and each chloroplast carries several genomes (Sharma and Sharma 2009). Therefore, these are targets for genome engineering, and optimized protocols for biomolecule delivery in the organelle are required.

Mitochondria is an essential organelle for metabolites, with applications in therapeutics and energy production, and has a significant influence on crop traits, which could be further enhanced by exogenous gene insertion into its genome. Despite the significant advancement of genome engineering tools in plants, biomolecule delivery in plant mitochondria remains challenging due to the double membrane and the small size of the organelle (Daniell and Chase 2004). Furthermore, these have a pore size of about 5–20 nm, and the inner mitochondrial membrane has very low permeability for hydrophilic molecules (Schwab et al. 2016). In one of the earliest studies reported, mitochondria were extracted from the wheat embryo, and foreign DNA was introduced through electroporation. The *coxII* gene under the control of a mitochondrial promoter was expressed successfully, and post-transcriptional processing was validated (Farre 2001). Earlier, transformation of mitochondria has been reported in *Chlamydomonas* and *yeast* through particle bombardment (Fox et al. 1988; Randolph-Anderson et al. 1993). However, conventional methods such as electroporation and biolistics are not very successful in mitochondrial transformation because of very low transformant recovery (Chuah et al. 2015). Yoshinumi and colleagues used peptide-based DNA carriers to deliver the exogenous genes into the mitochondrial genome of cotyledons and apical meristem tissues of *Arabidopsis* (Yoshizumi et al. 2018). A Cytcox-KH peptide from *Saccharomyces cerevisiae* having a mitochondrial targeting signal was fused with candidate DNA and used for delivery of DNA in mitochondria in *Arabidopsis* and tobacco. The mitochondria insertion vector contained a gene encoding for *GFP* and spectinomycin resistance under the control of the Cox2 (mitochondria-specific promoter) promoter and terminator so that transcription of genes is activated only in mitochondria, not in the nucleus. The size of the peptide-DNA complex was ~356 nm with a −25 mV surface charge. *Arabidopsis* 2-day-old seedling leaves were used for DNA-peptide complex infiltration by the syringe-based vacuum system. Samples were incubated for 2 days, and then genomic DNA was isolated for PCR analysis, in which 9 out of 10 samples showed positive amplification. A reporter gene assay was also performed for GFP expression. These results show the potential of peptide-based targeting for organelle transformation. Recently, Cytox-KH peptide was used to guide CNTs for nucleic acid delivery in mitochondria (Law et al. 2022). SWCNTs were crosslinked with the polymer and fused with Cytox-KH through cysteine residue, thereby creating the SWCNT-polymer-Cytox-KH complex. Plasmid DNA containing the GFP gene under the control of a mitochondrial-specific promoter was loaded on the CNT complex and was used for plant transformation. It could successfully deliver the exogenous DNA to *Arabidopsis* seedlings. Expression of the exogenous DNA was confirmed through western blotting as well as reporter protein analysis.

As have been used for mitochondria, peptide-based nanocomplexes can facilitate rapid and efficient delivery of genetic material into the chloroplasts too. However, the synthesis of peptides and corresponding biomacromolecules is a major challenge, and that heavily influences the production cost of this technology. NPs can provide a good alternative for delivering biomolecules in chloroplasts too. Santana et al. (2022) developed biocompatible NPs and used CDs and SWCNTs for the delivery of chemicals or foreign DNA, respectively, in chloroplasts. Chloroplast-targeting peptides were attached to CDs or SWCNTs, and agrochemicals or plasmid DNAs were loaded onto CDs and SWCNTs, respectively. This resulted in significant improvement in biomolecule delivery in plant cell organelles. Plants treated with these NMs had a lower percentage of dead cells, and also no permanent cell and chloroplast damage was observed (Santana et al. 2022). However, the mechanisms behind the successful translocation of these NPs across plant cell walls and membranes remain poorly understood.

In another study, Kwak and collaborators (2019) reported the utilization of lipid exchange envelope and penetration (LEEP) models to study NP-mediated delivery in chloroplasts. Various types of NPs, like AuNPs, streptavidin-quantum dots, or different types of polymer-wrapped SWCNTs, were used in this study. As the size of the NPs is crucial for delivery across the membrane or cell wall, the LEEP model was utilized to calculate the size of the NP for effective delivery. SWCNTs were surface functionalized with chitosan or lipid-PEG and used for chloroplast transformation of several plants. YFP transient expression was successfully observed in the chloroplast of all plant species after 3 days of infiltration (Kwak et al. 2019). It has been reported that the translocation of SWCNTs across plant cells and chloroplasts is accomplished by the mechanism of LEEP (Wong et al. 2016). The currently established protocols used for DNA delivery to chloroplasts use a microcarrier approach that is universal across the plant and algal species. Recently, a study by Newkrick and collaborators reported that carboxylated single-walled carbon nanotubes (COOH-SWCNTs) coated with DNA (DNA–PEI) covalently bonded to fluorophore successfully delivered the cargo into algae cells and chloroplasts. Fluorophore attached to DNA cargo has been visualized with high spatial resolution confocal microscopy imaging. Furthermore, it was concluded that fluorophores have been effectively expressed in the outer membrane of algae and co-localized with DNA-SWCNT in the parts of chloroplast (Newkirk et al. 2023). Thus, it can be concluded that

PEI-conjugated SWCNTs-mediated gene delivery systems have the potential to become a novel tool for gene transfer into the chloroplast.

Though NPs have been shown to pass through the plant cell wall and deliver the biomolecule cargo in the cell, the plant cell wall is a major hurdle to genome engineering in plants. Furthermore, cell wall permeability also depends upon the shape, size, and other properties of the nanocarriers. Various strategies are being explored to pass the cell wall barriers. Burlaka and colleagues coated cellulase enzyme on the NPs to digest the cell wall locally and make nanoholes. It was observed that the use of cellulase significantly improved the CNT-mediated transformation efficiency of plant cells (Burlaka et al. 2015). In another study, the cell wall was exposed to the zwitterionic liquid (ZIL) that works as a cellulose-dissociating agent and disrupts the cellulose network partially by cleaving the hydrogen bonds. It results in the loosening of the cell wall and provides an opportunity for nanocarriers of diverse shapes and sizes to pass through the cell wall. Miyamoto and colleagues used ZIL to loosen the cell wall in *Arabidopsis* seedling tissues and could deliver peptide-DNA complexes in the cells. Chloroplast transformation efficiency was significantly enhanced, thereby suggesting the cellulose-dissociating agent can improve NP-mediated biomolecule delivery across plant cell walls (Miyamoto et al. 2022).

8.5.2 Nanoparticle-Based Gene Editing

As discussed in the previous section, diverse types of NPs have been developed for the delivery of nucleic acids in plants, and NP-mediated gene delivery methods are being optimized for generating transgenic plants (Jat et al. 2021; Wu et al. 2023a). Programmable nucleases are being directly used for editing the genomes successfully in animal or microbial cells; however, because of the presence of cell walls, it is difficult to deliver them directly in plant cells. Though these have been delivered directly in the naked plant cells successfully, i.e., in protoplasts, regeneration of edited plants from protoplasts is a challenge. Therefore, genes encoding nucleases are generally integrated into the nuclear genome of plant cells, and transgenic plants expressing the nucleases as well as gRNA are generated. Later, the T-DNA cassette is removed through segregation while growing the next generation. However, this is beneficial only in the case of plants that can be propagated sexually. In the plants that are propagated asexually, T-DNA cannot be removed through segregation methods.

Several proof-of-concept studies have been published that report that NPs can be used to deliver DNA or other biomolecule cargo across the cell wall in plant cells without the use of any external stimuli like biolistics (Cunningham et al. 2018). NPs such as nanotubes or mesoporous silica NPs, which have high cargo-carrying capabilities, have opened up new opportunities for the delivery of programmable nucleases and other required components directly in the plant cells. The use of pre-assembled Cas9-gRNA ribonucleoproteins rules out the insertion of recombinant DNA in the host genome. RNA-guided nucleases can edit the target locations on the chromosomes and are also cleared from the system by endogenous proteases. Similarly, Cas9 RNP cargo can be attached to the nanoparticle surface through covalent interactions and can be cleaved to release at the target locations. Enzymatically cleavable linkers can be introduced in the cargo for their attachment to the NP surface. Martin-Ortigosa and coworkers used MSNs functionalized with gold NPs to deliver protein and nucleic acids together in the plant cells and successfully confirmed the delivery of the protein component and expression of the gene in the plant cells (Martin-Ortigosa et al. 2012). Later, they transferred the "Cre recombinase enzyme" to transgenic maize cells that had a DNA fragment flanked by loxP recognition sites. Cre enzyme was successfully delivered in the maize cells, with a reported recombination frequency of about 20% (Martin-Ortigosa et al. 2012). Preassembled Cas9-nucleases and gRNA were also transfected into the protoplasts of various plants, including Arabidopsis, tobacco, and rice. This resulted in targeted genome modifications that were stably maintained in plants that were regenerated from these transfected protoplasts and were successfully transmitted to germline cells. Similarly, preassembled Cas9-gRNA was transfected directly into zygotes through PEG-Ca^{2+} mediated transfection, or into embryonic cells of maize and wheat through biolistics, and plants with targeted mutations were successfully regenerated (Toda et al. 2019). However, the percent success was very low when it was directly delivered to the embryonic cells. As Cas9-gRNA is directly delivered in the plant cells, it does not involve the integration of recombinant DNA in the host genome and its expression. Therefore, neither codon optimization for Cas9 nor the use of specific regulatory sequences, such as species-specific promoters, would be required.

NPs like carbon nanotubes or MSNs could be a potential candidate for delivery of preassembled complexes in plant cells, as these have very high cargo-carrying capacity and cargo is also protected. Zhang et al. (2020) used a DNA nanostructure-based biomolecule delivery system in plants. DNA nanostructures of various sizes, shapes, stiffness, or compactness were created, and nucleic acid and protein cargo were conjugated. These DNA nanostructures, along with the cargo, are successfully internalized into the mature plant cells across the cell wall. Furthermore, high DNA nanostructures or stiffer single-walled carbon nanotubes (SWCNT) were constructed and tested for their ability to internalize into the plant cells. Despite the larger size of the nanostring-SWCNT conjugates, they could successfully internalize into the mature plant cells, thereby suggesting that these can deliver preassembled complexes of programmable nucleases along with other required components (Zhang et al. 2020). These DNA nanostructures and their conjugates were used to deliver siRNAs specific to the GFP gene. siRNA delivered through this method could successfully silence the GFP expression in tobacco leaves.

Demirer and colleagues developed a carbon nanotube NP-mediated delivery system for plasmid DNA in intact plant cells. It will allow the expression from the Cas9 coding plasmid DNA without its integration into the host genome (Demirer et al. 2019). SWCNTs with specialized surface chemistries for DNA delivery were used to deliver nuclease-encoding plasmids in plant cells. It resulted in transient expression of the Cas9 nucleases that caused permanent mutations in the host genome. It not only enables genome editing without the stable expression of the nuclease from the host genome but also significantly reduces off-target editing and toxicity (Zhang et al. 2016). Similarly, the capabilities of several different types of NPs, like gold NPs, nanocapsules, DNA nanoclews, hydrogels, etc., to facilitate the delivery of preassembled gene editing complexes have been studied. Quantum dots, metal, or silica NPs have been shown to directly enter the plant cells and therefore can be used to deliver gene-editing cargo in the plant cells for the production of transgene-free crops (Vats et al. 2022). Doyle et al. (2019) used carbon dots to deliver plasmid DNA into the plant cells. When plasmid-coated carbon dots were sprayed on the leaves, Cas9 and gRANs were synthesized in the plant cells that successfully edited the target genes. However, nanoparticle uptake and their transport in the plants may vary among different tissues, and therefore, optimization of the NP delivery in a species-specific manner may be required (Doyle et al. 2019).

8.5.3 Nano-Based Plant Regeneration

In vitro, plant tissue culture that involves dedifferentiation of explant tissues to form a callus and molecular reprogramming of these dedifferentiated cells to regenerate complete plants is very important for genetic engineering-mediated improvement of plants. Therefore, the availability of an optimized regeneration protocol for a plant species is essential for biotechnological interventions in that species. However, many crops of commercial importance, like cereals, millets, medicinal plants, etc., are recalcitrant to regeneration, and this limits the applications of genetic engineering in these species. In vitro regeneration of complete plants depends upon several factors like nutrition components in the media; type, combination, and concentration of various phytohormones; quality and duration of light; humidity, and temperature.

Along with their role in various aspects of plant growth and development, nanomaterials are being investigated for their role in in vitro plant regeneration. Several reports have been published that show a positive nano-enabled outcome; for instance, Chutipaijit and Sutjaritvorakul (2017) evaluated the effect of zinc oxide (ZnO) and titanium oxide (TiO_2) on in vitro rice regeneration. Various concentrations of these NPs were supplemented in the regeneration media, and it was observed that their use significantly improved regeneration from rice explants. ZnO at a concentration of 5 mg/L and TiO_2 at a concentration of 20 mg/L showed maximum callus induction and significant improvement over control treatments, i.e., without NPs. It was also observed that the impact of TiO_2 NPs on regeneration was greater than the ZnO NPs. Use of TiO_2 NPs at a concentration of 20 mg/L produced the highest regeneration frequency, i.e., 67%. Similarly, the application of ZnO NPs during in vitro cultures improved organogenesis frequency in Agave (Canales-Mendoza et al. 2023). In another study, the impact of TiO_2 NPs on in vitro regeneration was evaluated in indica rice cultivars, i.e., Suphanburi1 and Suphanburi90. Various concentrations of TiO_2 NPs ranging from 10 to 50 mg/L were used in the media. Though the callus induction was maximum when media was supplemented with 50 mg/L of NPs, the regeneration frequency as well as the number of plantlets produced per callus was highest when media was supplemented with 40 mg/L of TiO_2 NPs (Chutipaijit and Sutjaritvorakul 2018).

Silver NPs have also been frequently evaluated for their effect on in vitro plant regeneration, and significant improvement has been reported. Hoai and colleagues have explored the effect of AgNPs on the indirect embryogenesis and maturation of somatic embryos. In this study, in vitro root explants and ex vitro leaves of purple passion fruit (*Passiflora edulis Sims f. edulis*) were observed under the impact of auxins, cytokinins, and AgNPs. It was observed that BAP associated with 1.5 mg/L AgNPs had a higher regeneration response of the explants. In addition, indirect somatic embryos were observed from leaf explants via callus formation, which imprints a positive effect of AgNPs. Moreover, with the addition of 3.0 mg/L AgNPs, significant proliferation, and somatic embryo maturation have been observed from the thin cell layer internode explant. The results revealed a reliable plant regeneration process through direct and indirect SEs from leaf explants that can be applied for the commercial propagation of purple passion fruit. From these results, it can be concluded that AgNPs can be used in the regeneration of purple passion fruit that can enhance commercial propagation of the same (Hoai Phong et al. 2023). Decontamination of the explants for in vitro culture is also crucial. NP can play an important role in tissue sterilization. In a similar study, AgNPs have been used for explant surface sterilization and in vitro growth and acclimatization of *Limonium sinuatum.* During the shoot multiplication and rooting stages of plantlets, endogenous hormones modulating the effect of AgNPs were recorded. In this comparative study, the results showed that explants treated with 200 mg/L AgNPs for 20 min not only helped in the disinfection process but also significantly improved shoot induction compared to explants treated with 1000 mg/L $HgCl_2$ for 5 min. Noteworthy, 0.4 mg/L AgNPs-derived plantlets showed better acclimatization compared to the control plantlets grown in greenhouse conditions. These results highlight the perspective of nanoparticles in enhancing plant growth, regeneration, and resilience (Cuong et al. 2023). Similarly, the application of AgNPs enhanced the shoot proliferation in *Gaillardia pulchella* as well as the health of the shoots. Shoots produced on the media supplemented with NPs were sturdy, healthy, and accumulated a higher amount of green pigment compared to the control plants. Furthermore, AgNPs promoted root induction and helped in the acclimatization of plantlets (Manokari et al. 2023). Similarly, Andujar and coworkers used AgNPs successfully for the enhancement of dipentene production in Stevia. Though AgNPs did not have any impact on shoot proliferation up to 12 mg/L of NPs, shoot multiplication was decreased when media was supplemented with >12 mg/L of AgNPs. Silver NPs may enhance antioxidant activity and can protect plants from oxidative damage. Antioxidant properties of AgNPs were observed in boysenberry plants under PEG-mediated drought stress, which reduces the growth of roots and shoots. It was observed that shoot and root growth in media supplemented with 4% and 8% PEG and AgNPs was indistinguishable from that of control plants grown on PEG- and AgNP-free media. The results indicated that the addition of AgNPs enhanced antioxidant activity in boysenberry plants and improved shoot production under drought stress (Şener and Sayğı 2023). Furthermore, NM-based NCs may help in the controlled release of the various regulators in the media, potentially regulating tissue proliferation in vitro. Overall, we are beginning to understand the role of NPs as media supplements for in vitro plant tissue culture, which may play a key role in optimizing efficient protocols for callus proliferation, organogenesis, somatic embryogenesis, or shoot regeneration in commercial and recalcitrant crops.

8.6 Major Challenges

The introduction of NPs as a vehicle for biomolecule delivery across the cell wall in plant cells is changing the landscape of plant genetic engineering strategies. NPs hold enormous potential for DNA delivery; however, to use them to their full potential, we need to work on how NPs interact with biological systems. Very little is known about how NPs internalize through the cell wall? and how do NP properties like size, shape, tensile strength, compactness, etc., affect the internalization process? Whether NP-DNA complex passage through the cell wall requires energy or can it pass via simple diffusion? The selection of the dose of the NPs is another area that directly impacts NP interaction with plant cells. If the concentration of the NPs or the material used for functionalization is low, it may result in ineffective functionalization or weak interactions between the NPs and the DNA. Therefore, optimization of the NP dose is crucial, and it may change with the type of NM and the size or shape of the NPs (Vats et al. 2022).

Nucleic acid loading on the NP surface and their release inside the plant cells has a direct effect on the transformation efficiency. NP surface can be customized for charge and chemical composition and therefore can be customized to bind biomolecules of diverse shapes and sizes. Therefore, it is important to understand the nature of the materials, i.e., composition, surface charge, etc., used for NP surface functionalization and their interaction with the biomolecules. Most of the nanoparticle-mediated plant transformations have been used for transient gene expression, but for genetic engineering stable integration is required (Wu et al. 2023b). DNA release from NPs in the plant cell and its subsequent transfer to the nucleus is a complex process. As different biomolecules do not bind with the NP surface with equal affinity, their release may not be with the same efficiency in the cell. Therefore, the efficiency of the same NPs to deliver the cargo may differ depending upon the nature of the cargo. The size range of the exogenous DNA that can be delivered using NPs is not known (Lv et al. 2020). Furthermore, the mechanism of how DNA moves to the nucleus and integrates into the host genome is not clear.

The stability of NP colloids directly depends upon the ionic strength of the buffer, osmoticum, and other parameters of plant cell culture buffers. The compatibility of plant cell culture buffers with individually dispersed and colloidal stable forms of NPs is very low, and hence NP activity is affected. Therefore, to optimize reliable conditions that can maintain NP stability, they should be explored. Although

NP-mediated gene delivery opens the gate for very innovative transformations, the efficiency of NP-mediated gene delivery to plants is still relatively low compared to other biological systems. Therefore, modifications in the tools and optimization procedures should be explored for successful nano-based gene delivery to plants.

8.7 Conclusion

Plant genetic engineering-mediated crop improvement involves the development of high-yielding, nutrient, and stress-tolerant crop varieties and is directly dependent upon the capability to deliver molecules like DNA, RNA, and proteins in the plant cells. NP-mediated delivery of biomolecules to living cells has emerged as a promising tool and can play an important role in the development of designer crops. Though NP-mediated DNA delivery in plant cells is still in its infancy due to the presence of rigid cell walls, several proof-of-concept studies published in the last decade show that NPs can carry DNA cargo across plant cell walls. Several innovative strategies, like the use of cellulase or cell wall-losing chemicals with NPs, have been designed for improving the NP transport across cell walls and are expected to improve the transformation procedures. The capability of the MSN or CNTs to deliver CRISPR/Cas9-mediated gene editing components directly into the intact plant cells can provide breakthrough opportunities for plant genome engineering without the integration of CRISPR/Cas9 expression cassettes in the host genome. However, several technical challenges need to be addressed. To realize the full potential of NPs for delivering biomolecules in plants, a thorough understanding of the NP interactions with the cell wall, the release of the cargo in the cell, and the impact of NPs on the cellular process is required. Overall NP-mediated gene delivery has enormous potential and can have an effect on revolutionizing the agriculture sector in an effective, robust, and sustainable manner.

References

Ahmar S, Gill RA, Jung K-H, Faheem A, Qasim MU, Mubeen M, Zhou W (2020) Conventional and molecular techniques from simple breeding to speed breeding in crop plants: recent advances and future outlook. Int J Mol Sci 21(7):2590. https://doi.org/10.3390/ijms21072590

Ahmed S, Gao X, Jahan MA, Adams M, Wu N, Kovinich N (2021) Nanoparticle-based genetic transformation of Cannabis sativa. J Biotechnol 326:48–51. https://doi.org/10.1016/j.jbiotec.2020.12.014

Alshammari BH, Lashin MMA, Mahmood MA, Al-Mubaddel FS, Ilyas N, Rahman N, Sohail M, Khan A, Abdullaev SS, Khan R (2023) Organic and inorganic nanomaterials: fabrication, properties and applications. RSC Adv 13(20):13735–13785. https://doi.org/10.1039/D3RA01421E

Altpeter F, Baisakh N, Beachy R, Bock R, Capell T, Christou P, Daniell H, Datta K, Datta S, Dix PJ, Fauquet C, Huang N, Kohli A, Mooibroek H, Nicholson L, Nguyen TT, Nugent G, Raemakers K, Romano A, Somers DA, Stoger E, Taylor N, Visser R (2005) Particle bombardment and the genetic enhancement of crops: myths and realities. Mol Breed 15:305–327. https://doi.org/10.1007/s11032-004-8001-y

Ambreen T, Kim MH (2020) Influence of particle size on the effective thermal conductivity of nanofluids: a critical review. Appl Energy 264:114684. https://doi.org/10.1016/j.apenergy.2020.114684

Amreddy N, Babu A, Muralidharan R, Munshi A, Ramesh R (2017) Polymeric nanoparticle-mediated gene delivery for lung cancer treatment. Top Curr Chem 375(2):35. https://doi.org/10.1007/s41061-017-0128-5

Andrade Â, Ferreira R, Fabris J, Domingues R (2011) Coating nanomagnetic particles for biomedical applications. Biomedical Engineering-Frontiers and Challenges, PP157-176. https://doi.org/10.5772/19519

Ansari MJ, Kadhim MM, Hussein BA, Lafta HA, Kianfar E (2022) Synthesis and stability of magnetic nanoparticles. BioNanoScience 12:627–638. https://doi.org/10.1007/s12668-022-00947-5

Báez-Cruz R, Baptista LA, Ntim S, Manidurai P, Espinoza S, Ramanan C, Cortes-Huerto R, Sulpizi M (2021) Role of pH in the synthesis and growth of gold nanoparticles using L-asparagine: a combined experimental and simulation study. J Phys Condens Matter 33(25):254005. https://doi.org/10.1088/1361-648X/abf6e3

Begum SLR, Jayawardana NU (2021) A review of nanotechnology as a novel method of gene transfer in plants. J Agric Sci-Sri Lanka 16(2):300–316. https://doi.org/10.4038/jas.v16i2.9336

Behrooz, Darbani S, Farajnia M, Toorchi S, Zakerbosta S, Noeparvar C, Neal S (2008) DNA-Delivery Methods to Produce Transgenic Plants Biotechnology (Faisalabad) 7(3):385–402. https://doi.org/10.3923/biotech.2008.385.402

Berry CC (2009) Progress in functionalization of magnetic nanoparticles for applications in biomedicine. J Phys D Appl Phys 42(22):224003. https://doi.org/10.1088/0022-3727/42/22/224003

Bhardwaj P, Singh B, Behera SP (2021) Green approaches for nanoparticle synthesis: emerging trends. In: Nanomaterials. Elsevier, pp 167–193. https://doi.org/10.1016/B978-0-12-822401-4.00015-5

Bharti C, Gulati N, Nagaich U, Pal A (2015) Mesoporous silica nanoparticles in target drug delivery system: a review. Int J Pharm Investig 5(3):124. https://doi.org/10.4103/2230-973x.160844

Burlaka OM, Pirko YV, Yemets AI, Blume YB (2015) Application of carbon nanotubes for plant genetic transformation, Tsitol Genet, pp 233–255. PMID: 26841488

Cai Y, Liu Z, Wang H, Meng H, Cao Y (2024) Mesoporous silica nanoparticles mediate SiRNA delivery for long-term multi-gene silencing in intact plants. Adv Sci 11(9). https://doi.org/10.1002/advs.202301358

Caizer (2016) Nanoparticle size effect on some magnetic properties. In: Aliofkhazraei, M.(eds) Handbook of Nanoparticles Nanoparticle. Springer, Cham 475–519. https://doi.org/10.1007/978-3-319-15338-4_24

Canales-Mendoza AI, Villanueva-Ibáñez M, Tovar-Jiménez X, Álvarez-Cervantes J (2023) Effect of the addition of fungal extracellular biogenic zinc oxide nanoparticles on the in vitro multiplication of Agave salmiana shoots. Plant Cell Tissue Organ Cult 155(2):479–491. https://doi.org/10.1007/s11240-023-02589-1

Chandrakala V, Aruna V, Angajala G (2022) Review on metal nanoparticles as nanocarriers: current challenges and perspectives in drug delivery systems. Emerg Mater 5:1593–1615. https://doi.org/10.1007/s42247-021-00335-x

Chandrasekaran R, Rajiv P, Abd-Elsalam KA (2020) Carbon nanotubes: plant gene delivery and genome editing. In: Carbon nanomaterials for agri-food and environmental applications. Elsevier, pp 279–296

Chang FP, Kuang LY, Huang CA, Jane WN, Hung Y, Hsing YIC, Mou CY (2013) A simple plant gene delivery system using mesoporous silica nanoparticles as carriers. J Mater Chem B 1(39): 5279–5287. https://doi.org/10.1039/c3tb20529k

Cheng M, Lowe BA, Spencer TM, Ye X, Armstrong CL (2004) Invited review: factors influencing agrobacterium-mediated transformation of monocotyledonous species. In Vitro Cell Dev Biol Plant 40:31–45. https://doi.org/10.1079/IVP2003501

Chuah JA, Yoshizumi T, Kodama Y, Numata K (2015) Gene introduction into the mitochondria of Arabidopsis thaliana via peptide-based carriers. Sci Rep 5. https://doi.org/10.1038/srep07751

Chutipaijit S, Sutjaritvorakul T (2018) Titanium dioxide (TIO 2) nanoparticles induced callus induction and plant regeneration of Indica rice cultivars (Suphanburi1 and Suphanburi90). Digest journal of nanomaterials and biostructures. 13(4): 1003–1010

Chutipaijit S, Sutjaritvorakul T (2017) Application of nanomaterials in plant regeneration of rice (Oryza sativa L.). In: Materials today: proceedings. Elsevier, pp 6140–6145. https://doi.org/10.1016/j.matpr.2017.06.107

Colilla M, Baeza A, Vallet-Regí M (2015) Mesoporous silica nanoparticles for drug delivery and controlled release applications. In: The sol-gel handbook. Wiley Blackwell, pp 1309–1344. https://doi.org/10.1002/9783527670819.ch42

Cunningham FJ, Goh NS, Demirer GS, Matos JL, Landry MP (2018) Nanoparticle-mediated delivery towards advancing plant genetic engineering. Trends Biotechnol 36:882–897. https://doi.org/10.1016/j.tibtech.2018.03.009

Cuong DM, Mai NTN, Tung HT, Khai HD, Luan VQ, Phong TH, Van The VB, Phuong HTN, Van Binh N, Tan Nhut D (2023) Positive effect of silver nanoparticles in micropropagation of Limonium sinuatum (L.) Mill. 'White'. Plant Cell Tissue Organ Cult 155(2):417–432. https://doi.org/10.1007/s11240-023-02488-5

Daniell H, Chase C (2004) Molecular biology and biotechnology of plant organelles, Chloroplasts and Mitochondria. Springer Nature. https://doi.org/10.1007/978-1-4020-3166-3

Demirer GS, Zhang H, Goh NS, Pinals RL, Chang R, Landry MP (2020) Carbon nanocarriers deliver siRNA to intact plant cells for efficient gene knockdown. Sci Adv 6(26):eaaz0495. https://doi.org/10.1126/sciadv.aaz0495

Demirer GS, Zhang H, Matos JL, Goh NS, Cunningham FJ, Sung Y, Chang R, Aditham AJ, Chio L, Cho MJ, Staskawicz B, Landry MP (2019) High aspect ratio nanomaterials enable delivery of functional genetic material without DNA integration in mature plants. Nat Nanotechnol 14(5):456–464. https://doi.org/10.1038/s41565-019-0382-5

Ding Y, Jiang Z, Saha K, Kim CS, Kim ST, Landis RF, Rotello VM (2014) Gold nanoparticles for nucleic acid delivery. Mol Ther 22(6):1075–1083. https://doi.org/10.1038/mt.2014.30

Doyle C, Higginbottom K, Swift TA, Winfield M, Bellas C, Benito-Alifonso D, Fletcher T, Galan MC, Edwards K, Whitney HM (2019) A simple method for spray-on gene editing in planta. https://doi.org/10.1101/805036

Dunbar T, Tsakirpaloglou N, Septiningsih EM, Thomson MJ (2022) Carbon nanotube-mediated plasmid DNA delivery in rice leaves and seeds. Int J Mol Sci 23(8):4081. https://doi.org/10.3390/ijms23084081

Farooq N, Ather L, Shafiq M, Nawaz-ul-Rehman MS, Haseeb M, Anjum T, Abbas Q, Hussain M, Ali N, Asad Abbas SAA, Mushtaq S, Haider MS, Sadiq S, Shahid MA (2022) Magnetofection approach for the transformation of okra using green iron nanoparticles. Sci Rep 12(1):16568. https://doi.org/10.1038/s41598-022-20569-x

Farre J-C (2001) Gene expression in isolated plant mitochondria: high fidelity of transcription, splicing and editing of a transgene product in electroporated organelles. Nucleic Acids Res 29(12):2484–2491. https://doi.org/10.1093/nar/29.12.2484

Fox, T. D., Sanford, J. C., & McMullin, T. W. (1988). Plasmids can stably transform yeast mitochondria lacking endogenous mtDNA. Proceedings of the National Academy of Sciences of the United States of America, 85(19), 7288–7292. https://doi.org/10.1073/pnas.85.19.7288

Furuhata Y, Sakai A, Murakami T, Morikawa M, Nakamura C, Yoshizumi T, Fujikura U, Nishida K, Kato Y (2019) A method using electroporation for the protein delivery of Cre recombinase into cultured Arabidopsis cells with an intact cell wall. Sci Rep 9(1):2163. https://doi.org/10.1038/s41598-018-38119-9

Gade A, Ingle P, Nimbalkar U, Rai M, Raut R, Vedpathak M, Jagtap P, Abd-Elsalam KA (2023) Nanofertilizers: the next generation of agrochemicals for long-term impact on sustainability in farming systems. Agrochemicals 2(2):257–278. https://doi.org/10.3390/agrochemicals2020017

Gao F, Gu Z (2015) Melting temperature of metallic nanoparticles. In: Aliofkhazraei, M. (eds) Handbook of nanoparticles. Springer, Cham, pp 1–25. https://doi.org/10.1007/978-3-319-15338-4_6

Ghaffari A, Abazari M, Moghimi HR (2023) Wound healing and nanotechnology: opportunities and challenges. In: Bioengineered nanomaterials for wound healing and infection control. Elsevier, pp 115–174

Ghaghelestany AB, Jahanbakhshi A, Taghinezhad E (2020) Gene transfer to German chamomile (L chamomilla M) using cationic carbon nanotubes. Sci Hortic 263:109106. https://doi.org/10.1016/j.scienta.2019.109106

Giraldo JP, Landry MP, Faltermeier SM, McNicholas TP, Iverson NM, Boghossian AA, Reuel NF, Hilmer AJ, Sen F, Brew JA, Strano MS (2014) Plant nanobionics approach to augment photosynthesis and biochemical sensing. Nat Mater 13(4):400–408. https://doi.org/10.1038/nmat3890

Golestanipour A, Nikkhah M, Aalami A, Hosseinkhani S (2018) Gene delivery to tobacco root cells with single-walled carbon nanotubes and cell-penetrating Fusogenic peptides. Mol Biotechnol 60(12):863–878. https://doi.org/10.1007/s12033-018-0120-5

Guisbiers G, Mejía-Rosales S, Leonard Deepak F (2012) Nanomaterial properties: size and shape dependencies. J Nanomater 2012(1). https://doi.org/10.1155/2012/180976

Gur T (2022) Green synthesis, characterizations of silver nanoparticles using sumac (Rhus coriaria L.) plant extract and their antimicrobial and DNA damage protective effects. Front Chem 10. https://doi.org/10.3389/fchem.2022.968280

Hajiahmadi Z, Shirzadian-Khorramabad R, Kazemzad M, Sohani MM (2019) Enhancement of tomato resistance to Tuta absoluta using a new efficient mesoporous silica nanoparticle-mediated plant transient gene expression approach. Sci Hortic 243:367–375. https://doi.org/10.1016/j.scienta.2018.08.040

Hao Y, Yang X, Shi Y, Song S, Xing J, Marowitch J, Chen J, Chen J (2013) Magnetic gold nanoparticles as a vehicle for fluorescein isothiocyanate and DNA delivery into plant cells. Botany 91(7):457–466. https://doi.org/10.1139/cjb-2012-0281

Haydar MS, Ghosh D, Roy S (2024) Slow and controlled release nanofertilizers as an efficient tool for sustainable agriculture: recent understanding and concerns. Plant Nano Biol 7:100058. https://doi.org/10.1016/j.plana.2024.100058

Hoai Phong T, Hieu T, Thanh Tung H, Thi Nhu Mai N, Dac Khai H, Manh Cuong D, Quoc Luan V, Ba Nam N, Tan Nhut D (2023) Somatic embryogenesis as potential method for commercial propagation in Passiflora edulis Sims f. edulis - an important horticultural crop.Scientia horticulturae 316:112020 https://doi.org/10.1016/j.scienta.2023.112020

Hui Y, Yi X, Hou F, Wibowo D, Zhang F, Zhao D, Gao H, Zhao CX (2019) Role of nanoparticle mechanical properties in cancer drug delivery. ACS Nano 13:7410–7424. https://doi.org/10.1021/acsnano.9b03924

Issa B, Obaidat IM, Albiss BA, Haik Y (2013) Magnetic nanoparticles: surface effects and properties related to biomedicine applications. Int J Mol Sci 14:21266–21305. https://doi.org/10.3390/ijms141121266

Jat SK, Bhattacharya J, Sharma MK (2020) Nanomaterial based gene delivery: a promising method for plant genome engineering. J Mater Chem B 8:4165–4175. https://doi.org/10.1039/D0TB00217H

Jat SK, Bhattacharya J, Sharma MK (2021) Nanomaterial based gene delivery: A promising method for plant genome engineering. J Mater Chem B 8:4165–4175

Jia R, Zhao H, Huang J, Kong H, Zhang Y, Guo J, Huang Q, Guo Y, Wei Q, Zuo J, Zhu YJ, Peng M, Guo A (2017) Use of RNAi technology to develop a PRSV-resistant transgenic papaya. Sci Rep 7(1):12636. https://doi.org/10.1038/s41598-017-13049-0

Kanu GA, Parambath JBM, Abu Odeh RO, Mohamed AA (2022) Gold nanoparticle-mediated gene therapy. Cancers (Basel):14(21), 5366; https://doi.org/10.3390/cancers14215366

Khan I, Awan SA, Rizwan M, Hassan ZU, Akram MA, Tariq R, Brestic M, Xie W (2022) Nanoparticle's uptake and translocation mechanisms in plants via seed priming, foliar treatment, and root exposure: a review. Environ Sci Pollut Res 29(60):89823–89833. https://doi.org/10.1007/s11356-022-23945-2

Khan I, Saeed K, Khan I (2019) Nanoparticles: properties, applications and toxicities. Arab J Chem 12:908–931. https://doi.org/10.1016/j.arabjc.2017.05.011

Khodakovskaya MV, de Silva K, Nedosekin DA, Dervishi E, Biris AS, Shashkov EV, Galanzha EI, Zharov VP (2011) Complex genetic, photothermal, and photoacoustic analysis of nanoparticle-plant interactions. Proc Natl Acad Sci 108(3):1028–1033. https://doi.org/10.1073/pnas.1008856108

Kumar A, Zhang X, Liang XK (2013) Gold nanoparticles: Emerging paradigm for targeted drug delivery system. Biotechnology Adv 31:593–606

Kwak SY, Lew TTS, Sweeney CJ, Koman VB, Wong MH, Bohmert-Tatarev K, Snell KD, Seo JS, Chua NH, Strano MS (2019) Chloroplast-selective gene delivery and expression in planta using chitosan-complexed single-walled carbon nanotube carriers. Nat Nanotechnol 14(5):447–455. https://doi.org/10.1038/s41565-019-0375-4

dos Santos OAL, dos Santos MS, Antunes Filho S, Backx BP (2024) Nanotechnology for the control of plant pathogens and pests. Plant Nano Biol 8:100080. https://doi.org/10.1016/j.plana.2024.100080

Law SSY, Liou G, Nagai Y, Giménez-Dejoz J, Tateishi A, Tsuchiya K, Kodama Y, Fujigaya T, Numata K (2022) Polymer-coated carbon nanotube hybrids with functional peptides for gene delivery into plant mitochondria. Nat Commun 13(1):2417. https://doi.org/10.1038/s41467-022-30185-y

Li RN, Da XH, Li X, Lu YS, Gu FF, Liu Y (2021) Functionalized magnetic nanoparticles for drug delivery in tumor therapy. Chin Phys B 30 017502. https://doi.org/10.1088/1674-1056/abb3e6

Ling D, Hackett MJ, Hyeon T (2014) Surface ligands in synthesis, modification, assembly and biomedical applications of nanoparticles. Nano Today 9:457–477. https://doi.org/10.1016/j.nantod.2014.06.005

Liu GC, Xing D (2009) A non-invasive and rapid seed vigor biosensor based on quantitative measurement of superoxide generated by aleurone cell in intact seeds. Biosens Bioelectron 24(6):1537–1542. https://doi.org/10.1016/j.bios.2008.06.040

Liu Y, Yang F, Jing X, Liu X, Wang G, Jian-Ping A, Huo H, Zhang Z, Wang XF, Zhang S, You C (2023) A biomimetic nanoparticle for pDNA delivery and expression in plant cells in a pH-dependent manner. ACS Agric Sci Tech 3(8):631–641. https://doi.org/10.1021/acsagscitech.3c00068

Loo YY, Rukayadi Y, Nor-Khaizura M-A-R, Kuan CH, Chieng BW, Nishibuchi M, Radu S (2018) In vitro antimicrobial activity of green synthesized silver nanoparticles against selected gram-negative foodborne pathogens. Front Microbiol 9. https://doi.org/10.3389/fmicb.2018.01555

Low L-Y, Yang S-K, Kok D-XA, Ong-Abdullah J, Tan N-P, Lai K-S (2018) Transgenic plants: gene constructs, vector and transformation method. In: New visions in plant science. InTech https://doi.org/10.5772/intechopen.79369

Lv Z, Jiang R, Chen J, Chen W (2020) Nanoparticle-mediated gene transformation strategies for plant genetic engineering. Plant J 104(4):880–891. https://doi.org/10.1111/tpj.14973

Madhu, Sharma A, Kaur A, Antil D, Singh SP, Upadhyay SK (2023) Plants as bioreactors. In: Plants as bioreactors for industrial molecules. Wiley, pp 1–27. https://doi.org/10.1002/9781119875116.ch1

Mahmood MA, Naqvi RZ, Rahman SU, Amin I, Mansoor S (2023) Plant virus-derived vectors for plant genome engineering. Viruses 15(2):531. https://doi.org/10.3390/v15020531

Mahmoud LM, Kaur P, Stanton D, Grosser JW, Dutt M (2022) A cationic lipid mediated CRISPR/Cas9 technique for the production of stable genome edited citrus plants. Plant Methods 18(1): 33. https://doi.org/10.1186/s13007-022-00870-6

Maiti D, Tong X, Mou X, Yang K (2019) Carbon-based nanomaterials for biomedical applications: a recent study. Front Pharmacol 9:1401. https://doi.org/10.3389/fphar.2018.01401

Majeed N, Panigrahi KCS, Sukla LB, John R, Panigrahy M (2020) Application of carbon nanomaterials in plant biotechnology. In: Materials today: proceedings. Elsevier, pp 340–345. https://doi.org/10.1016/j.matpr.2020.01.618

Manokari M, Raj MC, Dey A, Faisal M, Alatar AA, Joshee N, Shekhawat MS (2023) Silver nanoparticles improved morphogenesis, biochemical profile and micro-morphology of Gaillardia pulchella Foug cv. 'Torch Yellow'. Plant Cell Tissue Organ Cult (PCTOC) 155(2):433–445. https://doi.org/10.1007/s11240-023-02502-w

Martin-Ortigosa S, Valenstein JS, Lin VSY, Trewyn BG, Wang K (2012) Gold functionalized mesoporous silica nanoparticle mediated protein and DNA codelivery to plant cells via the biolistic method. Adv Funct Mater 22(17):3576–3582. https://doi.org/10.1002/adfm.201200359

Mehta M, Bui TA, Yang X, Aksoy Y, Goldys EM, Deng W (2023) Lipid-based nanoparticles for drug/gene delivery: an overview of the production techniques and difficulties encountered in their industrial development. ACS Mater Au 3(6):600–619. https://doi.org/10.1021/acsmaterialsau.3c00032

Meyers B, Zaltsman A, Lacroix B, Kozlovsky SV, Krichevsky A (2010) Nuclear and plastid genetic engineering of plants: comparison of opportunities and challenges. Biotechnol Adv 28:747–756. https://doi.org/10.1016/j.biotechadv.2010.05.022

Miyamoto T, Tsuchiya K, Toyooka K, Goto Y, Tateishi A, Numata K (2022) Relaxation of the plant cell wall barrier via Zwitterionic liquid pretreatment for micelle-complex-mediated DNA delivery to specific plant organelles. Angew Chem 134(32). https://doi.org/10.1002/ange.202204234

Mohamed MA, Mohamed AEMA, Abd-Elsalam KA (2019) Magnetic nanoparticles in plant protection: promises and risks. In: Nanotechnology in the life sciences. Springer, pp 225–246. https://doi.org/10.1007/978-3-030-16439-3_12

Mohanpuria P, Rana NK, Yadav SK (2008) Biosynthesis of nanoparticles: Technological concepts and future applications. Journal of Nanoparticle Research 10:507–517

Mohapatra J, Xing M, Liu JP (2019) Inductive thermal effect of ferrite magnetic nanoparticles. Materials 12(19):3208. https://doi.org/10.3390/ma12193208

Naphatsamon U, Ohashi T, Misaki R, Fujiyama K (2018) The production of human β-glucocerebrosidase in Nicotiana benthamiana root culture. Int J Mol Sci 19(7):1972. https://doi.org/10.3390/ijms19071972

Newkirk GM, Jeon S-J, Kim H-I, Sivaraj S, De Allende P, Castillo C, Jinkerson RE, Giraldo JP (2023) DNA delivery by high aspect ratio nanomaterials to algal chloroplasts. Environ Sci Nano 10(10):2890–2903. https://doi.org/10.1039/D3EN00268C

Nomura K, Komamine A (1986) Embryogenesis from microinjected single cells in a carrot cell suspension culture. Plant Sci 44(1):53–58. https://doi.org/10.1016/0168-9452(86)90168-8

Ozyigit (2020) Gene transfer to plants by electroporation: methods and applications. Mol Biol Rep. 47(4):3195–3210. https://doi.org/10.1007/s11033-020-05343-4. PMID: 32242300

Paine JA, Shipton CA, Chaggar S, Howells RM, Kennedy MJ, Vernon G, Wright SY, Hinchliffe E, Adams JL, Silverstone AL, Drake R (2005) Improving the nutritional value of Golden Rice through increased pro-vitamin A content. Nat Biotechnol 23(4):482–487. https://doi.org/10.1038/nbt1082

Park C-W, Choi J-Y, Son Y-J, Kim D-H, Li H, Liang W, Lee C, Jung K-H, Kim Y-J (2024) Magnetofected pollen gene delivery system could generate genetically modified Cucumis sativus. Hortic Res 11. https://doi.org/10.1093/hr/uhae179

Rai M, Deshmukh S, Gade A, Abd-Elsalam KA (2012) Strategic nanoparticle-mediated/gene transfer in plants and animals-a novel approach. Current Nanoscience 8(10):170-179. https://doi.org/10.2174/1573413711208010170

Rajkumari N, Alex S, Soni KB, Anith KN, Viji MM, Kiran AG (2021) Silver nanoparticles for biolistic transformation in Nicotiana tabacum L. 3 Biotech 11(12):497. https://doi.org/10.1007/s13205-021-03043-9

Randolph-Anderson BL, Boynton JE, Gillham NW, Harris EH, Johnson AM, Dorthu M-P, Matagne RF (1993) Further characterization of the respiratory deficient dum-1 mutation of Chlamydomonas reinhardtii and its use as a recipient for mitochondrial transformation. Mol Gen Genet 236–236(2–3):235–244. https://doi.org/10.1007/BF00277118

Saleh TA (2022) Characterization and description of adsorbents and nanomaterials. In: Interface Science and Technology, Volume 34, pp 199–232. https://doi.org/10.1016/B978-0-12-849876-7.00008-7

Salih SJ, Mahmood WM (2023) Review on magnetic spinel ferrite (MFe2O4) nanoparticles: from synthesis to application. Heliyon 9(6):e16601. https://doi.org/10.1016/j.heliyon.2023.e16601

Santana I, Jeon SJ, Kim HI, Islam MR, Castillo C, Garcia GFH, Newkirk GM, Giraldo JP (2022) Targeted carbon nanostructures for chemical and gene delivery to plant chloroplasts. ACS Nano 16(8):12156–12173. https://doi.org/10.1021/acsnano.2c02714

Schwab F, Zhai G, Kern M, Turner A, Schnoor JL, Wiesner MR (2016) Barriers, pathways and processes for uptake, translocation and accumulation of nanomaterials in plants – critical review. Nanotoxicology 10:257–278. https://doi.org/10.3109/17435390.2015.1048326

Şener S, Sayğı H (2023) The role of silver nanoparticles in response of in vitro boysenberry plants to drought stress. Horticulturae 9(11). https://doi.org/10.3390/horticulturae9111177

Sharma AK, Sharma MK (2009) Plants as bioreactors: recent developments and emerging opportunities. Biotechnol Adv 27(6):811–832. https://doi.org/10.1016/j.biotechadv.2009.06.004

Singh H, Kumar A, Bansod BS, Singh T, Thakur A, Singh T, Sharma J (2018) Enhanced moisture sensing properties of a nanostructured ZnO coated capacitive sensor. RSC Adv 8(7):3839–3845. https://doi.org/10.1039/c7ra10917b

Sircaik S, Dhiman K, Gambhir G, Kumar P, Srivastava DK (2021) Transgenic implications for biotic and abiotic stress tolerance in agricultural crops. In: Agricultural biotechnology: latest research and trends. Springer, Singapore, pp 185–221

Sun D, Hussain HI, Yi Z, Rookes JE, Kong L, Cahill DM (2016) Mesoporous silica nanoparticles enhance seedling growth and photosynthesis in wheat and lupin. Chemosphere 152:81–91. https://doi.org/10.1016/j.chemosphere.2016.02.096

Sun X-D, Ma J-Y, Feng L-J, Duan J-L, Xie X-M, Zhang X-H, Kong X, Ding Z, Yuan X-Z (2023) Magnetite nanoparticle coating chemistry regulates root uptake pathways and iron chlorosis in plants. Proc Natl Acad Sci 120(27). https://doi.org/10.1073/pnas.2304306120

Thomas M, Klibanov AM (2003) Conjugation to gold nanoparticles enhances polyethylenimine's transfer of plasmid DNA into mammalian cells. Proc Natl Acad Sci 100(16):9138–9143. https://doi.org/10.1073/pnas.1233634100

Tiwari M, Mishra AK, Chakrabarty D (2022) Agrobacterium-mediated gene transfer: recent advancements and layered immunity in plants. Planta 256(2):37. https://doi.org/10.1007/s00425-022-03951-x

Toda E, Koiso N, Takebayashi A, Ichikawa M, Kiba T, Osakabe K, Osakabe Y, Sakakibara H, Kato N, Okamoto T (2019) An efficient DNA- and selectable-marker-free genome-editing system using zygotes in rice. Nat Plants 5(4):363–368. https://doi.org/10.1038/s41477-019-0386-z

Torney F, Trewyn BG, Lin VSY, Wang K (2007) Mesoporous silica nanoparticles deliver DNA and chemicals into plants. Nat Nanotechnol 2(5):295–300. https://doi.org/10.1038/nnano.2007.108

Toyoda H, Matsuda Y, Utsumi R, Ouchi S (1988) Intranuclear microinjection for transformation of tomato callus cells. Plant Cell Rep 7(5):293–296. https://doi.org/10.1007/BF00269921

Vats S, Kumawat S, Brar J, Kaur S, Yadav K, Magar SG, Jadhav PV, Salvi P, Sonah H, Sharma S, Deshmukh R (2022) Opportunity and challenges for nanotechnology application for genome editing in plants. Plant Nano Biol 1:100001. https://doi.org/10.1016/j.plana.2022.100001

Vijayakumar PS, Abhilash OU, Khan BM, Prasad BLV (2010) Nanogold-loaded sharp-edged carbon bullets as plant-gene carriers. Adv Funct Mater 20(15):2416–2423. https://doi.org/10.1002/adfm.200901883

Wang, Huang J, Zhang M, Wang Y, Wang H, Ma Y, Zhao X, Wang X, Liu C, Huang H, Liu Y, Lu F, Yu H, Shao M, Kang Z (2020a) Carbon dots enable efficient delivery of functional DNA in plants. ACS Appl Bio Mater 3(12):8857–8864. https://doi.org/10.1021/acsabm.0c01170

Wang X, Xie H, Wang P, Yin H (2023) Nanoparticles in plants: uptake, transport and physiological activity in leaf and root. Materials 16(8):3097. https://doi.org/10.3390/ma16083097

Wang Z, Zhang Z, Zheng D, Zhang T, Li X, Zhang C, Yu R, Wei J, Wu Z (2022) Efficient and genotype independent maize transformation using pollen transfected by DNA-coated magnetic nanoparticles. J Integr Plant Biol 64(6):1145–1156. https://doi.org/10.1111/jipb.13263

Wang Z-P, Agro-Biotechnology B, Center R, Zhang Z-B, Zhang C, Yin L-F, Yu R, Wei J-H, Wu Z-Y (2020b) Efficient and genotype independent maize pollen transfection mediated by magnetic nanoparticles. https://doi.org/10.21203/rs.3.rs-97984/v1

Wong MH, Misra RP, Giraldo JP, Kwak S-Y, Son Y, Landry MP, Swan JW, Blankschtein D, Strano MS (2016) Lipid exchange envelope penetration (LEEP) of nanoparticles for plant engineering: a universal localization mechanism. Nano Lett 16(2):1161–1172. https://doi.org/10.1021/acs.nanolett.5b04467

Wu K, Xu C, Li T, Ma H, Gong J, Li X, Sun X, Hu X (2023a) Application of nanotechnology in plant genetic engineering. Int J Mol Sci 24(19):14836. https://doi.org/10.3390/ijms241914836

Wu K, Xu C, Li T, Ma H, Gong J, Li X, Sun X, Hu X (2023b) Application of nanotechnology in plant genetic engineering. Int J Mol Sci 24(19):14836. https://doi.org/10.3390/ijms241914836

Wu Q, Miao WS, Zhang YD, Gao HJ, Hui D (2020) Mechanical properties of nanomaterials: a review. Nanotechnol Rev 9(1):259–273. https://doi.org/10.1515/ntrev-2020-0021

Wu W, Wu Z, Yu T, Jiang C, Kim W-S (2015) Recent progress on magnetic iron oxide nanoparticles: synthesis, surface functional strategies and biomedical applications. Sci Technol Adv Mater 16(2):023501. https://doi.org/10.1088/1468-6996/16/2/023501

Yadav S, Jat SK, Bhattacharya J, Sharma MK (2023) Nanotechnology mediated gene transfer in plants: a novel approach. In: Nanotechnology in agriculture and agroecosystems. Elsevier, pp 141–168. https://doi.org/10.1016/B978-0-323-99446-0.00005-2

Yong J, Zhang R, Bi S, Li P, Sun L, Mitter N, Carroll BJ, Xu ZP (2021) Sheet-like clay nanoparticles deliver RNA into developing pollen to efficiently silence a target gene. Plant Physiol 187(2):886–899. https://doi.org/10.1093/plphys/kiab303

Yoshizumi T, Oikawa K, Chuah JA, Kodama Y, Numata K (2018) Selective gene delivery for integrating exogenous DNA into plastid and mitochondrial genomes using peptide-DNA complexes. Biomacromolecules 19(5):1582–1591. https://doi.org/10.1021/acs.biomac.8b00323

Yu-qin F, Lu-hua L, Pi-wu W, Jing Q, Yong-ping F, Hui W, Jing-ran S, Chang-li L (2012) Delivering DNA into plant cell by gene carriers of ZnS nanoparticles. Them. res. Chinese Universities 28(4): 672–676

Zeleňáková A, Zeleňák V, Beňová E, Kočíková B, Király N, Hrubovčák P, Szűcsová J, Nagy, Klementová M, Mačák J, Závišová V, Bednarčík J, Kupčík J, Jacková A, Volavka D, Košuth J, Vilček (2024) The surface modification of the silica-coated magnetic nanoparticles and their application in molecular diagnostics of virus infection. Sci Rep 14(1):14427. https://doi.org/10.1038/s41598-024-64839-2

Zhang B, Huang S, Meng Y, Chen W (2023) Gold nanoparticles (AuNPs) can rapidly deliver artificial microRNA (AmiRNA)-ATG6 to silence ATG6 expression in Arabidopsis. Plant Cell Rep 42(7):1191–1201. https://doi.org/10.1007/s00299-023-03026-5

Zhang H, Cao Y, Xu D, Goh NS, Demirer GS, Cestellos-Blanco S, Chen Y, Landry MP, Yang P (2021) Gold-nanocluster-mediated delivery of siRNA to intact plant cells for efficient gene knockdown. Nano Lett 21(13):5859–5866. https://doi.org/10.1021/acs.nanolett.1c01792

Zhang H, Demirer GS, Zhang H, Ye T, Goh NS, Aditham AJ, Cunningham FJ, Fan C, Landry MP (2019) DNA nanostructures coordinate gene silencing in mature plants. Proc Natl Acad Sci 116(15):7543–7548. https://doi.org/10.1073/pnas.1818290116

Zhang H, Zhang H, Demirer GS, González-Grandío E, Fan C, Landry MP (2020) Engineering DNA nanostructures for siRNA delivery in plants. Nat Protoc 15(9):3064–3087. https://doi.org/10.1038/s41596-020-0370-0

Zhang J, Lei W, Meng Y, Zhou C, Zhang B, Yuan J, Wang M, Xu D, Meng X, Chen W (2022) Expression of PEI-coated gold nanoparticles carrying exogenous gene in periwinkle mesophyll cells and its practice in huanglongbing research. iScience 25(6):104479. https://doi.org/10.1016/j.isci.2022.104479

Zhang Y, Liang Z, Zong Y, Wang Y, Liu J, Chen K, Qiu J-L, Gao C (2016) Efficient and transgene-free genome editing in wheat through transient expression of CRISPR/Cas9 DNA or RNA. Nat Commun 7(1):12617. https://doi.org/10.1038/ncomms12617

Zhao X, Meng Z, Wang Y, Chen W, Sun C, Cui B, Cui J, Yu M, Zeng Z, Guo S, Luo D, Cheng JQ, Zhang R, Cui H (2017) Pollen magnetofection for genetic modification with magnetic nanoparticles as gene carriers. 3(12):956-964. https://doi.org/10.1038/s41477-017-0063-z

Zolghadrnasab M, Mousavi A, Farmany A, Arpanaei A (2021) Ultrasound-mediated gene delivery into suspended plant cells using polyethyleneimine-coated mesoporous silica nanoparticles. Ultrason Sonochem 73:105507. https://doi.org/10.1016/j.ultsonch.2021.105507

Chapter 9
Role of Nanotechnology in Combating Climate Change

Sachit Agarwal and Vinay Kumar

Abstract Nanotechnology nowadays is emerging as a solution for addressing climate change. By using the unique properties of nanoparticles (NPs), researchers have enhanced the productivity of renewable energy systems and remediation systems, as well as increased the crop yield in agriculture. Nanomaterials (NMs) such as MOFs, nanotubes, and perovskites have enhanced the system by increasing their properties such as absorption and adsorption, and in agriculture, nanofertilizers and nanopesticides have increased the yield and also promoted soil health. Nanosensors being employed for the monitoring of climate change could help in policymaking. Although nanotechnology has potential, it still has some limitations that need to be addressed to utilize its full potential. This chapter explores how nanotechnology is being applied and what are the challenges it faces.

Keywords Nanotechnology · Climate change · Renewable energy · Sustainable agriculture · Nanoremediation · Carbon capture

9.1 Introduction

Climate change is presently one of the most difficult challenges in the twenty-first century. One of the main reasons is human activity which has increased over the last hundred years. This has caused an increase in the levels of atmospheric gases like CO_2 which causes a rise in global temperature causing global warming. Addressing global warming is the main priority which is the main reason for climate change. Over the last few decades, the Earth's surface temperature has been increasing at an alarming rate. According to the recent data from *NOAA (National Centers for Environmental Information)*, 2024 became the warmest year with a global temperature of 1.29 °C (2.32 °F) surpassing the value of the year 2023 which had a global temperature of 0.10 °C (0.18 °F). A major reason for this increase in trends is the

S. Agarwal · V. Kumar (✉)
Department of Botany, Central University of Punjab, Ghudda, Bathinda, India
e-mail: vinay@cup.edu.in; vinayihbt@gmail.co.in

S. K. Yadav, A. Kumari (eds.), *Nanobiotechnology in Agri-food Sector*, Sustainability Sciences in Asia and Africa, https://doi.org/10.1007/978-981-96-7756-6_9

accumulation of greenhouse gas emissions (GHGs). These can be natural such as CO_2 and CH_4 or industrial such as SF_6 and hydrofluorocarbons (HFCs). The combination of these gases contributes to ozone layer depletion causing an increase in global warming and resulting in widespread environmental harm. The impacts can be changes in season patterns, glaciers melting, pollution, flooding, drought, soil erosion, coastal erosion, and a decline in crop yields (Chausali et al. 2023). This direct relation between the rising temperatures and the increasing GHGs shows the importance of mitigation of climate change. To solve the problem of these emissions, global energy has started to transition towards lower-carbon energy systems, which could help in lowering our carbon footprint. The use of renewable sources such as solar, wind, geothermal, and tidal is one way through which we can minimize our dependency on fossil fuels as compared to fossil fuels these are not finite and can also help in reducing GHGs. Another way is to adopt sustainable practices of agriculture which means to limit the resource use keeping in mind for future generations. Although renewable energy has enormous potential still it is difficult to widely adapt at a large scale due to technical reasons like integration with compatible renewable in the energy sector and economic reasons mainly for developing countries. Furthermore, the solar and wind energy which are expected to account for 70% of electricity generation by 2050 requires a large-scale project and requires a risk assessment (Adanma and Ogunbiyi 2024). Here, nanotechnology emerges to give innovative solutions to solve these problems making a path for a sustainable future.

Nanotechnology is simply referred to as the manipulation of matter. It deals with the making of nanomaterials or nanoparticles (NPs). Nanoparticles compared to bulk particles are smaller in size ranging between 1 and 100 nm and have unique characteristics. Due to the reduction in size, there is a change in the chemical and physical characteristics of the material. Utilizing these nanoparticles (NPs) through nanoscale devices, nanotechnology addresses a variety of issues in diverse sectors. For example, nanomaterials (NMs) are used in the making of solar cells making them thin, and flexible and enhancing their efficiency through the absorption of sunlight. Similarly, in the wind and tidal energy sector, NMs are used in making turbines that are anti-corrosive and are lighter in weight aiming to increase their life span. In the agriculture sector, NMs are used in fertilizers and pesticides enhancing their efficacy (Lameirinhas et al. 2022; Liu et al. 2023; Arora et al. 2022).

This chapter will document the role of nanotechnology in mitigating climate change through its various roles in different sectors, like in renewable energy sources in terms of scalability and cost-effectiveness. The additional role of nanotechnology in sustainable agriculture, which is also important to solve the population hunger, with limited resources. More, applications like monitoring of climate and environmental remediation also help to create a more sustainable future which can help in mitigating climate change.

9.2 Reasons for Selecting Nanotechnology

Nanotechnology research to fight against climate change has increased significantly in the twenty-first century. Due to its extremely small size and high surface area-to-volume ratios, it can speed up chemical reactions. This unique property helps NPs to absorb pollutants more effectively. Nanoscale filters have been developed to remove toxins and pollutants from fuels which further improves the fuel quality and reduces harmful emissions. Because of this increased efficiency, nanotechnology is an alternative to fossil fuels through progress in renewable sources of energy and helps in lowering consumption which ultimately helps in lowering greenhouse gas emissions (Vasantha et al. 2021). In the agriculture sector, through nanotechnology, the development of nanopesticides and nanofertilizers ensures precise nutrient delivery while minimizing overuse and greenhouse gas emissions. Nanosensors have also been made that monitor soil health and help farmers practice resilient farming practices (Yang and Duncan 2021).

These innovations show the manipulative potential of nanotechnology across different sectors. By providing both sustainable and feasible solutions, nanotechnology has the potential to revolutionize different sectors.

9.3 Applications of the Nanotechnology Against Climate Change

Advancements in nanotechnology are having a positive impact on the environment. These creative innovations in nanotechnology have given solutions to some of the most environmental challenges. Nanotechnology gives sustainable solutions to many problems like wastewater treatment, reducing greenhouse gas emissions, fuel problems, and many other such problems that contribute to climate change in our day-to-day lives (Fig. 9.1).

9.3.1 Nanotechnology in Renewable Energy

An important step in addressing the issue of climate change is reducing the consumption of fossil fuels that is to switch from nonrenewable to renewable energy. Natural gas, metals, minerals, oils, and coals are some of the nonrenewable sources that are currently being used to meet the demand for energy production. But these are limited in nature. However, issues like its cost-effectiveness and energy efficiency have prevented switching to renewable sources. In this scenario, nanotechnology has provided many ways to sustainably use renewable energy systems. By using the unique property of NMs, researchers have enhanced the production efficiency of many things like energy storage devices (e.g. supercapacitors), solar cells, and

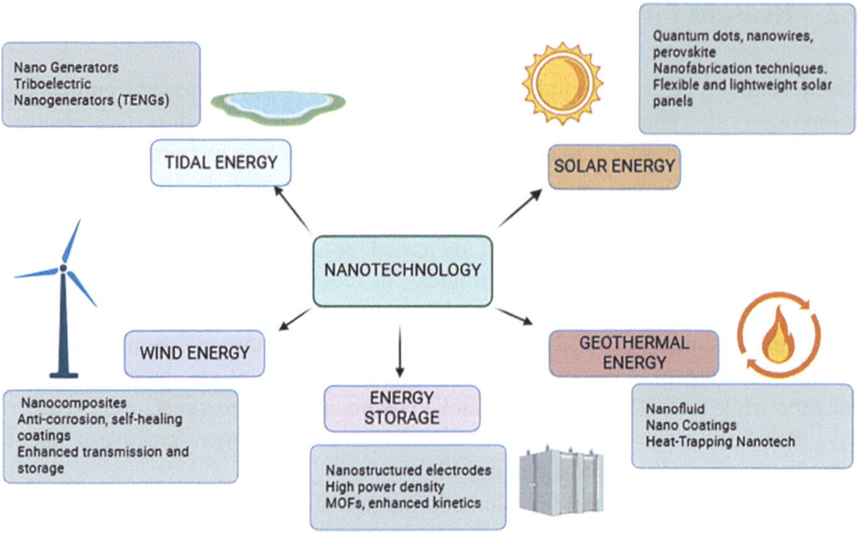

Fig. 9.1 Nanotechnology application in diverse sectors

hydrogen fuel cells (Simpa et al. 2024). These advancements in nanotechnology have made the implementation of renewable energy sources much easier. This section will explore how nanotechnology is helping to make different renewable energy sources cheaper, more reliable, and more efficient. In Fig. 9.1, an overview of the use of nanotechnology is shown across various sectors.

9.3.1.1 Nanotechnology in Solar Energy

Among the various options of renewable sources solar energy is one of the promising energy sources and a solution to combating climate change. However, despite the benefits its efficiency and cost-effectiveness continue to be challenges limiting their widespread adoption. Nanotechnology addresses these limitations to enhance solar efficiency. Through the development of NMs like Carbon Quantum Dots (CQDs) and Graphene Quantum Dots (GQDs), researchers have made many advances regarding energy conversion and charge efficiency (Ghosh et al. 2020). Recently, the integration of nanotechnology with nanofluid cooling systems and the incorporation of citrate-stabilized and PVP-stabilized silver NPs have improved the overall efficiency of solar systems by leveraging the unique thermal and optical properties of NMs (Rajukkannu et al. 2025).

Furthermore, nanotechnology plays a role in advancing photovoltaic cells, also known as solar cells, which are devices that convert light energy into electrical energy. The conventional first-generation solar cells were made of silicon, a semiconductor that faced several limitations, like efficiency limits which accounted for around 29.4% of material cost. Through NM enhancement, solar PV got higher efficiency which allowed them to capture a broad spectrum of sunlight, low-cost production due to simple and less energy-intensive manufacturing, and most importantly reduced environmental impact (Lameirinhas et al. 2022). Carbon-based materials like graphene and NMs like quantum dots and nanowires are commonly used to enhance solar cells. Quantum dots (QDs) possess unique optical and electrical properties that include a tunable bandgap energy that maximizes energy capture and conversion by allowing absorption at a wide range of wavelengths from the solar spectrum. Unlike, conventionally used single-layer designs, QDs allow for the development of advanced tandem solar cells, where several layers are arranged in such a way as to absorb light wavelengths increasing the energy output (Ghosh et al. 2020).

Plasmonic nanoparticles are usually made up of gold (Au) or silver (Ag) and have a phenomenon known as localized surface plasmon resonance (LSPR). When these are exposed to light, they produce a collective oscillation at a specific light wavelength. Due to this interaction, it strengthens the electromagnetic field close to the nanoparticle surface, which later has many benefits especially in solar energy harvesting, sensing, and imaging (Alkhalayfeh et al. 2021). Recently, many advances have been made in this technology which is expanding its application in the development of solar cell types, including organic, inorganic, dye-sensitized, plasmonic enhanced photocatalysis, and perovskite solar cells. Silver (Ag) nanoparticles are incorporated in photoelectrodes to improve efficiency in CdS/CdSe co-sensitized solar cells (Mandal 2022).

Nanotechnology has not only played a role in enhancing energy efficiency but also reduced manufacturing costs so that the product made can be scalable. Advanced fabrication methods such as nanoprint lithography, enable faster production compared to conventional methods. Furthermore, NMs help in the substitution of costly materials like silver in a conductive paste. This overall reduces the cost and enhances the efficiency. These are some of the breakthroughs through the nanotechnology in the solar system which have not only improved the efficiency and reduced costs but also expanded the uses of solar energy applications across many sectors.

9.3.1.2 Nanotechnology in Wind Energy

Wind energy is a renewable energy mainly produced by converting the kinetic energy into electric energy. The kinetic energy is converted into rotational energy which is made by the air motion which in turn rotates wind turbines' rotor blades. This rotational energy is then transferred via a shaft to a generator which generates electrical energy. Significant progress has been made in wind energy since 2020, including the installation of larger and more stronger turbines with an output of up to 12 MW in offshore systems. The main challenge in this is the efficiency and longevity of the wind turbines that are often faced with harsh environments both offshore and onshore. Innovations through the inclusion of NPs make the novel approach through the use of nanocoating, nanolubricants, and nanofluids discussed below are playing the role of improving turbine lifespan with their efficiency (Simpa et al. 2024).

Wind turbines are often faced with harsh environments like low temperatures and cold climate create ice on rotor blades which increases the weight resulting in greater strain on the blades and impaired function of equipment or causing corrosion, ultimately leading to higher energy consumption and lower output and increase in the maintenance cost. Nanocoatings can be incorporated into the materials to give turbine blades self-healing, anticorrosion, and antiicing properties. When tested, on the aluminum surface which is exposed to chloride-rich atmospheres mainly in offshore wind farms, where turbines are faced with high humidity, and salt-laden air, metal oxide nanoparticles have shown a noticeable slowdown in corrosion advancement. This is due to the nanostructure's innate capacity to strengthen the coating structure, which reduces stress and prevents crack formation (Aljibori et al. 2023). In an experiment done by B. Liu et al. (2023), $MoS_2/ZnO/PDMS$ nanocoating shows superhydrophobic characteristics, which helps in resisting the ice formation.

Nanolubricants enhance the performance and life span of the mechanical systems when they are used in turbine gearboxes and generators that often experience random conditions and harsh weather. In these environments, poor lubrication in the gearbox can cause wear and tear failure due to friction leading to gearbox failure. Additionally, friction causes energy loss in the form of heat which also shows how lubrication is important. Conventional lubricants face several limitations including limited load-carrying capacity which can lead to component wear and failure under high stress. Additionally, they can deposit on the components and form sludge which would affect the performance of the gearbox. These problems were solved by the integration of NPs in the lubricants. NPs such as graphene, molybdenum disulfide, or carbon nanotubes were incorporated into the lubricants which enhanced the thermal conductivity, also reduced friction which resulted in less wear and tear which further improved the performance efficiency and life span (Jason et al. 2020).

Nanofluids are a mixture of a base fluid (like water, oil, diesel, and ethylene glycol) and a small number of nanosized particles like metal or metal oxides with average sizes below 100 nm which are utilized not only in this wind energy sector but also in many other daily aspects of daily human activity like medical, power, and chemical engineering. These are heat transfer fluid that acts as a coolant system in any machinery. Compared to a traditional cooling system, these have improved thermal efficiency along with reducing the required surface area (Aglawe et al. 2021). NPs like Al_2O_3, TiO_2, and Fe_3O_4 have been added to a base oil like water which has been demonstrated to increase the efficiency of the thermal fluid. A study by Salim et al. (2024) showed that just a 1% volume fraction of Al_2O_3/water nanofluid had increased the thermal efficiency by 2.09% in a copper N-shaped double-pipe heat exchanger. This shows the potential of nanofluids in reducing the waste of energy in the form of heat by achieving higher thermal efficiency.

The integration of nanotechnology into wind energy systems has significantly improved the efficiency, lifespan, and overall performance of the system which plays a critical role in reducing carbon emissions and combating climate change.

Furthermore, this has also minimized the usage of resources in the making of the system which has made it more reliable, sustainable, and cost-effective energy production.

9.3.1.3 Nanotechnology in Tidal Energy

Tidal energy is also another untouched resource. Oceans cover a massive part of the earth and can be used as an alternative to fossil fuels and offer various forms of renewable energy sources. These include thermal energy that can be derived from the temperature difference in water layers due to seasonal changes, kinetic energy that can be harnessed from the movement of water during tides and waves which are formed by the movement of celestial bodies like the moon, the sun, chemical energy from the seawater components, and biological energy from the biomass. Together, these are a massive unused resource for sustainable power generation (Khan et al. 2022a, b). Although not much is emphasized on tidal energy as compared to the other sources, advancement is still being done. For instance, projects are underway to increase global power generation with the target of 300GW of ocean energy capacity by 2050 (Garanovic 2023).

Nanotechnology enhances the tidal energy systems through the nanoscale generators and triboelectric nanogenerators (TENGs), which helps in improving the energy output. An innovative nanoscale generator named "E8-PowerBuoys" has shown the efficient harness of tidal and wave energy tested in various aquatic systems like saline water. These new-generation generators are made for easy installation and maintenance ensuring durability and lifespan (Suwanapingkarl and Srivallop 2020). Triboelectric nanogenerators (TENGs) are capable of harvesting energy from waste energy which is also of low-frequency vibrations like human walking and ocean waves. These have enhanced the power output compared to conventional technologies (Gonçalves et al. 2023).

Similar to wind energy systems, nanolubricants and nanofluids are used in the same way in tidal energy systems to increase the efficiency of systems. Although nanotechnology has immense potential, its effects are still limited studies on the marine environment, and also their long-term sustainability remains significant.

9.3.1.4 Nanotechnology in Geothermal Energy

Geothermal energy is the heat energy stored in the Earth's interior that comes from its chemical and physical processes. Although present in vast quantities, its uneven distribution and the depth at which it is present make it commercially challenging to extract. Geothermal energy, for long a time, has been used to generate power by using it in the form of natural steam and hot water (Jayakumari et al. 2023). According to the report of the International Renewable Energy Agency (IRENA), as of 2023, the globally installed geothermal power generation has reached up to 16,335 MW, at a growth rate annually of about 3.5%. Although there is an increase in energy generation, it still only shows a small part of the world's total renewable energy capacity. The main challenges it faces are associated with the cost and hazards associated with deep drilling. Nanotechnology here plays an important role in providing solutions to these challenges. One application of nanostructures is to refine special fluid blends like alkanes or other nanofluids which are currently used in geothermal power plants to drive turbines to depths of the earth. As the interior of the Earth is very hot, these nanofluids act as a cooling system for them (Abdalla et al. 2020). These enhance the thermal conductivity and the stability of drills. Nanocoatings also enhance the ground heat exchangers (GHEs) which are used to extract geothermal energy. Nanocoatings can protect against corrosion and can increase their lifespan (Jayakumari et al. 2023). NMs like metal-organic frameworks (MOFs) are used in heat-trapping nanotechnology which enhances the geothermal efficiency by using carbon-based nanotrap solution which reduces the chance of earthquakes during the operations. NPs like magnetic iron oxides or functionalized silica NPs used in heap leaching techniques which make the extraction of valuable minerals easier and reduce the extraction costs (Bhattacharjee and Maiti 2021).

9.3.1.5 Nanotechnology in Energy Storage

In sustainable development, the use of limited resources is required to minimize the exploitation, and a medium of energy storage is required whenever we capture surplus energy from solar, wind, and tidal or from any form and release it during a period of high demand or when there is insufficient energy generation in the grid. In recent times, nanotechnology incorporation has further improved the energy storage technology by increasing storage capacity, transmission from the device or by increasing its charging capabilities. For instance, electrodes improved by nanostructures give better ionic transport and electronic conductivity by increasing the surface area for diffusion of nanostructures faster compared to the traditional Li-ion batteries which gives us a more efficient and durable energy storage solution in the long term (Li et al. 2021). This section will tell how nanotechnology played a transformative in this sector and also its potential to meet the low-carbon future demands.

Battery

A battery is an electrochemical device that stores chemical energy and releases it as electrical energy during discharge. An ideal battery should have high storage capacity, shelf-life, and low cost. The development of battery materials (anodes, cathodes, and electrolytes) depends upon the dimension alteration by NMs to improve the conductivity efficiency and reduce the wasteful reactions that occur (Hsu et al. 2024). Nanoscale materials have played a significant role in improving the performance of batteries. Smaller sizes of NPs facilitate better ionic transport which leads to faster ion movement and overall improved battery efficiency. For instance, lithium phosphorous oxynitride (LiPON) shows high conductivity when compared with conventional electrolytes. Additionally, a smaller size increases the surface contact area between the electrode and electrolyte. This allows for an increase in sites for activation of lithium-ion intercalation, which boosts the capacity of the electrodes. For example, cobalt sulfide (CoS) nanosheets are being used in conjugation with an inorganic electrolyte that creates a large surface area contact leading to better contact between the electrode and electrolyte. Nanoscale materials like lithium titanate ($Li_4Ti_5O_{12}$) are used for improving structural stability during volume changes reducing the risk of cracking and preventing degradation of performance due to volume change thereby increasing the life span of batteries. Some materials like SnO_2 are inactive in their bulk form but become active at the nanoscale showing enhanced lithium-ion interaction, overall enhancing the battery performance (Parameswaran et al. 2022; Julien and Mauger 2024).

NPs integration in the battery technology has overall enhanced the energy storage capabilities, its structural stability significantly increasing the lifespan which is crucial for renewable energy systems, where it can help in balancing the supply and demand of energy generation.

Electric Vehicle Batteries

Nowadays, there is a rise in the production of electric vehicles (EVs) to reach the goal of net zero carbon emissions. These are expected to reduce emissions as they do not depend on conventional sources of fuel (Gnanavendan et al. 2024). Nanotechnology enhances this technology by targeting critical challenges faced like energy density, lifespan, and thermal management, making EVs more feasible. Silicon-based NMs increase the energy density compared to the conventional use of graphite. Other materials like carbon nanotubes and graphene improve charge/discharge rates increasing the overall battery efficiency (Davies 2024; Kumar et al. 2024).

These are some of the advancements done by nanotechnology in EVs through the more efficient, lighter, and long-lasting batteries. With continuous progress, other problems like long charging times in EVs, high production costs, or any other problems could be overcome. This could give us a bit way off from fossil fuels.

Supercapacitors

Supercapacitors, also known as ultracapacitors are energy storage devices that are intermediate between traditional capacitors and batteries. Supercapacitors store energy electrostatically through an electrode–electrolyte interface as compared to batteries which use chemical reactions. Through this mechanism, they provide high power density, quick charge-discharge cycles, and enhancement in their lifespan due to the minimal degradation over time. The integration of NMs like carbon nanotubes, graphene, and transition metal oxides has overall enhanced the properties of supercapacitors. Materials like graphene and carbon nanotubes have played a role in enhancing the charge-discharge cycles that make them reliable for applications that require frequent power surges like those used in improving the efficiency of solar cells and fuel cells in rapid energy delivery (Mangaiyarkkarasi 2024). Nowadays, computational methods like machine learning (ML) and many more are being used to model complex nanoparticles in supercapacitors. For instance, ML has been used to predict the supercapacitor performance when integrated with graphene oxide nanorings (Yogesh et al. 2024).

Supercapacitors still have some issues like lower energy densities when compared with batteries which are being currently addressed by rapid advancement in nanotechnology. These innovations may help supercapacitors to become more scalable and an efficient solution for a wider range of storage energy applications.

Hydrogen Energy Production and Storage

Hydrogen energy is another alternative that holds a promising potential for addressing climate change and is also a cleaner source of energy than fossil fuels. Due to its properties like high density and potential for less CO_2 emissions, it is suitable for sustainable energy systems. Its adaptation could reduce global carbon emissions by 830 million tonnes annually (Sadeq et al. 2024). The use of NMs can enhance the production of hydrogen energy such as utilizing nanocatalysts in the process of photocatalysis for the making of H_2 energy. Nanostructures such as Cu-$ZnIn_2S_4$ nanosheets combined with carbon dots are found to have increased the rate of photocatalytic efficiency, and through this, they also made a sustainable process for hydrogen generation (Chen et al. 2024).

Another component of the hydrogen energy system is hydrogen storage which enables efficient transportation and optimum utilization of hydrogen as a clean energy source. Conventional techniques of hydrogen storage include storing hydrogen in either gaseous or liquid forms which has many drawbacks like energy leakage and safety concerns. NMs offer solutions to these problems by enhancing the efficiency and safety of hydrogen storage to facilitate the transition to cleaner energy sources. Solid-state hydride NMs are used to prevent high dehydrogenation temperatures and slow absorption rates through nanointerface techniques. For instance, hydrides like magnesium hydride (MgH_2) are engineered which improve hydrogen absorption and desorption kinetics significantly reducing the energy required for storage (Huang et al. 2024).

These are some of the advancements that show the role of nanotechnology in hydrogen energy production and storage. However, some challenges still remain which are its scalability, and its safety concerns linked to hydrogen explosive nature. By using the unique properties of nanomaterials, researchers and engineers are finding a way to address these issues. Some key innovations done by nanotechnology are listed in Table 9.1.

Table 9.1 Some key innovation done by nanotechnology

Innovations	Components	Applications	Impact
Quantum dots (QDs)	Cadmium (Cd), Selenium (Se), Lead (Pb)	Advanced photovoltaic systems in solar cells.	Enhanced light absorption and energy conversion efficiency.
Perovskite nanostructures	Lead (Pb), Iodine (I)	Perovskite-based solar cells.	Higher photovoltaic efficiency and low-cost manufacturing.
Nano-coatings	Silicon Dioxide (SiO_2), Aluminium Oxide (Al_2O_3)	Protection for wind and water turbines	Enhanced durability and resistance to harsh environment.
Nanostructured membranes	Cerium (Ce), Titanium (Ti)	Hydrogen fuel cells	Improved ion transport and energy efficiency.
Carbon nanotubes (CNTs)	Carbon (C)	Supercapacitors and energy storage devices.	Higher energy density and faster charge-discharge cycles.
Thermoelectric nanomaterials	Bismuth (Bi), Tellurium (Te)	Waste heat recovery systems.	Conversion of industrial waste heat into usable energy.
Graphene-based materials	Carbon (C)	Lightweight materials for electric vehicles and batteries.	Reduced energy consumption in transportation.
Nano catalysts for Hydrogen Production	Platinum (Pt), Nickel (Ni), Iron (Fe)	Water splitting for green hydrogen production.	Increased catalytic efficiency, reducing fossil fuel dependence.

9.3.2 *Carbon Capture and Sequestration*

Carbon capture and sequestration is a process of capturing and storing the atmospheric carbon dioxide (CO_2) emitted by industrial processes and power plants before reaching the atmosphere which helps in reducing greenhouse gas emissions which overall helps to mitigate climate change. The captured carbon is then transported and stored in geological storages like saline aquifers or depleted oil and gas reservoirs that can be used for CO_2 storage, in deep down oceans where it can be dissolved in seawater, or just immediately utilized in various applications like in building materials production of some chemical fuels (Liu et al. 2022).

Nanotechnology plays an important role in developing more efficient processes for enhancing carbon sequestration and capture methods which is crucial for addressing climate change. Through nanoscale engineering, we get to know that CO_2 diffusion and adsorption are affected by pore size and shape smaller the pore size greater the preference for CO_2. For instance, metal-organic frameworks (MOFs) are used as they have a tunable pore size, making them effective for CO_2 adsorption (Lee et al. 2023). More applications of NMs across various sectors include gas stream separation where advanced NMs like MOFs and graphene oxide-based membranes are used for the removal of CO_2 gas from the gas streams (Alguacil 2023).

For the development of novel methods of greener energy sources, advancements in nanotechnology play a critical role. However, there are still some challenges that need to be addressed before they can be commercialized for public use. Problems like high production costs, shorter lifespan, and the integration of NMs into current infrastructure still need to be addressed to make full use of the potential of nanotechnology in mitigating climate change.

9.3.3 Nanotechnology in Sustainable Agriculture

Sustainable agriculture means meeting the needs of a growing population for food with the simultaneous preservation of environmental resources. Many of the developing countries are dependent on agriculture for their livelihoods. So, it is important to have sustainable practices in meeting the long term. However, there are several issues that the agriculture sector must face like ever-changing season patterns due to global warming, a decline in the crop yield, reduction in essential nutrients and minerals in the soil, inadequate water supply, crop diseases, and lack of manpower. According to a statistical study done by combining multiple datasets, it is estimated that global food demand will increase by 35% to 56% between 2010 and 2050 under various socio-economic and climate change scenarios (Van Dijk et al. 2021). Although agriculture practices introduced during the Green Revolution have increased our production yields, these are found to have certain drawbacks on health and environmental consequences due to excessive fertilizer practices. It does not only raise concerns about their presence in the food chain at various levels but also their presence in the groundwater contaminating them. This has led to finding other alternatives for eco-friendly practices. Nanotechnology is one of the solutions to these problems. NMs are applied in a small amount of dosage which saves many resources. Its applications range from using precise delivery systems using nanofertilizers and nanopesticides for reducing chemical run-off to developing nanosensors that enable monitoring of soil health and food security and sustainable agriculture practices that can help in minimizing the effect of climate change on agriculture productivity (Singh et al. 2021). The limitations of NPs in the agriculture sector are still unknown. For instance, there is limited knowledge about how NPs are absorbed in plants or what is their mechanism of translocation within the plant tissue and their effect is still unknown on humans in the long term. One study on wheat leaves found that inorganic NPs like ZnO enter through the stomata. Fe_3O_4 NPs enter through roots consequently enhancing the root defenses. Trace elements like selenium in its inorganic form selenate (SeO_4^{2-}) or selenite (SeO_3^{2-}) can be absorbed by the roots and transported to their aerial part (Quintarelli et al. 2024). These studies show the importance of the application of NPs while reducing the reliance on fertilizers and pesticides.

9.3.3.1 Plant Nutrition

Plants require both micro- and macronutrients for their proper growth. Macronutrients are those that are needed in large quantities such as carbon, hydrogen, oxygen, nitrogen, phosphorous, potassium, calcium, magnesium, and sulfur. A combination of these makes complex biological compounds such as carbohydrates, proteins, and nucleic acids. Micronutrients are those that are needed in smaller quantities such as minerals and vitamins. The main elements of micronutrients are zinc, copper, iron, boron, manganese chlorine, and molybdenum. For instance, Mn and Cu act as a co-factor that plays an important role in metabolic reactions. Their deficiency or toxicity causes many diseases in plants which ultimately leads to a decrease in yield. To solve these problems without relying on chemicals, NPs are being utilized in plant nutrition mainly through the development of nanofertilizers and nanobiofertilizers. These can enhance plants' nutrition by enhancing nutrient delivery by making capsules that have essential nutrients which can be released slowly having a sustained release of nutrients. Through nanofertilizers, they enhance the plant's resistivity to adverse conditions, and also, they help in the synthesis of nanobiofertilizers (NBFs). For instance, *Azospirillum* a beneficial bacterium combined with Fe is used to make nanofertilizers. This overall helps in enhancing soil health and also sustaining biodiversity (Babu et al. 2021; Arora et al. 2022).

These points highlight the applications and address issues of nanofertilizers. This shows the importance of understanding and further progressing in nanotechnology. NFs and NBFs are eco-friendlier than traditional fertilizers due to low dosage requirements further reducing the chemical load on the soil. Furthermore, more study is required to determine the optimal dosage for the NFs to minimize the potential harmful effects.

9.3.3.2 Nanoparticles in Weed and Pest Management

In current times, agrochemicals are used to protect crops from biotic factors like weeds and pests like pathogens or insects. Conventional pesticide practices such as integrated pest management have shown limitations like environmental pollution through chemicals causing a loss in soil fertility, development of pesticide resistance by the pests, and the harmful effects on humans and animals causing a loss in biodiversity. For instance, one study found that these have poor dispersion rates, with only 30% of the crops targeted. Furthermore, conventional herbicides face challenges like contamination of groundwater, and their inability to penetrate seed coats or dormant seeds (Yousef et al. 2023). Nanotechnology offers an alternative to these limitations. Wang et al. (2022) had shown that the integration of nanotechnology enhances the dispersion rate and improved the effectiveness by 31.5% and 18.9% when compared with conventional pesticides. By utilizing NPs, the required ingredients are encapsulated ensuring minimum wastage of sources and also by ensuring precise pest targeting. Nanoherbicides also known as "smart herbicides" directly deliver these ingredients to the required site which directly works at the molecular level, ensuring maximum efficiency. Due to their minimum concentration, they provide a highly effective and sustainable alternative.

Although nanotechnology has the potential for reducing our reliance on chemical pesticides and herbicides which helps in combating climate change, it also has many risks associated with it. Nanoformulations might have long-term release activity, but their small size might cause NPs to accumulate at the tropic level causing the increased risk of bioaccumulation. Other studies have also highlighted dose-dependent infection of nanoherbicides which cause blocking of vascular bundles in plants and reducing their pollination. Despite these challenges, a study by He et al. (2023) has shown that formulations are less likely to cause environmental pollution as compared to conventional pesticides as they are released in small amounts and have reduced losses.

9.3.4 Environmental Remediation

In today's world, climate change is affecting living conditions due to the pollutants released by industrial growth and urbanization. This has caused many problems like the increase in air pollution from automobiles and sludge formations from the untreated water which also causes contamination in groundwater. So, remedial processes are done for the removal of pollutants from the environment. This is defined as environmental remediation. All remediation techniques are ultimately used to lower the risks of environmental pollution. However, these remediation techniques still face many problems. Like in water formation during the treatment, sludge formation can occur which can further increase the contaminant, another one is the requirement of high energy which further creates a paradox of sustainability.

Nanotechnology offers a solution to these problems and helps in addressing climate change. Using the unique properties of NMs like their high surface area can increase the absorption capabilities like those used in air filters or in soil remediation by giving nano fertilizers to plants to enhance their absorption (Fig. 9.2). This section will explore in different ways how nanotechnology is being used in treatments reducing their impact on climate.

9.3.4.1 Nanotechnology in Water Treatment

Due to industrial activities, our water bodies are getting heavily polluted around the world. Water remediation is important for mitigating climate change. Conventional methods used for water remediation faced several limitations high energy input and lack of efficiency in removing pollutants that are present in trace amounts. These processes also often generate sludge which causes even more pollution if not treated. Nanotechnology advancements in water remediation techniques have been made to solve these limitations to mitigate climate change. Nanoscale materials like

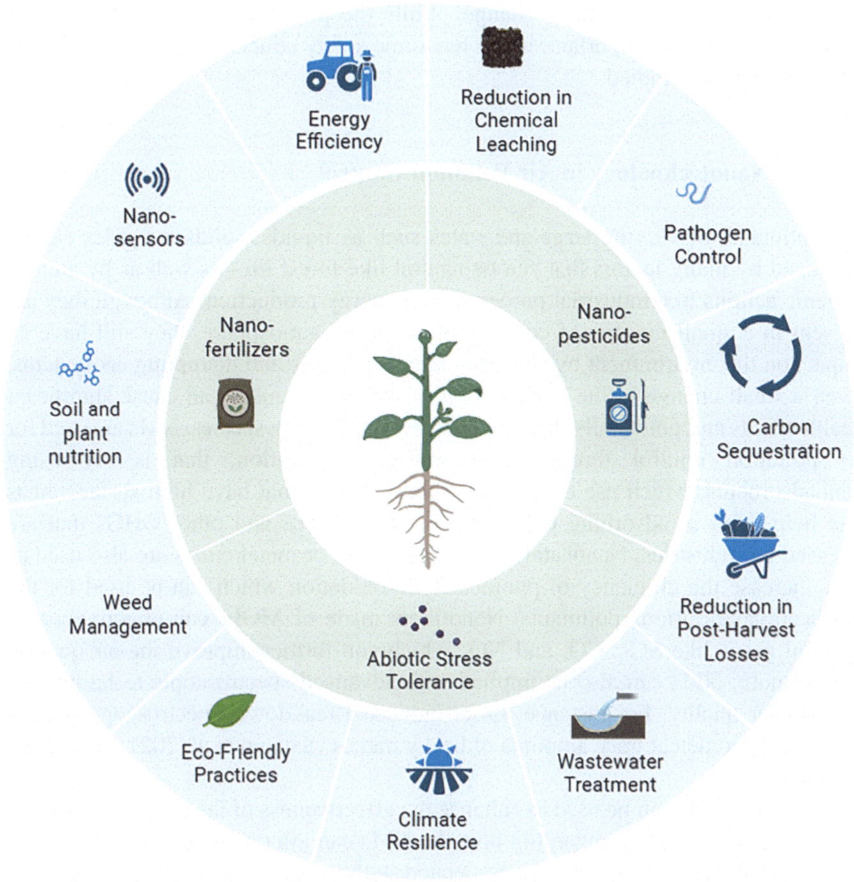

Fig. 9.2 Nanotechnology used in enhancing the plant health in different ways promoting sustainable agriculture

nanomembranes, carbon nanotubes (CNTs), graphene oxide, and metal-oxide NPs are used due to their unique property of small size and high surface area. These enhance the absorption of pollutants like heavy metals and even microplastics in the water treatment process. CNTs and graphene are used for the removal of toxic pollutants like mercury and arsenic. Industrial wastewater mainly contains pollutants like oil, dyes, and drugs which are difficult to remove by conventional methods. So, nanomaterials like silica and nanoclays are used to enhance the absorption of these materials. Nanotechnology also improves energy efficiency by increasing the efficiency of reverse osmosis membranes which enable the production of freshwater mitigating water scarcity in a growing population through lower carbon footprints (Adityawarman et al. 2024; Satyam and Patra 2024).

By solving problems like improving water quality treatment and desalination, nanotechnology not only solves the water scarcity problem but also indirectly solves the problem of global climate change. While the potential of nanotechnology in water remediation is important, it still has some safety concerns regarding its usage which need to be studied.

9.3.4.2 Nanotechnology in Air Pollution Control

Air pollutants of varying sizes and states such as liquids, solids, or gases can be produced by many factors that can be natural like forest fires as well as by anthropogenic actions like industrial processes and energy production. Although they are present in a small number of concentrations in the atmosphere, they still have an impact on the environment by changing season patterns and disrupting ecosystems. Even a small change in the composition of the environment can cause significant health effects and potentially threaten the survival of many species. NMs are used for air pollution control through their various innovation, that is, by using nanoadsorbents, which use carbon nanotubes (CNTs) that have high surface areas that help them in adsorbing pollutants like CO_2, CH_4, and other GHGs that are released by industries. Nanocatalysts mainly metals or metaloxides are also used as they increase the efficiency of photocatalytic oxidation which can be used for the purification of various pollutants. Nanofilters made of MOFs can effectively trap harmful gases like SO_2, CO, and NO_x which can further improve the air quality. Furthermore, NMs can also be implied with advanced spectroscopic techniques to monitor air quality. For instance, Laser-induced Breakdown Spectroscopy (LIBS) can be used to detect trace amounts of heavy metals (Saleem et al. 2021; Attia et al. 2024).

Although NMs can be used to enhance the effectiveness of the process, still there is limited understanding regarding how the NMs can interact with pollutants and if there are any environmental risks associated with them. Some might even become toxic due to their reactions, causing further health risks. Therefore, a study is required for the safe and successful implementation of NMs for air pollution control.

9.3.4.3 Nanotechnology in Soil Remediation

Soil is one of the vital components, and in recent times, soil contamination is a significant concern due to the release of toxic compounds from human activities like the use of extensive chemicals due to fertilizers and waste from industrial processes. These have resulted in the accumulation of pollutants causing loss of soil fertility resulting in ultimately loss of biodiversity as well as causing risks to human health through bioaccumulation. Nanotechnology utilizes NPs to remove pollutants like pesticides and persistent organic pollutants (POPs) for soil remediation (Table 9.2). Iron NPs, carbon nanotubes, and metal oxide NPs are used for remediation as they have high adsorption capacity and can also be used as photocatalysts to speed up the breakdown of organic pollutants like pesticides into less harmful products. These allow the 100% removal of pollutants within a short period of time (Raj et al. 2024).

Table 9.2 Nanotechnology in environmental remediation

Nanomaterial	Application Area	Advantages	Examples
Nano Zero-Valent Iron (nZVI)	Groundwater and soil remediation	High reactivity, cost-effective, and scalable	Used for treating chlorinated solvents and heavy metals
Zinc Oxide Nanoparticles (ZnO)	Water disinfection and degradation of pollutants	Antibacterial and photocatalytic properties	Used for dye degradation and microbial control
Titanium Dioxide Nanoparticles (TiO2)	Water treatment and pollutant degradation	Photocatalytic activity, works under UV light	Effective for degrading organic pollutants like dyes and pesticides
Copper-Based Nanoparticles	Industrial waste water and soil treatment	High catalytic efficiency, cost effective	For removal of nitrates and phosphate
Silver Nanoparticles	water and soil treatment	Effective at low concentrations	Applied in drinking water purification and soil microbial control
Magnetic Nanoparticles (e.g., Fe_3O_4)	Remediation of contaminants of water and soil	Easy recovery via magnetic fields, high surface area	Removal of arsenic, heavy metals and oil spills
Nano-Ceria (CeO_2)	Antioxidant applications in water and air systems	Redox properties, versatile catalytic activity	Pollutant removal and oxidative stress mitigation

Innovative applications of NPs when applied with hyperaccumulator plants can enhance the phytoremediation activity of these plants to boost their absorption ability. Metal-organic frameworks (MOFs) are mainly used for these purposes. These can also be used with the microbes to enhance the biodegradation process. Another application of NPs is their regeneration property. This is essential for the reusing of NPs which reduces the manufacturing cost and can ensure long-term viability for remediation (Inobeme et al. 2022).

Although nanotechnology has significant potential, it still faces several limitations. One of the major concerns as already discussed is the toxicity. Another is the biocompatibility study which is still limited. Regulations regarding the use of NMs, for safety standards, are still under study for reducing uncertainties. Solving these problems can unlock the full potential of nanotechnology for the efficient use of NMs for sustainable remediation.

9.3.5 Environmental Climate Monitoring

Real-time collection of data on contaminated regions is important as it provides informed decision-making and makes effective policies to solve it. Nowadays, satellite images are generated of contaminated regions like polluted water bodies, land degradation, and tracking of air pollution levels. This satellite technology helps in collecting these geographical data aiding in creating efficient strategies. Satellite remote sensing doesn't measure these variables directly but through electromagnetic radiation. Then, some complex models are used to calculate the desired parameter. NMs are integrated with satellite technology to enhance this step. These improve the input of data through the making of advanced nanosensors which with improved accuracy provide relevant data on the contaminated sites (Ferral et al. 2024). NMs have unique properties such as improved reactivity and provide novel optical, magnetic, and electronic properties when integrated due to their high surface-to-volume ratio. This made it possible to create selective sensors for the detection of contaminants. NMs like quantum dots and carbon-based structures use a process known as the confinement effect that is it amplifies the light absorption and scattering, of a pollutant which makes it traceable even at a minimum amount (Fu et al. 2024). These nanosensors can also be used for the management of agriculture practices through real-time monitoring of soil health, moisture, and nutrients, reducing resource waste and environmental impact. They reduce the need for resources indirectly like pesticides by detecting the diseases early on. Through this, they promote biodegradable and eco-friendly activities (Yang and Duncan 2021).

However, several challenges must be addressed to fully realize the potential of nanoenabled environmental monitoring. Issues such as sensor stability, reproducibility, scalability, and cost-effectiveness require continued research and innovation. Furthermore, concerns regarding the environmental fate and toxicity of nanomaterials highlight the need for rigorous risk assessment and regulatory oversight. Overcoming these challenges will be pivotal in advancing nanotechnology's role in creating sustainable and effective solutions for environmental monitoring and protection.

9.4 Challenges Associated with Nanotechnology

Nanotechnology has come in trend nowadays for addressing climate change that gives solutions across diverse sectors due to its unique property of small size and having a high-surface area ratio like in renewable energy sources, energy storage, sustainable agriculture, environmental remediation by improving the efficiency and further reducing the greenhouse gas emissions by minimizing their number of emissions. However, the use of nanomaterials still faces many challenges during their implication. Furthermore, the lack of regulations about the usage raises concerns about its long-term impact.

9.4.1 Environmental Risks

Although nanomaterial usage has several benefits it also has drawbacks that need to be considered before usage. First, is the waste generated during the synthesis of nanomaterials, as it requires high energy during its production. As NMs are used in small dosages their overall concentrations are still below harmful levels but point sources of discharges like sewage may pose risk in its increased concentrations. Still, the long-term effects of NMs are still unknown due to limited research on them (Quintarelli et al. 2024).

9.4.2 Economic Challenges

Nanotechnology has several pros but the economic feasibility is still an issue. Lower costs of production will make the manufacturing process easy and also reduce the investment risks. But currently, the manufacturing needs specialized equipment making the scalability limited. Furthermore, resources used in the manufacturing of NMs could also have their own negative effects which makes a sustainability issue. Developing countries couldn't employ nanotechnology as it is too expensive for them. Overcoming these economic barriers through the lowering of manufacturing process, cooperation between countries could ensure the adoption of technology and this can also help in mitigation of climate change. For instance, nanomaterials used in the manufacturing of energy storage devices are now being integrated with EVs launched in many countries nowadays as these are becoming affordable due to bringing down the costs which also reduces the emissions of GHGs. This highlights the importance of scalability and cost considerations.

9.4.3 Technical Challenges

Although the study of NMs is advancing rapidly, it still faces several challenges like the complexity of production, their performance, safety, and lastly the long-term impact they cause. The production of NMs needs skill-intensive labor as it requires interdisciplinary research knowledge like chemistry, materials science, and fabrication techniques. Although nanotechnology is being used for various purposes nowadays, still there are limited studies on the long-term impact on the environment. These studies could help us in making policies if necessary. Along with these challenges education regarding the NMs to stakeholders is critical for dispelling the fear and misconception about it to promote the acceptance of NMs (Huang et al. 2024).

9.5 Future Direction

NMs are playing an important role in enhancing technologies that can help in climate change mitigation through various means. Currently, the integration of AI and Machine Learning (ML) with nanotechnology is done for the simulation to test the compatibility of NPs with the required system which also reduces the trial and error. By using nanotechnology across diverse sectors sustainable practices can be effectively applied that could help in mitigating climate change.

9.6 Conclusion

Nanotechnology has helped in rapid advancement in many sectors that can play a role in mitigating climate change through various means. In the renewable energy sector, nanotechnology has enhanced energy efficiency from different sources like solar energy, wind energy, tidal energy, and geothermal energy. Furthermore, it has also enhanced the energy transportation and storage system which has reduced the wasteful energy. This has helped in reducing the GHG emissions in the environment. Also, through the agriculture sector, nanotechnology has helped in enhancing the plant yield which can also improve the biodiversity by minimizing the damage on it. Indirectly targeted the food scarcity problem that we might face due to the increasing population.

However, despite its potential, nanotechnology still faces many challenges such as environmental risks, technical risks, and scalability risks which can be further solved by having a balance between sustainability and development. By solving such problems by further advancements, nanotechnology can help in climate change mitigation efforts which could bring a greener and more sustainable future.

References

Abdalla AM, Elnaghi BE, Hossain S, Dawood M, Abdelrehim O, Azad AK (2020) Nanotechnology utilization in energy conversion, storage and efficiency: a perspective review. Adv Energy Convers Mater:30–54. https://doi.org/10.37256/aecm.11202075

Adanma NUM, Ogunbiyi NEO (2024) Assessing the economic and environmental impacts of renewable energy adoption across different global regions. Eng Sci Technol J 5(5):1767–1793. https://doi.org/10.51594/estj.v5i5.1154

Adityawarman D, Lugito G, Kawi S, Wenten IG, Khoiruddin K (2024) Advancements and future trends in nanostructured membrane technologies for seawater desalination. Desalination:118390. https://doi.org/10.1016/j.desal.2024.118390

Aglawe K, Yadav R, Thool S (2021) Preparation, applications and challenges of nanofluids in electronic cooling: a systematic review. Mater Today Proc 43:366–372. https://doi.org/10.1016/j.matpr.2020.11.679

Alguacil FJ (2023) Nanomaterials for CO2 capture from gas streams. Separations 11(1):1. https://doi.org/10.3390/separations11010001

Aljibori H, Alamiery A, Kadhum A (2023) Advances in corrosion protection coatings: a comprehensive review. Int J Corros Scale Inhib 12(4). https://doi.org/10.17675/2305-6894-2023-12-4-6

Alkhalayfeh MA, Aziz AA, Pakhuruddin MZ, Katubi KMM (2021) Recent advances of perovskite solar cells embedded with plasmonic nanoparticles. Phys Status Solidi A 218(17). https://doi.org/10.1002/pssa.202100310

Arora S, Murmu G, Mukherjee K, Saha S, Maity D (2022) A comprehensive overview of nanotechnology in sustainable agriculture. J Biotechnol 355:21–41. https://doi.org/10.1016/j.jbiotec.2022.06.007

Attia YA, Ezet AE, Saeed S, Galmed AH (2024) Nano carbon-modified air purification filters for removal and detection of particulate matters from ambient air. Sci Rep 14(1):621. https://doi.org/10.1038/s41598-023-50902-x

Babu S, Singh R, Yadav D, Rathore SS, Raj R, Avasthe R, Yadav S, Das A, Yadav V, Yadav B, Shekhawat K, Upadhyay P, Yadav DK, Singh VK (2021) Nanofertilizers for agricultural and environmental sustainability. Chemosphere 292:133451. https://doi.org/10.1016/j.chemosphere.2021.133451

Bhattacharjee I, Maiti DK (2021) Nanotools and nanodevices in geothermal energy. In: Elsevier eBooks, pp 507–518. https://doi.org/10.1016/b978-0-12-821709-2.00022-0

Chausali N, Saxena J, Prasad R (2023) Nanotechnology as a sustainable approach for combating the environmental effects of climate change. J Agric Food Res 12:100541. https://doi.org/10.1016/j.jafr.2023.100541

Chen Q, Mao B, Liu Y, Zhou Y, Huang H, Wang S, Li L, Yan W, Shi W, Kang Z (2024) Designing 2D carbon dot nanoreactors for alcohol oxidation coupled with hydrogen evolution. Nat Commun 15(1):8052. https://doi.org/10.1038/s41467-024-52406-2

Davies EL (2024) Innovations in energy through nanomaterials enhancing solid-state battery safety and efficacy. Highl Sci Eng Technol 116:250–256. https://doi.org/10.54097/fjrk7k69

Ferral A, Bonansea M, Scavuzzo CM, Nemiña F, Paci MB, Ramirez JC, Sepúlveda B, Fraxedas J, Esplandiu MJ (2024) Bringing satellite and nanotechnologies together: unifying strengths against pollution and climate change. Front Nanotechnol 6. https://doi.org/10.3389/fnano.2024.1332820

Fu Y, Liu T, Wang H, Wang Z, Hou L, Jiang J, Xu T (2024) Applications of nanomaterial technology in biosensing. J Sci Adv Mater Devices 9(2):100694. https://doi.org/10.1016/j.jsamd.2024.100694

Garanovic A (2023) IEA-OES unveils international roadmap for 300GW of ocean energy by 2050. Offshore Energy. https://www.offshore-energy.biz/iea-oes-unveils-international-roadmap-for-300gw-of-ocean-energy-by-2050/

Ghosh D, Sarkar K, Devi P, Kim K, Kumar P (2020) Current and future perspectives of carbon and graphene quantum dots: from synthesis to strategy for building optoelectronic and energy devices. Renew Sust Energ Rev 135:110391. https://doi.org/10.1016/j.rser.2020.110391

Gnanavendan S, Selvaraj SK, Dev SJ, Mahato KK, Srii Swathish R, Sundaramali G, Oussama A, Azab M (2024) Challenges, solutions and future trends in EV-technology: a review. IEEE Access 12:17242–17260

Gonçalves I, Rodrigues C, Ventura J (2023) Sea state adaptation enhances power output of triboelectric nanogenerators for tailored ocean wave energy harvesting. Adv Energy Mater 14(2). https://doi.org/10.1002/aenm.202302627

He J, Li J, Gao Y, He X, Hao G (2023) Nano-based smart formulations: a potential solution to the hazardous effects of pesticide on the environment. J Hazard Mater 456:131599. https://doi.org/10.1016/j.jhazmat.2023.131599

Hsu C, Ajaj Y, Ghadir GK, Al-Tmimi HM, Alani ZK, Almulla AA, Hussein MA, Al-Tameemi AR, Mahmoud ZH, Mustafa MA, Kianfar F, Habibzadeh S, Kianfar E (2024) Rechargeable batteries for energy storage: a review. e-Prime 8:100510. https://doi.org/10.1016/j.prime.2024.100510

Huang X, Auffan M, Eckelman MJ, Elimelech M, Kim J, Rose J, Zuo K, Li Q, Alvarez PJJ (2024) Trends, risks and opportunities in environmental nanotechnology. Nat Rev Earth Environ 5(8): 572–587. https://doi.org/10.1038/s43017-024-00567-5

Inobeme A, Mathew JT, Adetunji CO, Ajai AI, Inobeme J, Maliki M, Okonkwo S, Adekoya MA, Bamigboye MO, Jacob JO, Eziukwu CA (2022) Recent advances in nanotechnology for remediation of heavy metals. Environ Monit Assess 195(1):111. https://doi.org/10.1007/s10661-022-10614-7

IRENA and IGA (2023) Global geothermal market and technology assessment, International Renewable Energy Agency, Abu Dhabi; International Geothermal Association, The Hague

Jason YJJ, How HG, Teoh YH, Chuah HG (2020) A study on the tribological performance of nanolubricants. PRO 8(11):1372. https://doi.org/10.3390/pr8111372

Jayakumari ABN, Malik NG, Mittal G, Martelo D, Kale N, Paul S (2023) A review on geothermal heat exchangers: challenges, coating methods, and coating materials. Coatings 13(12):1988. https://doi.org/10.3390/coatings13121988

Julien CM, Mauger A (2024) Fabrication of Li4Ti5O12 (LTO) as anode material for Li-ion batteries. Micromachines 15(3):310. https://doi.org/10.3390/mi15030310

Khan F, Pandey P, Upadhyay TK (2022a) Applications of nanotechnology-based agrochemicals in food security and sustainable agriculture: an overview. Agriculture 12(10):1672. https://doi.org/10.3390/agriculture12101672

Khan MZA, Khan HA, Aziz M (2022b) Harvesting energy from ocean: technologies and perspectives. Energies 15(9):3456. https://doi.org/10.3390/en15093456

Kumar P, Channi HK, Babbar A, Kumar R, Bhutto JK, Khan TMY, Bhowmik A, Razak A, Wodajo AW (2024) A systematic review of nanotechnology for electric vehicles battery. Int J Low-Carbon Technol 19:747–765. https://doi.org/10.1093/ijlct/ctae029

Lameirinhas RAM, Torres JPN, De Melo Cunha JP (2022) A photovoltaic technology review: history, fundamentals and applications. Energies 15(5):1823. https://doi.org/10.3390/en15051823

Lee G, Ahmed I, Jhung SH (2023) CO2 adsorption using functionalized metal–organic frameworks under low pressure: contribution of functional groups, excluding amines, to adsorption. Chem Eng J 481:148440. https://doi.org/10.1016/j.cej.2023.148440

Li L, Liu W, Dong H, Gui Q, Hu Z, Li Y, Liu J (2021) Surface and Interface engineering of nanoarrays toward advanced electrodes and electrochemical energy storage devices. Adv Mater 33(13):e2004959. https://doi.org/10.1002/adma.202004959

Liu N, Wang Y, Bai Q, Liu Y, Wang P, Xue S, Yu Q, Li Q (2022) Road life-cycle carbon dioxide emissions and emission reduction technologies: a review. J Traffic Transport Eng 9(4):532–555. https://doi.org/10.1016/j.jtte.2022.06.001

Liu B, Liu Z, Li Y, Feng F (2023) A wind tunnel test of the Anti-Icing properties of MOS2/ZNO hydrophobic Nano-Coatings for wind turbine blades. Coatings 13(4):686. https://doi.org/10.3390/coatings13040686

Mandal P (2022) Application of Plasmonics in solar cell efficiency improvement: a brief review on recent Progress. Plasmonics 17(3):1247–1267. https://doi.org/10.1007/s11468-022-01616-9

Mangaiyarkkarasi J (2024) Nanostructural innovations. In: Advances in chemical and materials engineering book series, pp 51–74. https://doi.org/10.4018/979-8-3693-5320-2.ch003

Parameswaran AK, Azadmanjiri J, Palaniyandy N, Pal B, Palaniswami S, Dekanovsky L, Wu B, Sofer Z (2022) Recent progress of nanotechnology in the research framework of all-solid-state batteries. Nano Energy 105:107994. https://doi.org/10.1016/j.nanoen.2022.107994

Quintarelli V, Hassine MB, Radicetti E, Stazi SR, Bratti A, Allevato E, Mancinelli R, Jamal A, Ahsan M, Mirzaei M, Borgatti D (2024) Advances in Nanotechnology for Sustainable Agriculture: a review of climate change mitigation. Sustain For 16(21):9280. https://doi.org/10.3390/su16219280

Raj P, Ghosh A, Ravi N, Jadav SK, Patra H, Shehnaz N, Panotra N, Sharma V, Sharma R, Pandey SK, Singh BV (2024) Nanomaterials for sustainable soil remediation and contaminant

immobilization. Asian J Soil Sci Plant Nutr 10(4):684–697. https://doi.org/10.9734/ajsspn/2024/v10i4439
Rajukkannu S, Bunpheng W, Dhairiyasamy R, Gopinath V (2025) Efficiency improvement in silicon and perovskite solar cells through nanofluid cooling using citrate and PVP stabilized silver nanoparticles. Sci Rep 15(1):833. https://doi.org/10.1038/s41598-025-85374-8
Sadeq AM, Homod RZ, Hussein AK, Togun H, Mahmoodi A, Isleem HF, Patil AR, Moghaddam AH (2024) Hydrogen energy systems: technologies, trends, and future prospects. Sci Total Environ 939:173622. https://doi.org/10.1016/j.scitotenv.2024.173622
Saleem H, Zaidi SJ, Ismail AF, Goh PS (2021) Advances of nanomaterials for air pollution remediation and their impacts on the environment. Chemosphere 287:132083. https://doi.org/10.1016/j.chemosphere.2021.132083
Salim A, Jihan JI, Das BK, Ahmed R (2024) Thermal performance investigation of N-shape double-pipe heat exchanger using Al2O3, TiO2, and Fe3O4-based nanofluids. Energy Sci Eng 12(3):1072–1090. https://doi.org/10.1002/ese3.1698
Satyam S, Patra S (2024) Innovations and challenges in adsorption-based wastewater remediation: a comprehensive review. Heliyon 10(9):e29573. https://doi.org/10.1016/j.heliyon.2024.e29573
Simpa NP, Solomon NNO, Adenekan NOA, Obasi NSC (2024) Nanotechnology's potential in advancing renewable energy solutions. Eng Sci Technol J 5(5):1695–1710. https://doi.org/10.51594/estj.v5i5.1137
Singh S, Sangwan S, Sharma P, Devi P, Moond M (2021) Nanotechnology for sustainable agriculture: an emerging perspective. J Nanosci Nanotechnol 21(6):3453–3465. https://doi.org/10.1166/jnn.2021.19012
Suwanapingkarl P, Srivallop K (2020) Reviews existing technologies and proposes 'E8-PowerBuoys' Nano-scale generator of tidal and wave energy for river and ocean. In: International Conference on Power, Energy and Innovations (ICPEI), Chiangmai, Thailand, pp 85–88. https://doi.org/10.1109/icpei49860.2020.9431402
Van Dijk M, Morley T, Rau ML, Saghai Y (2021) A meta-analysis of projected global food demand and population at risk of hunger for the period 2010–2050. Nat Food 2(7):494–501. https://doi.org/10.1038/s43016-021-00322-9
Vasantha V, Sharvari S, Alfia N, Praveen N (2021) Application of nanotechnology toward improved production of sustainable bioenergy. Nanomaterials:445–479. https://doi.org/10.1016/b978-0-12-822401-4.00010-6
Wang D, Saleh NB, Byro A, Zepp R, Sahle-Demessie E, Luxton TP, Ho KT, Burgess RM, Flury M, White JC, Su C (2022) Nano-enabled pesticides for sustainable agriculture and global food security. Nat Nanotechnol 17(4):347–360. https://doi.org/10.1038/s41565-022-01082-8
Yang T, Duncan TV (2021) Challenges and potential solutions for nanosensors intended for use with foods. Nature Nanotechnol 16(3):251–265. https://doi.org/10.1038/s41565-021-00867-7
Yogesh GK, Nandi D, Yeetsorn R, Wanchan W, Devi C, Singh RP, Vasistha A, Kumar M, Koinkar P, Yadav K (2024) A machine learning approach for estimating supercapacitor performance of graphene oxide nano-ring based electrode materials. Energy Adv 4:119. https://doi.org/10.1039/d4ya00577e
Yousef HA, Fahmy HM, Arafa FN, Allah MYA, Tawfik YM, Halwany KKE, El-Ashmanty BA, Al-Anany FS, Mohamed MA, Bassily ME (2023) Nanotechnology in pest management: advantages, applications, and challenges. Int J Trop Insect Sci 43(5):1387–1399. https://doi.org/10.1007/s42690-023-01053-z

Chapter 10
Role of Nanotechnology in Preventing Post-harvest Losses of Agriproducts

Rajni Garg, Ansar Anjum, Rishav Garg, and Nnabuk Okon Eddy

Abstract The United Nations has set a goal of eradicating hunger by the year 2030 in its sustainability goals; yet, achieving global food security seems to be a foolhardy goal. The root cause is not only a decline in crop production but also the loss at many stages of food handling, transportation, and storage, commonly termed post-harvest loss. The major food spoilage is primarily due to temperature variations, changes in humidity, and infestation by pests resulting in the perishing of the food. This spoilage reduces the quality, nutritional value, and shelf life of food, resulting in economic loss. Although conventional food packaging is used to avoid spoilage, it is not successful in combating post-harvest loss. Nanotechnology is a potentially useful approach to mitigate these issues, providing sustainable and environmentally friendly agricultural practices. This chapter explores the use of various nanomaterials (NMs) as nanocarriers, nanoemulsions, and nanoencapsulation to enhance the harvest yield and prevent post-harvest loss. Active and smart food packaging with the integration of NMs having antioxidant, antibacterial, and/or barrier protection potential increases the shelf life of the enclosed products. The chapter also highlights the associated challenges and future prospects of the application of nanotechnology in handling post-harvest loss.

Keywords Active packaging · Smart packaging · Post-harvest loss · Shelf life · Nanomaterials

R. Garg (✉)
Department of Applied Sciences and Humanities, Galgotias College of Engineering & Technology, Greater Noida, Uttar Pradesh, India
e-mail: rajnigarg@science.org.in

A. Anjum
Department of Applied Chemistry, Delhi Technological University, Delhi, India

R. Garg
Department of Civil Engineering, Galgotias College of Engineering & Technology, Greater Noida, Uttar Pradesh, India

N. O. Eddy
Department of Pure and Industrial Chemistry, University of Nigeria, Nsukka, Enugu State, Nigeria

S. K. Yadav, A. Kumari (eds.), *Nanobiotechnology in Agri-food Sector*, Sustainability Sciences in Asia and Africa, https://doi.org/10.1007/978-981-96-7756-6_10

10.1 Introduction

Food is the basic need of every living organism to sustain its metabolic activities. Since the evolution of human civilization, agriculture has had prime significance as human beings are primarily dependent upon various agricultural sources as their prime source of food. Various food processing and packaging industries obtain their raw materials from the agricultural harvest (Kavakebi et al. 2021). However, post-harvest loss around the world is a significant concern leading to a decrease in overall food availability. The presiding global food crisis is anticipated to escalate with the current rate of increase in population, which is expected to reach more than 9 billion people by 2050 (Guan et al. 2023). Almost 33% of the global post-harvest amounting to billions of money is getting lost when there is already demand for more food production. This post-harvest loss is in terms of both quality and quantity, making unintended reductions in food availability for human consumption (Hu et al. 2024).

Post-harvest loss can be due to various factors, including ineffective processing techniques, inadequate storage facilities, and incorrect handling of the produce leading to environmental degradation due to the attack of air, moisture, pests, and microbes (Fig. 10.1). The advent of nanotechnology has transformed the agriculture sector by providing solutions to these critical issues (Yang et al. 2023). Efficient food processing, proper handling, and storage with active, smart, and intelligent packaging help to maintain the food quality and enhance its shelf life thereby efficiently reducing post-harvest loss, and are termed as sustainable agricultural practices (Hezma et al. 2019). Various NMs such as Ag (Shankar et al. 2021), ZnO, and TiO_2 NPs (Azizi-Lalabadi et al. 2020) have been explored recently to protect pre-harvest to post-harvest and enhance overall food production, quality, and availability.

Nano-formulations like nanocarriers, nanoemulsions, and nanoencapsulation not only assist in maintaining the quality of agricultural goods but also help in the enhancement of overall agricultural yield. Nano-formulated agrochemicals like pesticides and fertilizers are used to reduce pest infestations and thus increase crop productivity (Tymczewska et al. 2022). Nanoencapsulation is used to deliver targeted nano-formulated pesticides and fungicides that reduce the overuse of chemicals and their harmful environmental impact. Throughout the journey of agri-food products from production to packing, transportation to storage, and selling, nanosensors are used to monitor a variety of chemical, physical, and biological aspects (Hu et al. 2024). These nanosensors are used in smart packaging to detect spoilage and contamination for the real-time monitoring of the packaged products and help in the early diagnosis of the issue. Nanosensors are also capable of detecting diseases caused by various pathogens after harvest, as well as adulteration, allowing timely treatment and reducing losses (Hassan et al. 2024).

NMs-based packaging materials and coatings can extend the shelf life of the enclosed product by providing barrier protection and controlling the amount of moisture, oxygen, and other harmful gases around the food product (Singh and

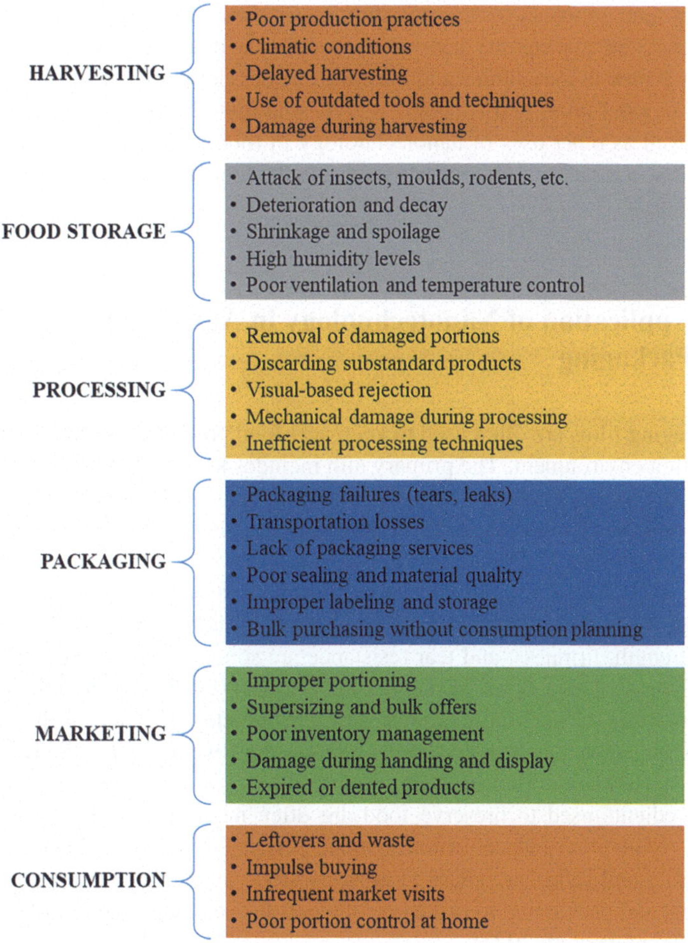

Fig. 10.1 Causes for agri-food loss

Sahareen 2017). NPs can also absorb or neutralize ethylene gas, delaying ripening and senescence (Indumathi et al. 2019). The incorporation of silver NPs in the packaging films limits the growth of harmful microorganisms, thereby reducing food deterioration and improving food safety (Mihaly Cozmuta et al. 2018). Silica NPs-integrated food packaging provides a barrier to excess moisture and helps in reducing food spoilage due to the hydrolysis of enzymes and growth of microorganisms (Videira-Quintela et al. 2022), while nano-cellulose-based coatings help in the retention of water in chopped fruits, nuts, and vegetables (Mihaly Cozmuta et al. 2018). NMs-based packaging also has the potential to maintain the right temperature during storage and transportation. NMs can also aid food safety by detecting the presence of contaminants and reducing microbial growth.

This chapter emphasizes the possibilities of nanotechnology in agriculture to indicate that the appropriate use of NMs offers a practical path for addressing important issues in agriculture and forward our development toward a sustainable and food-secure future. In this regard, active, smart/intelligent, biodegradable packaging as well as other uses of nanotechnology in food packaging have been investigated. Also, a toxicity study of NMs for use in the food and agriculture industry has been included.

10.2 Application of Nanotechnology in Agri-Food Packaging

The packaging films are often used to keep agri-food products safe from the outside deteriorating environment. The primary aim includes keeping the product safe from the attack of oxygen, light, moisture, and microbes in the surroundings that can cause product spoilage (Yang et al. 2023). Films with excellent barrier protection and the potential of blocking UV radiation enhance the service life of the enclosed product (Shankar et al. 2018). Packaging papers should be strong and flexible enough to handle manufacturing, packaging, storage, transportation, and handling stresses. Tensile strength, firmness, and tear resistance are some of the mechanical qualities of films (Li et al. 2017). Some packaging films are made to be see-through or clear so the customer can see the product. Some films are opaque, which means they don't let light into the package and protect items that are sensitive to light (Mihaly Cozmuta et al. 2018). Packaging films must be strong and leak-proof to stop contamination as well. Ingredients used to preserve food are often made to prevent the growth of microbes, oxidation, and control enzymatic activity (Singh and Sahareen 2017). Food quality and safety are checked at many steps, such as during travel, storage, processing, and packaging, to make sure it meets the standards and is still safe to eat (Videira-Quintela et al. 2022).

Accordingly, the packaging is broadly classified into active, smart, and biodegradable packaging. Active packaging involves the addition of certain chemicals to the packaging film or container to enhance the longevity of the enclosed product (Mihaly Cozmuta et al. 2018). The additives are selected to enhance the mechanical (Nguyen et al. 2021), barrier (Cheng et al. 2021), and optical qualities (Indumathi et al. 2019) of the packaging material. The various additives include natural polymers (Azizi-Lalabadi et al. 2020), essential oils (Shankar et al. 2021), plant extracts (Tavana et al. 2024), phytochemicals (Hassan et al. 2024), and NPs (Li et al. 2017). These additives are used in composite/blend form and infused with the packaging material. On the other hand, intelligent packaging materials are used to know about the condition of the packaged product. These materials react to changes in the packaged product, such as changes in quality, contamination, oxidation, or the stage of ripeness. These materials may convey information about the status of the item within without coming into direct touch with it and thus make it much simpler to monitor the quality and safety of food. Alterations in temperature, concentrations of gasses, allergens, and differences in nutritional content can also be confirmed. Using these methods can help find and identify different bacteria, such as microbes and mushrooms, in kept food (Tonezzer et al. 2022).

The usual plastic-based packaging can't be decomposed easily by physical, chemical, or biological ways, which leads to a buildup of waste leading to a lot of environmental and health issues. Research has shown that putting trash on city streets and roads can damage the sewage system, causing it to get clogged up and overflow (Garg et al. 2022a). The health of aquatic creatures is seriously threatened by the release of large amounts of plastic trash into sea and freshwater environments. Although burning is the easiest method to get rid of waste, the process of burning things releases harmful gases into the air, which then makes the air quality worse. In addition, this release of toxic gases worsens global warming (Amin et al. 2022). More and more people around the world are worried about the problems with getting rid of trash and the damage that many manmade plastics do to the environment and public health. This means that there is a greater need to look into other materials that are better for the earth. In this regard, biodegradable packaging has become a good option for plastic packaging (Indumathi et al. 2019). The biodegradable materials break down by the action of microbes like bacteria, fungi, and viruses and are generally obtained from renewable natural sources. As biodegradable plastics break down, they release simple substances like water, carbon dioxide, inorganic chemicals, and biomass (Sobhan et al. 2021). They have slightly different tensile strength, thermal properties, and barrier protection as compared to conventional plastic. Biodegradable packaging is now being used in the agri-food sector (Mathew et al. 2019). Nanocomposites of biodegradable materials that can kill microbes and/or protect cells from damage are now used in biodegradable packaging. Especially, biogenic NMs have been embedded within the packaging films and containers without exposure to the enclosed product (Echegoyen and Nerín 2013).

10.2.1 Nanotechnology for Enhanced Food Quality and Shelf Life

The food industry is highly concerned with reducing post-harvest loss as well as maintaining the shelf life and quality of food products. Perishable foods such as fruits, vegetables, seafood, and meat are highly susceptible to spoilage and need special protective measures. Different NMs have been incorporated into natural compound-based films to improve the shelf life of various perishable foods and maintain their nutritional quality (Cheng et al. 2021). Zinc oxide NPs and neem oil have been used to obtain chitosan bio-nanocomposites for the packaging of carrots. The composite film has improved thickness, transparency, and elongation. It also exhibited increased barrier qualities, tensile strength, and decreased water entry.

Table 10.1 Applications of nanomaterials in improving the shelf life of food products

Materials used	Findings/Applications	Reference
Ag NPs/Cellulosic material	Retained nutritional content and extended shelf life of the enclosed vegetable by improving antimicrobial action and maintaining moisture.	Singh and Sahareen (2017)
TiO_2 NPs/Chitosan/Cellulose active papers	The film showed 79.25% yeast/mold inhibition after 6 months of storage, extending the shelf life of walnut kernel.	Mihaly Cozmuta et al. (2018)
Ag NPs/essential oils/ Chitosan film	Extended shelf life of strawberries with gamma irradiation to exhibit improved antibacterial resistance and reduced weight loss.	Shankar et al. (2021)
Ag NPs/agar/maltodextrin polyvinyl alcohol coating	Reduction in acidity loss, weight loss, pH variations, and softening of bananas during storage at 25 °C, extended shelf life by 5 days.	Nguyen et al. (2021)
Ag NPs/Pectin/pullulan film	Improved antimicrobial activity against foodborne pathogens and enhanced barrier properties.	Lee et al. (2019)
Ag NPs/PLA film	Improved transmission rates of oxygen and moisture without impacting the strength and stability of the film. Enhanced anti-oxidation, antibacterial, and anti-ultraviolet properties extend the shelf life of strawberries by 7 days.	Cheng et al. (2021)
ZnO and TiO2 NPs supported on 4A zeolite/PVA/gelatin films	Doubled the shelf life of shrimp from 6 to 12 days by inhibiting bacterial growth.	Azizi-Lalabadi et al. (2020)
ZnO NPs/PVA/arginine/ chitosan film	Exhibited superior mechanical properties, water, and oxygen barrier.	Yang et al. (2023)
ZnO NPs/chitosan/cellulose acetate phthalate films	Optimal hydrophobicity and oxygen permeability extended the shelf life of black grape fruits.	Indumathi et al. (2019)
ZnO NPs/PLA films	Improved firmness, phenolic content, color retention, and sensory quality of apples.	Li et al. (2017)
ZnO NPs/PLA films	Reduced gas permeability and improved tensile properties and antimicrobial activity against E. coli.	Marra et al. (2016)
ZnO NPs/PVA/chitosan film	The 10–15 wt% ZnO NPs showed the greatest thermal stability, mechanical strength, and antibacterial activity.	Hezma et al. (2019)

Additionally, the film was able to inhibit the attack of *Escherichia coli* (Sanuja et al. 2015).

Coating food is a very old concept in food preservation with the purpose of improving barrier protection against the harmful environment, chemicals, pests, and physical damage (Nguyen et al. 2021). It also inhibits the growth of microorganisms and prevents early dehydration and ripening of the enclosed food, thereby preserving the food quality and enhancing its shelf life. The shelf life of Le Conte pears was extended by 105 days by using an edible film (Table 10.1). The film was infused with beeswax, chitosan, and pollen grains and used as a coating. The study reported an increased water contact angle of these coatings. As a result, the coating limited the entry of moisture and efficiently regulated the amount of moisture around the pear (Sultan et al. 2021). Various traditional food packaging materials are non-biodegradable and pose environmental concerns. Nanotechnology has offered innovative solutions regarding nano-coatings with antimicrobial, barrier protection, and sensory properties (Lee et al. 2019).

Researchers have also investigated the application of bioactive substances such as essential oils (Shankar et al. 2021) and plant extracts (Bahrami et al. 2019) in improving and maintaining the nutritional value of food. These compounds are encapsulated in the nano-form in the biopolymers to obtain edible coatings for fruits and vegetables. A nanoemulsion of ginger essential oil and carnauba wax was used to obtain hydroxypropyl methylcellulose-based nanocomposites. The nano-product was used in coating for preserving and improving the quality of papayas. The coating delayed the ripening process, prevented color changes, and reduced the weight loss of the enclosed fruit in cold storage (Miranda et al. 2022). Nano-encapsulated tarragon essential oil was incorporated in a chitosan-gelatin edible coating. This coating was then applied to pork slices to prevent lipid oxidation and microbial action during refrigeration. The coating exhibited improved antibacterial, antioxidant, and sensory potential. The study reported an increase in the shelf life of the pork slice by up to 12 days (Zhang et al. 2020).

These coatings help to preserve food for a longer period and also improve its nutritional content. Thus, it prevents food wastage due to spoilage during storage. Titanium dioxide NPs and litchi peels were blended into chitosan films. The active film lowered the activity of polyphenol oxidase, limited the electrolyte leakage, and inhibited the production of malondialdehyde. This action delayed the browning process of the packaged apples and also prevented weight loss over a more than 180-day period (Liu et al. 2021). Qiu et al. explored a nanoemulsion of gelatin-chitosan, cinnamon essential oil, and rosemary extract to increase the shelf life of chicken patties (Qiu et al. 2022). Likewise, Cui et al. (2022) prepared xanthan gum-based nanoliposomes containing essential oil extract from *Litsea cubeba.* The product was used to coat salmon to prevent degradation by *Vibrio parahaemolyticus* and oxidation during cold preservation. Additionally, the preservation was extended to several days by this coating. Essential oil nanoemulsions have also been used to prolong the shelf life of meat of turkey (Milani et al. 2020), rainbow trout (Shokri et al. 2020), and lamb loin (Hasani-Javanmardi et al. 2021).

10.2.2 Nanotechnology for Protection Against Microbes

The most common NPs used in active nanopacking include metals like silver (Ag) (Youssef et al. 2014), gold (Au) (Kavakebi et al. 2021), and copper (Cu) (Ahmed et al. 2018), along with metal oxides like zinc oxide (ZnO) (Priyadarshi and Negi 2017), titanium dioxide (TiO_2) (Li et al. 2018), and silicon oxide (SiO_2) (Videira-Quintela et al. 2022), with photocatalytic properties. The active nanopackages can manage and absorb carbon dioxide, ethylene, moisture, oxygen, water vapor, and other harmful gases. These packages also improve microbiological safety and act as thermal insulators (Marra et al. 2016). Including antibacterial potential against gram-negative bacteria (*Escherichia coli* ATCC 8739, *Salmonella typhimurium* ATCC 14028, *Pseudomonas aeruginosa* ATCC 9027, and *Klebsiella pneumoniae* ATCC 10031) compared to gram-positive bacteria (*Bacillus cereus* ATCC 1247, *Staphylococcus aureus* ATCC 25923, *Streptococcus pneumoniae* ATCC 49615, and *Listeria monocytogenes* ATCC 7644), NPs have shown the potential to essentially stop the growth of several bacteria (Bahrami et al. 2019) and fungi (Youssef et al. 2014).

Although the exact mechanism for the antibacterial action of AgNPs is still unclear. Yet, nanosilver, being highly stable and able to fight a broad spectrum of diseases, is used extensively in food packaging (Bhardwaj et al. 2023). It is considered that Ag^{+} ions are released from silver NPs near the bacterial cells that absorb these free silver ions. As a result, the membrane permeability in the cytoplasm activates, and the ions invade the cell. These ions hinder the respiratory enzymes and produce reactive oxygen species (ROS) that further impede the synthesis of adenosine triphosphate (ATP) (Lata et al. 2022). ROS generation damages the bacterial proteins, mitochondrial enzymes, and DNA, restricting the cell metabolism that can cause bacterial death. This antimicrobial action of AgNPs has been used to provide protection against microbes and extend the shelf life of various food products (Garg et al. 2022b). Ginger extract-mediated Ag NPs were reinforced in PVA-montmorillonite K10 clay to obtain biodegradable and active nanocomposite pouches for improved packaging of chicken sausages in the refrigerator, extending their shelf life with protection from microbes (Mathew et al. 2019). Many forms of natural fibers like pullulans, chitosan, and PLA can carry stable Ag and ZnO NPs. Hybrid materials produced by mixing various fibers and NPs have shown encouraging results in fighting dangerous microbes identified in enclosed food products (Roy et al. 2021).

The nanocomposite films have shown antibacterial action against *Escherichia coli* and *Listeria monocytogenes* with the improvement of moisture and UV-light barrier (Shankar et al. 2018). Low-concentration essential oil-supported Ag NPs were incorporated in pullulan films that exhibited antibacterial effects against *Staphylococcus aureus* and *Listeria monocytogenes*. The enclosed poultry and meat remained safe from microbial contamination, thereby guaranteeing their freshness and nutritional worth (Morsy et al. 2014). AgNPs/PVA films proved more effective in preventing *mesophilic bacteria* action and prolonged freshness of *Oncorhynchus mykiss* fillets (Kavakebi et al. 2021). Hybrid Ag/Au NPs/Chitosan films showed antimicrobial action against fungi, yeast, and gram-negative bacteria as well as gram-positive bacteria (Youssef et al. 2014). PVA films using ZnO NPs/nutmeg oil/starch showed improved *S. typhimurium* resistance. This difference stems from gram-negative bacteria having a different cellular structure than gram-positive bacteria. Fascinatingly, gram-positive bacterial cell walls consist of several layers of peptidoglycan, which presents a difficulty for silver NPs trying to get through (Jayakumar et al. 2019).

With considerable promise as a dependable antibacterial agent fit for many medicinal and food-related uses, Cu NPs have become very popular and are chosen over Ag NPs in active nanopackaging even if copper lacks significant biocidal qualities, as they are more reasonably priced, compatible with polymers, and provide a greater spectrum of physical characteristics (Saravanakumar et al. 2020). Strong antibacterial qualities of copper have been demonstrated, therefore fighting several types of dangerous diseases and other microorganisms. The impact of low-density polyethylene/copper nanocomposites on *E. coli* has been investigated on gram-positive and gram-negative food-deteriorating microorganisms, extending the shelf life of Peda (Lomate et al. 2018).

PLA films with ZnO NPs and chitosan films have been used to improve the shelf quality and life (Li et al. 2017). The goal was to keep apples fresher for longer during storage. Recent investigations have revealed variable degrees of enhanced barrier for moisture, oxygen, and pathogens in ZnO NPs films. Fascinatingly, the hybrid ZnO/TiO_2 composites have been shown to be most effective in almost doubling the shelf life of shrimp (Azizi-Lalabadi et al. 2020). The mechanical studies have demonstrated that the incorporation of these NPs in natural polymers improves tensile properties, providing mechanical strength (Marra et al. 2016) and thermal stability (Hezma et al. 2019). PLA films functionalized with Ag NPs have been used to improve the anti-oxidation and anti-UV properties, drastically lower the count of microbes that prolonged the shelf life of strawberries, and stabilize the barrier protection of the films (Cheng et al. 2021). Table 10.2 illustrates the significant antimicrobial action of nanocomposite-based films for their suitable application in food packaging.

10.2.3 Nanotechnology in Sensing

Food product quality and lifetime depend much on appropriate temperature and humidity regulation in storage. Nanotechnology-based transducers have transformed our methods of monitoring food product properties and quality. Real-time monitoring made possible by this modern technology guarantees that the food's physical, chemical, and biological features are kept in good condition during transit and storage (Lata et al. 2022). Nanosensors (Table 10.3) can offer important insights to avoid spoiling and decrease post-harvest loss by tracking many criteria, including pH, moisture, temperature, deteriorating chemicals, and microbial growth. The application of nanosensors to detect bacteria, including *Escherichia coli* (Roy et al. 2021), *Salmonella* spp. (Mathew et al. 2019), and *Listeria monocytogenes* (Morsy et al. 2014), in food products has made great advancements. The possible integration of nanosensors into food packages has been proven as a significant step to lower post-harvest losses. It also highlights the critical application of NMs in the agri-food industry (Patri et al. 2019). Various NMs with antioxidant and antibacterial properties have been integrated into food packaging. These integrations help to maintain the quality and extend the shelf life of preserved food. Especially,

Table 10.2 Antimicrobial action of some nanocomposite-based films for food packaging applications

Materials	Applications	References
Ag/Au NPs/Chitosan films	Inhibited growth of *Staphylococcus aureus* (gram positive), *Pseudomonas aeruginosa* (gram-negative bacteria), *Aspergillus niger* (fungi), and *Candida albicans* (yeast).	Youssef et al. (2014)
Ag NPs/PVA film	Antibacterial action against *mesophilic bacteria* and prolonging the freshness of *Oncorhynchus mykiss* fillet, extended shelf life up to 7 days.	Kavakebi et al. (2021)
Ag NPs/Essential Oil/Pullulan films	Antimicrobial activity against *Staphylococcus aureus* and *Listeria monocytogenes* contamination in poultry and meat.	Morsy et al. (2014)
Ag NPs/pullulan and Au NPs/pullulan composites	Inhibition of quorum sensing of *Escherichia coli* ATCC 25922, *Staphylococcus aureus* ATCC 25923, *Streptococcus mutans* ATCC 25175, *Salmonella enterica serotype typhmurium* SL 1344, *Bacillus thuringiensis,* and *Pseudomonas aeruginosa* ATCC 27853.	Ghaffarlou et al. (2022)
ZnO NPs/nutmeg oil/Starch/PVA films	Enhanced mechanical, water barrier, and antimicrobial properties. Inhibited *S. typhimurium* with visual pH sensing.	Jayakumar et al. (2019)
Ag NPs/PVA/clay pouches	Improved barrier properties, mechanical strength, and water resistance. Potent antibacterial action against *Salmonella Typhimurium* and *S. aureus* for meat packaging.	Mathew et al. (2019)
ZnO NPs/PLA films	Enhanced antibacterial efficacy against *Escherichia coli* and *Listeria monocytogenes* in packaged minced fish paste.	Shankar et al. (2018)
Tragacanth/Hydroxypropyl methylcellulose/Beeswax/Ag NPs Film	Antibacterial potential against gram-negative bacteria (*Escherichia coli* ATCC 8739, *Salmonella typhimorum* ATCC 14028, *Pseudomonas aeruginosa* ATCC 9027, and *Klebsiella pneumoniae* ATCC 10031) compared to gram-positive bacteria (*Bacillus cereus* ATCC 1247, *Staphylococcus aureus* ATCC 25923, *Streptococcus pneumoniae* ATCC 49615, and *Listeria monocytogenes* ATCC 7644).	Bahrami et al. (2019)
ZnO NPs/Pullulan/Chitosan film	Demonstrated superior antioxidant properties and bactericidal efficacy against *E. coli* and *L. monocytogenes*.	Roy et al. (2021)
ZnO NPs/chitosan film	Enhanced mechanical properties, thermal stability, and antibacterial action against *Bacillus subtilis*, and *Escherichia coli*.	Priyadarshi and Negi (2017)
ZnO NPs/PLA films	Improved the UV-light barrier, water vapor barrier, and antibacterial properties against *Escherichia coli* and *Listeria monocytogenes*.	Shankar et al. (2018)

(continued)

Table 10.2 (continued)

Materials	Applications	References
ZnO NPs/PVA/gelatin film	Effectively inhibited the growth of gram-positive bacterial strains.	Tymczewska et al. (2022)
ZnO NPs/PVA/Chitosan film	Improved mechanical strength, reduced moisture, and water vapor permeability, with color changes allowing fish fillet freshness monitoring.	Amaregouda et al. (2023)
ZnO NPs/Chitosan film	Enhanced water vapor permeability and demonstrated fair antibacterial action against *E. coli.*	Bajpai et al. (2010)

nanocomposites can be blended as preservatives and antibacterial agents with nondegradable or biodegradable packing polymers (Priyadarshi and Negi 2017).

Humidity indicators are among the specialized technologies developed to monitor and control the storage conditions of packaged agri-food products (Calabretta et al. 2023). Researchers have developed a unique edible humidity indicator using two food-grade caseinate/glycerol-based polymer films with distinct ratios of casein, a milk protein, in the presence of glycerol as a plasticizer. The flat indicator exhibited irreversible self-bending and rolling to report humidity exposure of milk powder and vitamin C tablet (Zhang et al. 2023). Kafirin-PEO electrospun bio-nanocomposites containing beetroot extract have been used to obtain pH indicators to perform image processing based on pH sensing. The indicator exhibited a significantly reversible color with a pH change from 1 to 10 (Tavana et al. 2024). Likewise, visual pH sensing was obtained through the fabrication of a starch-PVA composite polymeric film with the incorporation of ZnO NPs in the presence of essential extracts from nutmeg and jamun. The film exhibited a color change of orange (alkaline pH), dark wheat (neutral pH), and blackish brown (acidic pH) (Jayakumar et al. 2019).

Environment-sensitive nanosensors have proven to be beneficial in reducing food deterioration. Nanotechnology has helped researchers to create time-temperature indicators capable of efficiently tracking and analyzing temperature variations throughout transit and storage. It is especially crucial in the preservation of delicate and perishable food products (Cutter 2002). Metal nanosensors have become very common as a kind of sensor because of their amazing sensitivity and stability. Researchers nowadays are measuring and identifying various contaminants using nanosensors. For instance, an Au NPs-based label-free electrochemical immunosensor has been fabricated using a functionalized nanofiber membrane composed of polyvinyl alcohol and gelatin for the ultrasensitive detection of quinalphos, an insecticide (Hu et al. 2024). The nanozyme activity of Au NPs was employed to fabricate a paper-based microfluidic chip that was integrated with a smartphone platform-based visual detection of ascorbic acid in samples. This portable colorimetric sensor was quite effective with a very low detection limit (Guan et al. 2023). A bio-based enzymatic biosensor for the detection of monosodium glutamate in tomato soup has been obtained through a hybrid system. A poly (amidoamine) generation 1 dendrimer-modified montmorillonite was incorporated

Table 10.3 Performance metrics of nanosensors

Sensor type	Materials used	Target analyte	Performance metrics	Study
Label-free electrochemical immunosensor	Polyvinyl alcohol/gelatin-gold nanoparticles (VA/G-AuNPs)	Quinalphos (insecticide)	Linear range: 0.06–1000 ng/mL; Detection limit: 50.74 pg/mL	Hu et al. (2024)
Enzymatic biosensor using electrospun nanofibers	PCL–CHIT/PAMAMG1–Mt/GluOx (PCL, chitosan, PAMAM, montmorillonite)	Monosodium Glutamate (MSG)	Linear range: 0.0025–0.175 mM; detection limit: 1.045 μM	Atilgan et al. (2023)
Electrospun fiber mats for food packaging	Chitosan-carrageenan (CH–KC) + Malva sylvestris extract	Freshness of carp fillets	Color change as pH indicator; monitors fish freshness	Shavisi (2024)
Kafirin-PEO electrospun bio-nanocomposite pH indicator	Red beetroot extract	pH indicator	Image processing-based pH monitoring	Tavana et al. (2024)
Bacterochromic cellulose biosensor	Cellulose nanofibers immobilized with red cabbage anthocyanin	Bacterial flora detection	Color shift (595 nm to 448 nm) indicating bacterial activity	Hassan et al. (2024)
Paper-based sensor for biogenic amines detection	Paper-based sensor integrated into smart packaging	Biogenic amines (putrescine)	Detection limit: 0.1 mM of putrescine in meat samples	Calabretta et al. (2023)
Porphyrin-based sensor foil	Polyethylene foil with dispersed silica and active sensing molecules	Biogenic amines in fish	Real-time detection of spoilage	Altmann et al. (2023)
Electrochemical sensor for antioxidant detection	Manganese dioxide and electrically reduced graphene oxide (ERGO)	Tert-butylhydroquinone (TBHQ)	Detection limit: 0.8 μM in edible oils	Cao et al. (2019)
Paper-based colorimetric sensor with smartphone	Portable paper-based system	Ascorbic acid	Detection limit: 0.406 μmol/L	Guan et al. (2023)
Biodegradable and conducting biosensor	Biochar/PLA/Ag NPs	NH_3	Detection limit: 5–60 ppm	Sobhan et al. (2021)
Nanocomposite sensor film		Hg^{2+} ions	Detection limit: 0.9–1200 ppb	Sahu et al. (2017)

(continued)

Table 10.3 (continued)

Sensor type	Materials used	Target analyte	Performance metrics	Study
	Sodium Alginate/PVA/Ag NPs			
Screen-printed electrodes-based sensor	Ag NPs/Chitosan embedded graphite	Nitrite levels in commercial milk powder	Detection limit: 30–1140 μM	(Patri et al. 2019)
Enzyme loaded strip-based sensor	ZnO NPs/Chitosan/ coconut fiber	Polyamine levels in fruits and vegetables	Detection limit: 0.01–5.0 mM	Hooda (2018)
Amperometric detection-based sensor film	ZnO NPs/xanthine oxidase/ chitosan/carbon nanotube/ polyaniline film	Xanthine in fish meat	Detection limit:	Devi et al. (2012)

in the poly-ε-caprolactone and chitosan-based electrospun nanofibers that were conjugated with glutamate to fabricate electrodes with a significant detection limit (Atilgan et al. 2023).

A ratiometric electrochemical sensor composed of a manganese dioxide-incorporated electrochemically reduced graphene oxide nanocomposite film was fabricated for the detection of tert-butylhydroquinone in edible oils with an efficient detection limit (Cao et al. 2019). The concentration of biogenic amines, present in certain foods, increases in spoiled food and is considered an indicator of food quality and freshness. A user-friendly aglycone genipin-based colorimetric sensing paper was integrated into smart packaging to monitor the concentration of putrescine in the packaged commercial chicken meat. The detection was monitored via image acquisition of visual color change through a smart camera (Calabretta et al. 2023). In another study, mesoporous silica was dispersed in polyethylene after functionalization with porphyrin to obtain a sensor foil by thermal extrusion. This foil was used for real-time detection of biogenic amines and microbial contamination in the packaged fish fillets (Altmann et al. 2023). An electrospinning technique was used to encapsulate the extract of *Malva sylvestris* into chitosan-carrageenan fibers to obtain fiber mats that were used to monitor the freshness of packaged silver carp fillets. The sensor displayed a visual color change as red (pH $>$ 3), purple (pH 4–6), dark blue (pH $=$ 7), green (pH 8–10), and brown (pH 11–12).

Food safety testing is the focus of constant research as different microbes and contaminants produce various foodborne diseases. Scientists are concerned about the possible health hazards connected to foodborne diseases brought on by different bacteria. Rapid, precise, and reasonably priced detection techniques are desperately needed (Hassan et al. 2024). Microbial detection initiatives increasingly aim for NMs to use their special catalytic, magnetic, electrical, and luminescent properties instead of depending on conventional microbiological analytical methods (Lata et al. 2022). The development of speedier, more accurate, and more reasonably priced diagnostic tests is a constant endeavor to assist this continuing war. Nanotechnology field researchers have been committed to create very efficient systems with traits like sensitivity, speed, cost, robustness, and repeatability (Hu et al. 2024). A thorough assessment of nanosensors used to track microbial growth just came out from a recent study. A bacterochromic textile sensor was fabricated to monitor hypsochromic shift for gram-negative bacteria (595–448 nm) and gram-positive bacteria (595–478 nm) affected medium. Anthocyanin extracted from red cabbage and potassium aluminum sulfate were immobilized as NPs on the surface of electrospun cellulose acetate nanofibers. The system exhibited significant and reversible colometric alterations (Hassan et al. 2024).

Over their storage and transit, these nanosensors track the condition of perishable goods like meat and dairy (Shavisi 2024). The nanosensors can also be easily included in food packaging to identify the presence of volatile organic compounds released by bacteria. One-pot synthesis was used to imprint Ag NPs on the thin nanocomposite films of polyvinyl alcohol and sodium alginate. The films were used to selectively detect Hg^{2+} in the aqueous phase (Sahu et al. 2017). A biodegradable and conductive composite material was fabricated using Ag NPs, polylactic acid, and activated biochar. This nanomaterial was used to develop a biosensor for detecting ammonia in spoiled food products (Sobhan et al. 2021). An amine-containing nanoporous covalent organic polymer was designed. It showed exceptional sensitivity and selectivity to detect ammonia at room temperature with significant response (Sharma et al. 2024). Likewise, nanosheets of vanadium pentaoxide were used to design a resistive nanosensor. The nanosensor helped in ammonia detection and monitoring of food spoilage (Van Duy et al. 2023). Trace-level detection of nitrite in commercial milk samples was achieved through graphite screen-printed electrodes that were coated with Ag NPs embedded in chitosan nanocomposites (Patri et al. 2019). Polyamines in locally purchased fresh fruits (orange, pear, mango, and banana) and vegetables (green peas, soybean, cauliflower, potato, mushroom, and broccoli) were detected using a polyamine-sensitive strip. This strip was obtained by immobilizing oxidases onto ZnO NPs that were further embedded into chitosan and coconut fibers (Hooda 2018). Xanthine in fish meat was detected using a xanthine-sensitive electrode that was obtained by immobilizing oxidase onto a ZnO NPs-embedded chitosan film electrodeposited on the polyaniline and multiwalled carbon nanotube electrode (Devi et al. 2012).

10.3 Toxicological Analysis

The use of nanotechnology in the agri-food sector is being applied in crop production, food processing, packaging, and storage. By use of active, smart, and biodegradable packaging, the food sector employs nanoscience to improve food security by means of detection of adulteration or diseases in food at several phases—

including production, processing, storage, packaging, and transportation. However, it has also raised safety concerns about the application of nanotechnology in the food sector. Specific study results have connected the intake of NPs to several biological consequences, including the induction of protein denaturation, oxidative stress response, and DNA damage to aquatic organisms (Tubatsi et al. 2022). However, some researchers have linked the continuous use of NMs to damage to the various body systems, including respiratory, gastrointestinal, and cardiovascular systems (Garg et al. 2021).

The possible negative consequences of NPs in food, such as their carcinogenicity, teratogenicity, or genotoxicity, have been highlighted in various studies providing insights on the characterization of dangers, evaluation of exposure, physicochemical characteristics of NMs, and their potential consequences. An NM's degree of danger is determined in part by its chemical composition, physical characteristics, tissue interactions, and exposure level. Even then, little research on possible health consequences has been carried out (Eddy et al. 2024). The possible safety for the use of NMs in the agri-food sector has to be assessed considering many elements. Initially, a material's physicochemical characteristics help to explain why it belongs in the NP category. For an NM, the nanoscale features are attributed to its special qualities, and it might lose its nano effects and behave like its parent material, whether via disintegration or degradation, and inflict negative effects on tissues and organs. However, the Food and Drug Administration (FDA) lists a safe limit of ZnO and TiO_2 for use in food contact items, cosmetics, and medications as well as for general consumption (U.S. Food and Drug Administration 2019). Its amazing self-disinfecting ability has been a central focus in the search for hygienic design goals for food processing and packaging surfaces (Musial et al. 2020).

10.4 Future Prospective

Food nanotechnology has come a long way, especially with respect to its safety and its influence on civilization. It is imperative to investigate thoroughly their possible human harm before nano-packaged foods are commercially introduced. Still absent are rules for the safety evaluation of nano-packaged foods (Echegoyen and Nerín 2013). Furthermore, the environmental damage resulting from the discharge of contaminants results from the chemical production of NPs applied utilizing different chemical techniques. Thus, it is very important to have a regulatory framework and bio-safety protocols for manufacturing, packaging, and consuming nano-packaged foods (Ashfaq et al. 2022). The development of nanopackaging materials should also consider the migration possibility of the nano-scaled chemicals into food. Migration into the food matrix might be viewed as a mass transfer mechanism in which smaller molecular mass particles are trapped in packaging before being released into the food item (Gallocchio et al. 2016). It is very critical to explore the possible problems that can arise when NPs, particularly those with finer size and lower concentration, might spread from polymer-based food packaging to the enclosed product (Li et al. 2018).

Table 10.4 Migration potential of NPs in packaging material

Packaging material	Results	References
Polypropylene/polyolefin/Ag NPs/containers and bags (Kinetic Go Green Basic Nanosilver Food Storage container, Oso Fresh Food Storage container, and FresherLonger™ Plastic Storage bags)	Migration of silver ions in commercial food containers lesser than the permissible limits set by European legislation.	Echegoyen and Nerín (2013)
Ag NPs/FreshLonger™ Plastic Storage bags	Migration of ions in chicken meatballs below regulatory limits.	Gallocchio et al. (2016)
TiO_2 and Ag NPs/PLA film	Migration of ions in cottage cheese below regulatory limits.	Li et al. (2018)
ZnO NPs/PBAT/TPS biopolymer film	Migration of ions for more than 1% ZnO NPs in case of acidic food contact and 5% for aqueous food products.	Bumbudsanpharoke et al. (2024)
Cellulose nanocrystals/PLA-based films	Migration below the permissible limits.	Mondal et al. (2024)
Ag NPs/LDPE films	No significant migration.	da Costa et al. (2024)
TiO_2 and ZnO NPs/LDPE film	Migration of ions in minced beef within safe limits.	Kavuncuoglu et al. (2023)
TiO_2 NPs/PET composite films for Laminated Steel Cans	Migration of ions within safe limits.	Cui et al. (2023)

Various studies have used efficient analytical methods to track and identify the migration of NPs to food from packing materials, as illustrated in Table 10.4.

10.5 Conclusion

Post-harvest loss may result from several sources, including suboptimal processing methods, insufficient storage infrastructure, and improper handling of food, which can lead to environmental deterioration caused by exposure to air, moisture, pests, and bacteria. The emergence of nanotechnology has revolutionized the agricultural industry by offering answers to these significant challenges. Efficient food processing, appropriate handling, and storage utilizing active, smart, and intelligent packaging contribute to preserving food quality and extending shelf life, hence effectively minimizing post-harvest waste. Diverse NMs have been investigated in sustainable agriculture practices to regulate pre-harvest and post-harvest processes, therefore augmenting total food production, protection, and accessibility. Packaging materials and coatings derived from NMs can prolong the shelf life of contained food by offering barrier protection and regulating moisture, oxygen, and ethylene concentrations surrounding the food product. Nonetheless, research lacking any NP movement yields insights about the detection limit of their evaluative approach. The safety of food-engineered NPs should be assessed on a case-by-case basis rather than providing a broad classification as safe or unsafe. Therefore, it is essential to assess each nano-engineered food comprehensively to ensure its safety and achieve individual approval.

References

Ahmed J, Arfat YA, Bher A, Mulla M, Jacob H, Auras R (2018) Active chicken meat packaging based on polylactide films and bimetallic Ag–Cu nanoparticles and essential oil. J Food Sci 83: 1299–1310. https://doi.org/10.1111/1750-3841.14121

Altmann A, Eden M, Hüttmann G, Schell C, Rahmanzadeh R (2023) Porphyrin-based sensor films for monitoring food spoilage. Food Packag Shelf Life 38:101105. https://doi.org/10.1016/j.fpsl.2023.101105

Amaregouda Y, Kamanna K, Kamath A (2023) Multifunctional bionanocomposite films based on chitosan/polyvinyl alcohol with ZnO NPs and Carissa carandas extract anthocyanin for smart packaging materials. ACS Food Sci Technol 3:1411–1422. https://doi.org/10.1021/acsfoodscitech.3c00065

Amin M, Attia MM, Agwa IS, Elsakhawy Y, El-Hassan KA, Abdelsalam BA (2022) Effects of sugarcane bagasse ash and nano eggshell powder on high-strength concrete properties. Case Stud Constr Mater 17:e01528. https://doi.org/10.1016/j.cscm.2022.e01528

Ashfaq A, Khursheed N, Fatima S, Anjum Z, Younis K (2022) Application of nanotechnology in food packaging: Pros and Cons. J Agric Food Res 7:100270. https://doi.org/10.1016/j.jafr.2022.100270

Atilgan H, Unal B, Yalcinkaya EE, Evren G, Atik G, Ozturk Kirbay F, Kilic NM, Odaci D (2023) Development of an enzymatic biosensor using glutamate oxidase on organic–inorganic-structured, electrospun nanofiber-modified electrodes for monosodium glutamate detection. Biosensors 13:430. https://doi.org/10.3390/bios13040430

Azizi-Lalabadi M, Ehsani A, Ghanbarzadeh B, Divband B (2020) Polyvinyl alcohol/gelatin nanocomposite containing ZnO, TiO2 or ZnO/TiO2 nanoparticles doped on 4A zeolite: microbial and sensory qualities of packaged white shrimp during refrigeration. Int J Food Microbiol 312:108375. https://doi.org/10.1016/j.ijfoodmicro.2019.108375

Bahrami A, Rezaei Mokarram R, Sowti Khiabani M, Ghanbarzadeh B, Salehi R (2019) Physico-mechanical and antimicrobial properties of tragacanth/hydroxypropyl methylcellulose/beeswax edible films reinforced with silver nanoparticles. Int J Biol Macromol 129:1103–1112. https://doi.org/10.1016/j.ijbiomac.2018.09.045

Bajpai SK, Chand N, Chaurasia V (2010) Investigation of water vapor permeability and antimicrobial property of Zinc Oxide nanoparticles-loaded chitosan-based edible film. J Appl Polym Sci 115:674–683. https://doi.org/10.1002/app.30550

Bhardwaj S, Lata S, Garg R (2023) Application of nanotechnology for preventing postharvest losses of agriproducts. J Hortic Sci Biotechnol 98:31–44. https://doi.org/10.1080/14620316.2022.2091488

Bumbudsanpharoke N, Nurhadi RP, Chongcharoenyanon B, Kwon S, Harnkarnsujarit N, Ko S (2024) Effect of migration on the functionality of zinc oxide nanoparticle in polybutylene adipate terephthalate/thermoplastic starch films: a food simulant study. Int J Biol Macromol 263:130232. https://doi.org/10.1016/j.ijbiomac.2024.130232

Calabretta MM, Gregucci D, Desiderio R, Michelini E (2023) Colorimetric paper sensor for food spoilage based on biogenic amine monitoring. Biosensors 13:126. https://doi.org/10.3390/bios13010126

Cao W, Wang Y, Zhuang Q, Wang L, Ni Y (2019) Developing an electrochemical sensor for the detection of tert-butylhydroquinone. Sensors Actuators B Chem 293:321–328. https://doi.org/10.1016/j.snb.2019.05.012

Cheng J, Lin X, Wu X, Liu Q, Wan S, Zhang Y (2021) Preparation of a multifunctional silver nanoparticles polylactic acid food packaging film using mango peel extract. Int J Biol Macromol 188:678–688. https://doi.org/10.1016/j.ijbiomac.2021.07.161

Cui H, Yang M, Shi C, Li C, Lin L (2022) Application of xanthan-gum-based edible coating incorporated with Litsea cubeba essential oil nanoliposomes in salmon preservation. Food Secur 11:1535. https://doi.org/10.3390/foods11111535

Cui J, Chen J, Gu J, Hong X, Hong M, Wu Z, Hu Y (2023) Migration of TiO2 from PET/TiO2 composite films used for polymer-laminated steel cans in acidic solution. Coatings 13:887. https://doi.org/10.3390/coatings13050887

Cutter CN (2002) Microbial control by packaging: a review. Crit Rev Food Sci Nutr 42:151–161. https://doi.org/10.1080/10408690290825493

da Costa BS, Pereira VAC, Prado ACF, Tobias TJ, Paris EC, Ferreira MD (2024) Antimicrobial potential of linear low-density polyethylene food packaging with Ag nanoparticles in different carriers (Silica and Hydroxyapatite). J Microbiol Methods 217–218:106873. https://doi.org/10.1016/j.mimet.2023.106873

Devi R, Yadav S, Pundir CS (2012) Amperometric determination of xanthine in fish meat by zinc oxide nanoparticle/chitosan/multiwalled carbon nanotube/polyaniline composite film bound xanthine oxidase. Analyst 137:754–759. https://doi.org/10.1039/c1an15838d

Echegoyen Y, Nerín C (2013) Nanoparticle release from nano-silver antimicrobial food containers. Food Chem Toxicol 62:16–22. https://doi.org/10.1016/j.fct.2013.08.014

Eddy NO, Garg R, Ukpe RA, Akpanudo NW, Abugu H, Garg R, Anjum A, Anand B (2024) Ethical and environmental implications associated with the application of nanotechnology. In: Smart and sustainable applications of nanocomposites. pp 274–299

Gallocchio F, Cibin V, Biancotto G, Roccato A, Muzzolon O, Losasso C, Simone B, Manodori L, Fabrizi A, Patuzzi I, Ricci A (2016) Testing nano-silver food packaging to evaluate silver migration and food spoilage bacteria on chicken meat. Food Addit Contam 33:1063–1071. https://doi.org/10.1080/19440049.2016.1179794

Garg R, Rani P, Garg R, Eddy NO (2021) Study on potential applications and toxicity analysis of green synthesized nanoparticles. Turkish J Chem 45:1690–1706. https://doi.org/10.3906/kim-2106-59

Garg R, Garg R, Okon Eddy N (2022a) Handbook of research on green synthesis and applications of nanomaterials. IGI Global

Garg R, Rani P, Garg R, Khan MA, Khan NA, Khan AH, Américo-Pinheiro JHP (2022b) Biomedical and catalytic applications of agri-based biosynthesized silver nanoparticles. Environ Pollut 310:119830. https://doi.org/10.1016/j.envpol.2022.119830

Ghaffarlou M, İlk S, Hammamchi H, Kıraç F, Okan M, Güven O, Barsbay M (2022) Green and facile synthesis of pullulan-stabilized silver and gold nanoparticles for the inhibition of quorum sensing. ACS Appl Biol Mater 5:517–527. https://doi.org/10.1021/acsabm.1c00964

Guan H, Du S, Han B, Zhang Q, Wang D (2023) A rapid and sensitive smartphone colorimetric sensor for detection of ascorbic acid in food using the nanozyme paper-based microfluidic chip. Lwt 184:115043. https://doi.org/10.1016/j.lwt.2023.115043

Hasani-Javanmardi M, Fallah AA, Abbasvali M (2021) Effect of safflower oil nanoemulsion and cumin essential oil combined with oxygen absorber packaging on the quality and shelf-life of refrigerated lamb loins. Lwt 147:111557. https://doi.org/10.1016/j.lwt.2021.111557

Hassan NF, Khattab TA, Fouda MMG, Abu Zaid AS, Aboshanab KM (2024) Electrospun cellulose nanofibers immobilized with anthocyanin extract for colorimetric determination of bacteria. Int J Biol Macromol 257:128817. https://doi.org/10.1016/j.ijbiomac.2023.128817

Hezma AM, Rajeh A, Mannaa MA (2019) An insight into the effect of zinc oxide nanoparticles on the structural, thermal, mechanical properties and antimicrobial activity of Cs/PVA composite. Colloids Surfaces A 581:123821. https://doi.org/10.1016/j.colsurfa.2019.123821

Hooda V (2018) Enzymes loaded chitosan/coconut fibre/zinc oxide nanoparticles strip for polyamine determination. Food Chem 239:1100–1109. https://doi.org/10.1016/j.foodchem.2017.07.057

Hu JC, Wen P, Wang Y, Yang JY, Xiao ZL, Xu ZL, Shen YD, Wang H, Hammock BD (2024) Fabrication of a label-free electrochemical immunosensor by functionalized nanofiber membrane for the ultrasensitive detection of quinalphos. Food Control 162:110423. https://doi.org/10.1016/j.foodcont.2024.110423

Indumathi MP, Saral Sarojini K, Rajarajeswari GR (2019) Antimicrobial and biodegradable chitosan/cellulose acetate phthalate/ZnO nano composite films with optimal oxygen permeability and hydrophobicity for extending the shelf life of black grape fruits. Int J Biol Macromol 132:1112–1120. https://doi.org/10.1016/j.ijbiomac.2019.03.171

Jayakumar A, Heera KV, Sumi TS, Joseph M, Mathew S, Praveen G, Nair IC, Radhakrishnan EK (2019) Starch-PVA composite films with zinc-oxide nanoparticles and phytochemicals as intelligent pH sensing wraps for food packaging application. Int J Biol Macromol 136:395–403. https://doi.org/10.1016/j.ijbiomac.2019.06.018

Kavakebi E, Anvar A, Ahari H, Motalebi A (2021) The effect of PVA coated film containing silver nanoparticles synthesized from aqueous Satureja rechingeri extract on shelf life of rainbow trout (Oncorhynchus mykiss) fillet. J Food Bioprocess Eng 4:135–145

Kavuncuoglu H, Yalcin H, Dogan M (2023) Development of (TiO2-ZnO)/LDPE based active nanocomposite films and detection of migration to minced beef during storage using response surface methodology. Food Chem 402:134278. https://doi.org/10.1016/j.foodchem.2022.134278

Lata S, Bhardwaj S, Garg R (2022) Nanomaterials for sensing and biosensing: applications in agri-food diagnostics. Int J Environ Anal Chem 00:1–18. https://doi.org/10.1080/03067319.2022.2115895

Lee JH, Jeong D, Kanmani P (2019) Study on physical and mechanical properties of the biopolymer/silver based active nanocomposite films with antimicrobial activity. Carbohydr Polym 224:115159. https://doi.org/10.1016/j.carbpol.2019.115159

Li W, Li L, Cao Y, Lan T, Chen H, Qin Y (2017) Effects of PLA film incorporated with ZnO nanoparticle on the quality attributes of fresh-cut apple. Nano 7:207. https://doi.org/10.3390/nano7080207

Li W, Li L, Zhang H, Yuan M, Qin Y (2018) Evaluation of PLA nanocomposite films on physicochemical and microbiological properties of refrigerated cottage cheese. J Food Process Preserv 42:1–9. https://doi.org/10.1111/jfpp.13362

Liu Z, Du M, Liu H, Zhang L, Xu X, Liu K, Tu J, Liu Q (2021) Chitosan films incorporating litchi peel extract and titanium dioxide nanoparticles and their application as coatings on watercored apples. Prog Org Coatings 151:106103. https://doi.org/10.1016/j.porgcoat.2020.106103

Lomate GB, Dandi B, Mishra S (2018) Development of antimicrobial LDPE/Cu nanocomposite food packaging film for extended shelf life of peda. Elsevier

Marra A, Silvestre C, Duraccio D, Cimmino S (2016) Polylactic acid/zinc oxide biocomposite films for food packaging application. Int J Biol Macromol 88:254–262. https://doi.org/10.1016/j.ijbiomac.2016.03.039

Mathew S, Mathew J, Radhakrishnan EK (2019) Biodegradable and active nanocomposite pouches reinforced with silver nanoparticles for improved packaging of chicken sausages. Food Packag Shelf Life 19:155–166. https://doi.org/10.1016/j.fpsl.2018.12.009

Mihaly Cozmuta A, Apjok R, Peter A, Mihaly Cozmuta L, Nicula C, Baia M, Vulpoi A (2018) Active papers coated with chitosan and containing TiO2 and Ag/TiO2 nanoparticles for increasing the shelf-life of walnut kernels. Cellulose 25:5205–5225. https://doi.org/10.1007/s10570-018-1925-x

Milani MA, Dana MG, Ghanbarzadeh B, Alizadeh A, Afshar PG (2020) Effect of novel bioactive coating enriched with nanoemulsion of mustard essential oil on the quality of Turkey meat. J Food Nutr Res 59:71–80

Miranda M, Sun X, Marín A, dos Santos LC, Plotto A, Bai J, Benedito Garrido Assis O, David Ferreira M, Baldwin E (2022) Nano- and micro-sized carnauba wax emulsions-based coatings incorporated with ginger essential oil and hydroxypropyl methylcellulose on papaya: preservation of quality and delay of post-harvest fruit decay. Food Chem X 13:100249. https://doi.org/10.1016/j.fochx.2022.100249

Mondal K, Soundararajan N, Goud VV, Katiyar V (2024) Cellulose nanocrystals modulate curcumin migration in PLA-based active films and its application as secondary packaging. ACS Sustain Chem Eng 12:9642–9657. https://doi.org/10.1021/acssuschemeng.4c00518

Morsy MK, Khalaf HH, Sharoba AM, El-Tanahi HH, Cutter CN (2014) Incorporation of essential oils and nanoparticles in pullulan films to control foodborne pathogens on meat and poultry products. J Food Sci 79:M675. https://doi.org/10.1111/1750-3841.12400

Musial J, Krakowiak R, Mlynarczyk DT, Goslinski T, Stanisz BJ (2020) Titanium dioxide nanoparticles in food and personal care products—what do we know about their safety? Nano 10:1–23. https://doi.org/10.3390/nano10061110

Nguyen TT, Huynh Nguyen T-T, Tran Pham B-T, Van Tran T, Bach LG, Bui Thi PQ, Ha Thuc CN (2021) Development of poly (vinyl alcohol)/agar/maltodextrin coating containing silver nanoparticles for banana (Musa acuminate) preservation. Food Packag Shelf Life 29:100740. https://doi.org/10.1016/j.fpsl.2021.100740

Patri SB, Adarakatti PS, Malingappa P (2019) Silver nanoparticles-chitosan composite embedded graphite screen-printed electrodes as a novel electrochemical platform in the measurement of trace level nitrite: application to milk powder samples. Curr Anal Chem 15:56–65. https://doi.org/10.2174/1573411014666180703142146

Priyadarshi R, Negi YS (2017) Effect of varying filler concentration on zinc oxide nanoparticle embedded chitosan films as potential Food packaging material. J Polym Environ 25:1087–1098. https://doi.org/10.1007/s10924-016-0890-4

Qiu L, Zhang M, Chitrakar B, Adhikari B, Yang C (2022) Effects of nanoemulsion-based chicken bone gelatin-chitosan coatings with cinnamon essential oil and rosemary extract on the storage quality of ready-to-eat chicken patties. Food Packag Shelf Life 34:100933. https://doi.org/10.1016/j.fpsl.2022.100933

Roy S, Priyadarshi R, Rhim JW (2021) Development of multifunctional pullulan/chitosan-based composite films reinforced with ZnO nanoparticles and propolis for meat packaging applications. Food Secur 10:2789. https://doi.org/10.3390/foods10112789

Sahu D, Sarkar N, Sahoo G, Mohapatra P, Swain SK (2017) Nano silver imprinted polyvinyl alcohol nanocomposite thin films for Hg2+ sensor. Sensors Actuators B Chem 246:96–107. https://doi.org/10.1016/j.snb.2017.01.038

Sanuja S, Agalya A, Umapathy MJ (2015) Synthesis and characterization of zinc oxide-neem oil-chitosan bionanocomposite for food packaging application. Int J Biol Macromol 74:76–84. https://doi.org/10.1016/j.ijbiomac.2014.11.036

Saravanakumar K, Sathiyaseelan A, Mariadoss AVA, Xiaowen H, Wang MH (2020) Physical and bioactivities of biopolymeric films incorporated with cellulose, sodium alginate and copper oxide nanoparticles for food packaging application. Int J Biol Macromol 153:207–214. https://doi.org/10.1016/j.ijbiomac.2020.02.250

Shankar S, Wang LF, Rhim JW (2018) Incorporation of zinc oxide nanoparticles improved the mechanical, water vapor barrier, UV-light barrier, and antibacterial properties of PLA-based nanocomposite films. Mater Sci Eng C 93:289–298. https://doi.org/10.1016/j.msec.2018.08.002

Shankar S, Khodaei D, Lacroix M (2021) Effect of chitosan/essential oils/silver nanoparticles composite films packaging and gamma irradiation on shelf life of strawberries. Food Hydrocoll 117:106750. https://doi.org/10.1016/j.foodhyd.2021.106750

Sharma M, Patel C, Sriram S, Mukherjee S, Das AK (2024) Tailoring NH3 sensing responsiveness through amine modification of imine-based nanoporous covalent organic polymer. ACS Appl Nano Mater 7:12926–12934. https://doi.org/10.1021/acsanm.4c01480

Shavisi N (2024) Electrospun fiber mats based on chitosan-carrageenan containing Malva sylvestris anthocyanins: physic-mechanical, thermal, and barrier properties along with application as intelligent food packaging materials. Int J Biol Macromol 266:131077. https://doi.org/10.1016/j.ijbiomac.2024.131077

Shokri S, Parastouei K, Taghdir M, Abbaszadeh S (2020) Application an edible active coating based on chitosan- Ferulago angulata essential oil nanoemulsion to shelf life extension of rainbow trout fillets stored at 4 °C. Int J Biol Macromol 153:846–854. https://doi.org/10.1016/j.ijbiomac.2020.03.080

Singh M, Sahareen T (2017) Investigation of cellulosic packets impregnated with silver nanoparticles for enhancing shelf-life of vegetables. LWT 86:116–122. https://doi.org/10.1016/j.lwt.2017.07.056

Sobhan A, Muthukumarappan K, Wei L, Zhou R, Ghimire N (2021) Development of a biosensor with electrically conductive and biodegradable composite by combinatory use of silver nanoparticles, novel activated biochar, and polylactic acid. J Electrochem Soc 168:107501. https://doi.org/10.1149/1945-7111/ac29dd

Sultan M, Hafez OM, Saleh MA, Youssef AM (2021) Smart edible coating films based on chitosan and beeswax-pollen grains for the postharvest preservation of Le Conte pear. RSC Adv 11: 9572–9585. https://doi.org/10.1039/d0ra10671b

Tavana F, Hematian Sourki A, Golmakani MT (2024) Enhancing the accuracy, speed, and efficiency of kafirin-PEO electrospun bio-nanocomposite pH indicators with red beetroot extract using image processing. Food Sci Nutr 12:2874–2885. https://doi.org/10.1002/fsn3.3968

Tonezzer M, Bazzanella N, Gasperi F, Biasioli F (2022) Nanosensor based on thermal gradient and machine learning for the detection of methanol adulteration in alcoholic beverages and methanol poisoning. Sensors 22. https://doi.org/10.3390/s22155554

Tubatsi, G., Kebaabetswe, L. P., & Musee, N. (2022). Proteomic evaluation of nanotoxicity in aquatic organisms: A review. Proteom 22:e2200008. https://doi.org/10.1002/pmic.202200008

Tymczewska A, Furtado BU, Nowaczyk J, Hrynkiewicz K, Szydłowska-Czerniak A (2022) Functional properties of gelatin/polyvinyl alcohol films containing black cumin cake extract and zinc oxide nanoparticles produced via casting technique. Int J Mol Sci 23:2734. https://doi.org/10.3390/ijms23052734

U.S. Food and Drug Administration (2019) FDA advances new proposed regulation to make sure that sunscreens are safe and effective. https://www.Fda.Gov/News-Events/Press-Announcements/Fda-Advances-New-Proposed-Regulation-Make-Sure-Sunscreens-Are-Safe-and-EffectiveFDANewsRelease

Van Duy L, Nguyet TT, Le DTT, Van Duy N, Nguyen H, Biasioli F, Tonezzer M, Di Natale C, Hoa ND (2023) Room temperature ammonia gas sensor based on p-type-like V2O5 nanosheets towards food spoilage monitoring. Nano 13:146. https://doi.org/10.3390/nano13010146

Videira-Quintela D, Guillén F, Martin O, Montalvo G (2022) Antibacterial LDPE films for food packaging application filled with metal-fumed silica dual-side fillers. Food Packag Shelf Life 31:100772. https://doi.org/10.1016/j.fpsl.2021.100772

Yang L, Ren L, Zhao Y, Liu S, Wang H, Gao X, Niu B, Li W (2023) Preparation and characterization of PVA/arginine chitosan/ZnO NPs composite films. Int J Biol Macromol 226:184–193. https://doi.org/10.1016/j.ijbiomac.2022.12.020

Youssef AM, Abdel-Aziz MS, El-Sayed SM (2014) Chitosan nanocomposite films based on Ag-NP and Au-NP biosynthesis by Bacillus Subtilis as packaging materials. Int J Biol Macromol 69: 185–191. https://doi.org/10.1016/j.ijbiomac.2014.05.047

Zhang H, Liang Y, Li X, Kang H (2020) Effect of chitosan-gelatin coating containing nano-encapsulated tarragon essential oil on the preservation of pork slices. Meat Sci 166:108137. https://doi.org/10.1016/j.meatsci.2020.108137

Zhang M, Arunachalam A, Perrin H, Polat S, Groenewold J, Mendes E, Eral HB (2023) An edible humidity indicator that responds to changes in humidity mechanically. ACS Appl Polym Mater 5:4780–4788. https://doi.org/10.1021/acsapm.3c00344

Chapter 11
Applications of Nanomaterials in Food Packaging

Ansar Anjum, Rajni Garg, Rishav Garg, and Nnabuk Okon Eddy

Abstract The packaging is essential for safeguarding food quality and enhancing safety and longevity. With an increasing population and rising consumer awareness, there is an increased demand for packaging that is clean, safe, cost-effective, and environmentally friendly. Nano-sized materials are a promising solution to meet these requirements. Nanotubes, nanofibers, and nanosheets have good physical, chemical, and biological properties that make them well-suited to improve the characteristics of packaging materials. Such materials are being used in active and smart packaging technologies to preserve food and monitor products in real time. This review explores the current applications of different nano-sized materials, such as metals and oxides of metal, carbon nanotubes, and graphene oxide nanoparticles, that are used in food packaging and also discusses their roles in improving barrier properties, antimicrobial activity, and shelf life. It also analyzes the challenges and limitations, including regulatory issues, scalability, and safety concerns, and suggests future research directions to maximize the potential of nano-sized materials in food packaging.

Keywords Nanomaterials · Active packaging · Smart packaging · Nanocomposites · Carbon based · Polymer based

---

A. Anjum
Department of Applied Chemistry, Delhi Technological University, Delhi, India

R. Garg (✉)
Department of Applied Sciences and Humanities, Galgotias College of Engineering & Technology, Greater Noida, Uttar Pradesh, India
e-mail: rajnigarg@science.org.in

R. Garg
Department of Civil Engineering, Galgotias College of Engineering & Technology, Greater Noida, Uttar Pradesh, India

N. O. Eddy
Department of Pure and Industrial Chemistry, University of Nigeria, Nsukka, Enugu State, Nigeria

S. K. Yadav, A. Kumari (eds.), *Nanobiotechnology in Agri-food Sector*, Sustainability Sciences in Asia and Africa, https://doi.org/10.1007/978-981-96-7756-6_11

11.1 Introduction

Food packaging protects food products from external factors and physical damage, ensures containment, traceability, convenience, and tamper resistance, and provides consumers with important information about ingredients and nutrition (Gaikwad et al. 2020). Largely, the goal of food packaging is to safely contain food and at the same time meet industrial standards and consumer expectations, ensuring food safety, cost-effectiveness, and minimal environmental impact (Silva et al. 2023). The food packaging should be safe, environmentally friendly, and affordable and made of high-quality materials to keep the food fresh without changing its physical, biological, and sensory properties. Thus, nanomaterials are important and are receiving more demand as a promising technology to address issues related to food safety, processing, packaging, and functional foods (Lata et al. 2022).

The materials with dimensions in the nanoscale range are called nanomaterials. The nanosize of particles leads to improved surface area and reactivity and modified properties, and they are added to tailor the conventional packaging materials (Sharma et al. 2024). The nanofibers, nanosheets, and nanocomposites are employed to transform and obtain advanced packaging materials with improved properties of superior barrier capabilities against oxygen and moisture, elasticity, tensile strength, temperature resistance, pH sensitivity, and humidity control that help in maintaining the quality and freshness of food and increase its longevity, thus enhancing the food value (Garg and Anjum 2024). Various nanomaterials have been prepared by reinforcing the polymer matrices with the incorporation of nanoscale fillers that improve strength, flexibility, thermal stability, and durability (Anjum et al. 2023). Also, the reinforcement helps to develop flexible packaging with upgraded barrier properties, biodegradability, and germicidal potential. Metal oxide nanoparticles such as titanium and zinc oxide enhance UV-blocking abilities, while carbon nanotubes and graphene oxide improve mechanical strength and function as an obstruction for gases (Nile et al. 2020). Silver nanoparticles are incorporated into packaging materials to introduce germicidal properties, extending the longevity of products. Layered silicates are effective toward the inclusion of fire-retardant characteristics in nano-sized composites utilized for enveloping food, thereby enhancing its ability to protect against fire (Bhardwaj et al. 2023).

The integration of nanomaterials in the packaging film is typically beneficial in preserving unpreserved foods, including fruits, fish, vegetables, meat, etc. This technology is commonly used in modern enveloping for minimally processed fresh products to extend their longevity by controlling or modifying the internal environment within the envelope and offering the suitable conditions for extended longevity (Lin et al. 2020). The fruits as well as vegetables are enclosed in packaging designed to regulate the atmosphere, which involves reducing the amount of oxygen and increasing the amount of carbon dioxide, which helps to maintain the produce's quality over a longer period (Peighambardoust et al. 2021). Additionally, this controlled environment inhibits the development of aerobic microorganisms by modifying the packaging atmosphere. The modified atmosphere packaging process reduces oxygen levels and controls water vapor to maintain ideal conditions to keep the food fresh. The gas mixture inside a package with fresh produce may differ from the surrounding air, but the pressure inside the package should be balanced with that outside (Sobhan et al. 2020). To achieve the equilibrium factors responsible, the respiration rate and mass of the product, thickness and surface area of packaging film, gas diffusion rates of the film, initial free volume and atmospheric composition inside the package, as well as external conditions including temperature and humidity, are important. These parameters are essential for preserving produce freshness and maintaining optimal storage conditions (Chausali et al. 2022).

In the chapter, the present state of understanding of applications of nano-sized materials in food enveloping has been explored, and challenges and prospects are discussed. The features and attributes of various types of nano-sized materials, including improved barrier properties, germicidal activity, and enhanced shelf life of food enveloping, have been included. Additionally, it also discusses the difficulties and constraints upon the use of nano-sized materials in food enveloping, regulatory issues, scalability, as well as safety concerns.

11.2 Types of Packaging

The application of nano-sized materials and technology in food packaging is aimed at nano-reinforcement by active and smart packaging using specialized nanomaterials to modify conventional polymer-based packaging materials (Lin et al. 2020). Active packaging is carried out by incorporating antimicrobial systems, oxygen scavengers, and immobilized enzymes that interact with food or the environment to contribute to food preservation (Table 11.1). While smart packaging focuses on the use of advanced technologies, such as sensors, RFID, and IoT connectivity, to provide real-time information and monitoring of the enclosed product (Wani et al. 2023).

Table 11.1 An overview of Packaging types (Ranjan et al. 2014)

Packaging type	Principle	Description	Nanomaterials used	Advantages
Active packaging	Interacts with package contents to improve shelf life or maintain quality during storage through various integrated elements.	Infusion of nanomaterials, e.g., ag, TiO_2	Ag, TiO_2	Enhanced antimicrobial, antioxidant, and UV-absorbing properties, aiming to improve food taste, freshness, and shelf life
Smart packaging	Focuses on using technology for enhanced functionalities beyond basic containment.	Integration of nanosensors, e.g., oxygen, freshness indicators	High variability	Real-time monitoring of food and safety assurance by detecting pathogens

11.2.1 Active Packaging

"Active packaging" is a contemporary method that is employed to extend the longevity of food items and functions by inhibiting the infiltration of microbes, absorption of moisture, oxidation, and premature ripening (Bhardwaj et al. 2023). This packaging includes the integration of compounds that can remove oxygen and absorb ethylene and antioxidants and can emit carbon dioxide and antimicrobial agents into the material used for packaging or the space available in the package to improve the characteristics of the polymer used (Slavin et al. 2017). The basic principle of working antimicrobial agents is either disruption of the cellular structure of microbes or inhibition of their metabolic pathways, thereby impeding their growth (Bansal et al. 2023). Additionally, nanoparticles can also generate reactive oxygen species, which interfere with DNA replication and ATP synthesis, ultimately leading to cellular damage or death (Wani et al. 2023).

Ethylene in gaseous form is an unsaturated hydrocarbon that accelerates the process of ripening and diminishes the quality of fruits and vegetables. Thus, active packaging that contains ethylene scavengers proves to be an assuring technology to extend the shelf life of perishable items (Alizadeh Sani et al. 2022). Ethylene absorbers trap to contain ethylene, while ethylene scavengers absorb water. Chitosan films that incorporate different concentrations of titanium dioxide (ranging from 0% to 2% w/w) were produced as well as assessed for their structural, mechanical, optical, and barrier characteristics (Lin et al. 2020). The nanocomposites of chitosan with TiO_2 demonstrated increased ethylene photodegradation efficiency as the TiO_2 content rose. Based on evaluations of water barrier properties, tensile strength, and photocatalytic degradation of ethylene, the chitosan film containing 1% TiO_2 proved to be the best choice and was chosen for germicidal testing. This composition exhibited germicidal activity against fungi (*Penicillium* and *Aspergillus*), positive-stain bacteria (*Staphylococcus aureus*), and negative-stain bacteria (*Escherichia coli, Pseudomonas aeruginosa,* and *Salmonella typhimurium*) (Gaikwad et al. 2020).

Additionally, a chitosan-TiO_2 nanocomposite film with a 1% TiO_2 concentration demonstrated photocatalytic degradation activity toward ethylene. At higher concentrations of TiO_2, spontaneous agglomeration of the nanoparticles was observed (Siripatrawan and Kaewklin 2018). To minimize inadvertent product waste, intelligent packaging concepts can be employed. However, these technologies offer more than just waste reduction benefits, as shown in Fig. 11.1. Effective combinations of nano-clay with starch-based polymers have the potential to create antimicrobial packaging systems due to improved mechanical properties. In a study, it was found that using 2.5% (w/w) montmorillonite clay and cornstarch films resulted in a 20% decrease in elongation at break, a 12% increase in tensile strength, and more than a doubling of Young's modulus compared to control films (Heydari et al. 2013). Similarly, another study demonstrated that adding 5% (w/w) MMT to starch-based films improved the tensile strength and decreased extension at the break by 15.8% and markedly enhanced Young's modulus by more than sixfold compared to

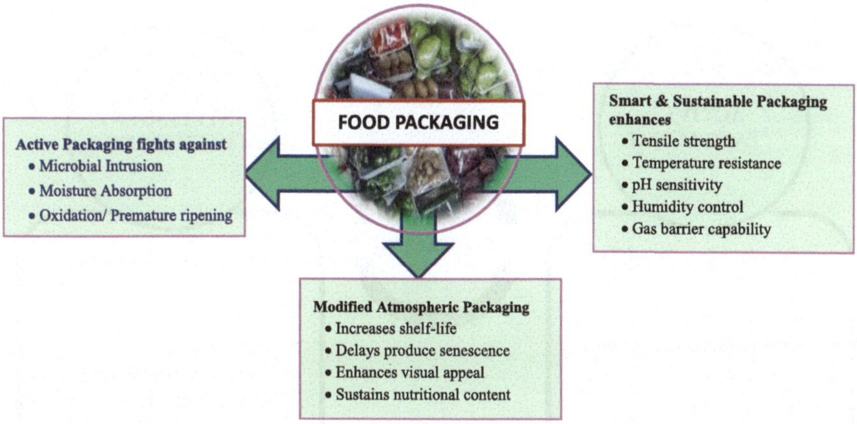

Fig. 11.1 Advancements in smart and sustainable packaging technology

controls (Cyras et al. 2008). Also, oxygen scavenger films incorporating TiO_2 in various polymers have been prepared (Xiao-e et al. 2004). These films are designed for a wide range of oxygen-sensitive products, and significant importance is being placed on the photocatalytic properties of nanocrystalline titania when exposed to ultraviolet (UV) radiation. However, a notable drawback of TiO_2 is its reliance on UVA light for activation due to its photocatalytic mechanism (Silva et al. 2023).

Avella et al. (2005) developed starch nanocomposite films using montmorillonite clay, noting significant improvements in modulus and tensile strength. The study highlighted that incorporating MMT into thermoplastic starch films resulted in a 107% increase in Young's modulus compared to controls. They observed variations in Young's modulus and tensile strength at different relative humidities, highlighting the complex impact of clay on the film of starch under varying environmental conditions.

11.2.2 Smart Packaging

It encompasses packaging utilizing technology to offer functionalities beyond just containing a product. It typically involves technology embedded in the packaging to sense or provide information about the product or brand. Smart packaging aims to enhance communication between producers and consumers by giving real-time information on food quality parameters. Nano-food additives, including colors, flavors, anti-caking agents, and gelling, agents have also been investigated (Nile et al. 2020). Furthermore, such systems commonly utilize nanosensors to determine the contaminants in food, with nanoparticles playing a crucial role in their development. Nanoparticle-based nanosensors offer significant potential for monitoring chemical, physical, and biological changes during the processing and preservation

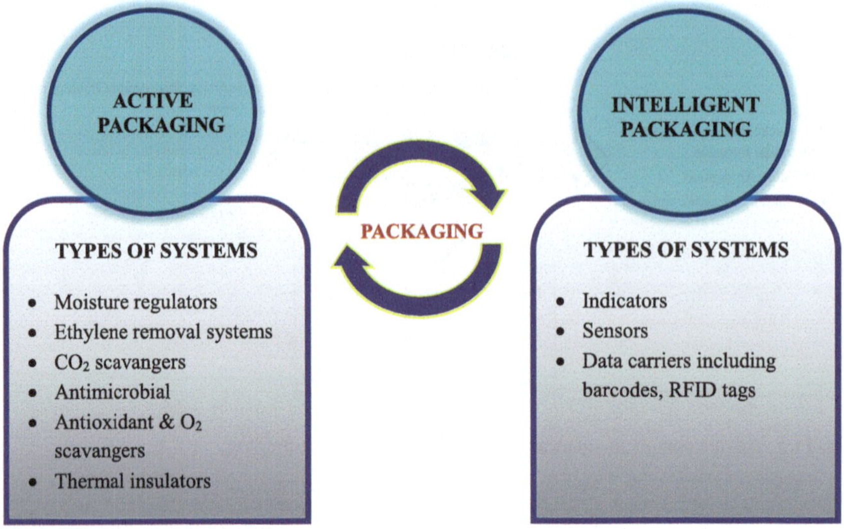

Fig. 11.2 Development of technology in packaging

of food. These sensors enable the detection of toxins and foodborne pathogens as well (Dias et al. 2013).

Smart and intelligent packaging systems help maintain food quality throughout the distribution process by capturing responses to internal and external environmental stimuli with the help of integrated sensors (Fig. 11.2). The common indicators that are used in food packaging help to monitor the integrity of packaging, freshness, and changes that occur over time and temperature. Keeping a check of such changes during production and throughout the supply chain is important to conserve the quality standards and extend the longevity of the product. Also, nano-barcodes made from nanoparticles are the identity tags within smart packaging solutions (Branton et al. 2009).

As society increasingly embraces sustainable practices, nanostructured materials have proved to help improve the accessibility and efficiency of green packaging solutions as we progress toward a more environmentally friendly planet (Yu et al. 2018). The key nanomaterials, including montmorillonite and kaolinite clays, zinc oxide, and silver nanoparticles, play significant roles in this advancement (Anjum et al. 2023). Montmorillonite creates strong films that effectively block oxygen and carbon dioxide, whereas the zinc oxide-coated silicate nanoparticles provide protective layers in sustainable packaging (Salmas et al. 2020). The silver nanoparticles possess strong germicidal properties that help in the longevity of food products (Oliani et al. 2022). Titanium dioxide is used to create barriers that protect against different substances, including oxygen and carbon dioxide (Sharma et al. 2020). These advancements underscore the important role of nanomaterials in improving standards for eco-friendly food packaging.

Furthermore, natural polysaccharides like starch (Adeyeye et al. 2023), chitosan (Sanuja et al. 2015), and derivatives of cellulose (Vipan et al. 2018) have antimicrobial properties and provide strength as well as resistance to moisture in packaging materials. Ecologically safe polymers like polyhydroxybutyrate, polylactic acid (Connolly et al. 2019), and polycaprolactone (Łukasiewicz et al. 2021) offer eco-friendly substitutes for traditional plastics. Additionally, edible films made of lipids, proteins, and polysaccharides can extend the longevity of food when used in packaging. The nano-sized composites integrate these materials, along with the antimicrobial agents, to prevent the spoilage of food and reduce packaging waste (Tabari 2018). UV absorption can affect the flavors, colors, and nutrients of milk-based products, which indicates the need for effective UV protection solutions. Similarly, crops need protection from direct UV radiation, making UV protective films essential in agriculture to preserve chlorophyll synthesis (Ahmed et al. 2017). Biopolymeric nanocomposites incorporating UV protective properties are being studied, employing biocompatible nanomaterials with optimal UV absorption (200–400 nm) to improve UV resistance. These advancements are crucial for maintaining the inherent physical and chemical properties and the biological attributes of food items, medications, and beverages during transit and storage (Peighambardoust et al. 2019).

11.3 Nanomaterials in Food Packaging

Nano-sized fillers are integrated into the polymer matrices to prolong the longevity of products while maintaining food safety and strengthening barrier defenses, specifically against oxygen and moisture. This type of packaging helps to enhance longevity and microbial safety and preserve the quality of food, with the addition of sensory attributes. These systems aim to replace conventional methods of food packaging and preservation that may involve excessive additives leading to undesirable effects (de Azeredo 2013).

Nanomaterials are increasingly being used in food enveloping for the longevity of packaged food products that is necessary due to concerns about food spoilage caused by inadequate barriers and active properties. The nano-sized materials that enhance the properties of composite film include zinc oxide (Salmas et al. 2020), titanium dioxide (Sharma et al. 2020), silicon dioxide (Pardo-Figuerez et al. 2018), Ag (Mohseni et al. 2020), and halloysite nano-clay (Abdullah and Dong 2019) owing to their small size and high dispersibility. These materials improve the tensile strength and elastic modulus, as well as reduce the concentration of moisture and gases, enhance barrier properties and antimicrobial functions, and extend the freshness of food products (Bhardwaj et al. 2023). Active packaging, which incorporates raw nanomaterials like nano-clay and essential oils, plays a significant role in imparting longevity to food items. The use of these nano-sized particles in food enveloping has shown a potential to reduce spoilage by preventing the propagation of harmful pathogens (Xiao-e et al. 2004). The incorporation of essential oils into nanocomposites improves their effectiveness, improving the longevity of food and preventing bacterial growth in high-moisture food items. The germicidal substances are often added to food packaging using casting or thermal processing methods. Casting involves applying a film-forming solution onto a surface, drying, and then removing the film. This method is commonly employed to prepare antimicrobial nanocomposite films for food packaging (Peighambardoust et al. 2021). Researchers have used different biopolymers, including chitosan (Lin et al. 2020), starch (Heydari et al. 2013), carboxymethyl cellulose (Tabari 2018), and gelatin, often combined with silver nanoparticles and nano-clay, to create highly efficient antimicrobial nanocomposite packaging films.

11.3.1 Clay Nanocomposites

The addition of nano-clay to polymer composites enhances their barrier properties by creating a complex pathway for gases, which slows down gas permeability. However, this also diminishes the biodegradability of the polymer, a process that can be quickened by enzymes (Anjum et al. 2024). Enzymes such as lipases, depolymerase, and proteinase K are commonly utilized to hasten degradation, primarily targeting the amorphous regions of the polymer, with increased effectiveness in the crystalline phase of PLA composites as degradation progresses (Connolly et al. 2019). Research shows that a masterbatch technique for the dispersion of nano-clay in polylactic acid through the melt mixing process in composite materials. Initially, the polylactic acid and masterbatch were loaded at a low speed of 300 rpm, gradually increasing to 2500 rpm. Both 2% and 4% w/w nano-clay concentrations show minimal temperature differences and incomplete disintegration within the studied time frames. Optical analysis indicated slow biodegradation with increasing nano-clay content, suggesting a longer film lifespan. Enzymes primarily degraded the surface, hindered by nano-clay, and influenced pH variations. Nano-clay effectively reduces the water vapor transmission rate, leading to prolonged biodegradation and enhanced film processability (Oliver-Ortega et al. 2021).

Another study aimed to develop operational characteristics of fish myofibrillar protein (FMP) films by adding MMT nano-clay and microbial transglutaminase (MTGase). The weight percent of MMT varied from 1% to 5% relative to the protein, and the effect of the additives on the physical, mechanical, and barrier properties of the FMP film was studied. The results demonstrated significant enhancements in water absorption, water vapor permeability, solubility, tensile strength, and elongation of the nanocomposites. These improvements were attributed to the exfoliation of MMT and the favorable interaction between fish myofibrillar protein and MTGase in the presence of MMT. When 3 wt% MMT was incorporated with MTGase, the tensile strength of the FMP film increased by about 62.3%. However, increasing the amount of MTGase further increased the tensile strength of the film, while adding more MMT did not contribute to further improvement (Rostamzad et al. 2016). The characteristics of films that are made using tilapia skin gelatin with hydrophobic and hydrophilic clay at different ranges (0–10%, w/w) were studied. It was found that the mechanical properties of the films improved when hydrophilic nano-clay (Cloisite Na^+) was added at 0.5–5% (w/w). The films with 1% (w/w) of Cloisite Na^+ and Cloisite 20A showed the lowest water vapor permeability ($P < 0.05$), and the yellowish hue of the films generally improved with higher concentrations of all nano-clay ($P < 0.05$). However, the transparency of the films decreased as the level of nano-clay incorporation was increased ($P < 0.05$). The scanning electron microscopy and wide-angle X-ray diffraction analyses indicate an intercalated/exfoliated structure of the nanocomposite gelatin-based films in both hydrophilic and hydrophobic nano-clay. SEM micrographs showed that the homogeneity and smoothness of the film surfaces decreased with the addition of both types of nano-clay. Differential scanning calorimetric and thermogravimetric analyses of gelatin-based films demonstrated that the use of nano-clay improved their stiffness and heat resistance. The type and concentration of nano-clay used had a notable influence on the characteristics of the tilapia skin gelatin films, affecting their mechanical strength, water vapor permeability, color properties, transparency, surface structure, and thermal stability (Nagarajan et al. 2014).

Lately, in a study, sodium montmorillonite (NaMt) and organo-montmorillonite were used as carriers for thyme, oregano, and basil essential oils in active low-density polyethylene (PE) films. The NaMt-based and organically modified montmorillonite films exhibit consistent antioxidant activity, retaining 50–70% of their antioxidant capacity even after 6 months. The combination of montmorillonite clay with essential oils also resulted in strong antimicrobial properties against pathogens, showing the potential as a natural antimicrobial agent. Furthermore, the study demonstrated that the incorporation of montmorillonite-based nanocomposites into packaging materials improved their capacity to act as an obstruction for oxygen and carbon dioxide, thereby extending the longevity of food products (de Oliveira et al. 2022). A comparison between organically modified montmorillonite-based low-density PE films and NaMt essential oil-based LDPE films shows that the OrgMt-based films exhibited superior tensile strength and barrier properties. This suggests their ability to be used as active packaging films. In testing with chicken breast fillets, these novel active packaging films effectively inhibited lipid oxidation, as determined by the thiobarbituric acid-reacting substances test (Giannakas 2020). The effectiveness of these blends for various uses was also studied, and it was found that they show promising roles as versatile, natural agents for diverse applications. Another study shows that the use of gelatin films containing nanoparticles, such as nano-clay and metal nanoparticles, is common in active packaging because of its strong antimicrobial properties. It is suggested that polyethylene nanocomposite films enhanced with nanoparticles like Cloisite 20A and Cloisite 30B significantly inhibit the growth of *S. aureus* and *E. coli* (Fasihnia et al. 2018). Also, Yang et al. created a film using polyvinyl alcohol, chitosan, and lignin nanoparticles for active packaging. This material reflects antioxidant and germicidal activity against the negative-stain bacteria *Erwinia carotovora subsp.* carotovora and *Xanthomonas arboricola*. This type of casing is suitable for preserving high-fat and fresh food (Yang et al. 2016).

Incorporating nanostructured nisin has the potential to create packaging with extended antimicrobial activity. This method involves gradually releasing the encapsulated peptide upon contact with food, effectively prolonging control over objectionable microorganisms. Enhanced germicidal efficiency is achieved through combining free nisin with liposome-encapsulated nisin within the template of polymer. Initially, the release of free antimicrobials significantly reduces bacterial contamination. This is followed by a sustained release of encapsulated substances, which continues to control the remaining bacteria, thereby extending shelf life (Imran et al. 2016). Gelatin and casein films were examined after integration with antimicrobial nanocomposites comprising nisin-loaded nanoliposomes and halloysite nano-clay reinforcement. Nanoliposomes with over 90% encapsulation efficiency for nisin and having approximately 150 nm size were integrated in the solutions of gelatin or casein. Halloysite was added at concentrations of 0.5 and 1.0 g/L, and the casting method was used to obtain the films. These films effectively inhibited foodborne pathogens *B. cereus, L. monocytogenes,* and *C. perfringens* when tested on milk agar as a food-simulating medium. The germicidal activity was compromised at the higher halloysite concentration (1.0 g/L), indicating the increased tortuosity in the diffusive path delayed nisin migration (Boelter and Brandelli 2016).

11.3.2 Nanoscale Polymer-Based Composites

Polymer nanocomposites are composed of a matrix made of polymer, nanofillers, plasticizers, and compatibilizers. The studies have demonstrated the positive impact of nanofillers on the barrier properties of polymer nanocomposites, making them promising materials for advanced food packaging. These materials have shown potential in preserving food quality and extending shelf life across a wide range of products, including processed meats, cheese, bakery items, confectionery, grains, boil-in-bag foods, fruit juices, dairy products, carbonated beverages, and beer bottles (Kanmani and Rhim 2014). Polymers like polylactic acid and polyhydroxyalkanoates are mixed with nanoparticles to produce biodegradable nanocomposites, contributing to environmental sustainability initiatives by offering substitutes for conventional plastic packaging. Polymer nanocomposites are advanced solid materials characterized by the dispersion of nanoscale fillers within a polymer matrix. Due to their nanosize, these composites exhibit significantly better mechanical, heat-related, visual, and physical-chemical characteristics as compared to pure polymers or usual microfilter composites at low filler loadings below 5 wt% (Silvestre et al. 2011).

Nanofillers enhance the properties essential for food packaging, but their adoption is hindered by cost and safety concerns. Polyamide 6, also known as thermoform nylon, is commonly used for making films, sheets, and bottles due to its affordability, transparency, strength, toughness, and thermoformability. However, nylon has high gas and water vapor permeability, so nylon nanocomposites with clay are being developed to improve barrier properties. Nylon-MXD6, a crystalline polyamide, is used in multilayer food packaging to extend shelf life by reducing CO_2 loss and oxygen ingress in carbonated beverages. These nanocomposites enhance barrier properties crucial for preserving food quality and shelf life, despite challenges in cost-effectiveness and regulatory acceptance (Kanmani and Rhim 2014).

11.3.3 Nanoscale Carbon-Based Composites

The composites comprising carbon nanotubes and graphene are highly prized due to their strong tensile strength and durability. They also contribute to improved thermal stability, which is crucial for maintaining the integrity of packaging in extreme conditions. The dependence of the food packaging industry upon petroleum-derived and synthetic materials contributes to environmental pollution (Lata et al. 2022). Therefore, carbon nanomaterials, such as carbon dots, graphene, and nanotubes, provide a sustainable alternative. These materials are created using eco-friendly processes and enhance food preservation and packaging. Carbon dots, known for their stability, water solubility, and surface functional groups like -COOH and -NH, exhibit powerful antibacterial properties (Bhardwaj et al. 2023). Recent research shows that the carbon dots are very effective against pathogens like *Staphylococcus aureus* and *Escherichia coli*, especially when used in formulations like Ag@carbon dot derived from *Cannabis sativa* (Raina et al. 2020). Although the carbon dots derived from natural sources like white mulberry also improve bacterial nanocomposite films, giving them UV-blocking and antimicrobial properties against *Listeria monocytogenes*. However, the preparation of these films involves optimization of the concentration of carbon dots and impregnation processes to enhance mechanical strength and antimicrobial efficacy, as seen through scanning electron microscopy (Salimi et al. 2021). Carbon nanomaterials present environmentally friendly solutions for enhancing food packaging integrity and offering improved mechanical properties and antimicrobial benefits, which are crucial for maintaining food safety and quality (Kousheh et al. 2020).

Carbon nanotubes exist in two primary forms: single-walled (SW) carbon nanotubes and multi-walled (MW) carbon nanotubes. SW carbon nanotubes are composed of a single layer of graphene rolled into a cylinder, while MW carbon nanotubes consist of multiple concentric layers held together by *van der Waals* forces. Recent studies have explored their antibacterial properties, such as the use of Ag-doped TiO_2 nanoparticles coated with SW carbon nanotubes and MW carbon nanotubes. This research demonstrated that SW carbon nanotubes—TiO_2/Ag exhibited greater toxicity against *E. coli* and *S. aureus* as compared to MW carbon nanotubes—TiO_2/Ag (Mohammad et al. 2019). However, these nanomaterials are also impregnated into food packaging. The functionalized carbon nanotubes reinforced poly(3-hydroxybutyrate-co-3-hydroxyvalerate) nanocomposites enhance the thermal and mechanical properties with an 88% increase in tensile strength and 172% in Young's modulus (Yu et al. 2014). Additionally, Dias et al. developed a film for packing pieces of cooked chicken meat using allyl isothiocyanate and carbon nanotubes, achieving a reduction in microbial growth for 40 days with allyl isothiocyanate concentrations >28% and carbon nanotube concentrations >0.02% (Dias et al. 2013). These innovations highlight the potential of carbon nanotubes to enhance food preservation and packaging efficiency.

On the other hand, graphene, a two-dimensional single-atom-thick structure derived from graphite, is made through mechanical or chemical exfoliation processes. Graphene oxide (GO) is synthesized using Hummer's method, while reduced graphene oxide (rGO) is derived from the reduction of GO, representing an intermediate between graphene and highly oxidized GO. Graphene-based nanomaterials are extensively researched for their antibacterial properties. Aunkor et al. (2020) conducted research on the germicidal properties of graphene oxide. Nanosheets against both positive-stain and negative-stain bacteria showed significant zones of inhibition, such as *Proteus mirabilis* (27 mm), *Pseudomonas aeruginosa* (38 mm), *Staphylococcus aureus* (38 mm), *Escherichia coli* (39 mm), *Serratia marcescens* (39 mm), and *Klebsiella pneumoniae* (41 mm). GO synthesized via the modified Hummers' method exhibited superior antibacterial efficacy compared to commercial antibiotics. Rago et al. (2015) demonstrated the germicidal activity of graphene nanoplatelets against *Streptococcus mutans*, noting that efficacy was influenced by nanoplatelet size and thickness. Smaller and thinner nanoplatelets showed stronger antibacterial effects against S. *mutans.* A silver-cellulose film modified with activated carbon (AgNPs/AC-CNF) was created using a casting method. Antimicrobial testing against *S. aureus* and *E. coli* showed that AgNP concentrations ranging from 140 to 450 ppm effectively inhibited bacterial growth on agar plates. Additionally, the film's electrical conductivity was evaluated using voltammetric studies that indicated an increase in peak current for CV (0.2 to 1.1 μA) and DPV (0.78 to 1.16 μA), while the electrical resistivity decreased (6.8 to 1.9 kΩ.cm) with an increase in the concentration of AgNPs. The findings demonstrate that the AgNPs/AC-CNF film exhibits promising antimicrobial and conductive properties, making it a viable option for enhancing food safety and quality through smart packaging applications (Sobhan et al. 2020).

11.4 Challenges, Safety, and Regulatory Considerations

There are many processes that are involved when gas molecules move through the polymer layer, including the adsorption and infiltration through the layers. The packaging material of food is prepared in such a way that it acts as a gas barrier (Ahmed et al. 2017). The infiltration of gaseous molecules is determined by taking into consideration the rate of diffusion and thickness of the tested film. Under comparable testing conditions, including temperature, relative moistness, and force, this metric allows for the comparison of barrier performance across various materials, although it is most accurate for thicker films (>10 μm) (Huang and Paul 2007). The absorbency, solubility, and movement into the films of polymer are determined experimentally through the isostatic and quasi-isostatic techniques, the water cup method, magnetic suspension balance, quartz crystal microbalance, and inverse gas chromatography, wherein each method has its specific application, including determination of absorbency, solubility, and movement into polymer films (Sonchaeng et al. 2018). Due to the diversity of methods used for such measurement, the researchers use varied units to express transmission rate or permeability, such as mol/m.s.Pa. This reflects the complexity and precision required in assessing barrier properties. The incorporation of nano-sized materials into food packaging raises several safety concerns that must be carefully addressed to ensure consumer protection and regulatory compliance.

(i) Migration of Nanomaterials: The primary concern in the preparation of packaging material is the movement of nano-sized materials from the packaging film into food products. The small size and increased area of surface of these materials cause higher mobility as compared to large-sized particles. This migration causes unintended ingestion of nanoparticles by the consumers. The biodegradable polymers like polyhydroxy alkanoates (Fabra et al. 2014), polyglycolic acid (Regubalan et al. 2022), polyvinyl alcohol (Yang et al. 2016), and thermoplastic starch (Avella et al. 2005) are viable options for barrier layers in packaging. However, their suitability for blown film processing, which is a common technique in packaging production, requires enhancement. Polyglycolic acid possesses effectiveness in enhancing gas and water barrier properties of polyethylene terephthalate flasks if employed as a medium barrier layer. This highlights its capability for enhanced overall performance, despite current challenges in processing methods like blown film (Yamane et al. 2014). The increased reactivity and potential to penetrate biological barriers have raised concerns about the influence of such materials on the health of humans. Investigations suggest that some nanoparticles are capable of exerting harmful effects concerning their structure, dimensions, and morphological characteristics (Xiao-e et al. 2004).

(ii) Environmental Impact: Nanoparticles enter the environment through methods of waste disposal and can accumulate in the ecosystems, due to which their disposal becomes a threat (Ahmad et al. 2024). The biodegradability of packaging materials prepared must adhere strictly to the standards depending on their design for factory composting, domestic composting, and marine environments. The European standard EN13432 specifically addresses biodegradability and verification of packaging. Also, the distinct criteria are established for domestic composting (European Bioplastics 2015). Many ASTM standards, including ASTM D7991, D7473, and D6691, are also tailored to claims of marine compostability (Wu et al. 2021).

(iii) Regulatory Challenges: The regulatory bodies have to face trials due to adequately ensuring the safety of nanomaterials because the existing regulations may not fully consider the distinct characteristics and possible risks associated with nanomaterials. Thus, there is a need for standardized testing protocols and valuation of all risks tailored specifically to nanotechnology (Singh et al. 2023).

(iv) Consumer Awareness: Effective communication of the advantages and possible hazards of using nanomaterials in packaging is essential. The purchasers should be made aware of the use of such materials in packaging, their safety evaluations, and any precautions necessary for handling and disposal (Khwaldia et al. 2010).

Thus, to adopt these safety concerns, a collaborative effort is required among stakeholders, including researchers, industry professionals, regulators, and consumer advocacy groups. Comprehensive risk assessments, transparent communication, and continuous monitoring of emerging scientific findings are crucial for safe and responsible use of such materials in the packaging of food products (Chausali et al. 2022).

11.5 Future Prospective

Introducing novel packaging materials, especially those incorporating nanomaterials, presents significant regulatory challenges that must be navigated carefully to ensure compliance and market acceptance.

(i) Lack of Specific Regulations: One major hurdle is the absence of specific regulations tailored to nanomaterials in food packaging. Existing regulatory frameworks may not encompass the unique properties and potential risks associated with nanotechnology.

(ii) Safety Assessment Requirements: Regulatory approval typically requires comprehensive safety assessments to evaluate the potential health risks posed by novel packaging materials. For nanomaterials, additional considerations such as toxicity, bioavailability, and environmental impact assessments may be necessary.

(iii) Standardization of Testing Protocols: The appraisal of the safe use of nanomaterials in food packaging is crucial but is lacking or underdeveloped. Establishing reliable methods for characterizing nanoparticles and assessing their behavior in food contact materials is essential for regulatory approval.

(iv) Global Harmonization: Differences in regulatory requirements across regions and countries can complicate the approval process for novel packaging materials. Harmonization efforts are needed to streamline regulatory pathways and facilitate market access for innovative technologies.

(v) Transparency and Communication: Transparent communication with regulatory authorities is vital to address their concerns and provide adequate data supporting the safety and efficacy of novel packaging materials. Clear documentation of risk assessments and compliance with regulatory requirements is essential for successful approval.

Navigating these regulatory hurdles requires collaboration between industry stakeholders, researchers, and regulatory agencies to develop guidelines and frameworks that ensure the safe introduction of innovative packaging technologies into the market. Adhering to regulatory requirements from the early stages of development and conducting robust safety assessments are critical steps toward gaining regulatory approval for novel packaging materials, including those incorporating nanomaterials.

11.6 Conclusion

The packaging of food products includes protection, containment, and information provision along with ensuring the quality and nutrient value of enclosed food products. Nanomaterials have been used as an innovative solution for packaging functional foods in the food industry, offering advanced packaging such as enhancement of strength, flexibility, durability, biodegradability, antimicrobial potential, thermal stability, barrier properties, and UV-blocking abilities. Studies reveal that some of the specific types of nanomaterials that have been used for these unique benefits include metal oxide nanoparticles for improved UV protection, carbon nanotubes and graphene oxide for enhanced mechanical strength and barrier performance, silver nanoparticles for antimicrobial properties, and layered silicates for flame-retardant properties. Despite the benefits, scalability, cost-effectiveness, safety concerns, and potential toxicity of nanomaterials, along with a lack of suitable regulatory and standardization protocols, pose challenges and limitations in their effective use in food packaging. Further investigations are required to address the challenges and limitations for developing safer, more effective, and sustainable food packaging solutions.

References

Abdullah ZW, Dong Y (2019) Biodegradable and water resistant poly(vinyl) alcohol (PVA)/starch (ST)/glycerol (GL)/halloysite nanotube (HNT) nanocomposite films for sustainable food packaging. Front Mater 6:1–17. https://doi.org/10.3389/fmats.2019.00058

Adeyeye SAO, Surendra Babu A, Guruprasath N, Sankar Ganesh P (2023) Starch nanocrystal and its food packaging applications. Curr Res Nutr Food Sci 11(1):01–21. https://doi.org/10.12944/CRNFSJ.11.1.1

Ahmad K, Li Y, Tu C, Guo Y, Yang X, Xia C, Hou H (2024) Nanotechnology in food packaging with implications for sustainable outlook and safety concerns. Food Biosci 58:103625. https://doi.org/10.1016/j.fbio.2024.103625

Ahmed J, Arfat YA, Al-Attar H, Auras R, Ejaz M (2017) Rheological, structural, ultraviolet protection and oxygen barrier properties of linear low- density polyethylene films reinforced with zinc oxide (ZnO) nanoparticles. Food Packag Shelf Life 13:20–26. https://doi.org/10.1016/j.fpsl.2017.04.005

Alizadeh Sani M, Maleki M, Eghbaljoo-Gharehgheshlaghi H, Khezerlou A, Mohammadian E, Liu Q, Jafari SM (2022) Titanium dioxide nanoparticles as multifunctional surface-active materials for smart/active nanocomposite packaging films. Adv Colloid Interf Sci 300: 102593. https://doi.org/10.1016/j.cis.2021.102593

Anjum A, Garg R, Kashif M, Eddy NO (2023) Nano-scale innovations in packaging: properties, types, and applications of nanomaterials for the future. Food Chem Adv 3:100560. https://doi.org/10.1016/j.focha.2023.100560

Anjum A, Gupta D, Singh B, Garg R, Pani B, Kashif M, Jain S (2024) Clay–polymer nanocomposites for effective water treatment: opportunities, challenges, and future prospects. Environ Monit Assess 196:666. https://doi.org/10.1007/s10661-024-12823-8

Aunkor MTH, Raihan T, Prodhan SH, Metselaar HSC, Malik SUF, Azad AK (2020) Antibacterial activity of graphene oxide nanosheet against multidrug resistant superbugs isolated from infected patients. R Soc Open Sci 7:200640. https://doi.org/10.1098/rsos.200640

Avella M, De Vlieger JJ, Errico ME, Fischer S, Vacca P, Volpe MG (2005) Biodegradable starch/clay nanocomposite films for food packaging applications. Food Chem 93:467–474. https://doi.org/10.1016/j.foodchem.2004.10.024

Bansal H, Singh S, Sharma A, Singh J, Mehta SK (2023) Polymer nano-composite coatings and films: modern insights and emerging strategies to lengthen the lifespan of fruits and vegetables. Food Mater Res 3. https://doi.org/10.48130/FMR-2023-0023

Bhardwaj S, Lata S, Garg R (2023) Application of nanotechnology for preventing postharvest losses of agriproducts. J Hortic Sci Biotechnol 98:31–44. https://doi.org/10.1080/14620316.2022.2091488

Boelter JF, Brandelli A (2016) Innovative bionanocomposite films of edible proteins containing liposome-encapsulated nisin and halloysite nanoclay. Colloids Surfaces B 145:740–747. https://doi.org/10.1016/j.colsurfb.2016.05.080

Branton D, Deamer DW, Marziali A, Bayley H, Benner SA, Butler T, Di Ventra M, Garaj S, Hibbs A, Huang X, Jovanovich SB, Krstic PS, Lindsay S, Ling XS, Mastrangelo CH, Meller A, Oliver JS, Pershin YV, Ramsey JM, Riehn R, Soni GV, Cossa VT, Wanunu M, Wiggin M, Schloss JA (2009) The potential and challenges of nanopore sequencing. In: Nanoscience and technology. Co-Published with Macmillan Publishers Ltd, UK, pp 261–268

Chausali N, Saxena J, Prasad R (2022) Recent trends in nanotechnology applications of bio-based packaging. J Agric Food Res 7:100257. https://doi.org/10.1016/J.JAFR.2021.100257

Connolly M, Zhang Y, Brown DM, Ortuño N, Jordá-Beneyto M, Stone V, Fernandes TF, Johnston HJ (2019) Novel polylactic acid (PLA)-organoclay nanocomposite bio-packaging for the cosmetic industry; migration studies and in vitro assessment of the dermal toxicity of migration extracts. Polym Degrad Stab 168:108938. https://doi.org/10.1016/J.POLYMDEGRADSTAB.2019.108938

Cyras VP, Manfredi LB, Ton-That M-T, Vázquez A (2008) Physical and mechanical properties of thermoplastic starch/montmorillonite nanocomposite films. Carbohydr Polym 73:55–63. https://doi.org/10.1016/j.carbpol.2007.11.014

de Azeredo HMC (2013) Antimicrobial nanostructures in food packaging. Trends Food Sci Technol 30:56–69. https://doi.org/10.1016/j.tifs.2012.11.006

de Oliveira LH, Trigueiro P, Souza JSN, de Carvalho MS, Osajima JA, da Silva-Filho EC, Fonseca MG (2022) Montmorillonite with essential oils as antimicrobial agents, packaging, repellents, and insecticides: an overview. Colloids Surfaces B 209:112186. https://doi.org/10.1016/j.colsurfb.2021.112186

Dias MV, de Soares NFF, Borges SV, de Sousa MM, Nunes CA, de Oliveira IRN, Medeiros EAA (2013) Use of allyl isothiocyanate and carbon nanotubes in an antimicrobial film to package shredded, cooked chicken meat. Food Chem 141:3160–3166. https://doi.org/10.1016/j.foodchem.2013.05.148

European Bioplastics (2015) Home composting. Factsheet

Fabra MJ, Lopez-Rubio A, Lagaron JM (2014) Nanostructured interlayers of zein to improve the barrier properties of high barrier polyhydroxyalkanoates and other polyesters. J Food Eng 127: 1–9. https://doi.org/10.1016/j.jfoodeng.2013.11.022

Fasihnia SH, Peighambardoust SH, Peighambardoust SJ (2018) Nanocomposite films containing organoclay nanoparticles as an antimicrobial (active) packaging for potential food application. J Food Process Preserv 42:e13488. https://doi.org/10.1111/jfpp.13488

Gaikwad KK, Singh S, Negi YS (2020) Ethylene scavengers for active packaging of fresh food produce. Environ Chem Lett 18:269–284. https://doi.org/10.1007/s10311-019-00938-1

Garg R, Anjum A (2024) Sustainable approach to protective nanocoatings. IGI Global

Giannakas A (2020) Na-montmorillonite vs. organically modified montmorillonite as essential oil nanocarriers for melt-extruded low-density poly-ethylene nanocomposite active packaging films with a controllable and long-life antioxidant activity. Nano 10. https://doi.org/10.3390/nano10061027

Heydari A, Alemzadeh I, Vossoughi M (2013) Functional properties of biodegradable corn starch nanocomposites for food packaging applications. Mater Des 50:954–961. https://doi.org/10.1016/j.matdes.2013.03.084

Huang Y, Paul DR (2007) Effect of film thickness on the gas-permeation characteristics of glassy polymer membranes. Ind Eng Chem Res 46:2342–2347. https://doi.org/10.1021/ie0610804

Imran M, Revol-Junelles A-M, Francius G, Desobry S (2016) Diffusion of fluorescently labeled bacteriocin from edible nanomaterials and embedded nano-bioactive coatings. ACS Appl Mater Interfaces 8:21618–21631. https://doi.org/10.1021/acsami.6b04621

Kanmani P, Rhim JW (2014) Properties and characterization of bionanocomposite films prepared with various biopolymers and ZnO nanoparticles. Carbohydr Polym 106:190–199. https://doi.org/10.1016/j.carbpol.2014.02.007

Khwaldia K, Arab-Tehrany E, Desobry S (2010) Biopolymer coatings on paper packaging materials. Compr Rev Food Sci Food Saf 9:82–91

Kousheh SA, Moradi M, Tajik H, Molaei R (2020) Preparation of antimicrobial/ultraviolet protective bacterial nanocellulose film with carbon dots synthesized from lactic acid bacteria. Int J Biol Macromol 155:216–225. https://doi.org/10.1016/j.ijbiomac.2020.03.230

Lata S, Bhardwaj S, Garg R (2022) Nanomaterials for sensing and biosensing: applications in agri-food diagnostics. Int J Environ Anal Chem 00:1–18. https://doi.org/10.1080/03067319.2022.2115895

Lin D, Yang Y, Wang J, Yan W, Wu Z, Chen H, Zhang Q, Wu D, Qin W, Tu Z (2020) Preparation and characterization of TiO2-Ag loaded fish gelatin-chitosan antibacterial composite film for food packaging. Int J Biol Macromol 154:123–133. https://doi.org/10.1016/j.ijbiomac.2020.03.070

Łukasiewicz S, Mikołajczyk A, Błasiak E, Fic E, Dziedzicka-Wasylewska M (2021) Polycaprolactone nanoparticles as promising candidates for nanocarriers in novel nanomedicines. Pharmaceutics 13:191. https://doi.org/10.3390/pharmaceutics13020191

Mohammad MR, Ahmed DS, Mohammed MKA (2019) Synthesis of Ag-doped TiO2 nanoparticles coated with carbon nanotubes by the sol–gel method and their antibacterial activities. J Sol-Gel Sci Technol 90:498–509. https://doi.org/10.1007/s10971-019-04973-w

Mohseni MS, Khalilzadeh MA, Mohseni M, Hargalani FZ, Getso MI, Raissi V, Raiesi O (2020) Green synthesis of Ag nanoparticles from pomegranate seeds extract and synthesis of Ag-Starch nanocomposite and characterization of mechanical properties of the films. Biocatal Agric Biotechnol 25:101569. https://doi.org/10.1016/j.bcab.2020.101569

Nagarajan M, Benjakul S, Prodpran T, Songtipya P (2014) Characteristics of bio-nanocomposite films from tilapia skin gelatin incorporated with hydrophilic and hydrophobic nanoclays. J Food Eng 143:195–204. https://doi.org/10.1016/j.jfoodeng.2014.06.038

Nile SH, Baskar V, Selvaraj D, Nile A, Xiao J, Kai G (2020) Nanotechnologies in food science: applications, recent trends, and future perspectives. Nano-Micro Lett 12:45. https://doi.org/10.1007/s40820-020-0383-9

Oliani WL, Pusceddu FH, Parra DF (2022) Silver-titanium polymeric nanocomposite non ecotoxic with bactericide activity. Polym Bull 79:10949–10968. https://doi.org/10.1007/s00289-021-04036-7

Oliver-Ortega H, Vandemoortele V, Bala A, Julian F, Méndez JA, Espinach FX (2021) Nanoclay effect into the biodegradation and processability of poly(lactic acid) nanocomposites for food packaging. Polymers (Basel) 13:2741. https://doi.org/10.3390/polym13162741

Pardo-Figuerez M, López-Córdoba A, Torres-Giner S, Lagaron JM (2018) Superhydrophobic Bio-Coating made by co-continuous electrospinning and electrospraying polyethylene terephthalate films proposed as easy emptying transparent food packaging. Coatings 8. https://doi.org/10.3390/coatings8100364

Peighambardoust SJ, Peighambardoust SH, Mohammadzadeh Pournasir N, Pakdel P (2019) Properties of active starch-based films incorporating a combination of ag, ZnO and CuO nanoparticles for potential use in food packaging applications. Food Packag Shelf Life 22: 100420. https://doi.org/10.1016/j.fpsl.2019.100420

Peighambardoust SH, Fasihnia SH, Peighambardoust SJ, Pateiro M, Domínguez R, Lorenzo JM (2021) Active polypropylene-based films incorporating combined antioxidants and antimicrobials: preparation and characterization. Food Secur 10:722. https://doi.org/10.3390/foods10040722

Rago I, Bregnocchi A, Zanni E, D'Aloia AG, De Angelis F, Bossu M, De Bellis G, Polimeni A, Uccelletti D, Sarto MS (2015) Antimicrobial activity of graphene nanoplatelets against Streptococcus mutans. In: 2015 IEEE 15th International Conference on Nanotechnology (IEEE-NANO). IEEE, pp 9–12

Raina S, Thakur A, Sharma A, Pooja D, Minhas AP (2020) Bactericidal activity of Cannabis sativa phytochemicals from leaf extract and their derived Carbon Dots and Ag@Carbon Dots. Mater Lett 262:127122. https://doi.org/10.1016/j.matlet.2019.127122

Ranjan S, Dasgupta N, Chakraborty AR, Melvin Samuel S, Ramalingam C, Shanker R, Kumar A (2014) Nanoscience and nanotechnologies in food industries: opportunities and research trends. J Nanopart Res 16:2464. https://doi.org/10.1007/s11051-014-2464-5

Regubalan B, Manibalan S, Pandit P (2022) Polyglycolic acid-based bionanocomposites for food packaging applications. pp 153–164

Rostamzad H, Paighambari SY, Shabanpour B, Ojagh SM, Mousavi SM (2016) Improvement of fish protein film with nanoclay and transglutaminase for food packaging. Food Packag Shelf Life 7:1–7. https://doi.org/10.1016/j.fpsl.2015.10.001

Salimi F, Moradi M, Tajik H, Molaei R (2021) Optimization and characterization of eco-friendly antimicrobial nanocellulose sheet prepared using carbon dots of white mulberry (<scp> *Morus alba* </scp> L.). J Sci Food Agric 101:3439–3447. https://doi.org/10.1002/jsfa.10974

Salmas C, Giannakas A, Katapodis P, Leontiou A, Moschovas D, Karydis-Messinis A (2020) Development of zno/na-montmorillonite hybrid nanostructures used for pvoh/zno/na-montmorillonite active packaging films preparation via a melt-extrusion process. Nano 10. https://doi.org/10.3390/nano10061079

Sanuja S, Agalya A, Umapathy MJ (2015) Synthesis and characterization of zinc oxide-neem oil-chitosan bionanocomposite for food packaging application. Int J Biol Macromol 74:76–84. https://doi.org/10.1016/j.ijbiomac.2014.11.036

Sharma R, Jafari SM, Sharma S (2020) Antimicrobial bio-nanocomposites and their potential applications in food packaging. Food Control 112:107086. https://doi.org/10.1016/j.foodcont.2020.107086

Sharma R, Lata S, Garg R (2024) Valorisation of agricultural waste and their role in green synthesis of value-added nanoparticles. Environ Technol Rev 13:40–59. https://doi.org/10.1080/21622515.2023.2283412

Silva MRF, Alves MFRP, Cunha JPGQ, Costa JL, Silva CA, Fernandes MHV, Vilarinho PM, Ferreira P (2023) Nanostructured transparent solutions for UV-shielding: recent developments and future challenges. Mater Today Phys 35:101131. https://doi.org/10.1016/j.mtphys.2023.101131

Silvestre C, Duraccio D, Cimmino S (2011) Food packaging based on polymer nanomaterials. Prog Polym Sci 36:1766–1782. https://doi.org/10.1016/j.progpolymsci.2011.02.003

Singh N, Gaur S, Chawla S, Singh S, Husen A (2023) Use of smart nanomaterials in food packaging. Adv Smart Nanomater their Appl: 233–245. https://doi.org/10.1016/B978-0-323-99546-7.00008-2

Siripatrawan U, Kaewklin P (2018) Fabrication and characterization of chitosan-titanium dioxide nanocomposite film as ethylene scavenging and antimicrobial active food packaging. Food Hydrocoll 84:125–134. https://doi.org/10.1016/j.foodhyd.2018.04.049

Slavin YN, Asnis J, Häfeli UO, Bach H (2017) Metal nanoparticles: understanding the mechanisms behind antibacterial activity. J Nanobiotechnol 15:65. https://doi.org/10.1186/s12951-017-0308-z

Sobhan A, Muthukumarappan K, Wei L, Van Den Top T, Zhou R (2020) Development of an activated carbon-based nanocomposite film with antibacterial property for smart food packaging. Mater Today Commun 23:101124. https://doi.org/10.1016/j.mtcomm.2020.101124

Sonchaeng U, Iñiguez-Franco F, Auras R, Selke S, Rubino M, Lim L-T (2018) Poly(lactic acid) mass transfer properties. Prog Polym Sci 86:85–121. https://doi.org/10.1016/j.progpolymsci.2018.06.008

Tabari M (2018) Characterization of a new biodegradable edible film based on Sago Starch loaded with Carboxymethyl Cellulose nanoparticles. Nanomedicine Res J 3:25–30. https://doi.org/10.22034/NMRJ.2018.01.004

Vipan B, Mahajan C, Tandon R, Kapoor S, Sidhu MK (2018) Natural coatings for shelf-life enhancement and quality maintenance of fresh fruits and vegetables-a review. J Postharvest Technol 06:12–26

Wani NR, Dar AH, Dash KK, Pandey VK, Srivastava S, Jan SY, Deka P, Sabahi N (2023) Recent advances in the production of bionanomaterials for development of sustainable food packaging: a comprehensive review. Environ Res 237:116948. https://doi.org/10.1016/j.envres.2023.116948

Wu F, Misra M, Mohanty AK (2021) Challenges and new opportunities on barrier performance of biodegradable polymers for sustainable packaging. Prog Polym Sci 117:101395. https://doi.org/10.1016/j.progpolymsci.2021.101395

Xiao-e L, Green ANM, Haque SA, Mills A, Durrant JR (2004) Light-driven oxygen scavenging by titania/polymer nanocomposite films. J Photochem Photobiol A Chem 162:253–259. https://doi.org/10.1016/j.nainr.2003.08.010

Yamane K, Sato H, Ichikawa Y, Sunagawa K, Shigaki Y (2014) Development of an industrial production technology for high-molecular-weight polyglycolic acid. Polym J 46:769–775. https://doi.org/10.1038/pj.2014.69

Yang W, Owczarek JS, Fortunati E, Kozanecki M, Mazzaglia A, Balestra GM, Kenny JM, Torre L, Puglia D (2016) Antioxidant and antibacterial lignin nanoparticles in polyvinyl alcohol/chitosan films for active packaging. Ind Crop Prod 94:800–811. https://doi.org/10.1016/j.indcrop.2016.09.061

Yu H-Y, Qin Z-Y, Sun B, Yang X-G, Yao J-M (2014) Reinforcement of transparent poly (3-hydroxybutyrate-co-3-hydroxyvalerate) by incorporation of functionalized carbon nanotubes as a novel bionanocomposite for food packaging. Compos Sci Technol 94:96–104. https://doi.org/10.1016/j.compscitech.2014.01.018

Yu H, Park JY, Kwon CW, Hong SC, Park KM, Chang PS (2018) An overview of nanotechnology in food science: preparative methods, practical applications, and safety. J Chem 2018:1. https://doi.org/10.1155/2018/5427978

Chapter 12
Role of Nanotechnology in Food Processing

Andersen Escobar Schlogl, Marília Magalhães Gonçalves, Brunna D'Onofre Couto, Rebeca Rocha Batista, Felipe Ramalho Estanislau, Liliane de Paula Gonçalves, George Lucas da Silva Magalhães, and Igor José Boggione Santos

Abstract Nanotechnology represents a revolutionary approach to addressing challenges in food processing and preservation, offering innovative solutions to enhance food quality, safety, and sustainability. This chapter explores the integration of nanotechnology in food science, focusing on its applications in production optimization, packaging, preservation, and contaminant detection. Nanomaterials such as nanoemulsions, nanoliposomes, and nanofibers play pivotal roles in improving bioavailability, nutrient delivery, and food shelf life. Additionally, nanostructured active and intelligent packaging materials exhibit antimicrobial properties and real-time monitoring capabilities, ensuring enhanced food safety and extended durability. The use of nanotechnology in unit operations, such as filtration and thermal processes, has improved efficiency and reduced environmental impact. Despite its numerous benefits, challenges such as consumer acceptance, regulatory complexities, and concerns regarding nanotoxicity persist. Future prospects indicate a

A. E. Schlogl
Postgraduate Program in Physics and Chemistry of Materials, Department of Natural Sciences (DCNAT) Federal University of São João del-Rei (UFSJ), São João del-Rei, Brazil

M. M. Gonçalves
Department of Chemistry, Biotechnology, and Bioprocess Engineering (DQBIO), Federal University of São João del-Rei (UFSJ), Ouro Branco, MG, Brazil

Department of Food Technology (DTA), Federal University of Viçosa, Viçosa, MG, Brazil

B. D'Onofre Couto · R. R. Batista · F. R. Estanislau
Department of Chemistry, Biotechnology, and Bioprocess Engineering (DQBIO), Federal University of São João del-Rei (UFSJ), Ouro Branco, MG, Brazil

L. de Paula Gonçalves · G. L. da Silva Magalhães
Department of Chemical Engineering (DEQUI), Federal University of São João del-Rei (UFSJ), Ouro Branco, MG, Brazil

I. J. B. Santos (✉)
Postgraduate Program in Physics and Chemistry of Materials, Department of Natural Sciences (DCNAT) Federal University of São João del-Rei (UFSJ), São João del-Rei, Brazil

Department of Chemistry, Biotechnology, and Bioprocess Engineering (DQBIO), Federal University of São João del-Rei (UFSJ), Ouro Branco, MG, Brazil
e-mail: igorboggione@ufsj.edu.br

S. K. Yadav, A. Kumari (eds.), *Nanobiotechnology in Agri-food Sector*, Sustainability Sciences in Asia and Africa, https://doi.org/10.1007/978-981-96-7756-6_12

growing role for nanotechnology in developing sustainable and efficient food systems, driving innovation in the global food industry.

Keywords Nanomaterials · Food processing · Food safety · Food contaminants · Nanotoxicity

12.1 Introduction

Considered the fifth cycle of technological innovation (Dutta 2011), nanotechnology encompasses a set of techniques and processes aimed at manipulating the morphology, geometry, and size of materials to obtain structures at the nanometric scale with distinct properties (Uskoković 2007). These materials can exhibit new characteristics or modifications to existing properties when not nanostructured. There are two main approaches to obtaining nanostructures: the bottom-up approach, which utilizes interactions of atoms and/or molecules to achieve nanometric size through chemical or biological synthesis, and the top-down approach, which fragments a material in its naturally macroscopic state using techniques such as grinding and physicochemical processes (Alencar et al. 2013; Garcia and Lagaron 2012). The chosen methodology for obtaining the nanostructure directly impacts the properties of the produced material, its quality, and manufacturing costs. These new properties generate technological advances in various sectors, including information technology, energy production, materials science, manufacturing, food, water, the environment, and safety. In the food sector, nanotechnology has developed a segment known as "nanofood" (Azevedo et al. 2014; Hebbar et al. 2020).

Nanofood refers to foods and/or food additives that utilize nanomaterials or processes containing nanotechnology to alter, enhance, or create properties and functionalities of the food (Meghani et al. 2020). These modifications can increase the bioavailability of nutrients, improve the taste, texture, and color of the food, extend the shelf life of food products, and monitor the quality status of the product. However, nanotechnology is not limited to enhancing the quality and safety of foods; it also offers solutions to some of the most pressing challenges in the food industry, such as reducing food waste and improving global nutrition (Hessel et al. 2023; Nile et al. 2020).

Among the main nanomaterials used in nanofoods are bioactive nanostructures, nanoemulsions, nanocarriers, and nanostructured packaging. Nanoparticles can be used as sources of bioactives such as vitamins, minerals, antioxidants, and essential fatty acids through the nanostructuring of bioactives, proteins, or metal oxides. These particles can perform nanoencapsulation of other food additives, providing protection against adverse conditions during processing and storage, thereby increasing their stability and bioavailability (Meghani et al. 2020; Sun et al. 2021). They also allow for the formation of nanoemulsions, improving the solubility and absorption of bioactive ingredients and stabilizing the created emulsion in products like beverages, sauces, butter, and mayonnaises (Oliveira Filho et al. 2022). Nanocarriers, nanometric structures such as liposomes and micelles, encapsulate nutrients and bioactive compounds, releasing them in a controlled manner within the body, which is particularly valuable for functional foods and dietary supplements where the efficacy of the bioactive compound is crucial (Meghani et al. 2020). Food packaging is another area where nanotechnology has shown great potential, with nanostructured packaging materials offering antimicrobial properties, resistance to gases and moisture, and increased food durability. These smart packaging solutions can, for example, contain indicators that show the freshness of the food or sensors that detect the presence of pathogens, ensuring greater food safety (Ahmad et al. 2023; Souza and Fernando 2016).

Despite its promise, nanofood faces some barriers that are slowing its widespread implementation in the consumer market. As high-tech and recent products, the global population still has a low level of understanding about nanotechnology, leading to reluctance toward nanofoods. This barrier to social acceptance of nanofoods is rooted in the fear of innovation, fear associated with the unknown, technological fears, concerns about the potential harm that nanotechnology might cause, rejection of highly industrialized products, and high demand for so-called "natural" foods with low processing levels and minimal addition of food additives. Educational and transparent campaigns about the benefits and safety of nanofoods are essential to gain consumer trust and reduce this barrier (Hasmin et al. 2021; Sodano et al. 2016).

As society strives to learn more about nanofoods, state governments are also developing new safety standards with tests and analysis protocols to ensure food safety for their populations. Regulatory agencies such as the Food and Drug Administration (FDA) in the USA and the European Food Safety Authority (EFSA) in Europe require extensive risk assessments to ensure that nanomaterials are safe for human consumption (Allan et al. 2021; Ilyas et al. 2021).

The scaling-up process for the production of nanofoods is another bottleneck for commercialization. The investment required to develop production infrastructure, purification, and quality analysis of the product poses significant technical and economic challenges. The production of nanomaterials, which can be food additives or the product itself, may present high costs, making the products more expensive for consumers; cost reduction through technological advances and economies of scale is necessary to make nanofoods competitive in the market (Herdiana et al. 2022; Jeevahan et al. 2020).

12.2 The Impact of Nanotechnology on the Efficiency of Industrial Processes

Processes in the food industry comprise numerous unit operations, from handling raw materials to obtaining finished products (Biswas et al. 2022). In this scenario, nanotechnology has emerged as a promising tool for improving the efficiency of many processes, acting on some important unit operations. The role of nanotechnology covers operations such as separation and purification, where nanofiltration and nanoadsorption techniques are used to recover valuable substances and remove impurities more effectively when operating on a nanometer scale (Assadpour et al. 2020; Bandini and Nataloni 2015; García Doménech et al. 2020; McClements 2020; Sánchez-Arévalo et al. 2023), food product concentration and clarification steps using nanofilters improve process yields. Thermal processing can be optimized with the use of nanofluids, resulting in lower energy expenditure (Assadpour et al. 2020).

Food preservation is also improved through encapsulation techniques with nanoparticles, which extend the shelf life of products (Mohammad et al. 2022) and the use of active and intelligent packaging made from nanomaterials (Alfei et al. 2020; Hamad et al. 2018; Neme et al. 2021).

The use of nanofiltration also plays a crucial role in the treatment of water, an essential input in the food industry, acting in the removal of contaminants and purification (García Doménech et al. 2020; Kim et al. 2021). In addition, nanocatalysts have been studied for the treatment of effluents from the food industry, which can reduce environmental impacts (Navia-Mendoza et al. 2021).

12.2.1 Optimization of Production Processes

The optimization of production processes has been improved by the use of nanotechnology in some unit operations along the production chain. In the stages of separation, purification, clarification, and concentration of compounds, nanofiltration has stood out due to its lower operating pressures, higher flows and retentions, continuous operation, ease of cleaning, and lower maintenance costs (Assadpour et al. 2020).

In previous studies, García-Martín et al. (2010) evaluated the use of nanofiltration to reduce the sugar concentration in wine musts to obtain wines with lower alcohol content. They obtained satisfactory results using only two stages of nanofiltration, optimizing the process compared to the most widely used method of de-alcoholization in the wine industry, which is carried out on the finished wine and involves several stages in a rotating conical column to first remove the aroma and then the alcohol and then re-add the flavors, making the process more expensive.

Bandini & Nataloni (2015) used nanofiltration to recover dextrose from industrial solutions containing different polysaccharides, testing various types of membranes to assess the feasibility of industrial application in the recovery of dextrose from mother liquors, a by-product of the industrial process. The results were satisfactory at an industrial level when using a polyamide membrane, achieving a dextrose purity in the permeate of over 97%.

Arend et al. (2017) carried out nanofiltration of strawberry juice to concentrate phenolic compounds using polyvinylidene difluoride (PVDF) membranes. This method proved to be effective in concentrating the main bioactive compounds in

Table 12.1 Applications of nanofiltration in the food industry

Method	Material	Application	References
Nanofiltration	Graphene, graphene oxide, boron nitride, carbon nanotubes, cellulose acetate, polysulfones, Polyvinylidene fluoride, the addition of ceramics such as zirconia, titania, and alumina	Water treatment	Domenech et al. (2020)
	Polyamide, Polysulfone	Removal of lactic acid from milk serums	Chandrapala et al. (2016)
	Polyamide, Polyethersulfone, Polypiperazinamide	Concentration of polyphenols in cranberry juice	Acosta et al. (2017)
	Polyvinylidene difluoride	Concentration of polyphenols in strawberry juice	Arend et al. (2017)
	Polyamide	Reducing sugar in musts to lower the alcohol content of wines	García-Martín et al. (2010)
	Polyamide	Dextrose extraction from industrial solutions containing different polysaccharides	Bandini and Nataloni (2015)
Organic solvent nanofiltration	Polyamide, silicone-coated polyamide, polydimethylsiloxane	Extraction and purification of polyphenols from olive pomace	Sánchez-Arévalo et al. (2023)

strawberry juice without causing changes in flavor, color, and phenolic compounds, preserving the quality of the juice.

In recent studies, Sánchez-Arévalo et al. (2023) investigated the application of organic solvent nanofiltration (OSN) on nine different types of membranes to purify and separate phenolic compounds in a model solution of wet olive pomace hydroalcoholic extract. The aim was to validate the industrial viability of purifying phenolic compounds from wet olive pomace, which are important as bioactive compounds for application in functional foods. Among the membranes tested, the DuraMem® 500, NFX e NF270 models showed the most satisfactory permeate flux results (10–100 $L \cdot h^{-1} \cdot m^{-2}$) compared to the other membranes studied. However, the hydrophilic NF270 membrane, with an active layer of cross-linked polyamide, stood out for its greater efficiency in the process, achieving permeate fluxes close to 100 $L\ h^{-1}\ m^{-2}$.

Other applications of nanofiltration in the food industry are used, and some of them are listed in Table 12.1.

12.2.2 Waste Reduction and Energy Efficiency

Nanotechnology is also being used to develop methods of reducing the waste generated by the food industry. These methods not only decrease environmental impact by reducing waste and pollution but also recover valuable compounds from rejects and by-products that can be reintegrated into industrial processes (Bandini and Nataloni 2015; García Doménech et al. 2020; McClements 2020; Sánchez-Arévalo et al. 2023).

Nanofiltration can be used to recover cellulose from food waste, which can be used in the production of functional nanofibers for bioactive components in the food industry (McClements 2020).

Nanocatalysts are being studied to treat wastewater contaminated by food dyes, dispose of effluents that are less toxic to the environment, and recover process water (Navia-Mendoza et al. 2021; Roy et al. 2021). Agri-food, industrial, and agricultural waste is being studied for the green synthesis of silver nanoparticles (AgNPs), zinc oxide nanoparticles (ZnONPs), used in antimicrobial coatings for food packaging, gold nanoparticles (AuNPs), palladium nanoparticles (PdNPs), platinum nanoparticles (PtNPs), titanium dioxide nanoparticles (TiO_2NPs), and iron oxide magnetic nanoparticles (IONPs) (Dhanker et al. 2022; Pushparaj et al. 2022).

The use of nanomaterials to produce active and intelligent packaging (Islam et al. 2022) reduces environmental impacts, as they replace plastic packaging made from synthetic polymers, which are harmful to the environment (McClements 2020).

Nanotechnology acts as a green technique not only to reduce waste but also to reduce energy costs in the food industry. For example, in the unitary separation operation, the use of nanofiltration enables lower energy expenditure due to fewer process steps and lower operating pressures (Wahab et al. 2024). Studies are being conducted to improve the efficiency of heat exchange in heat exchangers and fundamental equipment in various operations in the food industry, such as pasteurization, sterilization, fractionation, distillation, and concentration, using nanofluids. The application of these nanomaterials can help reduce energy consumption and shorten processing times due to their high thermal conductivity (Assadpour et al. 2020).

12.2.3 Applications in Preservation Technologies, Such as Pasteurization and Sterilization

Food waste is one of the recurring problems in food industries (Sridhar et al. 2020); approximately 30% of food is wasted due to the action of microorganisms and problems with shelf life. Food preservation techniques are therefore necessary (Chandrababu et al. 2024).

Among traditional preservation techniques, sterilization using chemical agents on food contact surfaces, such as chlorine, is widely used due to its low cost and effectiveness in cleaning. However, the reaction of active chlorine with organic matter can compromise its sanitizing action, resulting in a bacteriostatic state, in which reduced levels of bacterial metabolic activity still occur, increasing the risk of cross-contamination (Tao et al. 2021).

Preservation techniques using heat treatment, such as pasteurization and sterilization, are also used to inactivate pathogenic microorganisms in food (Chiozzi et al. 2022); however, these techniques can cause nutritional losses, energy waste, taste changes, and a reduction in the food matrix (Sridhar et al. 2020). Because of this, nanotechnology is a promising tool that can be integrated into preservation techniques; for example, in the case of food pasteurization, the encapsulation of food with nanotubes can prevent losses in food characteristics when subjected to pasteurization processes (Mohammad et al. 2022). Studies carried out by XU et al. (2020) combined the radiofrequency pasteurization technique with the use of ZnO nanoparticles and obtained sensory quality results for meat samples that were superior to the high-pressure steam sterilization method.

In addition, other processes use nanotechnology for food preservation. The nanofiltration technique acts as a sterilization technique, as it removes bacteria and other pathogens due to size exclusion (García Doménech et al. 2020). New methods of encapsulating bacteriocins, proteins produced by lactic acid bacteria with a bactericidal effect against other bacterial species, in nanoliposomes allow for a controlled release of antimicrobials. Nanoemulsions can be used to decontaminate equipment, and nanofibers containing bacteriocin have shown positive inhibitory activity against *Listeria innocua*, one of the pathogenic species that needs to be avoided in food industries (Chudasama and Goyary 2024; Pathakoti et al. 2017). Silver nanoparticles (AgNPs) are used as a powerful disinfectant (Deshmukh et al. 2019). In the future, nanoenzymes may be used to oxidize vital bacterial cell structures and eliminate biofilms from food contact surfaces (Wang et al. 2024a, b), and nanoengineered surfaces with antimicrobial nano-coatings of silver (AgNPs), titanium dioxide (TiO_2), and zinc oxide (ZnO) are used to protect against biofilm formation (Mohammad et al. 2022). Figure 12.1 shows a schematic of nanotechnology applications in food preservation techniques.

12.3 Applications of Nanotechnology in Food Processing

The insertion of products derived from nanotechnology in consumers' lives is likely to increase across the world in the coming years (Pushparaj et al. 2022). It is believed that nanotechnology can revolutionize the food system and has the potential to positively influence food science, as it can generate innovation in food texture, processability, taste, and stability during shelf life (Lamabam and Thangjam 2018). For example, protein- and lipid-based nanostructures are employed to bind and release nutrients in a controlled manner, while polymeric nanostructures with antimicrobial properties increase the stability and longevity of products (Pateiro et al. 2021; Namasivayam et al. 2023).

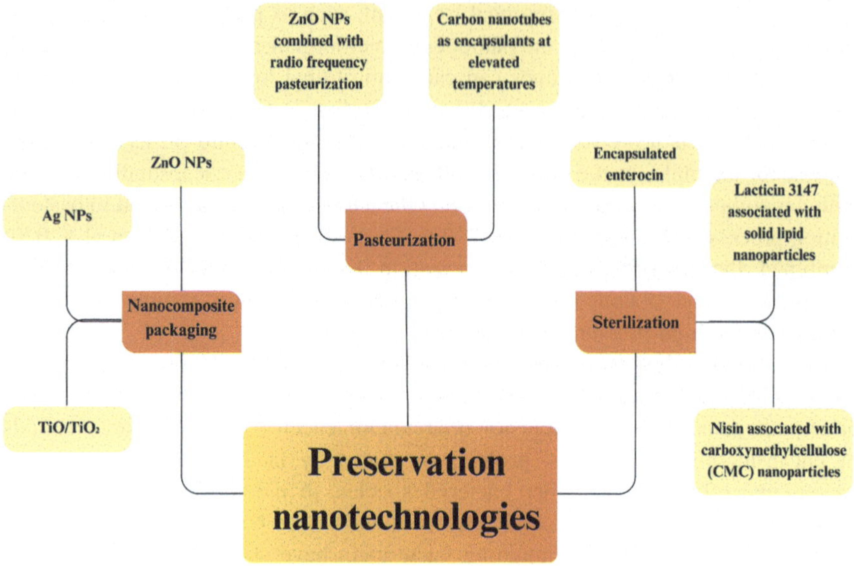

Fig. 12.1 Diagram of nanotechnology applications in food preservation

12.3.1 Improving the Texture and Flavor of Food

Nanostructures can improve food flavor and nutritional profile without altering sensory perception; however, studies on their sensory effects and consumer acceptance are still limited.

Preliminary reports suggest that nanofoods can reduce harmful compounds such as fat, sodium, and sugar while improving taste and texture (Nikravan et al. 2024; Sun et al. 2021). However, attention must be paid to changing the nutritional value, as several factors can affect this value (Szilagyi 2019), such as the presence of specific enzymes, such as lactase, which can affect the absorption of nutrients.

12.3.2 Encapsulation of Bioactive Ingredients

Food composition and molecular structures of foods also have a significant impact on nutrient release and absorption (Sabra et al. 2021; Turgeon and Brisson 2020). In this context, encapsulation allows the protection of specific nutrients during food processing, storage, and digestion, which allows for greater absorption by the human body. This facilitates the optimization of the absorption of essential nutrients such as calcium, fat-soluble vitamins, and high-biological-value proteins, all aimed at increasing your nutritional value (Jafari and McClements 2017; Jiang et al. 2023a, b).

The aforementioned vitamins and nutrients can become a bioactive ingredient added to food with the ability to treat or prevent diseases, being any substance that can be considered a food or part of a food and that provides health benefits (Dudeja and Gupta 2017).

Nanotechnology contributes to the advancement of encapsulation by providing new biocompatible nanostructures with a greater encapsulation capacity due to the high area/volume ratio. Thus, these nanocomposites will be recognized as GRAS—Generally Recognized As Safe—which is a designation of the *Food and Drug Administration* (FDA) for adding new substances to foods (Alu'datt et al. 2022).

The potential for encapsulating bioactives in nanostructures is proven, being a widely studied area and a strong alternative to optimize the bioavailability and stability of these molecules from processing to consumption, also improving their absorption in the gastrointestinal tract (Wang et al. 2024a, b; Azevedo et al. 2014).

Numerous techniques are used in the nanoencapsulation of bioactives via different approaches, such as nanoemulsions, nanogels, solid lipid nanoparticles, and nanoproteins. Each technique offers specific mechanisms to encapsulate bioactives with unique advantages differing from the offerings of others based on distinct applications in the food industry (Pateiro et al. 2021). Table 12.2 presents examples of different types of nanoencapsulation, along with their specific characteristics and benefits.

12.3.3 Increased Bioavailability of Nutrients

Nanotechnology appears in this scenario as a hopeful answer to increase the bioavailability of compounds, thus ensuring that important elements reach the human body systems easily (Ayala-Fuentes and Chavez-Santoscoy 2021; Pateiro et al. 2021; Adinepour et al. 2022). The bioavailability of a bioactive ingredient can be defined as the rate at which the bioactive is extracted from a product and the time it is available for action in the body. The bioavailability of a bioactive ingredient is considered to be high when it reaches effective concentrations in the blood and exerts its biological function (Jafari and McClements 2017).

Nanotechnology improves the bioavailability of encapsulated micronutrients by inducing physicochemical changes in materials that are reduced to the nanometric scale, in addition to increasing their surface área (Okagu et al. 2019). The nanoscale size allows faster digestion of the bioactive, easier penetration into mucous layers, and absorption by cells, optimizing the ability to perform the required biological function (Shafaei et al. 2017).

To verify the increase in bioavailability of the bioactive by encapsulation in nanostructures, in vitro experiments are performed that simulate the digestion stages of the human organism (Charoo et al. 2019). These simulated media are synthetic, with the appropriate pH range, and contain enzymes characteristic of both gastric and intestinal juices. Madalena and collaborators tested the in vitro digestion of the β-lactoglobulin nanostructure with encapsulated riboflavin for resistance to the pH of

Table 12.2 Main characteristics and benefits of nanoemulsions, nanogels, solid lipid nanoparticles, and nanoproteins for food

Type	Characteristics	Benefits	Examples	Reference
Nanoemulsion	High stability; transparency; Increased bioavailability of active compounds; droplet size in the range of 20–200 nm	Increases solubility in aqueous systems; Facilitates the formulation of transparent emulsions; Improves physicochemical, rheological, and sensory properties	Mayonnaise; Milk; Chantilly; Ice cream	Donsì (2018)
Nanoprotein	Protection against environmental factors (such as light and oxygen); Preservation of the quality of active ingredients; increased ingredient stability; extending the shelf life of food products	Improvement in the technical-functional properties of foods; Greater foam and emulsion stability; Low-calorie food formulations; Enhancement of color, flavor, texture, and activity of bioactive compounds	Soy protein; Whey protein	Biswas et al. (2022)
Nanogels	Hydrophilic three-dimensional structures; Composed of biopolymeric molecules; association by covalent bonds or non-covalent interactions.	Stabilizes sensitive compounds; Controls the release of nutrients; Facilitates incorporation into foods		Quazi & Park (2022)
Lipid nanoparticles	Produced from a solid matrix of lipids; They use solid lipids or a mixture of solid lipids; High stability of the lipid matrix	Improves the solubility and stability of bioactive compounds; Protects against degradation; Facilitates incorporation into food formulations		Ghosal and Bandara (2022)

the stomach and intestine. The nanostructure with bioactive was degraded only in the small intestine due to the action of bile salts, proving to be an excellent medium carrier of this bioactive, as it was resistant to the action of more acidic pH and digestive enzymes (Madalena et al. 2016).

In vivo studies are also conducted on animals, following the Animal Welfare Guidelines for Laboratory Research. One group of animals receives the bioactive of interest encapsulated in the nanostructure, and the other is a control group. The effect of the bioactive and the increase in bioavailability are verified by monitoring the gastric and intestinal juices of the animals, as well as by blood tests, depending on the objective of the study. After the study animals are killed, ex vivo analyses can be performed, such as imaging of the organs involved in the digestive system and evaluation of the release of the respective bioactive and physicochemical analyses of the macerated digestive organs. Fang and co-workers studied the release of insulin encapsulated in enteric nanospheres made from hydroxypropylmethylcellulose and demonstrated through an in vitro study that 20% of the insulin content was degraded in the acidic pH of the stomach and the rest was absorbed only in the large intestine. They conducted the in vivo study and showed that the nanosphere had a significant hypoglycemic effect in diabetic rats, which is a promising route for oral insulin delivery (Fang et al. 2019).

Nanostructures of dairy compounds, such as whey protein, can be used to encapsulate anthocyanin and folic acid and be used as dietary supplements (Assadpour et al. 2016). Studies demonstrate that the insertion of nanoprobiotics and nanoprebiotics in foods can promote the absorption of nutrients and even make the medication administration mechanism more effective (Dangi et al. 2023).

In all cases, in the development of functional ingredients using nanotechnology, research must be conducted to verify the interference of the food matrix in the stability and bioavailability of the active compound linked to the nanostructure. As well as whether the studied nanosystem could cause significant changes in the physical-chemical, texture, and sensory properties of foods (Biswas et al. 2022).

12.4 Nanotechnology in Food Preservation and Packaging

Several countries around the world are making considerable efforts to create and improve regulations to guarantee food safety for the population. However, the incidence of diseases caused by food-borne pathogens remains high. Therefore, the search for innovative alternatives that promote the safety of food products and food-related environments is becoming increasingly urgent (Mohammad and Ahmad 2024).

Nanotechnology has proved extremely attractive to the food industry, as it offers new approaches that differ from traditional techniques. This science has been applied in various sectors of the industry, such as in the creation of nanoemulsions, nanofibers, and nanoencapsulation, used to extend the shelf life of various foods (Meral et al. 2022; Yemmireddy et al. 2015).

Various factors affect food quality such as microbial contamination, oxidation, and lack of proper temperature control, among others (Sultana et al. 2022). Because of this, it is imperative to look for alternatives that promote the preservation and increase the shelf life of food products. Multiple methods have been developed as a solution to this problem, including refrigeration, the use of preservatives, vacuum preservation, and intelligent and active packaging, the latter being an innovative concept on the market (Correia et al. 2024; Qian et al. 2021).

Active packaging consists of including certain materials in the packaging to extend the shelf life of a specific product, play a key role in preserving food, and provide protection against external conditions. However, it is important to clarify that the concepts of active packaging and intelligent packaging are different. A smart package can be defined as a device capable of detecting and providing information about the properties and functionalities of food, for example, packages that have time and temperature indicators, dyes for detecting gases, controlling microorganisms, and monitoring product deterioration (Mane 2016; Mei and Wang 2020).

The application of nanotechnology in the food industry has increased the beneficial opportunities for packaging (Enescu et al. 2019). Due to their physical and chemical properties, nanomaterials can improve mechanical barriers, influence performance, and enhance the functionality of packaging. In addition, these materials can offer antimicrobial properties, UV protection, and increased heat resistance (Bajpai et al. 2018; He et al. 2019). Recent studies have investigated the use of nanoparticles and enzymes that convert molecular signals into visible information such as color, fluorescence, or electronic signals, detectable both visually and by specialized equipment (Niu et al. 2018).

However, although they have a variety of benefits, it is important to delve deeper into research related to the migration of these materials into food and their distribution in the human body to strengthen the knowledge base on the subject, contribute to the development of regulations and guidelines to guide the use of nanostructures in food packaging, and ensure the safety and viability of marketing these technologies (Mei and Wang 2020).

12.4.1 Nano-Coatings and Edible Films

In this context, the widespread use of nano-coatings has been noted in the area of food packaging. One example is the incorporation of active ingredients into packaging materials, giving products antimicrobial and antioxidant properties (Bahrami et al. 2020). In general, nano-coatings consist of ultra-thin layers on a nanometer scale, formed or deposited directly on the surface of a different material component, similar to the mechanical properties of the matrix. Usually, a component, for example, a chemical vapor deposition, is added to offer significant benefits and improvements, as well as increase the safety, quality, and durability of products (Gu et al. 2020; Meral et al. 2022).

This category includes edible coatings, which are a preservation method commonly applied as a thin layer on the surfaces of food products, such as fruits and vegetables, to extend their shelf life and delay ripening (Xing et al. 2019). This technology has gained popularity due to its ease of preparation, low cost, and lack of temperature control (Ch Momin et al. 2021; Jodhani and Nataraj 2021; Suhag et al. 2020).

These coatings are formulated by preparing solutions or emulsions from natural ingredients such as proteins, polysaccharides, and lipids, which are applied to food surfaces by dipping, spraying, or brushing. In addition, to ensure greater food quality and safety, it has become possible to add antimicrobial and antioxidant agents to coating materials (Mahela et al. 2020).

Nanotechnology has played an important role in this. Antimicrobial agents with sizes between 1 and 100 nm have been incorporated into edible coating materials, and studies have shown significant differences between traditional coatings and nano-coatings (Esmaeili et al. 2021; Xing et al. 2019).

Coating layers confer a specific chemical or physical function to a surface, such as a gas barrier, hydrophilic/hydrophobic or oleophobic properties, as well as improving insulating/conductive properties and corrosion resistance (Müller et al. 2017). The use of additives, in small quantities, can modify properties such as gloss, structure, surface tension, and resistance to ultraviolet rays (Suriati et al. 2022). Compared to traditional coatings, the use of nanomaterials results in improvements such as greater resistance to indentation, high elasticity, rapid drying, among others (Baig et al. 2021).

It is important to note that the effectiveness of antimicrobial activities in a coating depends on several factors, including the concentration and type of coating material, as well as the size, shape, release, interaction between them, surface structure, storage temperature, and other properties of the nanostructures (Sharma et al. 2019; Wei et al. 2014; Xing et al. 2011).

Factors related to the hydrophobicity of nanomaterials used to manufacture coatings should be considered as they are capable of conferring improvements in properties (Momin et al. 2021). For example, coatings developed from lipids have hydrophobic groups and are commonly used to confer antimicrobial activities, while those made from polymers, polysaccharides, or proteins (or a combination of these) have hydrophilic groups and are used in the formulation to promote better biodegradability, antimicrobial activity, and physical stability (Momin et al. 2021; Mahela et al. 2020). However, most of the materials used in edible coatings are hydrophilic, and therefore it is necessary to carry out studies regarding the migration of nanostructures into the food system due to dissolution in food (Mohammad and Ahmad 2024).

Chitosan nanoparticles, for example, are used extensively in these formulations because they can form films, are non-toxic, biodegradable, and have biocompatibility. In addition to demonstrating better mechanical properties and fungicidal/antimicrobial activity in coatings, they are also cheaper and more accessible when compared to other materials (Momin et al. 2021; Martinez et al. 2010). Thus, it can be seen that the food industry can benefit from the use of this technology by increasing food safety and extending the shelf life of its products. Researchers have studied the influence of increased food shelf life when nanoparticles are added to edible coatings and concluded that there was a significant increase in this parameter when compared to traditional coatings (Deng et al. 2017; Tahir et al. 2019).

In addition, the synergistic effect between nanostructures and coating substances plays an essential role and can significantly influence the effectiveness of microbial activity (Ribeiro et al. 2022; Wei et al. 2014). It occurs when two or more effects are combined, resulting in greater efficacy than when applied individually. The interaction of nanostructures with antimicrobials can optimize these properties due to the charge transfer between them (Ribeiro et al. 2022).

Table 12.3 Films with nanostructures used to extend the shelf life of fruits and vegetables

Nanostructure	Fruits and vegetables	Storage conditions	Results	References
Agar and zinc	Green grapes	Ambient conditions for 25 days	The application of the film made it possible to extend the shelf life of the green grapes by up to 21 days	Kumar et al. (2019)
Chitosan	Green pepper	Stored at 27 °C for 7 days	Color change was slower and remained uniform for up to 7 days	Priyadarshi et al. (2018)
Chitosan, soy protein, and stearic acid	Apple	Ambient conditions for 6 weeks	The shelf life of the apple was extended for about 6 weeks	Wu et al. (2017)
Pectin and manganese	Cherry tomatoes	Stored at 0 ± 2 °C and 61.2% relative humidity (RH) or 10 ± 0.5% °C and 90% RH	The film extended shelf life by up to 24 days in storage at 10 °C	Kumar et al. (2020a, b)
Silicon dioxide	Strawberry	Ambient conditions	The film extended the shelf life of the fruit by around 10 days and showed potential for future applications	Zhao et al. (2023)

Another alternative for food preservation is the application of films to inhibit or delay the deterioration of fresh products by minimizing enzymatic, chemical, biochemical, physical, and/or microbiological alterations. The most widely used techniques for manufacturing films in food products are solution molding, layer-by-layer assembly, and extrusion (Basumatary et al. 2022). The incorporation of nanostructures has helped to optimize the functionality of these films, improving antimicrobial and antioxidant properties and extending the shelf life of fruit and vegetables (Galus and Kadzińska 2015), as shown in Table 12.3.

It is therefore notable that the addition of nanoparticles to coatings and edible films offers the possibility of preserving a variety of foods, especially fresh fruits and vegetables, extending shelf life, improving quality and standardization, preventing the growth of microorganisms and fungi, minimizing economic losses for producers and traders, as well as promoting various benefits to the food sector (Mohammad and Ahmad 2024).

12.5 Food Safety and Quality Control

Food safety refers to the practice of measures during food processing to prevent changes or the presence of physical, chemical, or biological agents that could put the consumer's health or physical integrity at risk. This covers the entire production chain, from agricultural production to harvesting, processing, storage, transportation, distribution, and preparation.

The concept of quality control involves a set of procedures and measures designed to ensure that food products meet established quality standards and consumer satisfaction. In quality control, some aspects must be observed: establishing quality standards, carrying out inspection and testing, implementing systems for quality management, ensuring accurate labeling and adequate packaging, and monitoring consumer feedback.

Over time, the tools to secure safety and quality control in food production have evolved as new technologies have been developed. A current trend is the use of nanotechnology, which has enabled the development of innovative methods for detecting contaminants, extending the shelf life of products, and monitoring food quality in real time.

12.5.1 Advanced Methods for Contaminant Detection

Contaminants present in food are divided into biological, chemical, and physical (Fig. 12.2). Its presence is related to failures during processing and poses risks to consumer health and loss of product quality. Among the control measures that must be applied to avoid them are monitoring and carrying out tests for their detection. Many of these tests are time-consuming or require a large laboratory structure.

Techniques based on nanotechnology enable the detection of different compounds such as heavy metals, pathogens, pesticides, food allergens, and antibiotics during food processing, preventing product contamination. Some examples are molecular or immunological assays, electrochemical analysis, surface-enhanced Raman scattering (SERS), and colorimetry (Nam et al. 2022). These present advantages such as portability, specificity, speed, and precision (Zhou et al. 2023).

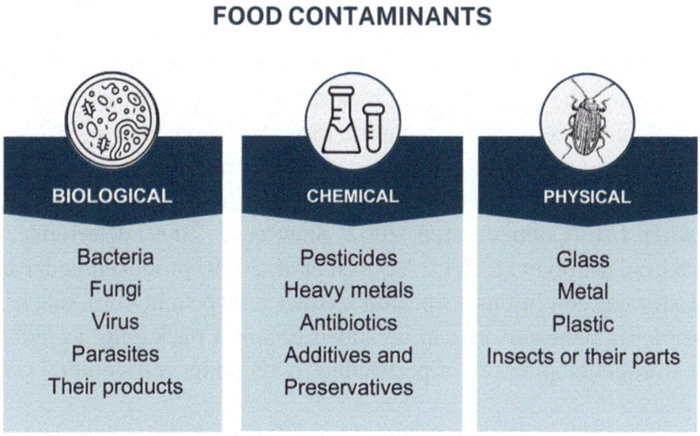

Fig. 12.2 Types of contaminants in food

Table 12.4 Applications of nanotechnology in the detection of chemical and biological contaminants in food

Detection technique	Contaminant	Reference
Immunosensor based on antibody-attached gold nanoparticles	Aflatoxin B1	Sharma et al. (2010)
Quantum dots linked to anti-*Salmonella* antibodies	*Salmonella*	Singh et al. (2018)
Colorimetric detection using gold nanoparticles coupled to aptamers.	Antibiotic tobramycin in milk and chicken eggs	Ma et al. (2018)
Electrochemical sensor composed of a glassy carbon electrode (GCE), gold nanoparticles, formaldehyde dehydrogenase, a nanocomposite membrane, and an ionic liquid	Formalin in fish	Bandyopadhyay et al. (2020)
Aptamer-based magnetic and upconversion nanoparticle sensor	*Escherichia coli* in pork	Li et al. (2020)
Antibody-based surface-enhanced Raman scattering (SERS) sensor	α-Lactalbumin in raw cow milk, ultra-heat-treated cow milk, walnut milk drink, and peanut milk drink	Zhang et al. (2021)

Table 12.4 presents some applications of nanotechnology in the detection of chemical and biological contaminants in food.

12.5.2 Real-Time Monitoring of Food Quality

Real-time food quality monitoring is an emerging technology for immediate and continuous detection of critical parameters that need to be controlled during food processing or storage to ensure food safety and quality. The main tools used are sensors, biosensors, and, recently, nanosensors.

The use of nanosensors in smart packaging makes it possible to track the environmental conditions in which the food is found while also being able to detect the first signs of spoilage (Pal 2017). To ensure the integrity of the packaging, especially regarding the entry of gases, nanosensors for detecting oxygen (Mills and Hazafy 2009; Sharma et al. 2017) or carbon dioxide (Borchert et al. 2013; Jung et al. 2012) can be used. In food spoilage, some compounds are generated, and nanosensors can be used to detect them. Some of these compounds are xanthine, generated when fish (Çubukçu et al. 2007; Shan et al. 2009) deteriorate, aflatoxin, and ochratoxin (Castillo et al. 2015; Yousefi et al. 2019) produced under inadequate drying and storage conditions in peanut extract and peanut-corn snacks. Furthermore, colorimetric nanosensors can be added to smart packaging to indicate variations in the storage temperature of perishable foods under refrigeration (Wang et al. 2017).

Nanosensors stand out among technologies for food quality control, especially when applied to smart packaging. They are advantageous over traditional sensors due to their sensitivity, portability, and flexibility of integration with different devices. However, some challenges are still faced in its application, such as difficulties in scaling up and a lack of nanotoxicological studies or those that verify the possibility of migration of nanostructures from packaging to food (Alfei et al. 2020; Kaur et al. 2024).

12.6 Regulatory Aspects and Legislation

With the advent of nanotechnology products entering the consumer market, new regulations are being implemented to ensure societal well-being. Nanotechnology is a highly impactful tool in the market; however, the health impacts of nanostructures on humans in the short, medium, and long term, as well as the environmental effects associated with the disposal of nanomaterial-containing products, are not yet fully understood (Jagtiani et al. 2022). To protect people and the environment, countries are establishing a set of norms and regulations for the use of nanostructures (Allan et al. 2021).

Global regulatory bodies have not yet reached a consensus on regulation. Each country and economic bloc is independently organizing its laws. These differences complicate the commercialization of nanostructured products between different economic blocs (Allan et al. 2021; Santos et al. 2024). However, there is a consensus that the closer the contact of nanostructures with consumers, the greater the number of tests and analyses required for product commercialization. Electronic products with nanostructures, due to their lack of direct contact with the final consumer, are subject to more lenient standards and conditions for application and commercialization (Allan et al. 2021). The scenario is different in the food industry, where nanostructures can be applied in various ways: as a chemical additive in the product, in filters to enhance filtration efficiency, in primary packaging to preserve food for longer, and in secondary and tertiary packaging. Each use presents a different level of proximity of the nanostructure to the consumer. For example, tertiary packaging has minimal contact between the nanostructure and both the food and the final consumer; meanwhile, nanostructures used as chemical additives in food have direct contact with the final consumer (Ghebretatios et al. 2021; Jagtiani et al. 2022).

A set of analyses is necessary to evaluate the toxicity of a nanomaterial. Some of the tests include migration tests, which monitor the release of nanostructures from packaging into the food matrix; cytotoxicity and genotoxicity tests, which aim to understand the relationship between concentration, exposure time, and the chemical structure of the applied nanostructure with possible damage to different types of human tissues; bioaccumulation tests, which analyze how the human body processes different nanostructures; and ecotoxicity tests, which verify possible environmental impacts (Ghebretatios et al. 2021; Vijayakumar et al. 2022).

However, determining the degree of toxicity of a nanostructure is not straightforward. The toxicity of a nanostructure varies depending on its size, crystallinity, chemical nature, geometry, concentration used, exposure time to cells, and the type

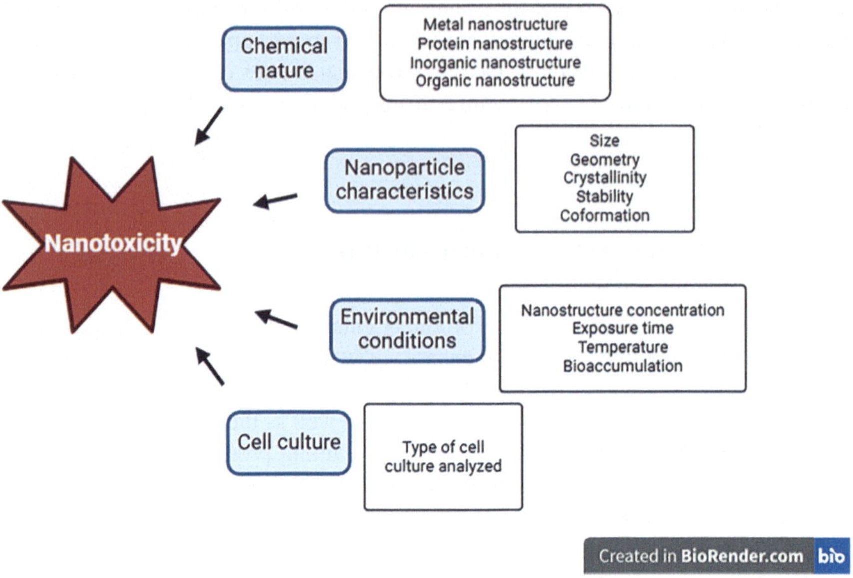

Fig. 12.3 Key variables influencing the nanotoxicity of a nanostructure

of cell exposed. Figure 12.3 illustrates the main variables influencing the nanotoxicity of a nanostructure.

These factors complicate the development of a unanimous study on the subject, forcing each country and economic bloc to conduct its studies and establish regulatory agencies. In 2022 alone, there were more than 2700 national nanotechnology standards (Database StatNano, 2024). The main recognized agencies are the European Chemicals Agency (ECHA), the European Food Safety Authority (EFSA), and the European Medicines Agency (EMA) in the European bloc; the Food and Drug Administration (FDA) in the USA; the Canadian Government's Chemical Management Plan (CMP) in Canada; the Japanese Ministry of Health, Labour and Welfare (MHLW), the Pharmaceutical and Medical Devices Agency (PMDA), the National Institute of Health Sciences (NIHS), and the Agency for Medical Research and Development (AMED) in Asia; and the National Health Surveillance Agency (ANVISA) in Brazil (Allan et al. 2021; Santos et al. 2024).

For the development of nanostructured food industry products, it is recommended to comply with local norms and legislation. For this purpose, it is advisable to refer to the following:

- Section 409 of the Federal Food, Drug, and Cosmetic Act (FDA)
- Registration with the Environmental Protection Agency (European Commission)
- EEC 89/109 (European Commission)
- Article 14, EC 450/2009 (European Commission)
- OECD TG318, TG412, and TG413
- Collegiate Board Resolution - RDC No. 326 (ANVISA)

12.7 Overview of the Global Nanotechnology Market and Practical Applications

Nanotechnology products are already a reality, with major companies incorporating them into their portfolios alongside the emergence of new technology-based startups (Chaudhry et al. 2017). Currently, there are 5631 companies with nanotechnology products, of which approximately 204 are in the food sector. The trend is toward an increased presence of nanotechnology in the market (StatNano 2023a, d).

Since the early 2000s, the number of patents granted for nanotechnology under ISO/TS 18110 has increased by 672.88%, rising from 1571 patents published annually worldwide by the European Patent Office (EPO) and the United States Patent and Trademark Office (USPTO) to 12,142 (StatNano 2023c, b). Although much data on investment values for technological development are scarce and incompletely reported, it is estimated that global investment in nanotechnology-related companies and technologies reaches US$ 250 billion, with China and the USA being the main players (Talebian et al. 2021). Due to these investments, the nanotechnology sector in the USA generated a gross revenue of US$ 904 million, in the EU US$ 941 million, US$ 1.14 billion in Asia in 2018 (Talebian et al. 2021). Market projections indicate that by 2024, the global nanotechnology market is expected to surpass US$ 125 billion (Rambaran and Schirhagl 2022). Figure 12.4 shows the growth in the number of patents granted from 2001 to 2023. It is

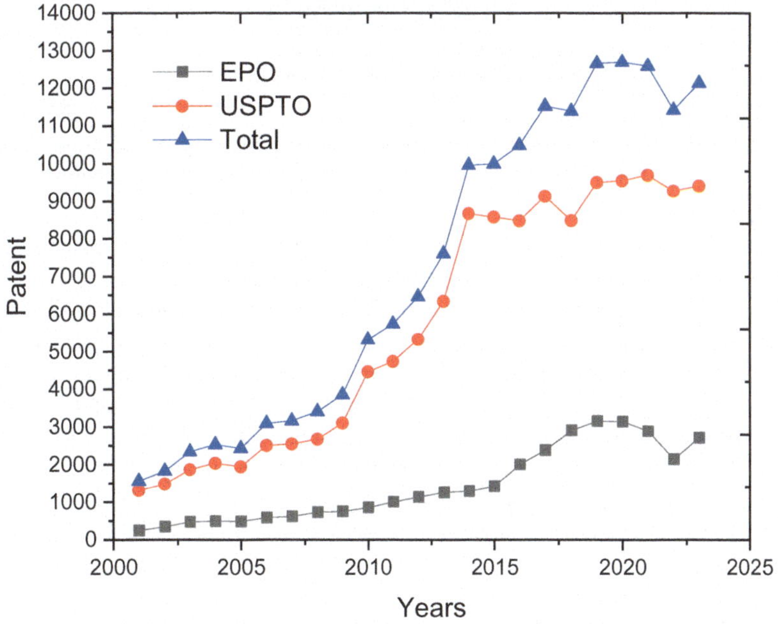

Fig. 12.4 Number of patents granted per year

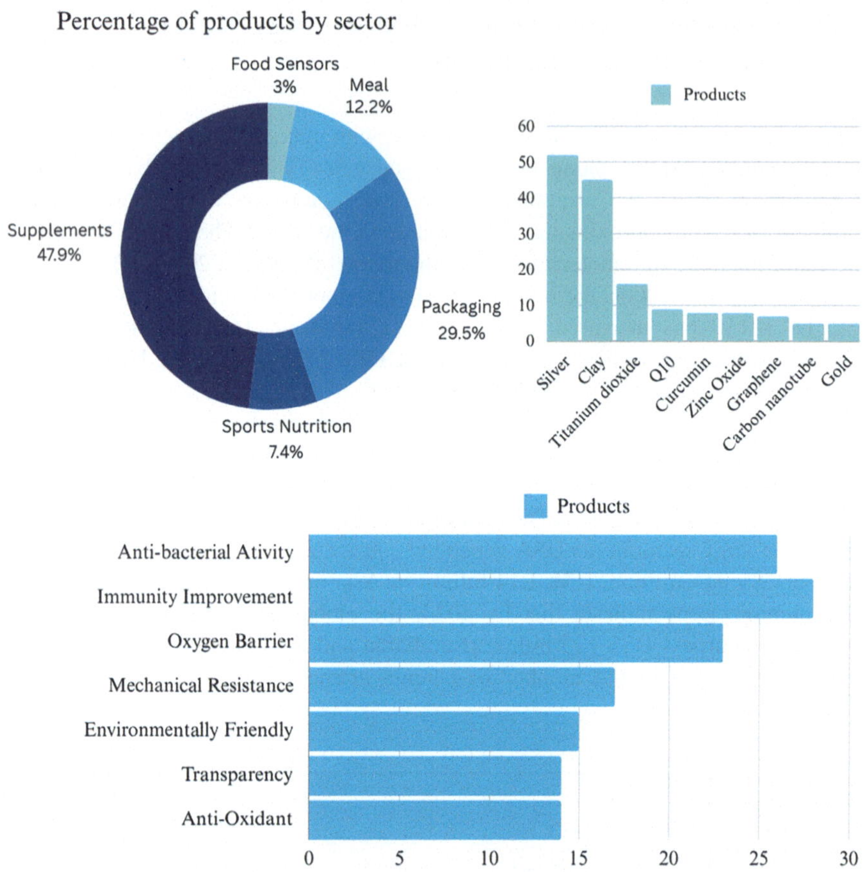

Fig. 12.5 Percentage of products marketed with nanomaterials by sector, most common nanomaterials used in food sector products, and most prominent properties of nanomaterials

noticeable that there is a stagnation and decline in granted patents from 2020 to 2023, a phenomenon attributed to the COVID-19 pandemic.

The StatNano database lists 434 nanotechnology products marketed by 204 companies across 32 countries. These are categorized into five main groups: Food Sensors, Meals, Packaging, Sports Nutrition, and Supplements. Figure 12.5 illustrates the percentage of products sold with nanomaterials by sector, along with the incidence of the most commonly used nanomaterials in the food sector and the most frequently exhibited properties of these nanomaterials (StatNano, 2023a, d).

The proportion of nanomaterial-containing products in the food sector remains relatively small compared to other sectors, such as electronics with 1962 products, medical with 1318 products, construction with 1170 products, and cosmetics with 1011 products. However, the outlook for the food sector is one of growth. The increasing global demand for food, driven by population growth and global events such as the COVID-19 pandemic, climate crises, water supply shortages, and wars, has stressed the global food production system (Numa et al. 2023). These concerning scenarios have triggered economic shifts toward new methods to enhance efficiency and safety in food production, processing, and logistics. In 2020 alone, it was estimated that US$ 31 billion was invested in Agri-FoodTech, with a total global investment of US$ 250 billion between 2016 and 2022 (Numa et al. 2023).

FoodTech is an interdisciplinary field focused on developing technologies across the entire food sector: production, preservation, processing, distribution, and food safety (Chaudhry et al. 2017; Sippel and Dolinga 2023). This sector includes various startups specializing in nanotechnology, nanobiotechnology, biotechnology, big data, and artificial intelligence (AI) (Sippel and Dolinga 2023), with investments increasing by 800%, from US$ 2.9 billion in 2012 to US$ 21.1 billion in 2020 (Sippel and Dolinga 2023).

12.8 Challenges and Future Perspectives

Nanotechnology has many applications in the food sector, as demonstrated in this chapter. At the same time, this emerging technology faces challenges both in research and development and in industrial applications. Researchers face challenges in developing the production process of nanocomposites, such as determining the production method, the concentration of the nanocomposites to be applied, its stability, its surface area, and its efficiency index for the proposed target (Meghani et al. 2020).

There are several concerns associated with the use of nanotechnology in the food industry, especially regarding the migration of nanocomposites into food products and the risk of toxicity, which can result in health problems and compromise food quality and standardization (Garcia et al. 2018; Störmer et al. 2017). In addition, there is the possibility of environmental impacts due to the possible formation of degraded by-products (Bystrzejewska-Piotrowska et al. 2009).

Therefore, it is extremely important to develop intelligent research models capable of measuring the possible toxicity of the nanocomposites, as well as other effects that may be caused by the application of nanotechnology, so that these effects can be mitigated and this technology can be effectively applied. Safely in the food sector (Chaudhry and Castle 2011). Furthermore, it is important to follow the regulations of the competent bodies to ensure the safety of the final product and the health of the consumer (Lamabam and Thangjam 2018).

However, the incorporation of nanomaterials into the food production process is increasing and has shown advantages over traditional methods, even proving suitable for food preservation (Du et al. 2023). In this sense, nanotechnology has the prospect of improving and incorporating indicators of color, temperature, time, and freshness; in addition to gas sensors, biosensors, and nanosensors for the detection of additives, adulterants, toxins, and/or contaminants; encapsulation of bioactive materials; development of active and intelligent food packaging; and other stages of the production process, which are viable in several studies previously carried out

(Bahrami et al. 2020; Mirza Alizadeh et al. 2021). Therefore, this area has great potential for exploration and promising application prospects. Thus, it is essential to consider studies on the interaction of food preservation materials interlinked with practical applications so that this technology can be commercialized safely and efficiently (Du et al. 2023; Mei and Wang 2020).

References

Acosta O, Vaillant F, Pérez AM, Dornier M (2017) Concentration of polyphenolic compounds in blackberry (Rubus Adenotrichos Schltdl.) juice by nanofiltration. J Food Process Eng 40(1): e12343. https://doi.org/10.1111/JFPE.12343

Adinepour F, Pouramin S, Rashidinejad A, Jafari SM (2022) Fortification/enrichment of milk and dairy products by encapsulated bioactive ingredients. Food Res Int 157:111212. https://doi.org/10.1016/j.foodres.2022.111212

Ahmad A, Qurashi A, Sheehan D (2023) Nano packaging-Progress and future perspectives for food safety, and sustainability. Food Packag Shelf Life 35:100997

Alencar MSMD, Bochner R, Dias MFF (2013) Nanotecnologia em Ciências da Saúde no Brasil: um olhar informétrico sobre os grupos de pesquisa

Alfei S, Marengo B, Zuccari G (2020) Nanotechnology application in food packaging: a plethora of opportunities versus pending risks assessment and public concerns. Food Res Int 137:109664. https://doi.org/10.1016/J.FOODRES.2020.109664

Allan J, Belz S, Hoeveler A, Hugas M et al (2021) Regulatory landscape of nanotechnology and nanoplastics from a global perspective. Regul Toxicol Pharmacol 122:104885

Alu'datt MH, Alrosan M, Gammoh S, Tranchant CC et al (2022) Encapsulation-based technologies for bioactive compounds and their application in the food industry: a roadmap for food-derived functional and health-promoting ingredients. Food Biosci 50:101971. https://doi.org/10.1016/J.FBIO.2022.101971

Arend GD, Adorno WT, Rezzadori K, Di Luccio M et al (2017) Concentration of phenolic compounds from strawberry (Fragaria X ananassa Duch) juice by nanofiltration membrane. J Food Eng 201:36–41. https://doi.org/10.1016/J.JFOODENG.2017.01.014

Assadpour E, Maghsoudlou Y, Jafari SM, Ghorbani M et al (2016) Optimization of folic acid nano-emulsification and encapsulation by maltodextrin-whey protein double emulsions. Int J Biol Macromol 86:197–207. https://doi.org/10.1016/J.IJBIOMAC.2016.01.064

Assadpour E, Dima C, Jafari SM (2020) Fundamentals of food nanotechnology. Handbook of Food Nanotechnology: Applications and Approaches, pp 1–35. https://doi.org/10.1016/B978-0-12-815866-1.00001-7

Ayala-Fuentes JC, Chavez-Santoscoy RA (2021) Nanotechnology as a key to enhance the benefits and improve the bioavailability of flavonoids in the food industry. Food Secur 10(11):2701. https://doi.org/10.3390/foods10112701

Azevedo MA, Bourbon AI, Vicente AA, Cerqueira MA (2014) Alginate/chitosannanoparticles for encapsulation and controlled release of vitamin B2. Int J Biol Macromol 71:141–146. https://doi.org/10.1016/J.IJBIOMAC.2014.05.036

Bahrami A, Delshadi R, Assadpour E, Jafari SM et al (2020) Antimicrobial-loaded nanocarriers for food packaging applications. Adv Colloid Interfac Sci 278:102140. https://doi.org/10.1016/j.cis.2020.102140

Baig N, Kammakakam I, Falath W, Kammakakam I (2021) Nanomaterials: a review of synthesis methods, properties, recent progress, and challenges. Mater Adv 2:1821. https://doi.org/10.1039/d0ma00807a

Bajpai VK, Kamle M, Shukla S, Mahato DK et al (2018) Prospects of using nanotechnology for food preservation, safety, and security. J Food Drug Anal 26(4):1201–1214. https://doi.org/10.1016/j.jfda.2018.06.011

Bandini S, Nataloni L (2015) Nanofiltration for dextrose recovery from crystallization mother liquors: a feasibility study. Sep Purif Technol 139:53–62. https://doi.org/10.1016/J.SEPPUR.2014.10.025

Bandyopadhyay K, Misra S, Bhattacharya S, Mukherjee A et al (2020) Review on utilization of biosensors for detection of adulteration in fish. Int J Modern Trend Sci Technol 6(10):50–55. https://doi.org/10.46501/IJMTST061009

Basumatary IB, Mukherjee A, Katiyar V, Kumar S (2022) Biopolymer-based nanocomposite films and coatings: recent advances in shelf-life improvement of fruits and vegetables. Crit Rev Food Sci Nutr 62(7):1912–1935. https://doi.org/10.1080/10408398.2020.1848789

Biswas R, Alam M, Sarkar A, Haque MI et al (2022) Application of nanotechnology in food: processing, preservation, packaging and safety assessment. Heliyon 8(11):e11795. https://doi.org/10.1016/J.HELIYON.2022.E11795

Borchert NB, Kerry JP, Papkovsky DB (2013) A CO2 sensor based on Pt-porphyrin dye and FRET scheme for food packaging applications. Sensors Actuators B Chem 176:157–165. https://doi.org/10.1016/J.SNB.2012.09.043

Bystrzejewska-Piotrowska G, Golimowski J, Urban PL (2009) Nanoparticles: their potential toxicity, waste and environmental management. Waste Manag 29(9):2587–2595. https://doi.org/10.1016/j.wasman.2009.04.001

Castillo G, Spinella K, Poturnayová A, Šnejdárková M et al (2015) Detection of aflatoxin B1 by aptamer-based biosensor using PAMAM dendrimers as immobilization platform. Food Control 52:9–18. https://doi.org/10.1016/J.FOODCONT.2014.12.008

Ch Momin M, Jamir AR, Ankalagi N, Bijaya Devi O et al (2021) Edible coatings in fruits and vegetables: a brief review. Pharma Innov J 10(7):71

Chandrababu V, Parameswaranpillai J, Gopi JA, Pathak C et al (2024) Progress in food packaging applications of biopolymer-nanometal composites—a comprehensive review. Biomater Adv 162:213921. https://doi.org/10.1016/J.BIOADV.2024.213921

Chandrapala J, Duke MC, Gray SR, Weeks M et al (2016) Nanofiltration and nanodiafiltration of acid whey as a function of pH and temperature. Sep Purif Technol 160:18–27. https://doi.org/10.1016/J.SEPPUR.2015.12.046

Charoo NA, Rahman Z, Khan MA (2019) Nanoparticles for improvement in oral bioavailability. Nanoarchitectonic Biomed:371–410. https://doi.org/10.1016/B978-0-12-816200-2.00006-2

Chaudhry Q, Castle L (2011) Food applications of nanotechnologies: an overview of opportunities and challenges for developing countries. Trends Food Sci Technol 22(11):595–603. https://doi.org/10.1016/J.TIFS.2011.01.001

Chaudhry, Q., Watkins, R., & Castle, L. (2017). Nanotechnologies in Food: What, Why and How? In Q. Chaudhry, L. Castle, & R. Watkins (Eds.), Nanotechnologies in Food (p. 0). The Royal Society of Chemistry. https://doi.org/10.1039/9781782626879-00001

Chiozzi V, Agriopoulou S, Varzakas T (2022) Advances, applications, and comparison of thermal (pasteurization, sterilization, and aseptic packaging) against non-thermal (ultrasounds, UV radiation, ozonation, high hydrostatic pressure) technologies in food processing. Appl Sci 12(4):2202. https://doi.org/10.3390/APP12042202

Chudasama M, Goyary J (2024) Nanostructured materials in food science: current progress and future prospects. Next Mater 5:100206. https://doi.org/10.1016/J.NXMATE.2024.100206

Correia LF, da Silva Pinho G, da Cruz Neves TJ, de Oliveira Vieira KC et al (2024) Nanotechnology innovation combined with bacteriocins as emerging strategy for the development of active and intelligent food packaging. Sustain Chem Pharm 39:101551. https://doi.org/10.1016/j.scp.2024.101551

Çubukçu M, Timur S, Anik Ü (2007) Examination of performance of glassy carbon paste electrode modified with gold nanoparticle and xanthine oxidase for xanthine and hypoxanthine detection. Talanta 74(3):434–439. https://doi.org/10.1016/J.TALANTA.2007.07.039

Dangi P, Chaudhary N, Chaudhary V, Virdi AS et al (2023) Nanotechnology impacting probiotics and prebiotics: a paradigm shift in nutraceuticals technology. Int J Food Microbiol 388:110083. https://doi.org/10.1016/J.IJFOODMICRO.2022.110083

Database StatNano. (2024, June 30). Number of national nanotechnology standards (Standard). https://Statnano.Com/Report/S105

Deng Z, Jung J, Simonsen J, Wang Y et al (2017) Cellulose nanocrystal reinforced chitosan coatings for improving the storability of postharvest pears under both ambient and cold storages. J Food Sci 82(2):453–462. https://doi.org/10.1111/1750-3841.13601

Deshmukh SP, Patil SM, Mullani SB, Delekar SD (2019) Silver nanoparticles as an effective disinfectant: a review. Mater Sci Eng C 97:954–965. https://doi.org/10.1016/J.MSEC.2018.12.102

Dhanker R, Rawat S, Chandna V, Deepa, Kumar R, Das S, Sharma A, Kumar V et al (2022) Recovery of silver nanoparticles and management of food wastes: obstacles and opportunities. Environ Adv 9:100303. https://doi.org/10.1016/J.ENVADV.2022.100303

Domenech NG, Purcell-Milton F, Gun'ko YK (2020) Recent progress and future prospects in development of advanced materials for nanofiltration. Mater Today Commun 23:100888

Donsì F (2018) Applications of nanoemulsions in foods. In: Nanoemulsions. Elsevier, pp 349–377. https://doi.org/10.1016/B978-0-12-811838-2.00011-4

Du H, Sun X, Chong X, Yang M et al (2023) A review on smart active packaging systems for food preservation: applications and future trends. Trends Food Sci Technol 141:104200. https://doi.org/10.1016/j.tifs.2023.104200

Dudeja P, Gupta RK (2017) Nutraceuticals. Food Safety in the 21st Century: Public Health Perspective, pp 491–496. https://doi.org/10.1016/B978-0-12-801773-9.00040-6

Dutta S (2011) Global innovation index 2011: accelerating growth and development. WIPO

Enescu D, Cerqueira MA, Fucinos P, Pastrana LM (2019) Recent advances and challenges on applications of nanotechnology in food packaging. A literature review. Food Chem Toxicol 134:110814. https://doi.org/10.1016/j.fct.2019.110814

Esmaeili M, Ariaii P, Nasiraie LR, Pour MY (2021) Comparison of coating and nano-coating of chitosan- Lepidium sativum seed gum composites on quality and shelf life of beef. J Food Meas Charact 15(1):341–352. https://doi.org/10.1007/s11694-020-00643-6

Fang Y, Wang Q, Lin X, Jin X et al (2019) Gastrointestinal responsive polymeric nanoparticles for Oral delivery of insulin: optimized preparation, characterization, and in vivo evaluation. J Pharm Sci 108(9):2994–3002. https://doi.org/10.1016/J.XPHS.2019.04.020

Galus S, Kadzińska J (2015) Food applications of emulsion-based edible films and coatings. Trends Food Sci Technol 45(2):273–283. https://doi.org/10.1016/j.tifs.2015.07.011

García Doménech N, Purcell-Milton F, Gun'ko YK (2020) Recent progress and future prospects in development of advanced materials for nanofiltration. Mater Today Commun 23:100888. https://doi.org/10.1016/J.MTCOMM.2019.100888

Garcia MDS, Lagaron JM (2012) Nanocomposites for food and beverage packaging materials. In: Nanotechnology in the food, beverage and nutraceutical industries. Elsevier, pp 335–361

Garcia CV, Shin GH, Kim JT (2018) Metal oxide-based nanocomposites in food packaging: applications, migration, and regulations. Trends Food Sci Technol 82:21–31. https://doi.org/10.1016/j.tifs.2018.09.021

García-Martín N, Perez-Magariño S, Ortega-Heras M, González-Huerta C, Mihnea M, González-Sanjosé ML, Palacio L, Prádanos P, Hernández A et al (2010) Sugar reduction in musts with nanofiltration membranes to obtain low alcohol-content wines. Sep Purif Technol 76(2):158–170

Ghebretatios, M., Schaly, S., & Prakash, S. (2021). Nanoparticles in the food industry and their impact on human gut microbiome and diseases. International Journal of Molecular Sciences, 22 (4), 1942

Ghosal A, Bandara N (2022) Lipid-based nanostructures in food applications. In: Food, medical, and environmental applications of nanomaterials. Elsevier, pp 113–128. https://doi.org/10.1016/B978-0-12-822858-6.00005-4

Gu Y, Xia K, Wu D, Mou J et al (2020) Technical characteristics and wear-resistant mechanism of nano coatings: a review. Coatings 10(3):233. https://doi.org/10.3390/coatings10030233

Hamad AF, Han JH, Kim BC, Rather IA (2018) The intertwine of nanotechnology with the food industry. Saudi J Biol Sci 25(1):27–30. https://doi.org/10.1016/J.SJBS.2017.09.004

Hasmin NA, Mahmood A, Azizun N, Abd Kadir NH et al (2021) Consumers' awareness and knowledge of nanofood in Malaysia. Soc Manag Res J 18(2):23–46

He X, Deng H, Hwang H (2019) The current application of nanotechnology in food and agriculture. J Food Drug Anal 27(1):1–21. https://doi.org/10.1016/j.jfda.2018.12.002

Hebbar U, Ranjan S, Dasgupta N, Mishra RK (2020) Nano-food engineering. Springer

Herdiana Y, Wathoni N, Shamsuddin S, Muchtaridi M (2022) Scale-up polymeric based nanoparticles drug delivery systems: development and challenges. OpenNano 7:100048

Hessel V, Escribà-Gelonch M, Schmidt S, Tran NN et al (2023) Nanofood process technology: insights on how sustainability informs process design. ACS Sustain Chem Eng 11(31): 11437–11458

Ilyas RA, Sapuan SM, Megashah LN, Ibrahim R et al (2021) Regulations for food packaging materials in bio-based packaging: material, environmental and economic aspects. CRC Press, pp 467–494

Islam F, Saeed F, Afzaal M, Ahmad A et al (2022) Applications of green technologies-based approaches for food safety enhancement: a comprehensive review. Food Sci Nutr 10(9):2855. https://doi.org/10.1002/FSN3.2915

Jafari SM, McClements DJ (2017) Nanotechnology approaches for increasing nutrient bioavailability. Adv Food Nutr Res 81:1–30. https://doi.org/10.1016/bs.afnr.2016.12.008

Jeevahan JJ, Chandrasekaran M, Venkatesan SP, Sriram V et al (2020) Scaling up difficulties and commercial aspects of edible films for food packaging: a review. Trends Food Sci Technol 100: 210–222

Jiang Y, Pan X, Yu T, Wang H (2023a) Intranasal administration nanosystems for brain-targeted drug delivery. Nano Res 16(12):13077–13099. https://doi.org/10.1007/s12274-023-6026-y

Jiang Y, Zhao G, Yang X, Fan F (2023b) Preparation and characterization of nano-SiO2-modified emulsified film and its application for strawberry preservation. Food Packag Shelf Life 40: 101181. https://doi.org/10.1016/j.fpsl.2023.101181

Jodhani KA, Nataraj M (2021) Synergistic effect of Aloe gel (Aloe vera L.) and lemon (citrus Limon L.) peel extract edible coating on shelf life and quality of banana (Musa spp.). J Food Meas Charact 15(3):2318–2328. https://doi.org/10.1007/s11694-021-00822-z

Jung J, Puligundla P, Ko S (2012) Proof-of-concept study of chitosan-based carbon dioxide indicator for food packaging applications. Food Chem 135(4):2170–2174. https://doi.org/10.1016/J.FOODCHEM.2012.07.090

Kaur G, Bhari R, Kumar K (2024) Nanobiosensors and their role in detection of adulterants and contaminants in food products. Crit Rev Biotechnol 44(4):547–561. https://doi.org/10.1080/07388551.2023.2175196

Kim IJ, Zhao W, Park JG, Meng Z (2021) Carbon nanotube filter for heavy metal ion adsorption. Ceram Int 47(23):33280–33285. https://doi.org/10.1016/J.CERAMINT.2021.08.230

Kumar S, Boro JC, Ray D, Mukherjee A et al (2019) Bionanocomposite films of agar incorporated with ZnO nanoparticles as an active packaging material for shelf life extension of green grape. Heliyon 5(6):e01867. https://doi.org/10.1016/j.heliyon.2019.e01867

Kumar N, Kaur P, Devgan K, Attkan AK (2020a) Shelf life prolongation of cherry tomato using magnesium hydroxide reinforced bio-nanocomposite and conventional plastic films. J Food Proc Preserv 44(4):e14379. https://doi.org/10.1111/jfpp.14379

Kumar P, Mahajan P, Kaur R, Gautam S (2020b) Nanotechnology and its challenges in the food sector: a review. Mater Today Chem 17:100332. https://doi.org/10.1016/J.MTCHEM.2020.100332

Lamabam SD, Thangjam R (2018) Progress and challenges of nanotechnology in food engineering. Impact Nanosci Food Indus:87–112. https://doi.org/10.1016/B978-0-12-811441-4.00004-2

Li H, Ahmad W, Rong Y, Chen Q et al (2020) Designing an aptamer based magnetic and upconversion nanoparticles conjugated fluorescence sensor for screening Escherichia coli in food. Food Control 107:106761. https://doi.org/10.1016/J.FOODCONT.2019.106761

Ma Q, Wang Y, Jia J, Xiang Y (2018) Colorimetric aptasensors for determination of tobramycin in milk and chicken eggs based on DNA and gold nanoparticles. Food Chem 249:98–103. https://doi.org/10.1016/J.FOODCHEM.2018.01.022

Madalena DA, Ramos ÓL, Pereira RN, Bourbon AI et al (2016) In vitro digestion and stability assessment of β-lactoglobulin/riboflavin nanostructures. Food Hydrocoll 58:89–97. https://doi.org/10.1016/J.FOODHYD.2016.02.015

Mahela U, Rana DK, Joshi U, Tariyal YS (2020) Nano edible coatings and their applications in food preservation. J Postharvest Technol 8(4):52–63

Mane KA (2016) A review on active packaging: an innovation in food packaging. Int J Environ Agric Biotechnol 1(3):544–549. https://doi.org/10.22161/ijeab/1.3.35

Martinez MR, Mihu MR, Tar M, Cordero RJB et al (2010) Demonstration of antibiofilm and antifungal efficacy of chitosan against candidal biofilms, using an in vivo central venous catheter model. J Infect Dis 201(9):1436–1440. https://doi.org/10.1086/651558

McClements DJ (2020) Nanotechnology approaches for improving the healthiness and sustainability of the modern food supply. ACS Omega 5(46):29623–29630. https://doi.org/10.1021/ACSOMEGA.0C04050

Meghani N, Dave S, Kumar A (2020) Introduction to nanofood. Nano-food. Engineering 1:1–23

Mei L, Wang Q (2020) Advances in using nanotechnology structuring approaches for improving food packaging. Ann Rev Food Sci Technol 11:339–364. https://doi.org/10.1146/annurev-food-032519-051804

Meral R, Ceylan Z, Kutlu N, Kılıçer A et al (2022) Antimicrobial nanocoating for food industry. In Handbook of Microbial Nanotechnology. https://doi.org/10.1016/B978-0-12-823426-6.00011-5

Mills A, Hazafy D (2009) Nanocrystalline SnO2-based, UVB-activated, colourimetric oxygen indicator. Sensors Actuators B Chem 136(2):344–349. https://doi.org/10.1016/J.SNB.2008.12.048

Mirza Alizadeh A, Masoomian M, Shakooie M, Zabihzadeh Khajavi M et al (2021) Trends and applications of intelligent packaging in dairy products: a review. Crit Rev Food Sci Nutr 62(2):383–397. https://doi.org/10.1080/10408398.2020.1817847

Mohammad ZH, Ahmad F (2024) Nanocoating and its application as antimicrobials in the food industry: a review. Int J Biol Macromol 254(2):127906. https://doi.org/10.1016/j.ijbiomac.2023.127906

Mohammad ZH, Ahmad F, Ibrahim SA, Zaidi S (2022) (2022) application of nanotechnology in different aspects of the food industry. Discov Food 2(1):1–21. https://doi.org/10.1007/S44187-022-00013-9

Müller K, Bugnicourt E, Latorre M, Jorda M et al (2017) Review on the processing and properties of polymer nanocomposites and nanocoatings and their applications in the packaging, automotive and solar energy fields. Nano 7(4):74. https://doi.org/10.3390/nano7040074

Nam NN, Do HDK, Trinh KTL, Lee NY (2022) Recent progress in nanotechnology-based approaches for food monitoring. Nano 12:4116. https://doi.org/10.3390/NANO12234116

Namasivayam SKR, Pandian UK, Chava V, Arvind Bharani RS et al (2023) Chitosan nanocomposite as an effective carrier of potential herbicidal metabolites for noteworthy phytotoxic effect against major aquatic invasive weed water hyacinth (Eichhornia crassipes). Int J Biol Macromol 226:1597–1610. https://doi.org/10.1016/j.ijbiomac.2022.11.272

Navia-Mendoza JM, Filho OAE, Zambrano-Intriago LA, Maddela NR et al (2021) Advances in the application of nanocatalysts in photocatalytic processes for the treatment of food dyes: a review. Sustain For 13:11676. https://doi.org/10.3390/SU132111676

Neme K, Nafady A, Uddin S, Tola YB (2021) Application of nanotechnology in agriculture, postharvest loss reduction and food processing: food security implication and challenges. Heliyon 7(12):e08539. https://doi.org/10.1016/J.HELIYON.2021.E08539

Nikravan L, Zamanpour S, Hashemi M, Marashi SMHM et al (2024) Application of nanotechnology to develop carrageenan-based films and coatings as carriers for essential oils. Curr Pharm Biotechnol 25. https://doi.org/10.2174/0113892010287269240222054020

Nile SH, Baskar V, Selvaraj D, Nile A et al (2020) Nanotechnologies in food science: applications, recent trends, and future perspectives. Nano-Micro Letters 12:1–34

Niu Q, Lan L, Li T, Guo Z et al (2018) A highly selective turn-on fluorescent and naked-eye colorimetric sensor for cyanide detection in food samples and its application in imaging of living cells. Sensors Actuators B Chem 276:13–22. https://doi.org/10.1016/j.snb.2018.08.066

Numa, I. A. N., Wolf, K. E., & Pastore, G. M. (2023). FoodTech startups: technological solutions to achieve SDGs. Food and Humanity, 1, 358–369

Okagu OD, Wang B, Acquah C, Udenigwe CC (2019) Protein-based nanodelivery systems for food applications. Encyclopedia of Food Chemistry, pp 719–726. https://doi.org/10.1016/B978-0-08-100596-5.21864-7

Oliveira Filho JGD, Silva GDC, Oldoni FCA, Miranda M et al (2022) Edible coating based on carnauba wax nanoemulsion and cymbopogon martinii essential oil on papaya postharvest preservation. Coatings 12(11):1700

Pal M (2017) Nanotechnology: a new approach in food packaging. J Food: Microbiol Safe Hygiene 2(2):1–2. https://doi.org/10.4172/2476-2059.1000121

Pateiro M, Gómez B, Munekata PES, Barba FJ et al (2021) Nanoencapsulation of promising bioactive compounds to improve their absorption, stability, functionality and the appearance of the final food products. Molecules 26(6):1547. https://doi.org/10.3390/molecules26061547

Pathakoti K, Manubolu M, Hwang HM (2017) Nanostructures: current uses and future applications in food science. J Food Drug Anal 25(2):245–253. https://doi.org/10.1016/J.JFDA.2017.02.004

Priyadarshi R, Sauraj Kumar B, Negi YS (2018) Chitosan film incorporated with citric acid and glycerol as an active packaging material for extension of green chilli shelf life. Carbohydr Polym 195:329. https://doi.org/10.1016/j.carbpol.2018.04.089

Pushparaj K, Liu WC, Meyyazhagan A, Orlacchio A et al (2022) Nano- from nature to nurture: a comprehensive review on facets, trends, perspectives and sustainability of nanotechnology in the food sector. Energy 240:122732. https://doi.org/10.1016/J.ENERGY.2021.122732

Qian M, Liu D, Zhang X, Yin Z et al (2021) A review of active packaging in bakery products: applications and future trends. Trends Food Sci Technol 114:459–447. https://doi.org/10.1016/j.tifs.2021.06.009

Quazi MZ, Park N (2022) Nanohydrogels: advanced polymeric nanomaterials in the era of nanotechnology for robust functionalization and cumulative applications. Int J Mol Sci 23(4):1943. https://doi.org/10.3390/ijms23041943

Rambaran, T., & Schirhagl, R. (2022). Nanotechnology from lab to industry–a look at current trends. Nanoscale Advances, 4(18), 3664–3675

Ribeiro AI, Dias AM, Zille A (2022) Synergistic effects between metal nanoparticles and commercial antimicrobial agents: a review. ACS Appl Nano Mater 5(3):524–532. https://doi.org/10.1021/acsanm.1c03891

Roy A, Elzaki A, Tirth V, Kajoak S et al (2021) Biological synthesis of nanocatalysts and their applications. Catalysts 11:1494. https://doi.org/10.3390/CATAL11121494

Sabra A, Netticadan T, Wijekoon C (2021) Grape bioactive molecules, and the potential health benefits in reducing the risk of heart diseases. Food Chem 12:100149. https://doi.org/10.1016/j.fochx.2021.100149

Sánchez-Arévalo CM, Croes T, Van der Bruggen B, Vincent-Vela MC et al (2023) Feasibility of several commercial membranes to recover valuable phenolic compounds from extracts of wet olive pomace through organic-solvent nanofiltration. Sep Purif Technol 305:122396. https://doi.org/10.1016/J.SEPPUR.2022.122396

Santos, I. J. B., Gonçalves, M. M., Schogl, A. E., de Paiva, G. M., & dos Reis Coimbra, J. S. (2024). Nanomaterial migration into the food matrix. Nanostructured Materials for Food Packaging Applications, 553–573

StatNano. (2023a, August 1). Nanotechnology Companies. https://Statnano.Com/Companies

StatNano. (2023b, August 1). Nanotechnology patents in EPO (Patent). Https://Statnano.Com/Report/S102

StatNano. (2023c, August 1). Nanotechnology patents in USPTO (Patent). Https://Statnano.Com/Report/S103

StatNano. (2023d, August 1). Nanotechnology Products Database. https://Product.Statnano.Com/

StatNano. (2023e, August 1). Nanotechnology publications (Article). https://Statnano.Com/Report/S29

Shafaei Z, Ghalandari B, Vaseghi A, Divsalar A et al (2017) β-Lactoglobulin: an efficient nanocarrier for advanced delivery systems. Nanomedicine 13(5):1685–1692. https://doi.org/10.1016/J.NANO.2017.03.007

Shan D, Wang Y, Xue H, Cosnier S (2009) Sensitive and selective xanthine amperometric sensors based on calcium carbonate nanoparticles. Sensors Actuators B Chem 136(2):510–515. https://doi.org/10.1016/J.SNB.2008.10.012

Sharma A, Matharu Z, Sumana G, Solanki PR, Kim CG, Malhotra BD et al (2010) Antibody immobilized cysteamine functionalized-gold nanoparticles for aflatoxin detection. Thin Solid Films 519(3):1213–1218. https://doi.org/10.1016/J.TSF.2010.08.071

Sharma C, Dhiman R, Rokana N, Panwar H (2017) Nanotechnology: an untapped resource for food packaging. Front Microbiol 8:243298. https://doi.org/10.3389/FMICB.2017.01735/BIBTEX

Sharma P, Panghal A, Gaikwad V, Jadhav S et al (2019) Nanotechnology: a boon for food safety and food defense. In Nanotechnology in the Life Sciences, pp 225–242. https://doi.org/10.1007/978-3-030-17061-5_9

Singh C, Ali MA, Reddy V, Singh D et al (2018) Biofunctionalized graphene oxide wrapped carbon nanotubes enabled microfluidic immunochip for bacterial cells detection. Sensors Actuators B Chem 255:2495–2503. https://doi.org/10.1016/J.SNB.2017.09.054

Sippel, S. R., & Dolinga, M. (2023). Constructing agri-food for finance: startups, venture capital and food future imaginaries. Agriculture and Human Values, 40(2), 475–488

Sodano V, Gorgitano MT, Verneau F, Vitale CD (2016) Consumer acceptance of food nanotechnology in Italy. Br Food J 118(3):714–733

Souza VGL, Fernando AL (2016) Nanoparticles in food packaging: biodegradability and potential migration to food—a review. Food Packag Shelf Life 8:63–70

Sridhar A, Ponnuchamy M, Kumar PS, Kapoor A (2020) Food preservation techniques and nanotechnology for increased shelf life of fruits, vegetables, beverages and spices: a review. Environ Chem Lett 19(2):1715–1735. https://doi.org/10.1007/S10311-020-01126-2

Störmer A, Bott J, Kemmer D, Franz R (2017) Critical review of the migration potential of nanoparticles in food contact plastics. Trends Food Sci Technol 63:39–50. https://doi.org/10.1016/j.tifs.2017.01.011

Suhag R, Kumar N, Petkoska AT, Upadhyay A (2020) Film formation and deposition methods of edible coating on food products: a review. Food Res Int 136:109582. https://doi.org/10.1016/j.foodres.2020.109582

Sultana A, Kathuria A, Gaikwad KK (2022) Metal-organic frameworks for active food packaging. A review. Environ Chem Letter 20(2):1479–1495. https://doi.org/10.1007/s10311-022-01387-z

Sun R, Lu J, Nolden A (2021) Nanostructured foods for improved sensory attributes. Trends Food Sci Technol 108:281–286. https://doi.org/10.1016/j.tifs.2021.01.011

Suriati L, Utama IMS, Harsojuwono BA, Gunam IBW (2022) Effect of additives on surface tension, viscosity, transparency and morphology structure of aloe vera gel-based coating. Front Sustain Food Syst 6. https://doi.org/10.3389/fsufs.2022.831671

Szilagyi A (2019) Digestion, absorption, metabolism, and physiological effects of lactose. In: Lactose. Elsevier, pp 49–111. https://doi.org/10.1016/B978-0-12-811720-0.00002-7

Tahir HE, Xiaobo Z, Mahunu GK, Arslan M et al (2019) Recent developments in gum edible coating applications for fruits and vegetables preservation: a review. Carbohydr Polym 224:115141. https://doi.org/10.1016/j.carbpol.2019.115141

Talebian, S., Rodrigues, T., Das Neves, J., Sarmento, B., Langer, R., & Conde, J. (2021). Facts and figures on materials science and nanotechnology progress and investment. ACS Nano, 15(10), 15940–15952

Tao M, Chen J, Huang K (2021) Bio-based antimicrobial delivery systems for improving microbial safety and quality of raw or minimally processed foods. Curr Opin Food Sci 41:189–200. https://doi.org/10.1016/J.COFS.2021.04.011

Turgeon SL, Brisson G (2020) Symposium review: the dairy matrix-bioaccessibility and bioavailability of nutrients and physiological effects. J Dairy Sci 103(7):6727–6736. https://doi.org/10.3168/jds.2019-17308

Usković V (2007) Nanotechnologies: what we do not know. Technol Soc 29(1):43–61

Wahab YA, Al-Ani LA, Khalil I, Schmidt S et al (2024) Nanomaterials: a critical review of impact on food quality control and packaging. Food Control 163:110466. https://doi.org/10.1016/J.FOODCONT.2024.110466

Wang YC, Lu L, Gunasekaran S (2017) Biopolymer/gold nanoparticles composite plasmonic thermal history indicator to monitor quality and safety of perishable bioproducts. Biosens Bioelectron 92:109–116. https://doi.org/10.1016/J.BIOS.2017.01.047

Wang T, Lai L, Huang Y (2024a) Nanozyme: an emerging tool for food packaging. Food Control 155:110104. https://doi.org/10.1016/J.FOODCONT.2023.110104

Wang Q, Xiang X, Chen B (2024b) Food protein-based nanotechnology: from delivery to sensing systems. Curr Opin Food Sci 56:101134. https://doi.org/10.1016/J.COFS.2024.101134

Wei X, Yang Z, Tay SL, Gao W (2014) Photocatalytic TiO_2 nanoparticles enhanced polymer antimicrobial coating. Appl Surf Sci 290:274–279. https://doi.org/10.1016/j.apsusc.2013.11.067

Wu T, Dai S, Cong X, Liu R et al (2017) Succinylated soy protein film coating extended the shelf life of apple fruit. J Food Proc Preserv 41(4). https://doi.org/10.1111/jfpp.13024

Xing Y, Xu Q, Che Z, Li X et al (2011) Effects of chitosan-oil coating on blue mold disease and quality attributes of jujube fruits. Food Function 2(8):466. https://doi.org/10.1039/c1fo10073d

Xing Y, Li W, Wang Q, Li X et al (2019) Antimicrobial nanoparticles incorporated in edible coatings and films for the preservation of fruits and vegetables. Molecules 24(9):1695. https://doi.org/10.3390/molecules24091695

Xu J, Zhang M, Cao P, Adhikari B (2020) Effect of ZnO nanoparticles combined radio frequency pasteurization on the protein structure and water state of chicken thigh meat. LWT 134:110168. https://doi.org/10.1016/J.LWT.2020.110168

Yemmireddy VK, Farrell GD, Hung YC (2015) Development of titanium dioxide (TiO2) nanocoatings on food contact surfaces and method to evaluate their durability and photocatalytic bactericidal property. J Food Sci 80(8):N1903–N1911. https://doi.org/10.1111/1750-3841.12962

Yousefi M, Orojzadeh P, Jafari SM (2019) Nanoencapsulation of food ingredients by dendrimers. Biopolymer Nanostructures for Food Encapsulation Purposes 1:607–625. https://doi.org/10.1016/B978-0-12-815663-6.00020-3

Zhang S, Jiang Z, Liang Y, Shen Y et al (2021) Effect of the duty cycle of flower-like silver nanostructures fabricated with a Lyotropic liquid crystal on the SERS Spectrum. Molecules 26(21):6522. https://doi.org/10.3390/MOLECULES26216522

Zhou YH, Mujumdar AS, Vidyarthi SK, Zielinska M et al (2023) Nanotechnology for food safety and security: a comprehensive review. Food Rev Intl 39(7):3858–3878

Chapter 13
Edible Films and Coatings for Fruits and Vegetables

Avnesh Kumari and Sudesh Kumar Yadav

Abstract Vegetable and fruit preservation has long been accomplished through the use of edible coatings and films. The literature review shows sufficient evidence that edible coatings and films have been used for centuries to preserve abundance of fruits and vegetables and extend their average shelf life. The mechanism by which remnants act to prolong the shelf life of fruits and vegetables beyond what is often expected is the monitoring influx and outflow of gases and moisture. Edible coatings and films composed of natural biopolymers can be used in place of nonbiodegradable and nonsustainable edible coatings. Therefore, a number of studies have been conducted for years to enhance edible coatings and films with efficient shelf life-extending properties while keeping in mind consumer safety and the environment. Protein-based edible films and coatings exhibit respectable mechanical strength, whereas polysaccharide-based films and coatings effectively block gases, but neither has moisture-protective properties. These deficiencies can be addressed by combining them with suitable hydrocolloids or lipids. The use of edible coatings and films has demonstrated a great deal of promise in extending the shelf life of fruits and vegetables, as well as being an environmentally responsible and sustainable method of food packaging.

Keywords Edible films · Coatings · Protein-based · Polysaccharide-based · Lipid-based · Nanomaterial-based

---

A. Kumari (✉)
Academy of Scientific and Innovative Research (AcSIR), CSIR- Human Resource Development Centre, Ghaziabad, Uttar Pradesh, India

Present Address: CSIR-Institute of Himalayan Bioresource Technology, Palampur, HP, India
e-mail: avnesh@ihbt.res.in

S. K. Yadav (✉)
Academy of Scientific and Innovative Research (AcSIR), CSIR- Human Resource Development Centre, Ghaziabad, Uttar Pradesh, India

Center of Innovative and Applied Bioprocessing (CIAB), Mohali, Punjab, India

Present Address: CSIR-Institute of Himalayan Bioresource Technology, Palampur, HP, India
e-mail: sudesh@ciab.res.in; director@ihbt.res.in

S. K. Yadav, A. Kumari (eds.), *Nanobiotechnology in Agri-food Sector*, Sustainability Sciences in Asia and Africa, https://doi.org/10.1007/978-981-96-7756-6_13

13.1 Introduction

Approximately 1.3 billion tons of food produced globally is wasted per annum, and perishable fruits and vegetables represent 40% of total waste (Qiaobin et al. 2023). Severe global food waste has been linked to major issues including starvation and fast climate change (Zhiqiang et al. 2021). Because fruits and vegetables have a limited shelf life, the majority of food waste is caused by their deterioration. Decomposing of fruits and vegetables involves dehydration, respiratory metabolism, microbial growth, and senescence (Zhiqiang et al. 2021). They are susceptible to microbial and fungal attacks and quickly become moldy if not stored properly (Lara et al. 2004).

Preservative addition, wax coating, irradiation, cold storage, heat treatment, and controlled environment storage are the most common methods for extending the shelf life of perishables (Pinela et al. 2016; Sivakumar et al. 2012). Food materials can have a thin layer of edible substances put on their surface to prolong their shelf life. The edible coatings and films market is expected to develop at a compound annual growth rate (CAGR) of 7.50% from 2022 to 2030, reaching USD 4.98 billion, from an estimated USD 2.6 billion in 2021.

Edible coatings can be consumed with food material, and they do not add unwanted unfavorable properties to the foodstuff (Manisha et al. 2021). Water, moisture, oxygen, carbon dioxide, and ethylene are all effectively blocked by edible coatings. In order to increase product shelf life and lower the chance of microbial development on food surfaces, edible coatings can also be loaded with active ingredients including antioxidants, anti-browning agents, flavors, colorants, spices, minerals, antimicrobials, and vitamins, among others (Manisha et al. 2021). Furthermore, an additional advantage of edible coating is the lessening of artificial packing waste since these coatings are composed of decomposable raw material (Dhall et al. 2013).

The US Food and Drug Administration has authorized the use of edible coatings in fruits and vegetables, which are regarded as meeting GRAS standards (Dhall et al. 2013; Martín-Belloso et al. 2009). Edible coatings can be made from substances that can form films. During the synthesis process, the film-forming material is scattered and dissolved using a solvent, such as water, alcohol, a mixture of water and alcohol, or a combination of other solvents. To create films, the film solution is then put into a container of the required form and dried at the appropriate temperature. Food can be coated with the film solutions by brushing, dipping, spraying, and then drying (Dhall et al. 2013).

By producing a semipermeable protective layer around the fruit's surface and altering the gaseous environment (O2 and CO2), the edible coatings extend the shelf life of fruits and vegetables. This consequently delays the ripening-related changes in the fruits by reducing the respiration rate and ethylene biosynthesis (Yadav et al. 2022).

Additionally, applying edible coatings to the surface of fruits and vegetables fills the pericarp's pores, which causes the stomata and lenticels to close. This delays the onset of physiological diseases, including a decrease in weight loss (Tokatli and Ashhan, 2021).

Edible coatings can be categorized based on the polymeric matrices which are used for synthesizing them. They are typically lipid-based (oil, waxes), protein-based (zein, gluten, gelatin), polysaccharide-based (starches, gums, chitosan), and composite-based (Kumar et al. 2019, 2022). Current edible coatings made of proteins and polysaccharides have low mechanical qualities, and lipids and waxes have inadequate water vapor barrier qualities. As a result, different approaches have been created to address these shortcomings.

13.2 Grouping of Edible Coatings

Edible coatings are classified on the basis of composition as polysaccharide-based, protein-based, lipid-based, and nanomaterial-based coatings (Table 13.1).

13.2.1 Polysaccharide-Based Coatings

Because of their wide availability, low cost, and lack of toxicity, polysaccharides—polymeric chains composed of monomers, i.e., mono- and disaccharides joined by glycosidic bonds—are utilized as biopolymers for edible coatings (Yu et al. 2023). Gellan gum, xanthan gum, chitosan, pullulan, carrageenan, cellulose, starch, alginate, pectin, guar gum, gum tragacanth, and gum are the commonly used polysaccharides in edible coatings (Sharma et al. 2024). Polysaccharide-based coatings are water-soluble, biodegradable, and of low cost. Polysaccharide-based coatings do not require organic solvents during application. In addition, polysaccharides have well-defined chemical structure that can be modified to allow tuning of their coating properties (Arnon et al. 2015). Arnon et al. have reported layer-by-layer coating of chitosan and carboxymethyl cellulose (CMC) on mandarins. They have found that this coating improved the firmness, gloss, and physiological quality of mandarins (Arnon et al. 2015). Another study reported the effectiveness of inexpensive edible coatings for enhancing the fruit gloss and improving postharvest quality of 'Valencia' oranges during storage (Khorram et al. 2017). Yet another study reported that the sodium alginate and pectin loaded with essential oils (EOs) reduced the microbial load, maintained all the nutritional parameters, and improved shelf life of strawberries (Guerreiro et al. 2015a, b). According to Castro Cegri et al., dextrin coatings preserved the quality of zucchini fruit at low temperatures by reducing weight loss, chilling damage, and oxidative stress. The addition of olive oil to dextrin coatings improved the induction of antioxidant enzymes and reduced the chilling injury during storage (Castro Cegri et al. 2023).

Table 13.1 Different edible coatings with additives/without additives used for coating fruits and vegetables

	Fruit/ vegetable	Additives	Advantages of coating technology and main results of study	Reference
Carboxymethyl cellulose	Apple slices	*Aloe vera*	Reduces PO and PPO involvement Reduced microbial burden; superior Hardness; protection against browning	Kumar et al. (2018)
Methyl cellulose	Sapota fruits	Palm oil	Reduces PO, PPO activity and discoloration; boosts ascorbic acid retention and anti-browning impact; Postpones the reduction in total phenolic content; increases the shelf life by three days	Vishvasrao et al. (2017)
Carboxymethyl cellulose	Strawberries	Lactobacillus plantarum	Decreases the pace at which yeasts and molds grow on strawberries' surface; boosts the fruit's probiotic properties	Khodaei et al. (2019)
Hydroxypropyl methyl cellulose	Hayward kiwis	Aloe vera gel and lemon essential oil	Lowers browning and weight loss while maintaining increased TSS, firmness, brightness, and greenness cut down on the microbial load	Passafiume et al. (2020)
Chitosan	Chitosan	Mango	Postpones the ripening process; raises TA and firmness of the fruit; lowers respiration rate, ethylene generation, and weight loss; preserves the ascorbic acid and AOC	Galus (2019)
Chitosan	Acetic or lactic acid	Blackberry	Antifungal effect against *Mucor racemosus*	Vilaplana et al. (2020)
Chitosan and cellulose nanofibers	Iron particles, curcumin	Kiwifruits	Lower stiffness and weight loss, as well as the low rate of transpiration	Ghosh et al. (2021)
Chitosan-pullulan	Pomegranate peel extract	Green bell pepper	Reduces color browning and weight loss while preserving stiffness, TPC, TFC, AOC, and sensory attributes	Kumar et al. (2021)
Chitosan and glycerol	Whey protein isolate	Strawberries	Reduce TA, TPC, DPPH, pH, weight loss, and color changes; enhance shelf life by 60%	Muley et al. (2020)
Chitosan	Cinnamon essential oil	Cucumber	Antifungal activity	Isturiz-Zapata et al. (2020)

(continued)

Table 13.1 (continued)

	Fruit/ vegetable	Additives	Advantages of coating technology and main results of study	Reference
Chitosan and alginate	Pomegranate peel extract	Capsicum	Reduce loss of ascorbic acid content, hardness, color, and weight	Nair et al. (2018)
Sodium alginate	Eugenol and citral	Arbutus unedo fruit	Reduces the growth of microorganisms and the loss of weight; enhances the physical, chemical, and biological characteristics, including color, hardness, AOC, and sensory aspects	Guerreiro et al. (2015a, b)
Sodium alginate	Lotus leaf extract	Goji berries	Decreases weight loss and decay rate while maintaining AA, TA, and TSS	Fan et al. (2019)
Gelatin	Ethanolic extract of propolis	Raspberries	Antifungal efficacy against strains of *Botrytis cinerea* and *Penicillium digitatum*; enhanced shelf life	Moreno et al. (2020)
Gelatin	Persian gum	Oranges	Reduces weight loss; reduces TA; increases TPC and AOC; preserves the firmness and gloss of the fruit	Khorram et al. (2017)
Arabic gum	Salicylic acid	Bananas	Reduces weight loss; enhances peel browning index and stiffness; preserves antioxidant activity	Alali et al. (2018)
Pectin and pullulan	*Vitis vinifera* grape seed extract	Peanuts	Decreases antibacterial activity against *Listeria monocytogenes* and *Escherichia coli*, as well as decreases lipid oxidation	Priyadarshi et al. (2022)

Polysaccharide-based edible coatings of alginate, cellulose, starch, and chitosan and carrageenan have shown useful effects on fruits and vegetables. According to Cruz-Monterrosa et al. (2023), they maintained the quality attributes, which included a longer shelf life, lower respiration rates, firmness, reduced weight loss, vitamins, preserved soluble solids, and bioactive compounds; they also reduced microbial growth, maintained antioxidant enzyme activities, postponed the degree of browning, and reduced overall food damage. Furthermore, the incorporation of EOs and plant extracts into edible coatings based on polysaccharides has shown remarkable antioxidant and antibacterial qualities (Cruz-Monterrosa et al. 2023). Therefore, when consumed with the food, polysaccharide-based edible coatings containing additional bioactives may contribute to food safety in addition to serving as an efficient preservation technique (Cruz-Monterrosa et al. 2023). Soto-Muñoz et al.

(2022) have used edible pregelatinized potato starch, glyceryl monostearate, glycerol emulsifiers as edible coating for maintaining the postharvest quality of cold-stored 'Fino' lemons.

13.2.2 Protein-Based Coatings

Animal-based proteins include collagen, keratin, casein, and gelatin, while plant-based proteins comprised of corn, wheat, soy, and so on have been used for preparing edible films and coatings (Armghan Khalid et al. 2022). Corn zein coatings have reduced water permeability (Hauzoukim et al. 2020), slowed down seed germination, and improved the shelf life of pears and apples (Park et al. 1996). Strawberries treated with soy protein-based edible coatings containing thymol maintained their appearance until 9 days after packing (Amal et al. 2010). According to a different study, fresh sliced apples' quality and shelf life were enhanced by edible coatings made of soy protein and ferulic acid (Alves et al. 2017). According to research by Tanada-Palmu & Grosso, applying wheat gluten to strawberries can slow down the senescence process, boost their firmness, stop moisture loss, delay fruit infection, and stop them from losing weight (Tanada-Palmu et al. 2005). Edible coverings made of whey protein have a lower oxygen permeability and are transparent, flavorless, colorless, and flexible (Hong et al. 2006). Edible coating based on whey protein and rice bran oil sustained the color, firmness, taste, and overall acceptability of the kiwifruit during 28 days of storage (Hassani et al. 2012). Protein-based edible coatings show good mechanical strength but poor gas blocking capabilities (Chettri et al. 2023). Fernando (2017) applied edible coating of casein-containing turmeric on carrots and witnessed that color, carotenoid concentration, texture, and antimicrobial properties remained acceptable for up to 10 days of storage. Another study reported that coating of the bell peppers with sodium caseinate delayed color change and moisture loss, and extended the storage life (Lerdthanangkul and Krochta 1996).

13.2.3 Lipid-Based Coatings

Edible films and coatings frequently contain lipids, which are abundant and inexpensive in nature (Devi et al. 2024). Several studies over the past few years have effectively identified lipid-based edible coating as a practical option for extending the postharvest shelf life of fruits and vegetables (Khorram et al. 2017; Milani et al. 2022; Devi et al. 2022). Lipid-based edible coatings such as shellac and wax reduce scratches and damages during handling, and improve fruit and vegetable gloss (Milani et al. 2022).

In a study, carnauba wax-based edible coatings containing oregano EO significantly reduced moisture loss and effectively delayed microbial spoilage in treated fruits and vegetables (Gutiérrez-Pacheco et al. 2020).

When applied to fresh strawberries, edible coatings based on candelilla wax demonstrated a strong ability to reduce the occurrence of strawberry weakening (Oregel-Zamudio et al. 2017). According to a different study, tangerine fruit's bioactive components and qualitative attributes were better preserved by an edible covering made of alginate and thyme EO. The study's findings demonstrated that adding thyme EO to an alginate-based edible coating improved its effectiveness in preserving the tangerine fruits' postharvest quality characteristics (Radi et al. 2023).

13.2.4 Nanomaterial-Based Coatings

Recently, NM-based edible coatings are also used for increasing the shelf life of fruits and vegetables. Nanoemulsions containing essential oils (EOs) have received considerable attention due to their antimicrobial properties. Sodium alginate-based edible coatings containing 0.5% nanoemulsions led to a notable reduction of *Salmonella typhimurium* and *Listeria monocytogenes* and extended the shelf life of fresh-cut pineapple (Prakash et al. 2020). Another study has reported alginate-based edible coatings containing thyme and oregano EOs in various concentrations. Although increasing the EO concentration extended the shelf life, the sensory panel's results were unfavorable because of the intense EO odor (Tabassum et al. 2020). Fresh-cut melons were coated with edible citral nanoemulsions embedded in carboxymethyl cellulose (CMC) or chitosan polysaccharides. The fruit benefited from improved appearance, extended storability, and excellent antibacterial protection thanks to the coatings made of nanoemulsions (Arnon-Rips et al. 2019). Yet, another study reported edible coating containing blend of silk fibroin, PVA, honey, and curcumin to enhance the shelf life of fruits and vegetables (Singh and Packirisamy, 2022). It was also documented that chitosan-clove-nisin edible coating increased the shelf life two times, prevented deterioration, and preserved the quality of food. Nanoemulsion-based edible coatings ensure improved quality and shelf life with reduced food waste and its environmental impact (Panwar et al. 2024).

13.3 Methods Used for the Preparation of Edible Coatings

Many methods are reported in the literature for the preparation of edible coatings.

13.3.1 Dipping

Using this technique, the fruit is submerged in a coating solution for a set amount of time to create a semipermeable layer that prevents moisture loss and gas transfer. First, for a length of time determined by the study, the sample is submerged in the coating and crosslinking solution. The average dipping duration is between 30 and 5 min. After that, it is removed to drain the excess solution. This method's main benefit is that even on difficult surfaces, the sample is equally covered from all sides. When coating chopped fruits, the layer-by-layer procedure is used since the fruit's stickiness prevents the coating from forming effectively (Pandey et al. 2022).

13.3.2 Spraying

It is one of the traditional techniques for coating food samples such that a semipermeable membrane can form. Using nozzles, the coating solutions are sprayed onto the samples in the form of droplets in this process. This system operates at high pressure (between 60 and 80 psi); thus, it doesn't need a lot of coating solution. A few benefits of this procedure are uniform coating solution dispersion over the fruit, coating in layers, and uniform coating thickness. There should be less viscosity in the coating mixture. It depends on a number factors, including the liquid's qualities (surface tension, density, and viscosity), the instrument's settings (nozzle design and spray angle), and the processing environment (temperature, pressure, and flow rate) (Pandey et al. 2022).

13.3.3 Vacuum Impregnation

Pressure is the single most significant aspect in the vacuum impregnation (VI) procedure because it is the key to this technique. This technique is typically employed to improve food components' vitamin and mineral content. The pressure is the only thing that differs from the typical dipping-draining procedure. The samples are partially immersed in the fluid and connected by vacuum pumps. Vacuum period, vacuum pressure, and atmospheric restoration time are crucial factors in the VI process (Pandey et al. 2022).

13.4 Edible Coatings for Shelf Life Improvement of Fruits and Vegetables

It is well acknowledged that the primary sources of vitamins, minerals, antioxidants, and fiber in the diets of consumers are fruits and vegetables. However, the main reason for their short shelf life is their high moisture content (75–95%), which also contributes to their rapid disintegration. The natural process of development, aging, and degradation that fruits and vegetables go through can be understood as a result of their living nature. Two categories of fruits are distinguished by the mechanisms involved in the ripening process. They are classified into climacteric and nonclimacteric fruits from this perspective (Miteluț et al. 2021). Compared to nonclimacteric fruits, fruits in the first category are more susceptible to microbial deterioration because they can ripen even after harvest and generate more ethylene. On the other hand, fruits in the second category do not continue to ripen after being harvested. For these food products to have a reasonable shelf life, proper coating and packaging methods are therefore crucial (Jafarzadeh et al. 2021). These days, fresh produce is frequently coated with a range of edible coatings and nanocomposite films to increase its shelf life. According to Guimaraes et al. (2018), these techniques are based on the same principles as modified environment techniques, which have shown promise in preserving the freshness of fruits and vegetables.

The way these edible films work is by creating a semipermeable safety layer around the fruit and vegetables, which significantly lessens the loss of quality characteristics. The comparatively large range of biopolymers, including xanthan gum, starch, pectin, carrageenan, alginate, and chitosan, which are widely employed to make edible films and food coatings, effectively preserves these qualities (Nair et al. 2020). The way these edible films work is by creating a semipermeable safety layer around the fruit and vegetables, which significantly lessens the loss of quality characteristics. The comparatively large range of biopolymers, including xanthan gum, starch, pectin, carrageenan, alginate, and chitosan, which are widely employed to make edible films and food coatings, effectively preserves these qualities (Nair et al. 2020).

Functionalized chitosan and chitosan nanoformulations are excellent choices for creating an edible coating for fruit products like chopped pineapple (Ghosh et al. 2019), chopped cherry tomatoes (Bitencourt et al. 2018), apricots (Zhang et al. 2018), peaches (Jiao et al. 2019), blueberries (Mannozzi et al. 2018), etc. Additionally, Sun et al.'s (2021) study investigated how three bio-based nanomaterial coatings composed of chitin, cellulose, and chitosan affected the shelf life of fruits such as kiwis, avocados, strawberries, bananas, nectarine, and apricots. The best antifungal activity was shown by chitosan nanofibers and wood nanocrystals. In addition, the coated strawberries' color changes and weight loss were cut in half when compared to the control samples (Sun et al. 2021). When the edible coatings are applied, respiration and senescence slow down without generating anaerobiosis, which can improve the fruit's external appearance and inhibit fungal proliferation (Mahajan et al. 2014).

The major goals of edible coatings, which have been evaluated by numerous researchers and used on a wide range of fruits, are to prolong the fruits' shelf life and preserve its biological characteristics. Studies on fruits, such as berries and tropical fruits (Jafarzadeh et al. 2021), papaya (Zillo et al. 2018; Mendy et al. 2019), goji (Fan et al. 2019), oranges (Parafati et al. 2016; Saberi et al. 2018; Rasouli et al. 2019), cherry (Dong et al. 2018), mango (Khaliq et al. 2016), strawberries (Dhital et al. 2017), and vegetables, such as tomatoes (de Jesús Salas-Méndez et al. 2019), carrots (Sanuja et al. 2015), and broccoli (Alvarez et al. 2013), have been conducted in significant quantities.

Pregelatinized potato starch along with sodium benzoate has been used as edible coating for improving the shelf life of Fino lemons (Soto-Muñoz et al. 2022). These coatings have significantly enhanced the postharvest quality of Fino lemons and also helped in maintaining the physiochemical quality parameters of Fino lemons (Soto-Muñoz et al. 2022). Another study reported that the coatings of lyophilized fish myofibrillar proteins, corn starch, and gum arabic delayed ripening, increased the commercial life of guavas, reduced postharvest losses, and added value to the fruit (Pereira et al. 2021).

Yet, another study reported that the coating of alginate and lemon essential oil inhibited fungal and bacterial growth after 15 days of storage, without changing quality parameters, and enhanced the shelf life of red raspberries (Gomes et al. 2017).

Gupta et al. have reported that coating treatment of mangoes with carboxymethyl cellulose, guar gum, and gum acacia showed higher values of ascorbic acid, total phenolic content, and total chlorophyll content during the storage period. This coating extended the shelf life along with preservation of quality and bioactive parameters of mango fruits (Gupta et al. 2023).

Osmunda japonica polysaccharide-incorporated CMC coatings reduced respiratory intensity, weight loss, and ascorbic acid loss in addition to raising titrable acidity and regulating malondialdehyde concentration. It also prevented tomato deterioration. Thus, the application of these CMC coatings preserved the quality and extended the shelf life of tomatoes after harvest (Zhang et al. 2019). Application of chitosan-based edible coating enriched with pequi peel extract on tomatoes improved fresh weight retention and reduced color change and delayed fungal growth (Breda et al. 2017). The application of edible coating of chitosan and cassava starch minimized weight loss and mold decay, and extended shelf life of black mulberries (Ojeda et al. 2021). Applying an edible covering made of fish protein hydrolysate on cherry tomatoes prevented the growth of yeasts and molds. However, the cherry tomatoes' ripening was not slowed down by the edible coating (de Quadros et al. 2020).

These studies' common findings demonstrated that whole fruit and vegetable samples coated or treated with edible coatings had better physical-chemical qualities (such as firmness, titratable acidity, and vitamin C content) and an extended shelf life (by reducing respiration rate and weight loss).

13.5 Edible Films and Coatings with Functional Additives

The edible coating materials' mechanical, sensory, and nutritional properties can all be improved using functional additions. It is apparent that edible coatings containing nanoparticles, essential oils, and other natural sources like plant extracts, antioxidants, antimicrobials, and antifungal agents have the potential to prolong citrus fruit shelf life by preserving their postharvest and physiochemical characteristics and inhibiting the growth of blue and green molds under storage conditions (Kumar et al. 2024).

The following developments in this subject are largely reflected in research aimed at enhancing edible coating performance with the addition of various functional ingredients (Riva et al. 2020).

Plasticizers are applied to coatings to improve flexibility and stop scorching, flaking, and cracking (glycerol, sorbitol, sucrose, mannitol, acetylated, monoglyceride, polyethylene glycol, and xylitol). Antioxidants (ascorbic acid, citric acid, and tocopherol) are added to coating matrices to prevent oxidative rancidity, degradation, and discoloration; emulsifiers (soy lecithin, stearic acid, and Tweens) and surfactants (Tweens) are added to coating matrices to improve coating adhesion; and antimicrobial agents (nisin, natamycin, phenolic compounds, natural seed extracts, and essential oils like cinnamaldehyde, and eugenol) are added to coating matrices to enhance their antimicrobial activity (Miteluț et al. 2021).

Utilizing the Box-Behnken design, alginate-based edible coatings enhanced with aloe vera were created with the goal of maximizing the minimal water vapor permeability. Titanium oxide nanoparticles (nTiO2) were mixed into the film at varying ratios to produce various films. The addition of titanium oxide nanoparticles enhanced the material's mechanical and antibacterial qualities. When alginate/*Aloe vera* film containing 5 wt% of TiO_2 was applied to tomatoes, shelf life experiments demonstrated strong resistance to weight loss and spoiling (Salama et al. 2020).

Samples of freshly cut apples were coated with anti-browning agents and edible coatings (carboxymethyl cellulose and *Aloe vera*) in varying proportions, either with one active component or both. Packed in polypropylene trays, the treated samples were kept at 5 ~ 2 ~ C. Throughout the course of the samples' storage, a number of characteristics were examined, including the following: microbiological indicators, ascorbic acid, total antioxidant, phenol, polyphenol oxidase, and peroxidase enzymes, and physical characteristics like color, physiological weight loss, and hardness. The edible covering material and anti-browning chemicals applied to the samples assisted in maintaining their quality. In the microbiological test, it was found that apple slices covered in *Aloe vera* and carboxymethyl cellulose had a much lower bacterial burden (Kumar et al. 2018).

The edible coatings of carrageenan, xanthan gum with 1% w/v lemongrass EO, and Arabic gum were investigated by Wan et al. (Wani et al. 2021). Over the course of 12 days, samples of strawberries treated with the three created edible coatings were subjected to postharvest quality assessments. The findings indicated that the coated strawberry samples had superior antioxidant activity, less weight loss, better ascorbic acid retention, and increased firmness. Throughout the storage period, the phenolic compounds and anthocyanin levels were retained in the edible films containing carrageenan gum. Additionally, the coatings containing carrageenan gum had the best outcomes in terms of preserving quality throughout storage.

Pomegranate seed oil is added to edible coatings of *Chlorella* sp. This helps to preserve the fruit by delaying the ripening of *Spondias tuberosa* and keeping it firmer, greener, and more turgid for longer. Furthermore, fruit quality, including vitamin C and phenolic components, is maintained by covering fruit with 2.0% *Chlorella* sp. in conjunction with pomegranate seed oil and storing it cold for 12 days at 14 ± 2 °C and 85 ± 5% relative humidity (Teodosio et al. 2021).

In a study conducted by Oyom et al. (2022), pear samples (*Pyrus bretschneideri* Rehd.) were coated with an edible formulation made of modified sweet potato starch and cumin essential oil. The coated samples were kept for 28 days at 25 °C. The coating technique utilized in the testing involved dipping the cleaned and sanitized samples into the coating formulations. The coating treatment was able to limit the respiration rate, delay weight loss, and maintain the firmness of the flesh after 21 days because of the modified starch coating's adhesiveness, which formed a thin layer on the pear and protected its pericarp and freshness.

Additionally, Passafiume et al. (2020) investigated the impact of three edible coatings on extending the shelf life and quality of fresh-cut Hayward kiwis. The coatings were based on hydroxypropyl methylcellulose, aloe vera gel, and lemon essential oil composite coating. Following ten days of chilling temperatures, the treated samples were subjected to a battery of chemical, microbiological, and sensory testing on days 2, 4, 7, and 10. The coatings that could be eaten were sprayed on. The coated samples maintained greater firmness, brightness, greenness, and soluble solid contents (SSC) while exhibiting less weight loss and browning, according to the data. Comparing the aloe vera gel and aloe vera gel + lemon essential oil samples to the control samples, the microbial burden was lowered.

Many studies have used plant extracts as additives in edible coatings. Plant extracts can improve the mechanical, sensory, and nutritional qualities of the original edible coating ingredients when added to coatings (Arnon-Rips et al. 2018). Pomegranate seed oil is a highly beneficial dietary component that is renowned for its abundance of antioxidants and phytochemicals, such as punicic acid. It is believed that the presence of phenolic chemicals, specifically orthodiphenols, is responsible for the advantageous qualities of extracts (Soleimanian et al. 2018). Emulsification of edible and biodegradable coatings can be aided by the nutritional and therapeutic qualities of pomegranate seed oil, as studies have shown (Sogut et al. 2019). Pomegranate seed oil can boost the creation of edible biofilms with better mechanical and barrier qualities, according to recent studies. The interactions between the coating's polysaccharides and proteins are responsible for these improvements (de Medeiros et al. 2021). A recent study has shown that Mexican lime fruit kept at 20 ± 2 °C and 50–60% relative humidity can have its shelf life extended by edible coatings made from wild sage seeds and pomegranate seed oil (Mohammadi et al. 2024). According to the study's findings, the treatments successfully stopped the fruit's weight and titratable acidity from declining. Over the course of its 24-day shelf life, the fruit's total phenolic, flavonoid, and antioxidant activity all showed a declining tendency (Mohammadi et al. 2024). On the other hand, during storage, the coated fruits—apart from pomegranate seed oil—showed increased levels of antioxidant capacity, peroxidase, and catalase activity. With the exception of PSO, the use of edible coatings improved the Mexican lime fruit's a* value, aesthetic appeal, and marketability (Mohammadi et al. 2024).

13.6 Conclusion

This analysis focuses on recently developed coatings used on fresh, minimally processed fruits and vegetables. The widespread use of lipid, protein, or nanomaterial as the primary ingredient in edible coatings, to which other useful additions like probiotics, antioxidant extracts, and essential oils are added, reflects the most recent advancements in this sector. In addition to their main function of forming a protective layer that prolongs shelf life, these functional additives offer consumers beneficial functional features like probiotic capabilities, antioxidant capacity, improved sensory qualities, and increased vitamin C content. Expanding the use of edible functional coatings for fruit and vegetable preservation may be made easier with a greater understanding of how these coatings work and how to market them to customers.

Furthermore, employing edible coatings and films to deliver bioactive substances and boost bioavailability has a lot of potential. The topic is always current, and experiments are absolutely required starting from the known data, even though there are many well-defined studies in this field and highly specific coating technology used for various fruits and/or vegetables with various proposed aims (ranging from improving shelf life to preserving a high nutritional value or increasing certain nutritional features of the products).

Acknowledgements The authors would like to thank the Director of CSIR-IHBT for providing necessary facilities for carrying out this work. CSIR-IHBT communication number of this chapter is 5729.

Author Contributions AK and SKY conceived the idea and wrote and edited the whole manuscript.

Ethical Declaration This declaration is not applicable to the current manuscript.

Conflict of Interest The authors have no conflict of interests.

Funding Financial support from the Council of Scientific and Industrial Research (MLP-207) is duly acknowledged.

References

Alali AA, Mohamed AA, Adel DAQ, Saleh AM (2018) Postharvest gum Arabic and salicylic acid dipping affect quality and biochemical changes of 'Grand Nain' bananas during shelf life. Sci Hortic 237:51–58

Alvarez MV, Alejandra GP, María DRM (2013) Antimicrobial efficiency of chitosan coating enriched with bioactive compounds to improve the safety of fresh cut broccoli. LWT-Food Sci Technol 50(1):78–87

Alves MM, Gonçalves MP, Cristina MRR (2017) Effect of ferulic acid on the performance of soy protein isolate-based edible coatings applied to fresh-cut apples. Lwt 80:409–415

Amal SHA, El-Mogy MMH, Aboul-Anean HE, Alsanius BW (2010) Improving strawberry fruit storability by edible coating as a carrier of thymol or calcium chloride. J Hortic Sci Ornam Plant 2(3):88–97

Armghan Khalid M, Bushra N, Farhan S, Muhammad A et al (2022) Edible coatings for enhancing safety and quality attributes of fresh produce: a comprehensive review. Int J Food Prop 25(1): 1817–1847

Arnon H, Rina G, Ron P, Elena P (2015) Development of polysaccharides-based edible coatings for citrus fruits: a layer-by-layer approach. Food Chem 166:465–472

Arnon-Rips H, Elena P (2018) Improving food products' quality and storability by using layer by layer edible coatings. Trends Food Sci Technol 75:81–92

Arnon-Rips H, Ron P, Elena P (2019) Enhancement of agricultural produce quality and storability using citral-based edible coatings; the valuable effect of nano-emulsification in a solid-state delivery on fresh-cut melons model. Food Chem 277:205–212

Bitencourt TA, Ana LFS, Mirna HRS, Odílio BGA (2018) Antifungal activity of chitosan and its quaternized derivative in gel form and as an edible coating on cut cherry tomatoes. J Agric Sci Belgrade 63(3):271–285

Breda CA, Daniella LM, Odilio Benedito GDA, Marta Cristina TD (2017) Effect of chitosan coating enriched with pequi (Caryocar brasiliense Camb.) peel extract on quality and safety of tomatoes (Lycopersicon esculentum Mill.) during storage. J Food Process Preserv 41(6):e13268

Castro-Cegrí A, Mariano OM, Sandra S, Fátima C et al (2023) Application of polysaccharide-based edible coatings to improve the quality of zucchini fruit during postharvest cold storage. Sci Hortic 314:111941

Chettri S, Sharma N, Ashish MM (2023) Edible coatings and films for shelf-life extension of fruit and vegetables. Biomater Adv 2:213632

Cruz-Monterrosa RG, Adolfo ARA, Ricardo MGR, María LZZ et al (2023) Application of polysaccharide-based edible coatings on fruits and vegetables: improvement of food quality and bioactivities. Polysaccharides 4(2):99–115

de Jesús Salas-Méndez E, Antonio V, Ana Cristina P, Lina FB et al (2019) Application of edible nanolaminate coatings with antimicrobial extract of Flourensia cernua to extend the shelf-life of tomato (*Solanum lycopersicum* L.) fruit. Postharvest Biol Technol 150:19–27

de Medeiros T, Albert EM, Bárbara GFLS, Jéssica Aline L, José Manoel FDLC, Elny Alves O, José Franciraldo DL, Railene Hérica CRA (2021) Preservation of Spondias tuberosa fruit with edible coatings based on Chlorella sp. enriched with pomegranate seed oil during storage. Food Bioproc Tech 14:2020–2031

de Quadros C d C, Karina OL, Caio Hendrix LB, Fabíola HDSF et al (2020) Effect of the edible coating with protein hydrolysate on cherry tomatoes shelf life. J Food Process Preserv 44(10): e14760

Devi LS, Kalita S, Mukherjee A, Kumar S (2022) Carnauba wax-based composite films and coatings: recent advancement in prolonging postharvest shelf-life of fruits and vegetables. Trends Food Sci Technol 129:296–305

Devi LS, Jaiswal AK, Jaiswal S (2024) Lipid incorporated biopolymer based edible films and coatings in food packaging: a review. Curr Res Foof Sci 8:100720

Dhall RK (2013) Advances in edible coatings for fresh fruits and vegetables: a review. Crit Rev Food Sci Nutr 53(5):435–450

Dhital R, Joshi P, Becerra-Mora N, Umagiliyage A et al (2017) Integrity of edible nano-coatings and its effects on quality of strawberries subjected to simulated in-transit vibrations. LWT 80: 257–264

Dong F, Wang X (2018) Guar gum and ginseng extract coatings maintain the quality of sweet cherry. Lwt 89:117–122

Fan XJ, Bin Z, He Y, Jun-Tao F et al (2019) Effect of lotus leaf extract incorporated composite coating on the postharvest quality of fresh goji (Lycium barbarum L.) fruit. Postharvest Biol Technol 148:132–140

Galus S (2019) Development of edible coatings in the preservation of fruits and vegetables. Polymers for Agri-Food Applications, pp 377–390

Ghosh T, Yoshikuni T, Katiyar V (2019) Influence of nontoxic magnetic cellulose nanofibers on chitosan based edible nanocoating: a candidate for improved mechanical, thermal, optical, and texture properties. J Agric Food Chem 67(15):4289–4299

Ghosh T, Kohei N, Katiyar V (2021) Curcumin doped functionalized cellulose nanofibers based edible chitosan coating on kiwifruits. Int J Biol Macromol 184:936–945

Gomes MDS, Maria DGC, Ana CGG, Adriana CG et al (2017) Effect of edible coatings with essential oils on the quality of red raspberries over shelf-life. J Sci Food Agric 97(3):929–938

Guerreiro AC, Custódia MLG, Maria LF, Maria GCM et al (2015a) The use of polysaccharide-based edible coatings enriched with essential oils to improve shelf-life of strawberries. Postharvest Biol Technol 110:51–60

Guerreiro AC, Custódia MLG, Maria LF, Maria GCM et al (2015b) The effect of alginate-based edible coatings enriched with essential oils constituents on Arbutus unedo L. fresh fruit storage. Postharvest Biol Technol 100:226–233

Guimaraes A, Luís A, Lorenzo MP, Miguel AC (2018) Edible films and coatings as carriers of living microorganisms: a new strategy towards biopreservation and healthier foods. Compr Rev Food Sci Food Saf 17(3):594–614

Gupta V, Meena NK, Sharma YK, Choudhary K (2023) Comparative study of different polysaccharide-based edible coatings on physicochemical attributes and bioactive compounds of mango cv. Dashehari fruits. eFood 4(1):e55

Gutiérrez-Pacheco MM, Luis AOR, Brenda ASE, Manuel RCV et al (2020) Individual and combined coatings of chitosan and carnauba wax with oregano essential oil to avoid water loss and microbial decay of fresh cucumber. Coatings 10(7):614

Hassani F, Farzaneh G, Majid J (2012) Edible coating based on whey protein concentrate-rice bran oil to maintain the physical and chemical properties of the kiwifruit (Actinidia deliciosa). Trakia Journal of Sciences 10(1):26–34

Hauzoukim SS, Mohanty B (2020) Functionality of protein-based edible coating. J Entomol Zool Stud 8(4):1432–1440

Hong SI, Krochta JM (2006) Oxygen barrier performance of whey-protein-coated plastic films as affected by temperature, relative humidity, base film and protein type. J Food Eng 77(3): 739–745

Isturiz-Zapata MA, Hernandez-Lopez M, Correa-Pacheco ZN, Barrera-Necha LL (2020) Quality of cold-stored cucumber as affected by nanostructured coatings of chitosan with cinnamon essential oil and cinnamaldehyde. LWT- Food Sci Technol 123:109089. https://doi.org/10.1016/j.lwt.2020.109089

Jafarzadeh S, Abdorreza MN, Ali S, Nazila OA et al (2021) Application of bio-nanocomposite films and edible coatings for extending the shelf life of fresh fruits and vegetables. Adv Colloid Interface Sci 291:102405

Jiao W, Chang Shu XL, Cao J, Fan X et al (2019) Preparation of a chitosan-chlorogenic acid conjugate and its application as edible coating in postharvest preservation of peach fruit. Postharvest Biol Technol 154:129–136

Khaliq G, Mahmud TMM, Ghazali HM, Phebe D et al (2016) Influence of gum arabic coating enriched with calcium chloride on physiological, biochemical and quality responses of mango (Mangifera indica L.) fruit stored under low temperature stress. Postharvest Biol Technol 111: 362–369

Khodaei D, Hamidi-Esfahani Z (2019) Influence of bioactive edible coatings loaded with Lactobacillus plantarum on physicochemical properties of fresh strawberries. Postharvest Biol Technol 156:110944

Khorram F, Asghar R, Seyyed MHH (2017) Shellac, gelatin and Persian gum as alternative coating for orange fruit. Sci Hortic 225:22–28

Kumar N (2019) Polysaccharide-based component and their relevance in edible film/coating: a review. Nutr Food Sci 49(5):793–823

Kumar P, Sethi S, Sharma RR, Singh S et al (2018) Improving the shelf life of fresh-cut 'Royal Delicious' apple with edible coatings and anti-browning agents. J Food Sci Technol 55:3767–3778

Kumar N, Ojha A, Upadhyay A, Singh R et al (2021) Effect of active chitosan-pullulan composite edible coating enrich with pomegranate peel extract on the storage quality of green bell pepper. LWT 138:110435

Kumar L, Dakuri R, Konala A, Kirtiraj KG (2022) Edible films and coatings for food packaging applications: a review. Environ Chem Lett 20:1–26

Kumar N, Upadhyay A, Shukla S, Bajpai VK et al (2024) Next generation edible nanoformulations for improving post-harvest shelf-life of citrus fruits. J Food Meas Charact 18(3):1825–1856

Lara I, Garcıa P, Vendrell M (2004) Modifications in cell wall composition after cold storage of calcium-treated strawberry (Fragaria× Ananassa Duch.) fruit. Postharvest Biol Technol 34:331–339

Lerdthanangkul S, Krochta JM (1996) Edible coating effects on postharvest quality of green bell peppers. J Food Sci 61(1):176–179

Mahajan PV, Oluwafemi JC, Singh Z, Christopher BW et al (2014) Postharvest treatments of fresh produce. Philos Trans R Soc A Math Phys Eng Sci 372:20130309

Manisha CM, Alemmoa RJ, Nikhil A, Takhellambam H et al (2021) Edible coatings in fruits and vegetables: a brief review. Pharm Innov J SP-10(7):71–78

Mannozzi C, Urszula T, Fabio C, Lorenzo S et al (2018) Effects of chitosan-based coatings enriched with procyanidin by-product on quality of fresh blueberries during storage. Food Chem 251:18–24

Martı́n-Belloso O, Alejandra Rojas-Graü M, Soliva-Fortuny R (2009) Delivery of flavor and active ingredients using edible films and coatings. Edible Films and Coatings for Food Applications, pp 295–313

Mendy TK, Misran A, Mahmud TMM, Ismail SI (2019) Application of Aloe vera coating delays ripening and extend the shelf life of papaya fruit. Sci Hortic 246:769–776

Milani JM, Nemati A (2022) Lipid-based edible films and coatings: a review of recent advances and applications. J Pack Technol Res 6(1):11–22

Miteluț AC, Elisabeta EP, Mihaela CD, Paul AP et al (2021) Latest developments in edible coatings on minimally processed fruits and vegetables: a review. Food Secur 10(11):2821

Mohammadi M, Somayeh R, Abbas R (2024) Enhancing shelf-life and quality of Mexican lime (Citrus aurantifolia cv.) fruit: utilizing edible coating from wild sage seeds enriched with pomegranate seed oils. J Food Meas Charact 18(1):331–344

Moreno MA, Ana MV, Ana-Rosa B, Catiana Z et al (2020) Antifungal edible coatings containing Argentinian propolis extract and their application in raspberries. Food Hydrocoll 107:105973

Muley AB, Singhal RS (2020) Extension of postharvest shelf life of strawberries (Fragaria ananassa) using a coating of chitosan-whey protein isolate conjugate. Food Chem 329:127213

Nair MS, Saxena A, Kaur C (2018) Characterization and antifungal activity of pomegranate peel extract and its use in polysaccharide-based edible coatings to extend the shelf-life of capsicum (Capsicum annuum L.). Food Bioproc Tech 11:1317–1327

Nair MS, Tomar M, Punia S, Kukula-Koch W et al (2020) Enhancing the functionality of chitosan- and alginate-based active edible coatings/films for the preservation of fruits and vegetables: a review. Int J Biol Macromol 164:304–320

Ojeda GA, Arias Gorman AM, Sgroppo SC, Zaritzky NE (2021) Application of composite cassava starch/chitosan edible coating to extend the shelf life of black mulberries. J Food Process Preserv 45(1):e15073

Oregel-Zamudio E, Angoa-Pérez MV, Oyoque-Salcedo G, Aguilar-González CN, Mena-Violante HG et al (2017) Effect of candelilla wax edible coatings combined with biocontrol bacteria on strawberry quality during the shelf-life. Sci Hortic 214:273–279

Oyom W, Hengping X, Zhitian L, Haitao L et al (2022) Effects of modified sweet potato starch edible coating incorporated with cumin essential oil on storage quality of 'early crisp'. Lwt 153: 112475

Pandey VK, Rayees UI, Rafeeya S, Dar AH (2022) A comprehensive review on the application of essential oils as bioactive compounds in Nano-emulsion based edible coatings of fruits and vegetables. Appl Food Res 2(1):100042

Panwar A, Kumar V, Dhiman A, Thakur P et al (2024) Nanoemulsion based edible coatings for quality retention of fruits and vegetables-decoding the basics and advancements in last decade. Environ Res 240:117450

Parafati L, Alessandro V, Cristina R, Gabriella C (2016) The effect of locust bean gum (LBG)-based edible coatings carrying biocontrol yeasts against Penicillium digitatum and Penicillium italicum causal agents of postharvest decay of mandarin fruit. Food Microbiol 58:87–94

Park HJ, Jong WR, Hong YL (1996) Edible coating effects on respiration rate and storage life of 'Fuji' apples and 'Shingo' pears. Food Sci Biotechnol 5(1):59–63

Passafiume R, Raimondo G, Giuseppe S, Vittorio F (2020) Effect of three different aloe vera gel-based edible coatings on the quality of fresh-cut Hayward kiwifruits. Food Secur 9(7):939

Pereira GVDS, Gleice VDSP, Luã CDO, Dilson Nazareno PC et al (2021) Rheological characterization and influence of different biodegradable and edible coatings on postharvest quality of guava. J Food Process Preserv 45(4):e15335

Pinela J, Barreira JCM, Barros L, Verde SC (2016) Suitability of gamma irradiation for preserving fresh-cut watercress quality during cold storage. Food Chem 206:50–58

Prakash A, Revathy B, Vellingeri V (2020) Citral nanoemulsion incorporated edible coating to extend the shelf life of fresh cut pineapples. Lwt 118:108851

Priyadarshi R, Zohreh R, Rhim JW (2022) Antioxidant pectin/pullulan edible coating incorporated with Vitis vinifera grape seed extract for extending the shelf life of peanuts. Postharvest Biol Technol 183:111740

Qiaobin H, Fang Z, Ngoc KL, Jerryck O (2023) Development of multifunctional nanoencapsulated trans-resveratrol/chitosan nutraceutical edible coating for strawberry preservation. ACS Nano 17:8586–8597

Radi M, Soraya S, Mehran S, Shima K et al (2023) Effect of Thymus vulgaris essential oil-loaded nanostructured lipid carriers in alginate-based edible coating on the postharvest quality of tangerine fruit. Food Bioproc Tech 16(1):185–198

Rasouli M, Mahmoud KS, Ramezanian A (2019) Inhibitory effect of salicylic acid and Aloe vera gel edible coating on microbial load and chilling injury of orange fruit. Sci Hortic 247:27–34

Riva SC, Umezuruike OO, Olaniyi AF (2020) Recent developments on postharvest application of edible coatings on stone fruit: a review. Sci Hortic 262:109074

Saberi B, Golding JB, Marques JR, Pristijono P, Chockchaisawasdee S, Scarlett CJ, Stathopoulos CE et al (2018) Application of biocomposite edible coatings based on pea starch and guar gum on quality, storability and shelf life of Valencia oranges. Postharvest Biol Technol 137:9–20

Salama HE, Mohamed SAA (2020) Optimized alginate and Aloe vera gel edible coating reinforced with nTiO2 for the shelf-life extension of tomatoes. Int J Biol Macromol 165:2693–2701

Sanuja S, Agalya A, Umapathy MA (2015) Synthesis and characterization of zinc oxide–neem oil–chitosan bionanocomposite for food packaging application. Int J Biol Macromol 74:76–84

Sharma S, Kohei N, Kumar S, Katiyar V (2024) Edible packaging to prolong postharvest shelf-life of fruits and vegetables: a review. Food chemistry. Advances 100711:100711

Singh DP, Packirisamy G (2022) Biopolymer based edible coating for enhancing the shelf life of horticulture products. Food Chem Mol Sci 4:100085

Sivakumar D, Van Deventer F, Terry LA, Polenta GA et al (2012) Combination of 1-methylcyclopropene treatment and controlled atmosphere storage retains overall fruit quality and bioactive compounds in mango. J Sci Food Agric 2:821–830

Sogut E, Ili Balqis AM, Nur Hanani ZA, Can SA (2019) The properties of κ-carrageenan and whey protein isolate blended films containing pomegranate seed oil. Polym Test 77:105886

Soleimanian Y, Hossein Goli SA, Varshosaz J, Francesca Maestrelli F et al (2018) Propolis wax nanostructured lipid carrier for delivery of β sitosterol: effect of formulation variables on physicochemical properties. Food Chem 260:97–105

Soto-Muñoz L, Martínez-Blay V, Pérez-Gago MB, Fernández-Catalán A, Argente-Sanchis M, Palou L et al (2022) Starch-based antifungal edible coatings to control sour rot caused by Geotrichum citri-aurantii and maintain postharvest quality of Fino lemon. J Sci Food Agric 102(2):794–800

Sun X, Qinglin W, David HP, Mary HF et al (2021) Comparative performance of bio-based coatings formulated with cellulose, chitin, and chitosan nanomaterials suitable for fruit preservation. Carbohydr Polym 259:117764

Tabassum N, Mohammad AK (2020) Modified atmosphere packaging of fresh-cut papaya using alginate based edible coating: quality evaluation and shelf-life study. Sci Hortic 259:108853

Tanada-Palmu PS, Carlos RFG (2005) Effect of edible wheat gluten-based films and coatings on refrigerated strawberry (Fragaria ananassa) quality. Postharvest Biol Technol 36(2):199–208

Teodosio AEMM, Railene HCRA, Bárbara GFLS, Jéssica AL et al (2021) Effects of edible coatings of Chlorella sp. containing pomegranate seed oil on quality of Spondias tuberosa fruit during cold storage. Food Chem 338:127916

Tokatlı K, Aslıhan D (2021) Influences of chitosan coatings on functional compounds of sweet cherries. J Food Sci Technol 58(5):1808–1818

Vilaplana R, Katherine G, Jessica G, Valencia-Chamorro S (2020) Chitosan coatings to control soft mold on fresh blackberries (Rubus glaucus Benth.) during postharvest period. *Sci Hortic* 262: 109049

Villafañe F (2017) Edible coatings for carrots. Food Rev Intl 33(1):84–103

Vishwasrao C, Ananthanarayan L (2017) Delayed post-harvest ripening-associated changes in Manilkara zapota L. var. Kalipatti with composite edible coating. J Sci Food Agric 97(2): 536–542

Wani SM, Amir G, Tehmeena A, Malik AR et al (2021) Effect of gum Arabic, xanthan and carrageenan coatings containing antimicrobial agent on postharvest quality of strawberry: assessing the physicochemical, enzyme activity and bioactive properties. Int J Biol Macromol 183:2100–2108

Yadav A, Kumar N, Upadhyay A, Sethi S et al (2022) Edible coating as postharvest management strategy for shelf-life extension of fresh tomato (*Solanum lycopersicum* L.): an overview. J Food Sci 87(6):2256–2290

Yu D, Indra BB, Kumar S, Fei Y et al (2023) Chitosan modified with bio-extract as an antibacterial coating with UV filtering feature. Int J Biol Macromol 230:123145

Zhang L, Chen F, Lai S, Wang H et al (2018) Impact of soybean protein isolate-chitosan edible coating on the softening of apricot fruit during storage. Lwt 96:604–611

Zhang X, Xiuling Z, Xiaochen L, Meiling D et al (2019) Effect of polysaccharide derived from Osmunda japonica Thunb-incorporated carboxymethyl cellulose coatings on preservation of tomatoes. J Food Process Preserv 43(12):e14239

Zhiqiang Zhou JM, Kun L, Wenwen Z, Kai L et al (2021) A plant leaf-mimetic membrane with controllable gas permeation for efficient preservation of perishable products. ACS Nano 15: 8742–8752

Zillo RR, Paula PMDS, Jacqueline DO, Eduardo MDG et al (2018) Carboxymethylcellulose coating associated with essential oil can increase papaya shelf life. Sci Hortic 239:70–77

Chapter 14
Nanotechnology in Horticulture Crops: Current Trends and Future Prospects

Avnesh Kumari and Sudesh Kumar Yadav

Abstract In horticulture crops, nanotechnology holds enormous promise. Horticultural crops' production, productivity, and shelf life can all be improved using nanotechnology. There is more potential for nanotechnology to extend the life and shelf life of horticulture crops. Packaging materials made from nanomaterials (NMs) help maintain the quality of horticulture produce. Additionally, the NMs-based packaging inhibits the growth of microorganisms in packed horticulture produce. Hence nanoparticles (NPs) have a great deal of promise for addressing and resolving post-harvest issues in horticulture crops.

14.1 Introduction

The synthesis of nanoparticles (NPs) for use in a variety of industries, including food (Kumari et al. 2014; Frewer et al. 2014), agriculture (Kumari et al. 2023), pharmaceuticals (Kumari et al. 2010), cosmetics (Raj et al. 2012), and more, is the focus of a new field of study called nanotechnology. NPs have a size in the nanoscale range and

A. Kumari (✉)
Academy of Scientific and Innovative Research (AcSIR), Academy of Scientific and Innovative Research (AcSIR) CSIR- Human Resource Development Centre, Ghaziabad, Uttar Pradesh, India

Present Address: Biotechnology Division, CSIR-Institute of Himalayan Bioresource Technology, Palampur, HP, India
e-mail: avnesh@ihbt.res.in

S. K. Yadav (✉)
Biotechnology Division, CSIR-Institute of Himalayan Bioresource Technology, Palampur, HP, India

Academy of Scientific and Innovative Research (AcSIR), Academy of Scientific and Innovative Research (AcSIR) CSIR- Human Resource Development Centre, Ghaziabad, Uttar Pradesh, India

Center of Innovative and Applied Bioprocessing (CIAB), Mohali, India
e-mail: sudesh@ciab.res.in; director@ihbt.res.in

S. K. Yadav, A. Kumari (eds.), *Nanobiotechnology in Agri-food Sector*, Sustainability Sciences in Asia and Africa, https://doi.org/10.1007/978-981-96-7756-6_14

their physiochemical properties differ from their counterparts, depending on the size and morphology (Mohanraj and Chen 2006; Stadler et al. 2018). Among all types of NPs, metal and metal oxide NPs have wide applications in biology, medicine, and engineering (Zhao et al. 2016). Many physical and chemical methods have been employed for the synthesis of metal and metal oxide NPs (Wang et al. 2007; Horwat et al. 2011). However, these conventional methods require a large amount of energy, release harmful and toxic chemicals, and involve expensive and complex equipment (Alsammarraie et al. 2018). Owing to this, these conventional methods have been replaced by green synthesis methods. Green synthesis methods use plant extracts and microbes to synthesize NPs (Ahmed et al. 2015; Mohanpuria et al. 2008). Till now, many metal and metal oxide NPs have been synthesized using extracts of different parts of plants (Devi et al. 2019; Chahardoli et al. 2018; Leili et al. 2018; Thovhogi et al. 2016; Sone et al. 2020) owing to their widespread accessibility, renewability, and environment-friendly nature (Vanlalveni et al. 2021).

Horticulture is a branch of science that deals with the cultivation and management of fruits, vegetables, flowers, and ornamental plants. Enhancing productivity and decreasing post-harvest losses by using nanotechnology approaches are considered the best strategies for ensuring food and nutritional security (Rana et al. 2021). The use of NPs in agriculture appears to be crucial for improving produce quality and production as well as lowering post-harvest losses in fruits and vegetables. According to published reports, microbiological and physiological processes can cause up to 30% of horticultural agricultural products to be lost. Using nanomaterials, we can significantly lower post-harvest losses to 5–10%, which would ultimately conserve a significant volume of fruits and vegetables (Rana et al. 2021).

The manufacture of metal and metal oxide nanoparticles has also been done using various plant parts, and ornamental plants have demonstrated good potential in this regard. Different types of metal NPs have also been used for improving the shelf life of ornamentals. In this chapter, we have tried to cover the latest developments in applications of nanotechnology in horticulture crops with special emphasis on ornamentals. We have covered the role of nanotechnology in enhancing the productivity, quality, and post-harvest losses of horticulture crops; explored various types of NPs and their applications in horticulture; discussed the use of nanotechnology in controlling fungal growth in horticulture crops; and examined nanomaterials (NMs)-based packaging solutions for horticulture crops.

14.2 Nanotechnology in Enhancing the Productivity, Quality, and Post-harvest Losses of Horticulture Crops

Crop productivity of horticulture crops has been enhanced by the use of nanofertilizers. Nanofertilizers are classified into two types based on the nutrition they provide to the crops, viz., macronutrient nanofertilizers and micronutrient nanofertilizers. Macronutrient nanofertilizers contain N, P, and K in encapsulated form within the NPs. Macronutrient nanofertilizers enhance the efficiency of N, P, and K and hence increase crop productivity. Micronutrient nanofertilizers such as copper (Cu), zinc (Zn), and iron (Fe) have shown beneficial effects on many horticultural crops, and they have been listed in Table 14.1.

Phosphorus, magnesium, sulfur, calcium, nitrogen, and potassium are important macronutrients that can be combined with NPs to provide plants with the maximum quantity of nutrients while lowering cultivation costs (Ditta and Arshad 2016). Numerous rulings have demonstrated the beneficial effects of macronanofertilizers on pollination, plant growth, and fruit yield and quality. These fertilizers have a variety of applications, including foliar spraying, injection, and soil application (Sharma et al. 2021). Numerous results have demonstrated the positive impact of macronanofertilizers on pollination, plant development, and fruit yield and quality. These fertilizers have a variety of applications, including soil application, foliar spraying, and injection (Sharma et al. 2021). In addition to an improvement in physicochemical properties, *Punica granatum* cv. Ardestani treated with nano-nitrogen by foliar application demonstrated a considerable increase in fruit output (17–44%) and the quantity of fruits produced per plant (15–38%), according to Davarpanah et al. (2017).

Furthermore, *Vitis vinifera* treated with nanosized calcite products showed decreased acidity, increased volume and length, total soluble solids, cluster number, and berry weight (Zahedi et al. 2020). A number of issues with conventional fertilizers, including as their cost-effectiveness and environmental impact, may be resolved by the creation of macro-nanofertilizers.

The development and yield of horticultural crops are improved by micronutrient nanofertilizers. The trace elements zinc, iron, boron, and copper found in these nanofertilizers are essential for a variety of plant physiological and metabolic functions (Ahmed et al. 2023).

14.3 Types and Possible Uses of Nanoparticles in Horticultural Crops

A variety of forms can be attributed to NPs according to their size, shape, chemical, biological, and physical characteristics. Moreover, quantum dots, nanoballs, nanorods, nanosheets, nanotubes, nanofibers, and aerogels are a few frequent

Table 14.1 Effect of micronutrients on various horticulture crops

Crops	Micronutrients encapsulated in NPs	Conc. (mg/ml)	Positive/negative impact on crops	References
Cucumber	Zn	1000	Inhibition of growth and deformation of the root tip	Zhao et al. (2014)
	Fe	50–2000	Increased biomass yield and activity of antioxidant enzymes	Moghaddas et al. (2017)
	Cu	1000	Growth reduction and enhances the activity of antioxidant enzymes	Kim et al. (2012)
Tomato	Zn	1	Decreases fungal growth	Khan et al. (2018)
	Cu	50–500	Enhances the firmness of fruits and antioxidant content	Ahmed et al. (2018a, b)
Garden pea	Zn	500	Reduces chlorophyll and hydrogen peroxide content	Nair et al. (2015)
	Fe	30–60	Enhances seed mass and chlorophyll content	Giorgetti et al. (2019)
	Cu	100–500	Reduces growth of plants and ROS production	Tripathi et al. (2017)
Lettuce	Fe	10–20	Decreased chlorophyll content and increased the activity of antioxidant enzyme	Trujillo-Reyes et al. (2014)
	Cu	1000	Growth reduction and enhancement of antioxidant activity of enzymes	Trujillo-Reyes et al. (2014)
Chili pepper	Zn	100–500	Enhances germination rate	Tantawy et al. (2015)
Coriander	Zn	400	Enhances pigment content	Pullagurala et al. (2018)
Onion	Zn	5–20	Decreases root growth	Ahmed et al. (2018a, b)
Squash	Cu	0–500	Increases concentration of ionic copper in media	Musante et al. (2012)
Radish	Cu	10–1000	Growth inhibition and DNA damage	Atha et al. (2012)
Bean	Cu	100–500	Causes nutrition imbalance	Alsaeedi et al. (2017)

categories of NPs (Hassan 2013; Nasrollahzadeh et al. 2019). The physical and chemical characteristics of NMs can be used to categorize them. These numerous forms and varieties of NPs have special qualities and are used in a variety of industries, including fruit crops, electronics, medicine, and agriculture.

14.3.1 Metal Nanoparticles

These are metal oxide or metal NPs, like titanium dioxide, silver, copper, zinc oxide, etc. Metal NPs possess antibacterial characteristics and can be employed as nanopesticides to manage plant diseases (Li et al. 2023). To improve nutrient uptake and raise crop productivity, they can also be applied as nanofertilizers (Qureshi et al. 2018). For instance, AgNPs are used as a nanopesticide in horticulture crops to manage plant diseases. AgNPs can be utilized to manage many bacterial and fungal diseases in horticulture crops, including powdery mildew in grapes because they have been demonstrated to have broad-spectrum antibacterial activities (Zhao et al. 2022).

AgNPs can be sprayed foliar on crops or mixed into the soil as a soil amendment or seed treatment (Mehmood 2018). In addition, crop output can be increased by applying zinc oxide NPs as a nanofertilizer to improve nutrient uptake. It has been demonstrated that ZnO NPs enhance crops' uptake of potassium, phosphorus, and nitrogen, increasing biomass and yield (Ahmed et al. 2023).

14.3.2 Polymer-Based Nanomaterials

According to Dutta et al. (2022), polymer-based NPs, which include chitosan, cellulose, and polyethylene glycol, can be employed as nanocarriers. They can also promote increased crop yields, improve nutrient absorption, and function as nanofertilizers (Dutta et al. 2022). Chitosan NPs, a polymer-based NM, have been researched for use in horticulture crops as a nanocarrier of nutrients, growth regulators, and insecticides. Chitosan NPs can be integrated into the soil as a soil amendment or loaded with different bioactive compounds and applied as a foliar spray to crops.

It has been demonstrated that Chitosan NPs improve crop productivity and nutrient uptake in a variety of crops, including tomato (Zhao et al. 2021) and strawberry (Gao et al. 2020; Guo et al. 2018; Ingle et al. 2022). The use of NMs in horticultural techniques has the potential to improve disease management and has long-lasting benefits on ecosystem functions and soil vitality (Dimkpa and Bindraban 2017; Duhan et al. 2017). Therefore, more study is required to fully understand the advantages and disadvantages of using NMs in horticulture as well as to create safe and effective applications.

14.3.3 Nanotechnology in Controlling Fungal Growth in Horticulture Crops

Fungal growths also affect the yield and growth of horticulture crops. Two pathogens, viz., *Botrytis cineria* and *Penicillium expansum*, cause maximum economic losses in horticulture crops. It is difficult to control this fungal growth because they have developed resistance to many traditional fungicides such as benzimidazoles and dicarboximides (Helmy et al. 2020). Zinc oxide NPs have shown better activity against *Penicillium expansum*. The effectiveness of ZnO NPs was evaluated against fungi viz., *Penicillium expansum*, *Alternaria alternata*, *Botrytis cinerea*, and *Rhizopus stolonifer*. ZnO NPs inhibited all the fungi at concentrations >6 mM and also caused morphological abnormalities in the fungal structures (Sardella et al. 2017).

Another study has reported that ZnO NPs have shown antifungal effects against *Aspergillus niger* and *Penicillin digitatum* for pre- and post-harvested fruits (Helmy et al. 2020). Many filamentous fungi produce mycotoxins, which have detrimental effects on human health. ZnO NPs have the ability to block and inhibit mycotoxins (Sun et al. 2018). Hassan et al. found that aflatoxin production was inhibited by the addition of 8 μg/mL of ZnO NPs.

At 10 μg/mL, ZnO NPs also showed inhibitory effects for fumonisin B1 and ochratoxin A. These results demonstrated that ZnO NPs' involvement in the food matrix may effectively prevent the growth of harmful fungi while also considerably reducing the amount of mycotoxins that toxic fungi create (Hassan et al. 2013). Table 14.2 lists the harmful fungi that ZnO NPs have demonstrated effectiveness against.

Table 14.2 List of pathogenic fungi against which ZnO NPs have shown efficacy

Particle size (nm)	Shape	Synthesis method	Fungus strain	References
8–23	Spherical, hexagonal, Rod shaped	Green	*Trichoderma harzianum*	Zaki et al. (2021)
–	Flower shaped	Chemical	*C. albicans*	Hui et al. (2016)
–	–	Microwave-assisted chemical	*Aspergillus niger*	Hassan et al. (2013)
Less than 50 nm	Spherical	Chemical	*Fusarium solani*	Malandrakis et al. (2021)
Less than 100	Spherical	Chemical	*Alternaria alternata*	Malandrakis et al. (2022)
55–85	Rod shaped	–	*Botrytis cinerea*	He et al. (2011)
40	–	Liquid-phase precipitation method	*Candida krusei*	Das et al. (2016)
50–100	–	Biosynthesis	*Trichophyton mentagrophytes*	Tiwari et al. (2017)
55–85	Rod shaped	Chemical	*Penicillium expansum*	Barad et al. (2017)

14.4 The Use of Nanotechnology to Improve Ornamental Production, Quality, Protection, and Post-harvest Losses

14.4.1 Green Synthesis of Nanoparticles Using Ornamentals

The concept of the green synthesis of metal NPs has been explored using many plants. Recently, many ornamentals have also been explored for the synthesis of metallic NPs (Kumar et al. 2009). *Rosmarinus officinalis* extracts (Fierascu et al. 2014) have been utilized for the synthesis of silver NPs. The appearance of a green color on the addition of rosemary extract to $AgNO_3$ aqueous solution led to the formation of colloidal silver NPs. These colors arise due to the excitation of surface plasmon vibrations of metal NPs.

14.4.2 Post-harvest Loss Reduction in Ornamentals

Kamiab et al. have used silver and silica NPs to enhance the shelf life of cut flower Lisianthus (*Eustoma grandiflora* cv. Echo). The treatment of Lisianthus cut flowers with 40 mg L^{-1} of both NPs enhanced the average vase life of cut flowers to 17 days, which was about 5 days in the control. NPs-treated flowers showed increased fresh weight, solution uptake, total chlorophyll, and total dissolved solids, particularly at greater concentrations. The application of silver and silica NPs decreased the proliferation at 40 mg L^{-1}. Overall the results suggested that nano-silver proved more effective for reducing ethylene content, and silicon NPs are applicable as an antimicrobial compound in combination with silver NPs to increase the vase life of Lisianthus flowers commercially (Kamiab et al. 2017). Another study reported the use of silica NPs to improve the shelf life of Lilium flowers. The shelf life of Lilium was extended by the foliar application of silica NPs, which was more successful than the soil treatment at boosting the plants' antioxidant system. The investigation found that the flowers had a longer shelf life, higher levels of Mg, P, and S, more photosynthetic pigments, and higher levels of glutathione peroxidase, phenols, and antioxidant activity. Consequently, using SiO_2 NPs to produce commercially valuable flowers could be a great way to enhance floral quality and extend flower shelf life (Sánchez-Navarro et al. 2021).

Carrillo-López et al. have reported the use of biosynthesized silver NPs for enhancing the vase life of Chrysanthemum cv. Puma. The study reported that a low concentration of silver NPs could enhance the inflorescence opening and augment the vase life of Chrysanthemum cv. Puma (Carrillo-López et al. 2016). Gupta et al. have reported that the shelf life of gladiolus was enhanced in the presence of sucrose-stabilized silver NPs (Gupta et al. 2022). The use of silver NPs to oriental lilies increased floral quantity, leaf greenness index, and biomass of leaves and bulbs. Furthermore, the AgNP-treated plants had longer flower

Table 14.3 Different concentrations of silver nanoparticles that have been used for improving the vase life of cut flowers

Concentration of AgNPs (mg/ml)	Cut flower	Application mode	Effects of treatment	References
5	Mokara Red orchid flower	Pulse	Enhanced vase life	Rahman et al. (2019)
5	Carnation	Solution	Inhibited ethylene synthesis	Koohkan et al. (2014)
10	Gerbera	Pulse	Water content and hydraulic conductance was retained	Motaghayer et al. (2019)
40	Lisianthus	Vase	Enhanced vase life and quality	Kamiab et al. (2017)
10	Chrysanthemum	Pulse	Reduced water stress	Kazemipour et al. (2013)

lifespans and started their anthesis sooner (Salachna et al. 2019). Another study also documented that the application of silver NPs on lily plants can contribute to evolving novel approaches to bulbous plant production and a real tactic to protect plants from ever-increasing land salinization (Byczyńska et al. 2023).

Cut flowers have two issues: vase life and quality. They also don't have a long shelf life after harvest because of sporadic water inflows brought on by microbial growth. The application of NPs can solve these issues. Applying silver nanoparticles to cut flowers prolonged their vase life by preventing bacterial growth and limiting ethylene generation by penetrating the stem's vascular tissues. Table 14.3 lists the various concentrations of silver nanoparticles that have been utilized in numerous experiments to extend the vase life of cut flowers.

14.4.3 Improvement of the Vase Life of Cut Flowers

Water absorption is positively impacted by nanotechnology because of the bactericidal properties of NPs (Liu et al. 2012). Reported an increase in vase life and demonstrated how applying 200 mg^{-1} of nanosilver to cut flowers of the cultivars Fiesta and Avalanche inhibited bacterial development and increased water uptake. Because of their special properties and large surface area, silver NPs have a stronger inhibitory effect on bacteria and other microorganisms in a variety of conditions. NPs have antimicrobial qualities, which make them useful in both medicine and water purification. According to Liu et al. (2012), when used in conjunction with sugar rather than alone, nano-silver particles produced the best results and extended the vases' life. When sugar and nano-silver are combined, the sugar serves as an energy source, and the nano-silver particles are responsible for retaining water. Kalayeh et al. also discovered that treating roses with nano-silver particles had improved results in terms of their vase life. It has been discovered that using nano-silver particles in conjunction with 8-hydroxyquinoline sulfate effectively increases the floral diameter and bud openings (Kalayeh et al. 2011).

To extend the vase life of different cut flowers, such as sweet peas (Elhindi 2012), gerberas (Solgi et al. 2009), roses (Abri 2013), lilies (Kazemi and Amari 2012), and carnations (Baker et al. 1985; Rahman et al. 2012), researchers found it fascinating to investigate the use of NPs in this area.

14.5 Nanomaterials Used in the Packaging of Horticulture Crops

In order to maintain the quality of horticultural cargo while pursuing economical and environmentally friendly packaging solutions, NMs are used in the storage and packing of horticultural commodities. According to Tissera et al. (2023), NMs are often inserted into polymer matrices, and the resulting polymer nanocomposites are utilized in packaging applications. Better-quality packaging, active packaging, and smart packaging are the three main categories. NMs in a polymer matrix enhance the gas barrier and mechanical characteristics of better packaging. Antimicrobials and other dynamic ingredients are included in the active packaging to improve the food product's safety, quality, and shelf life. Intelligent packaging involves nanosensors that can interact with internal and external environmental factors to sense the quality of the horticultural crops through their unique chemical and electro-optical properties (Tissera et al. 2023). Moreover, the ripening stage can also be monitored using sensors such as ethylene gas indicators, an intelligent packaging method. Nanosensors ensure that consumers buy fresh and tasty horticultural crops and diminish the occurrence of microorganisms (Emamhadi et al. 2020). The use of NMs in packaging can reduce the growth of many microorganisms, viz., *Lactobacillus plantarum*; *Escherichia coli*, and *Staphylococcus aureus*; Salmonella; *Alicyclobacillus acidoterrestris*; and aerobic psychrotrophic, molds, and yeasts (Emamhadi et al. 2020).

In this case, fresh orange juice's shelf life was extended by the application of antimicrobial nanocomposite materials like Ag and ZnO. High-performance, lightweight NMs present a fantastic opportunity to replace non-biodegradable plastic packaging materials in nano-packaging (Bahrami et al. 2020).

14.6 Conclusions

Nanotechnology holds significant potential in horticulture, where various NMs are employed to improve production quality and productivity while lowering fruit and vegetable deterioration after harvest. Utilizing the power of NPs and their distribution techniques, nanotechnology increases horticulture crop productivity. They lessen the excessive application of chemical pesticides and fertilizers. Because of their simplicity, affordability, and environmental friendliness, NMs can be created quickly, with minimal effort, and without endangering the environment. Nanotechnology helps to improve the quality and shelf life of both unprocessed and processed horticulture fruits, vegetables, and cut flowers.

Nanosensors are employed in crop pest diagnosis, nutrient needs analysis, pesticide residue detection, and soil moisture monitoring. However, the ecosystem and crop plants will suffer if NMs are used improperly. Therefore, correct usage of NMs based on research is required to increase the quality of food production and reduce post-harvest spoiling of horticulture products.

The availability of workable NPs and safety assessments of their field deployment are critical to ensuring food and nutritional security for the world's expanding population in a changing climate environment. Climate-smart horticulture, reduced post-harvest losses, and improved crop quality overall will come from increased usage of nanotechnology.

Acknowledgments Authors would like to acknowledge the Director, CSIR-IHBT, for providing infrastructure.

The CSIR-IHBT communication number of this chapter is 5726.

References

Abri FM, Ghasemnezhad RH, Bakhshi D (2013) Effect of ascorbic acid on vase life and petal senescence in cut rose flowers (Rosa hybrida) cv. 'Royal Class'. Amer-Eur J Agric Environ Sci 13(1):38–43

Ahmed S, Saiqa I (2015) Silver nanoparticles: one pot green synthesis using Terminalia arjuna extract for biological application. J Nanomed Nanotechnol 6(4:309

Ahmed B, Mohammad SK, Javed M (2018a) Toxicity assessment of metal oxide nano-pollutants on tomato (Solanum lycopersicon): a study on growth dynamics and plant cell death. Environ Pollut 240:802–816

Ahmed B, Mohammad S, Mohammad SK, Musarrat J (2018b) Chromosomal aberrations, cell suppression and oxidative stress generation induced by metal oxide nanoparticles in onion (*Allium cepa*) bulb. Metallomics 10(9):1315–1327

Ahmed R, Kamal Uddin M, Abdul Quddus M, Mohd Yusoff AS, Motalib Hossain MA, Ashfaqul Haque AN (2023) Impact of foliar application of zinc and zinc oxide nanoparticles on growth, yield, nutrient uptake and quality of tomato. Horticulturae 9(2):162

Alsaeedi AH, Hassan ER, Tarek A, Mohamed EG, Nevien E, Mahdi A (2017) Engineered silica nanoparticles alleviate the detrimental effects of Na+ stress on germination and growth of common bean (Phaseolus vulgaris). Environ Sci Pollut Res 24:21917–21928

Alsammarraie FK, Wei W, Peng Z, Azlin M, Mengshi L (2018) Green synthesis of silver nanoparticles using turmeric extracts and investigation of their antibacterial activities. Colloids Surf B: Biointerfaces 171:398–405

Atha DH, Huanhua W, Elijah JP, Danielle C, Holbrook RD, Pawel J, Dizdaroglu M, Xing B, Nelson BC (2012) Copper oxide nanoparticle mediated DNA damage in terrestrial plant models. Environ Sci Technol 46(3):1819–1827

Bahrami A, Rana D, Elham A, Seid MJ, Leonard W (2020) Antimicrobial-loaded nanocarriers for food packaging applications. Adv Colloid Interf Sci 278:102140

Baker JE, Chien YW, Daniel ET (1985) Delay of senescence in carnations by pyrazon, phenidone analogues, and tiron. HortScience 20:121–122

Barad SM, Roudbary ANO, Daryasari MP (2017) Preparation and characterization of ZnO nanoparticles coated by chitosan-linoleic acid, fungal growth and biofilm assay. Bratislavske Lekarske Listy 118(3):169–174

Byczyńska A, Agnieszka Z, Piotr S (2023) Colloidal silver nanoparticles enhance bulb yield and alleviate the adverse effect of saline stress on lily plants. J Ecol Eng 24(6)

Carrillo-López LM, Morgado-González A (2016) Biosynthesized silver nanoparticles used in preservative solutions for Chrysanthemum cv. Puma. J Nanomater 1:1769250

Chahardoli A, Karimi N, Ali F (2018) Nigella arvensis leaf extract mediated green synthesis of silver nanoparticles: their characteristic properties and biological efficacy. Adv Powder Technol 29(1):202–210

Das B, Khan MI, Jayabalan R, Behera SK, Yun SI, Tripathy SK, Mishra A (2016) Understanding the antifungal mechanism of Ag@ ZnO core-shell nanocomposites against Candida krusei. Sci Rep 6(1):36403

Davarpanah S, Ali T, Gholamhossein D, Mehdi A, Javier A, Reza K (2017) Effects of foliar nano-nitrogen and urea fertilizers on the physical and chemical properties of pomegranate (Punica granatum cv. Ardestani) fruits. HortScience 52(2):288–294

Devi HS, Muzaffar AB, Mohammad AS, Shazia P, Wani AH (2019) Green synthesis of iron oxide nanoparticles using Platanus orientalis leaf extract for antifungal activity. Green Proc Synth 8(1):38–45

Dimkpa CO, Bindraban PS (2017) Nanofertilizers: new products for the industry? J Agric Food Chem 66(26):6462–6473

Ditta A, Muhammad A (2016) Applications and perspectives of using nanomaterials for sustainable plant nutrition. Nanotechnol Rev 5(2):209–229

Duhan JS, Kumar R, Kumar N, Kaur P, Nehra K, Duhan S (2017) Nanotechnology: the new perspective in precision agriculture. Biotechnol Report 15:11–23

Dutta S, Pal S, Panwar P, Sharma RK, Bhutia PL (2022) Biopolymeric nanocarriers for nutrient delivery and crop biofortification. ACS omega 7(30):25909–25920

Elhindi KM (2012) Effects of postharvest pretreatments and preservative solutions on vase life longevity and flower quality of sweet pea (Lathyrus odoratus L.). Photosynthetica 50:371–379

Emamhadi MA, Mansour S, Mitra A, Yadolah F, Nguyen TTL, Amin MK (2020) Nanomaterials for food packaging applications: a systematic review. Food Chem Toxicol 146:111825

Fierascu RC, Ioana RB, Raluca S, Irina F, Rodica MI (2014) Characterization of silver nanoparticles obtained by using Rosmarinus officinalis extract and their antioxidant activity. Rev Roum Chim 59(3–4):213–218

Frewer LJ, Gupta N, George S, Fischer ARH, Giles EL, Coles D (2014) Consumer attitudes towards nanotechnologies applied to food production. Trends Food Sci Technol 40(2):211–225

Gao Y, Dan X, Dan R, Kaifang Z, Xiyu W (2020) Green synthesis of zinc oxide nanoparticles using Citrus sinensis peel extract and application to strawberry preservation: a comparison study. Lwt 126:109297

Giorgetti L, Carmelina S, Simonetta M, Lorenza B, Eliana T, Stefania B, Simona DG, Giovanna S, Luigi SDT, Monica RC (2019) An integrated approach to highlight biological responses of Pisum sativum root to nano-TiO2 exposure in a biosolid-amended agricultural soil. Sci Total Environ 650:2705–2716

Guo H, Jason CW, Zhenyu W, Baoshan X (2018) Nano-enabled fertilizers to control the release and use efficiency of nutrients. Curr Opin Environ Sci Health 6:77–83

Gupta Y, Biswas L, Kathpalia R (2022) Exploiting the potential of bio-synthesized silver nanoparticles to enhance the shelf life of Gladiolus. Indian J Biochem Biophys 59:486–490

Hassan AA, Howayda ME, Mahmoud HH (2013) Effect of zinc oxide nanoparticles on the growth of some mycotoxigenicmould. Stud Chem Process Technol 1(4):66–74

He L, Yang L, Azlin M, Mengshi L (2011) Antifungal activity of zinc oxide nanoparticles against Botrytis cinerea and Penicillium expansum. Microbiol Res 166(3):207–215

Helmy KG, Abir MP, Salah M (2020) Gamma radiation and polyvinyl pyrrolidone mediated synthesis of zinc oxide/zinc sulfide nanoparticles and evaluation of their antifungal effect on pre and post harvested orange and pomegranate fruits. Biocatal Agric Biotechnol 29:101728

Horwat D, Zakharov DI, Endrino JL, Soldera F, Anders A, Migot S, Karoum R, Vernoux P, Pierson JF (2011) Chemistry, phase formation, and catalytic activity of thin palladium-containing oxide films synthesized by plasma-assisted physical vapor deposition. Surf Coat Technol 205:S171–S177

Hui A, Junli L, Jianzhong M (2016) Synthesis and morphology-dependent antimicrobial activity of cerium doped flower-shaped ZnO crystallites under visible light irradiation. Colloids Surf A Physicochem Eng Asp 506:519–525

Ingle PU, Sudhir SS, Prashant RS, Suchitra SM et al (2022) Chitosan nanoparticles (ChNPs): a versatile growth promoter in modern agricultural production. Heliyon 8(11)

Iravani S (2011) Green synthesis of metal nanoparticles using plants. Green Chem 13(10): 2638–2650

Kalayeh SMO, Mostofi Y, Basirat M (2011) Study on some chemical compounds on the vase life of two cultivars of cut roses. J Ornam Horticultur Plant 1(2):123–128

Kamiab F, Sadegh SF, Elahe ZB (2017) Antimicrobial and physiological effects of silver and silicon nanoparticles on vase life of lisianthus (Eustoma grandiflora cv. Echo) flowers. Int J Hortic Sci Technol 4(1):135–144

Kazemi M, Ameri A (2012) Postharvest life of cut gerbera flowers as affected by Nano-silver and acetylsalicylic acid. Asian J Biochem 7:106–111

Kazemipour S, Davood H, Behzad K (2013) Effect of silver nanoparticles on the vase life and quality of cut chrysanthemum (Chrysanthemum morifolium L.) flower. Eu J Exp Biol 3:298–302

Khan M, Zaki AS (2018) Zinc oxide nanoparticles for the management of Ralstonia solanacearum, Phomopsis vexans and Meloidogyne incognita incited disease complex of eggplant. Indian Phytopathol 71:355–364

Kim S, Sooyeon L, Insook L (2012) Alteration of phytotoxicity and oxidant stress potential by metal oxide nanoparticles in Cucumis sativus. Water Air Soil Pollut 223:2799–2806

Koohkan F, Ahmadi N, Ahmadi SJ (2014) Improving vase life of carnation cut flowers by silver nano-particles acting as anti-ethylene agent. J Appl Hortic 16(3):210–214

Kumar V, Yadav SK (2009) Plant-mediated synthesis of silver and gold nanoparticles and their applications. J Chem Technol Biotechnol 84(2):151–157

Kumari A, Yadav SK (2014) Nanotechnology in agri-food sector. Crit Rev Food Sci Nutr 54(8): 975–984

Kumari A, Yadav SK, Yadav SC (2010) Biodegradable polymeric nanoparticles based drug delivery systems. Colloids Surf B: Biointerfaces 75(1):1–18

Kumari A, Rana V, Yadav SK, Kumar V (2023) Nanotechnology as a powerful tool in plant sciences: recent developments, challenges and perspectives. Plant Nano Biol 100046

Leili Mostafa MF, Bhatnagar A (2018) Green synthesis of nano-zero-valent iron from Nettle and Thyme leaf extracts and their application for the removal of cephalexin antibiotic from aqueous solutions. Environ Technol 39(9):1158–1172

Li Y, Zhang P, Mingshu L, Noman S, Muhammad A et al (2023) Application and mechanisms of metal-based nanoparticles in the control of bacterial and fungal crop diseases. Pest Manag Sci 79(1):21–36

Liu J, Kamani R, Daryl CJ, Shenggen H, Zhaoqi Z (2012) Effects of three different nano-silver formulations on cut Acacia holosericea vase life. Postharvest Biol Technol 66:8–15

Malandrakis AA, Nektarios K, Marianna A, Kalliope KP, Georgios T, Constantinos VC (2021) Metal nanoparticles: Phytotoxicity on tomato and effect on symbiosis with the Fusarium solani FsK strain. Sci Total Environ 787:147606

Malandrakis AA, Nektarios K, Constantinos VC (2022) Zinc nanoparticles: mode of action and efficacy against boscalid-resistant Alternaria alternata isolates. Sci Total Environ 829:154638

Mehmood A (2018) Brief overview of the application of silver nanoparticles to improve growth of crop plants. IET Nanobiotechnol 12(6):701–705

Moghaddasi S, Amir F, Amir HK, Karimzadeh F, Hamid RK, Reza K (2017) Bioavailability of coated and uncoated ZnO nanoparticles to cucumber in soil with or without organic matter. Ecotoxicol Environ Saf 144:543–551

Mohanpuria P, Rana NK, Yadav SK (2008) Biosynthesis of nanoparticles: technological concepts and future applications. J Nanopart Res 10:507–517

Mohanraj VJY, Chen JTJOPR (2006) Nanoparticles—a review. Trop J Pharm Res 5(1):561–573

Motaghayer MS, Majid A, Ali T (2019) Nanosilver, salicylic acid and essential oils effects on water relations of gerbera 'Rosalin' cut flowers. Adv Hortic Sci 33(2):271–282

Musante C, Jason CW (2012) Toxicity of silver and copper to Cucurbita pepo: differential effects of nano and bulk-size particles. Environ Toxicol 27(9):510–517

Nair PMG, Chung IM (2015) The responses of germinating seedlings of green peas to copper oxide nanoparticles. Biol Plant 59(3):591–595

Nasrollahzadeh, M, Atarod M, Sajjadi M, Mohammad Sajadi S , Zahra Issaabadi Z (2019) Plant-Mediated Green Synthesis of Nanostructures: Mechanisms, Characterization, and Applications. Intf Sci Technol 28:199–322

Pullagurala VLR, Ishaq OA, Rawat S, Kalagara S, Hernandez-Viezcas JA, Jose RPV, Jorge LGT (2018) ZnO nanoparticles increase photosynthetic pigments and decrease lipid peroxidation in soil grown cilantro (Coriandrum sativum). Plant Physiol Biochem 132:120–127

Qureshi A, Singh DK, Dwivedi S (2018) Nano-fertilizers: a novel way for enhancing nutrient use efficiency and crop productivity. Int J Curr Microbiol App Sci 7(2):3325–3335

Rahman MM, Ahmad SH, Lgu KS (2012) Psidium guajava and Piper betle leaf extracts prolong vase life of cut carnation (Dianthus caryophyllus) flowers. Sci World J 1:102805

Rahman MM, Ahmad SH, Mohamed MTM, Ab Rahman MZ (2019) Improving the vase life of cut Mokara red orchid flower using leaf extracts with silver nanoparticles. Proc Natl Acad Sci India Sect B Biol Sci 89:1343–1350

Raj S, Shoma J, Sumod US, Sabitha M (2012) Nanotechnology in cosmetics: opportunities and challenges. J Pharm Bioallied Sci 4(3):186

Rana RA, Siddiqui MN, Skalicky M, Brestic M, Hossain A et al (2021) Prospects of nanotechnology in improving the productivity and quality of horticultural crops. Horticulturae 7(10):332

Salachna P, Andżelika B, Agnieszka Z, Rafał P, Małgorzata M (2019) Stimulatory effect of silver nanoparticles on the growth and flowering of potted oriental lilies. Agronomy 9(10):610

Sánchez-Navarro JF, Yolanda González-García ABM, América Berenice Morales-Díaz SGM, Gregorio Cadenas-Pliego MDSGG, Antonio JM et al (2021) Silicon nanoparticles improve the shelf life and antioxidant status of lilium. Plants 10(11):2338

Sardella D, Ruben G, Vasilis PV (2017) Physiological effects and mode of action of ZnO nanoparticles against postharvest fungal contaminants. Food Res Int 101:274–279

Sharma S, Rana VS, Pawar R, Lakra J et al (2021) Nanofertilizers for sustainable fruit production: a review. Environ Chem Lett 19:1693–1714

Solgi M, Mohsen K, Toktam ST, Naderi R (2009) Essential oils and silver nanoparticles (SNP) as novel agents to extend vase-life of gerbera (Gerbera jamesonii cv.'Dune') flowers. Postharvest Biol Technol 53(3):155–158

Sone BT, Diallo A, Fuku XG, Gurib-Fakim A, Maaza M (2020) Biosynthesized CuO nano-platelets: physical properties & enhanced thermal conductivity nanofluidics. Arab J Chem 13(1):160–170

Stadler Lisa MH, Andreas Hartl SN, Massimo Colombo RZ, Petr Martin MB et al (2018) Recyclable magnetic microporous organic polymer (MOP) encapsulated with palladium nanoparticles and Co/C nanobeads for hydrogenation reactions. ACS Sustain Chem Eng 7(2): 2388–2399

Sun Q, Jianmei L, Tao L (2018) Zinc oxide nanoparticle as a novel class of antifungal agents: current advances and future perspectives. J Agric Food Chem 66(43):11209–11220

Tantawy AS, Salama YAM, El-Nemr MA, Abdel-Mawgoud AMR (2015) Nano silicon application improves salinity tolerance of sweet pepper plants. Int J Chem Tech Res 8(10):11–17

Thovhogi N, Park E, Manikandan E, Maaza M, Gurib-Fakim A (2016) Physical properties of CdO nanoparticles synthesized by green chemistry via Hibiscus Sabdariffa flower extract. J Alloys Compd 655:314–320

Tissera AN, Chanaka S, Nadeesh MA (2023) Emerging role of nanomaterials in storage and packaging of agricultural products. In: Engineered nanomaterials for sustainable agricultural production, soil improvement and stress management. Academic Press, pp 475–493

Tiwari N, Raksha P, Swapnil G, Aniket G, Rai M (2017) Biosynthesis of zinc oxide nanoparticles by petals extract of Rosa indica L., its formulation as nail paint and evaluation of antifungal activity against fungi causing onychomycosis. IET Nanobiotechnol 11(2):205–211

Tripathi DK, Singh S, Singh S, Srivastava PK et al (2017) Nitric oxide alleviates silver nanoparticles (AgNps)-induced phytotoxicity in Pisum sativum seedlings. Plant Physiol Biochem 110:167–177

Trujillo-Reyes J, Majumdar S, Botez CE, Peralta-Videa JR, Gardea-Torresdey JL (2014) Exposure studies of core–shell Fe/Fe3O4 and Cu/CuO NPs to lettuce (Lactuca sativa) plants: are they a potential physiological and nutritional hazard? J Hazard Mater 267:255–263

Vanlalveni C, Samuel L, Biswas A, Selvaraj M et al (2021) Green synthesis of silver nanoparticles using plant extracts and their antimicrobial activities: a review of recent literature. RSC Adv 11(5):2804–2837

Wang Y, Sean M, Shen R, Yang H (2007) Synthesis of iron oxide nanoparticles using a freshly-made or recycled imidazolium-based ionic liquid. Green Chem 9(10):1051–1056

Zahedi SM, Mahdieh K, da Silva JAT (2020) The use of nanotechnology to increase quality and yield of fruit crops. J Sci Food Agric 100(1):25–31

Zaki SA, Salama AO, Fawziah M, Marian MH et al (2021) Trichoderma harzianum-mediated ZnO nanoparticles: a green tool for controlling soil-borne pathogens in cotton. J Fungi 7(11):952

Zhao L, Jose RPV, Cyren MR, Jose AHV et al (2014) CeO2 and ZnO nanoparticles change the nutritional qualities of cucumber (Cucumis sativus). J Agric Food Chem 62(13):2752–2759

Zhao X, Wen L, Zhengqing C, Bing H, Qian T, Zhao D et al (2016) An overview of preparation and applications of stabilized zero-valent iron nanoparticles for soil and groundwater remediation. Water Res 100:245–266

Zhao X, Yu C, Huasheng L, Jinying L (2021) Influence of seed coating with copper, iron and zinc nanoparticles on growth and yield of tomato. IET Nanobiotechnol 15(8):674–679

Zhao X, Ruofei T, Jingyi Z, Yanlin L (2022) Multifunctional chitosan/grape seed extract/silver nanoparticle composite for food packaging application. Int J Biol Macromol 207:152–160

Chapter 15
Nanomaterials Used for Phytoremediation: A Review

Komal Sharma, Priya Saxena, and Alka Kumari

Abstract This chaper demonstrates an outline of the significance of nanomaterials (NMs) in the phytoremediation of heavy metals and metalloids. The paper also highlights the different types of nanoparticles (NPs), their synthesis methods, and characterization techniques. The aim of this review is to provide information about NMs' interaction with plant systems as well as some potential NPs used in the phytoremediation process. NMs have different surface structures and functions as nano-adsorbents, and their polymer-blended products are employed for the elimination of hazardous gases and adulterants (arsenic, manganese, nitrate, iron, and other heavy metals), organic entities (both aliphatic and aromatic), and biotic entities, including viruses, bacteria, and vermin. NMs possess properties like high surface area and associated high reactivity, which alleviate their performance in environmental restoration compared to other traditional techniques. Furthermore, the implementation of NMs in the monitoring of pollutants and accomplishments of nanostructured polymers in treating adulterants are also discussed, which provide a general idea to the future researchers. Besides these, the list of plant species and obtained NPs along with their utilization in the removal of different contaminants as well as their role in future research areas of nano-phytoremediation systems and prospects are briefly discussed in this paper.

Keywords Phytoremediation · Nanomaterials · Nano-phytoremediation · Characterization · Heavy metals

K. Sharma · A. Kumari (✉)
Department of Botany, University of Lucknow, Lucknow, Uttar Pradesh, India

P. Saxena
Department of Botany, University of Lucknow, Lucknow, Uttar Pradesh, India

Environmental Monitoring Division, FEST, CSIR- Indian Institute of Toxicology Research, Lucknow, Uttar Pradesh, India

S. K. Yadav, A. Kumari (eds.), *Nanobiotechnology in Agri-food Sector*, Sustainability Sciences in Asia and Africa, https://doi.org/10.1007/978-981-96-7756-6_15

15.1 Introduction

Several environmental challenges are piling up across the globe from widespread industrialization, urbanization, and modern agricultural practices that require immediate and sustained attention. The degradation of our environment, depletion of energy resources, and global energy consumption have become pressing issues that demand our unwavering attention. As we reflect on the historical Industrial Revolution, it is clear that our actions have resulted in unintended environmental harm, which, if left unaddressed, could have disastrous consequences. The extent of the damage inflicted on the environment has reached alarming levels, and failure to act expeditiously will expose us to perilous and horrifying outcomes in the future. Urgent action is required to address the environmental problems that are escalating globally for diverse reasons. The process of cleaning up pollutants in the environment—such as soil, water, and air—can take millions of years, but most pollutants are produced by industrial activities and vehicle emissions. This is a matter of great concern that necessitates immediate attention if we are to preserve our planet for future generations. We must act decisively and reduce our carbon footprint while seeking sustainable means to generate and employ energy (Pandit et al. 2022; Alebachew et al. 2023).

Human activities, like the burning of fossil fuels, municipal waste leaching into water bodies, ore mining, testing of nuclear weaponry, and using radioisotopes for medicinal and laboratory purposes, lead to environmental pollution on a large scale. Industrial waste, inclusive of colors, oils, and grease from mines, and landfill leachate, is hazardous in nature, and its untreated release in the environment is the primary reason behind chemical pollution. Chemical pollutants also contribute to air pollution, viz., episodic burning events cause the release of a particular chemical contaminant into the atmosphere. This may tend to be a significant environmental problem owing to increasing technical advancements and current expansions in the industrial sector. Pesticides and fertilizers applied to plants are transformed into toxic chemicals on interaction with humidity, sunlight, and other environmental factors, subsequently leading to adverse effects on plants and humans. In general, numerous factors, such as environmental pollution and rapidly deteriorating air and water quality, have progressively led to global warming, causing a rise in the earth's temperature. Therefore, environmental cleanup has become the most important duty for the survival of people and other species.

The swift expansion of agriculture and industries has resulted in a concerning rise in heavy metal pollution, jeopardizing both the ecosystem and food security. Furthermore, the disruption of natural habitats due to the surge in the global population has intensified heavy metal contamination on our planet (Masarovičová and Kráľová 2012). Environmental research has identified the prevention and management of heavy metal pollution as a key area of concern (Yang et al. 2022). These toxic metals are persistent and non-degradable, and their bioaccumulation and biomagnification along the food chain can have serious consequences on human health. The accumulation of heavy metals in the bodies of species at the top of the food chain is common, as they tend to be older, healthier, and have different dietary habits.

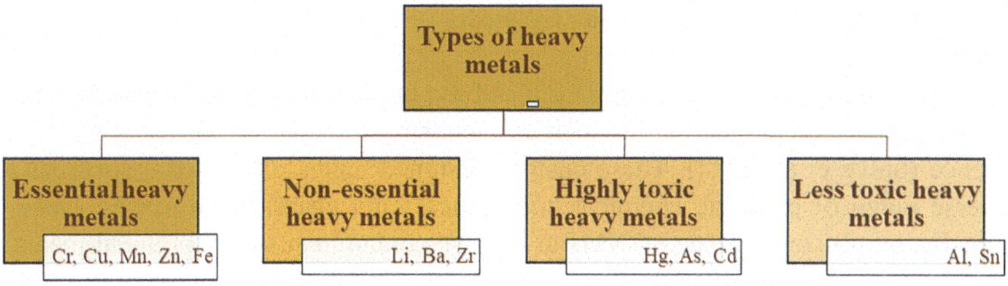

Fig. 15.1 Types of heavy metals. (Adapted from Mahmood et al. 2012)

Furthermore, heavy metals in the soil can damage the root system and performance of plants (Prakash 2023).

Heavy metals comprise a category of metals and metalloids characterized by atomic numbers exceeding 20 and a specific gravity greater than 5 g cm^{-3} (Ali and Khan 2018). These metals can be found in soil and can have toxic effects on organisms inhabiting the earth's surface, affecting their survival and functioning. Some of the most common heavy metals include arsenic, lead, cadmium, mercury, chromium, nickel, copper, and zinc. Heavy metals (Fig. 15.1.) can be classified into four categories depending on their use and toxicity: essential heavy metals (e.g., chromium, copper, manganese, zinc, and iron), non-essential heavy metals (e.g., lithium, barium, and zirconium), highly toxic heavy metals (e.g., mercury, arsenic, and cadmium), and less toxic heavy metals (e.g., tin and aluminum) (Mahmood et al. 2012). Some of these metals are essential micronutrients that are necessary for the proper functioning of the body. However, consuming more than the allowed amount of these metals can have harmful effects on the body and other organisms. Even at low concentrations, highly toxic metals can be extremely harmful.

Although soil has usual concentrations of heavy metals in different forms as ions, compounds, and complexes, several anthropogenic activities like metal excavation, hazardous wastes from various industries, and energy and fuel production cause a tremendous increment in heavy metal pollution of the environment (Li et al. 2018). Excessive amounts of heavy metals in the biosphere cause detrimental effects on human health and also reduce the productivity of plants by affecting the metabolic processes and soil-plant interactions (Rai et al. 2018). Consumption of heavy metals such as cadmium, chromium, lead, and mercury in elevated concentrations can cause severe diseases in both humans and animals.

Various remediation techniques have been presented and examined, but they are too expensive and make soil improper for plant growth. The technique of phytoremediation has turned out to be a promising strategy to rectify soils contaminated with heavy metals, which uses plants known as phyto-accumulators. Phytoremediation is an eco-friendly substitute for currently practiced physicochemical remediation methods, which are destructive to the environment.

Nowadays, rapid development in the area of nanotechnology provides solutions to different environmental challenges. Scientists and researchers of environmental sciences have also begun to exploit NMs in different industries, specifically in bioremediation, due to their peculiar structural and functional properties. So, the potentiality of NMs can furnish reclamation in deteriorating environmental conditions.

Nanotechnology refers to the creation and use of extremely small materials at the nanoscale across a range of fields such as chemistry, biology, physics, and material science (Salve et al. 2022). These custom-designed nanoscale materials possess a diverse range of physical, chemical, mechanical, and electrical properties. Nano-remediation utilizes engineered NMs to remove contaminants from polluted media, which are typically more effective and cost-efficient than traditional methods. NMs are also employed as useful catalysts, sensors, and adsorbents, despite the challenges associated with eliminating toxins from air and water (Roy et al. 2022). Furthermore, NMs serve as a vector for pest detection and eradication, energy storage, production, and conversion, as well as improving agricultural productivity, screening and diagnosis of disease, invention of newer drug delivery systems, food processing technologies, air pollution control, construction, and human health maintenance. NMs' finer electrical properties, notable surface-to-mass ratio, exalted catalytic activity, random placement of active sites, and numerous coating modification options make them stand out as potential remediating agents (Guerra et al. 2018a, b; Roy et al. 2022; Islam et al. 2022; Alebachew et al. 2023).

15.2 Nano-phytoremediation

Incorporation of nanotechnology in phytoremediation to clean up the environment is nano-phytoremediation (Aragay et al. 2012; Dhir 2021). Adulteration of air, water, and soil by heavy metal, organic, and inorganic pollutants, even in small amounts, can be remediated by nanoparticles because nanotechnology complements phytoremediation and speeds up its efficiency. NMs can be utilized for monitoring, recognizing, and removing pollutants (Kumar et al. 2021). It has been shown in recent studies that nano-sized zero-valent ions are capable of degrading organic contaminants such as atrazine, pesticides, and petroleum-based waste (Qu et al. 2013; Vyas et al. 2021). It is observed that in the size range of nanoparticles (NPs), materials possess far larger surface areas. Due to this, a greater quantity of a material can come into contact with neighboring materials, which affects reactivity. The use of NMs becomes an effective solution for phytoremediation of heavy metal pollution due to their high capability and cost-efficiency. Nano-phytoremediation is an emerging and promising strategy for environmental cleanup with the application of biosynthetic NPs. It is growing rapidly in the field of nanotechnology. Recently, researchers have been indulged in attaining knowledge about enhancing NM's ability and the processes for NM absorption and transportation for investigating toxic effects in the system. It combines the advantages of both nano-remediation and phytoremediation. By utilizing NMs in the remediation process, this approach offers the potential to eliminate organic and chemical contaminants, as well as organic pollutants, more efficiently. The conversion of harmful products facilitated by the high surface area and elevated surface energy of NMs is the causative factor behind such remediation approaches (Purushothaman and Kannan 2023).

15.3 Types of Nanoparticles Used for Phytoremediation

NPs can be broadly categorized into two major groups: organic and inorganic nanoparticles (Fig. 15.2.). The former encompasses carbon NPs, such as fullerenes, and those produced from organic waste materials like potato and orange peels. The latter includes magnetic, noble metal, and semiconductor NPs, such as gold, silver, titanium dioxide, and zinc oxide. According to Ruffini-Castiglione and Cremonini (2009), three categories of NPs have been identified: natural, incidental, and engineered. The first type includes volcanic and lunar dust, as well as mineral composites, while the second type includes products of anthropogenic activity such as exhaust from diesel engines, combustion of coal, and fumes from welding activities. The last type of NPs is metal-based materials, including quantum dots, nanogold, nano-zinc, nano-aluminum, TiO_2, ZnO, and Al_2O_3 (Srivastav et al. 2018).

Particles with dimensions of up to 100 nanometers in size are referred to as NPs, and these can be used to assist phytoremediation owing to certain physical properties as defined by Thangadurai et al. 2020:

(i) They assist gaseous phase interactions in the soil vadose zone/gas-liquid interactions in wastewater owing to their high surface-to-mass ratio. This helps to speed up the phytoremediation process.
(ii) New material can be prepared using very small changes to the size and shape of the original NM.
(iii) Transportation to the site via injection is easy owing to the very small size of the NP.

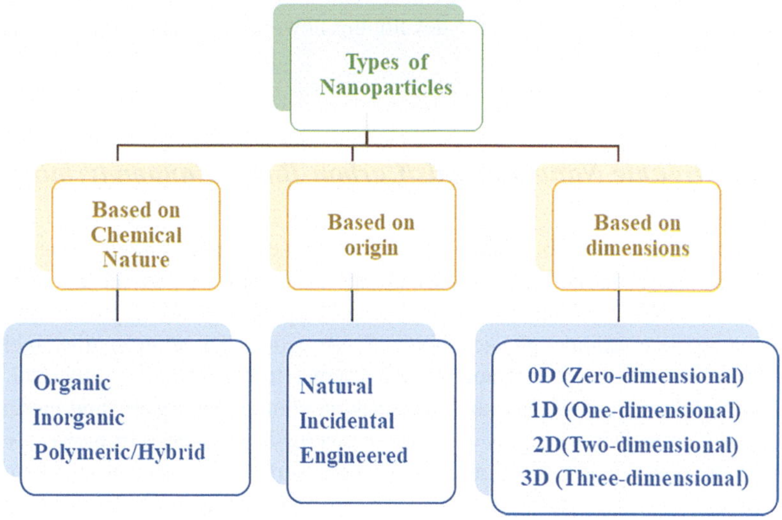

Fig. 15.2 Classification of nanoparticles

(iv) They are a relatively non-toxic and economic solution to traditional remediation practices.

NMs are usually of diverse types, based both on occurrence and chemical nature. Based on occurrence, Khan (2020) defined them as

(i) Naturally available, such as volcanic dust,
(ii) Those resulting from coal combustion or other related anthropogenic activities, and.
(iii) Engineered NPs.

Based on chemical properties, NMs are organic (with a carbon-based structure), inorganic, and polymer-based (Nwaadinigwe and Ugwu 2018; Guerra et al. 2018a, b; Thangadurai et al. 2020).

Generally, NMs are typically classified based on their chemical composition, shape, or number of dimensions at the nanoscale. The first classification is based on the NM's chemical nature, which can be organic, inorganic, polymeric, or hybrid. The second classification is based on the NM's shape, which can be NPs, nanorods, nanowires, quantum dots, and others. Finally, the third classification is based on the number of dimensions at the nanoscale (Kolahalam et al. 2019; Rahman et al. 2020). Nano-objects with three dimensions at the nanoscale are classified as 0D (e.g., NPs), while nano-objects with one and two dimensions at the nanoscale are classified as 1D (e.g., nanotubes and nanowires) and 2D (e.g., exfoliated clay layers and nanometric thin films) (Kolahalam et al. 2019; López-Villarreal et al. 2019; Rahman et al. 2020), respectively. Conversely, three-dimensional NMs are macroscopic materials that are composed of nano-objects, and their internal structure contains NMs. They are known as nanocomposites (Afreen et al. 2018; Sohail et al. 2019; Villalobos et al. 2019a, b; Palencia and García-Quintero 2023). The classification of NMs is essential for their design, manufacture, and application. Understanding the classification of NMs is crucial for developing new applications and technologies that can benefit various industries.

15.3.1 Organic Nanomaterials/Carbon-Based Nanomaterials

Since carbon is the most abundant element found on earth and is highly biocompatible, it can form varying numbers of compounds with various other elements. Carbon-based NPs become the most abundantly produced and historically used NPs due to their unique physicochemical properties. These properties include significant tensile strength, extraordinary resistance to corrosion, high thermal and electrical conductivity, stability, and versatility in eliminating organic and inorganic pollutants. These NMs are highly effective in treating heavy metals from wastewater due to their exceptional adsorption capacity, large surface area, and strong chemical reactivity. Moreover, carbon-based NMs are environmentally friendly, cost-effective, and easy to produce in large quantities, making them a promising alternative technology for the treatment of industrial wastewater (Astefanei et al. 2015; Dhiman et al. 2021). The two main types of carbon-based NMs are carbon nanotubes and graphene-based NMs. Other widely used carbon NMs are activated carbon, fullerene C60, fullerene C540, single-walled nanotubes, multi-walled nanotubes, etc. (Guerra et al. 2018a, b; Thangadurai et al. 2020). Such materials are highly appreciated due to their high surface area, heat and electrical conductivity, low toxicity, and environment-friendly nature. Fluoride removal from wastewater is reported using graphene, as, on the other hand, modified graphene can remediate gaseous pollutants and heavy metals too (El-Ramady et al. 2017). ZnO-graphene, TiO_2-graphene, and CdS-graphene nanocomposites may be used to treat benzene and other such organic contaminants (Guerra et al. 2018a, b). For a single open-ended carbon nanotube, absorption can occur at four surfaces, and it is therefore reported to be used extensively in heavy metal remediation.

15.3.2 Inorganic Nanomaterials

Inorganic NMs are either metal-based or metal oxide-based substances primarily used to remediate metallic contaminants. The most commonly used metal oxide NPs are silver and titanium oxides. Although silver oxide nanoparticles are known for their antimicrobial properties against both Gram-positive and Gram-negative bacteria, titanium oxide nanoparticles are mainly utilized for dealing with organic pollutants (Thangadurai et al. 2020). Also, binary mixed titanium oxide and silver oxide NMs are shown to have activity against a variety of pollutants. Shape-controlled, stable, and monodisperse inorganic NMs can be synthesized via thermal decomposition and/or reduction, hydrothermal extraction, and co-precipitation (Guerra et al. 2018a, b). Tsai et al. (2016) reported beneficial features of mesoporous silica, such as high surface area, huge pore volumes, and surface modification ability, that make it an excellent absorbent to be used as a remediator for gaseous and organic as well as metallic contaminants.

Poguberovic et al. (2016) synthesized nanoscale zero-valent iron from high-oxidant oak and mulberry leaf extracts, which causes the conversion of elemental Fe to nanoscale zero-valent iron (Machado et al. 2013), and used it for the removal of Ni^{2+} and Cu^{2+} from wastewater. Using such natural materials for NM synthesis is preferable owing to lesser environmental toxicity. Nanoscale zero-valent iron (nZVI) is also reported to be used by Kanel et al. 2005–2006, for the removal of arsenic (III) and arsenic (V) in colloidal solutions. Wang and Zhang 1997; Jiamjitrpanich et al. 2012) have also demonstrated a chemical method of preparation of nZVI using sodium borohydride and iron (III) chloride.

15.3.3 Polymer-Based Nanomaterials

Polymer-based NMs are colloidal bioactive solids that have dimensions on the order of 10 to 500 nm. The high thermal and mechanical stability, recyclability, and durability of synthetic NMs (Guerra et al. 2018a, b), in addition to the ability to be engineered with desirable properties (Tungittiplakorn et al. 2005), make polymeric NMs an excellent choice for remediation of chemicals, gases, and heavy metal contaminants.

Chitosan, alginate, starch, cellulose, and chondroitin sulfate are some natural polymeric substances, while chemically synthesized polymers encompass polyester, polyvinylpyrrolidone (PVP), polyurethane (PU), silicone rubber, and polyvinyl alcohol (Guerra et al. 2018a, b; Thangadurai et al. 2020). For the remediation of environmental pollutants, nanoscale zeolites, metal oxides, carbon nanotubes, and fibers are employed (Karn et al. 2009). Various NPs and their synthesis methods are reported in Table 15.1.

Table 15.1 Nanomaterial categories, synthesis formats, and illustrations

Nanomaterials	Synthesis formats	Illustration Rizwan et al. 2014
Metal NPs	Photochemical Electrochemical Biochemical Thermochemical	Ag, Au, Co, Cu, Ir, Ni, Pd, Pt, Rh, CdSe, CdTe, CoNi, Cu_3Au, FeNi, ZnS
Carbon NMs	Arc-discharge Laser ablation Chemical vapor deposition	Cylindrical nanotube (SWNT, MWNT) Fullerenes
Metal oxide NPs	Hydrothermal Solvothermal, Sol-gel Reverse micelles method Electrochemical deposition	ZnO, Fe_2O_3, Fe_3O_4, MgO, $BaCO_3$, $BaSO_4$, TiO_2
Polymer NMs	Electrochemical polymerization	Nanowires made of polypyrrole, polyaniline, and poly (3,4-ethylenedioxythiophene) dendrimers (PAMAM)
Nanocomposite	Innovative methods	A nanocomposite made up of polyethylene oxide and polyethyleneimine; CNT epoxy composites include hydrocarbon polymer composites, conjugated polymer composites, CNTs mixed with polycarbonates, fluoropolymers, polyethylene glycol, polyester polyamides, and several others
Bio-nanomaterials	Biological	Viruses, plasmids, and protein NPs

15.4 Synthesis, Uptake, and Characterization of NPs

Nanotechnology and nanoscience have emerged as promising research fields that offer exciting opportunities to generate and tailor NMs with unique properties, sizes, and chemical compositions. Researchers worldwide are increasingly focusing on developing reliable experimental processes to advance these fields and unlock their full potential. Studies have shown that these efforts have already yielded significant progress and promise to revolutionize numerous industries and applications. The production of NMs is achieved through two principal methods: top-down and bottom-up (Fig. 15.3.). The former approach involves the breakdown of bulk materials into smaller components using mechanical, chemical, or other forms of energy. This approach is utilized to miniaturize components and increase the efficiency of machines and devices. The top-down approach is predominantly implemented in the electronics industry. It is worth noting that the top-down approach to NM synthesis can lead to the production of particles with a broad dimension distribution and impurities in the resulting NPs. Defects in surface morphology are a significant issue linked to this method, as they can greatly influence the physical and chemical properties as well as the surface chemistry of materials synthesized biologically at the nanoscale. The orthodox top-down approach is widely critiqued for causing grave crystallographic damage to the

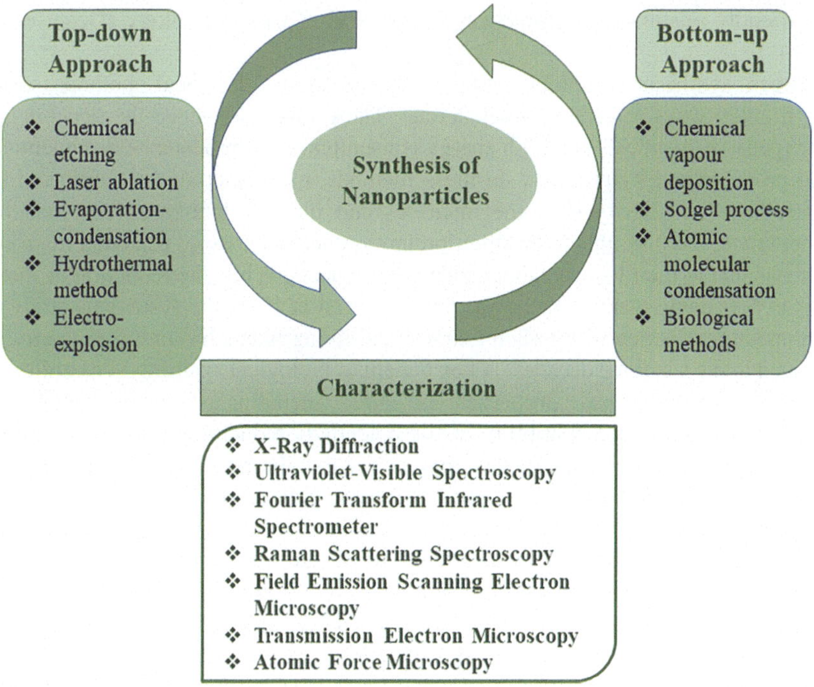

Fig. 15.3 Synthesis and characterization of nanoparticles (NPs)

prepared designs, as evidenced in recent studies. Therefore, caution must be exercised when employing the top-down approach, and alternative synthesis methods should be considered to obtain high-quality NMs with desired properties (Jalilian et al. 2020; Dejen et al. 2023). The bottom-up approach, on the other hand, entails the assembly of individual atoms or molecules to form NPs. This procedure depends on various elements, including the ability of self-assembly and physical and chemical interactions like hydrogen and ionic bonds, van der Waals forces, and water-mediated hydrogen bonds, which contribute to the organization of primary building blocks into discernible structures.

Recent advancements have seen biological synthesis gaining popularity as a superior alternative to traditional physical and chemical methods for producing nanoparticles (NPs). This bioreduction method utilizes various bioactive substances, such as plant extracts and microorganisms. Recognized as an eco-friendly or "green" synthesis approach, biosynthesis is currently considered the most environmentally sound, economical, and efficient method for NM production. There are three primary types of bio-assisted synthesis methods: (1) microbial-based synthesis: utilizing microorganisms as templates, (2) biomolecule-based synthesis: employing biomolecules as templates and (3) plant extract-based synthesis: involving different parts of plant extracts. Each of these methods has unique benefits and drawbacks, and careful consideration should be given to selecting the appropriate method based on the specific application and desired outcomes. Overall, the use of biological synthesis methods represents a promising avenue for the development of sustainable and eco-friendly manufacturing practices in the field of nanotechnology (Dejen et al. 2023).

Green synthesis techniques, often referred to as biological methods, offer eco-friendly ways to create nanomaterials. These methods reduce the dependence on expensive chemicals and high energy consumption and produce environmentally safe products and by-products. In these methods, nanomaterials are fabricated by controlling the activity of some microbes and plants. Plant-based methods for nanoparticle (NP) synthesis are often preferred over those using cells and microorganisms due to their lower maintenance requirements. While the reduction of metal ions to NPs using plant extracts was first observed in 1900, the systematic study of nanometal production using plant extracts and living plants has only been actively pursued in the last three decades. A key benefit of biological synthesis is its ability to enhance the biological properties of NPs, including their bioavailability, bioreactivity, and biocompatibility, while potentially reducing their toxicity. However, a significant drawback of this approach is that the resulting NPs may exhibit lower stability compared to those produced through traditional physical and chemical methods.

Nanoparticles (NPs), due to their diminutive size, can infiltrate plant roots by means of osmotic pressure, capillary forces, or by penetrating directly through the epidermal layers. The dimension of NPs plays a critical role in their entry into the cell membrane. Upon entering the cell wall, NPs move through the apoplast to reach the vascular system. The transport of NPs through the Casparian strip to the vascular system occurs through symplast movement. . NPs can bind to carrier proteins on the cell membrane, create pores, and pass through them. They can also move between cells through specialized channels called plasmodesmata (Tripathi et al. 2017; Pérez-de-Luque 2017). NPs can be absorbed by various plant structures, including the cytoplasm, nucleus, epidermis, and other cellular organelles. If NPs are absorbed by the seed coat, they can enter the seeds through spaces between cells or by diffusing through the cotyledon (Ali et al. 2021). Applying nanomaterials to leaves allows them to enter through stomata or the leaf's outer surface (through the cuticle). NPs larger than 10 nm can move through both the cell walls (apoplastic pathway) and within the cells (symplastic pathway). NPs between 10 and 50 nm typically follow the symplastic route. NPs between 50 and 200 nm generally move through the apoplastic path.

Once NPs have been fabricated, it is crucial to characterize them to gain a detailed understanding of their structure, composition, uniformity, charge, and other properties. There are several techniques available for characterizing NPs. These techniques include transmission electron microscopy (TEM), scanning transmission electron microscopy (STEM), scanning electron microscopy (SEM), electron energy loss spectroscopy (EELS), X-ray powder diffraction (XRD), and X-ray photoelectron spectroscopy (XPS) (Mourdikoudis et al. 2018).

The process by which plants uptake NPs is dependent on a complex interplay of several factors, including nanoparticle size, type, and chemical composition. Of these factors, NPs' size is the most critical determinant of whether the NPs can successfully enter the plant and move from the roots to other parts of the plant (Schwab et al. 2016). The uptake process of NMs is also influenced by the penetration mechanism. Once the NPs are inside the plant roots, they can move into the plant tissues through two pathways: apoplastic transport, which takes place outside the plasma membrane within xylem vessels, and symplastic transport, which involves the movement of water between the cytoplasm and sieve plates. Environmental conditions also play an important role in the uptake of NPs. For instance, temperature affects growth substances, which can, in turn, impact root length. Root structure also varies under field conditions compared to greenhouse conditions. The effectiveness of phytoremediation, specifically phytoextraction, is dependent on the ability of specific plants to hyperaccumulate particular contaminants. (Srivastav et al. 2018).

15.5 Factors Influencing Nano-phytoremediation of Pollutants

The uptake and distribution of contaminants within living plants are contingent upon various factors, including the characteristics of the pollutants, plants, and environmental conditions. Water solubility, vapor pressure, and molecular weight are the physical and chemical properties of a compound, which provide insight into its behavior and potential interactions with other substances in the environment.

Further, environmental factors, such as temperature, pH, organic matter, and soil moisture content, play a significant role in determining the fate and transport of chemical compounds in the environment. The type of root system and enzymes present in plants are additional characteristics that have an impact on the absorption and metabolism of chemical compounds by plants. Thus, a comprehensive understanding of the properties and characteristics of chemical compounds is imperative to predict and mitigate their potential environmental impacts (Srivastav et al. 2018).

Nanoparticles application in phytoremediation through different procedures are reported in Table 15.2.

15.6 Nanomaterials Interaction with Plant Systems

Nanomaterials with various shapes and sizes can be employed for environmental remediation since nanoparticles can penetrate contamination areas that microparticles are unable to access, and they exhibit greater reactivity toward redox-active pollutants. Uptake of nanoparticles in plants relies on different properties like physical-chemical composition, surface morphology, solubility, reactivity, and, most importantly, the concentration of nanoparticles, which varies from plant to plant (Fig. 15.4.). The efficient absorption of nanomaterial by plants is also determined by the nanomaterial penetration process. NPs size, surface area, plant species, contaminant concentrations, etc. are the key factors that define nanomaterial interaction with the plant system and the degree of remediation achieved. Upon entry into the plant roots, there are two routes that a contaminant may take: one is apoplastic transport, where the transport takes place outside the plasma membrane and xylem vessels (Sattelmatcher 2001; Srivastav et al. 2018), and the other is symplastic transport, which involves transport via cytoplasm and sieve plates (Roberts 2003; Srivastav et al. 2018).

Different factors, such as temperature and various soil properties, can influence root growth. It is essential to recognize that root architecture in field settings differs from that in greenhouse environments. The effectiveness of phytoremediation, particularly phytoextraction, relies on the use of hyperaccumulators that are specific to the contaminants (Srivastav et al. 2018). Additionally, engineered nanomaterials (NMs) are very effective at mitigating the negative impacts of metal contamination in plants. Because of their smaller dimensions and greater surface area, NMs have a heightened attraction to metals and can easily access contaminated areas in habitats affected by metals. NPs used in phytoremediation are listed in Table 15.3.

Many physiological and morphological changes occur in plants due to NPs' interaction, and plant response against these changes largely depends on the NPs' type and dose and also on the plant species involved. It is found in recent studies that various NPs positively affect the development and growth of plants.

Silver NPs efficiently (>90%) enhanced the production of phytohormones ABA and GA in plants, which increases plant endurance against stresses, and salicylic acid NPs enhanced the phytoremediation efficiency of plants under arsenic (As) stress.

Table 15.2 Application of different nanoparticles in remediation through different procedures. (Adapted from Guerra et al. 2018a, b; Tripathi et al. 2018 and Khan and Murphy 2023)

Type	Remediation process involved	Examples	Size	Application in remediation
Metal and metal oxides	Adsorption and redox reaction	Zero-valent iron NPs (nZVI)	50–200 nm	Detoxification of pesticides, herbicides, organochlorine, polychlorinated biphenyls, etc.
		Silver and gold NPs	10–20 nm	Water disinfection and kill pathogens
		Titanium oxide (TiO_2)	~ 5 nm	Dye and organic pollutant removal from textile wastewater, water purification, and pathogen inactivation
		Zinc oxide (ZnO)	~ 20 nm	Removal of dyes from wastewater
		Bimetallic NPs	20–40 nm	Decontaminate water and soil by removing and degrading chlorinated and brominated contaminants
Carbon-based nanomaterials	Absorption, adsorption, membrane filtration, antimicrobial agents, and environmental sensors	Carbon nanotubes	1–30 μm	Employed in water treatment and filtration processes, exhibits sorption capacity for organophosphates and heavy metals, facilitates the degradation of PFAs and herbicides, and aids in the recycling of electronic, computer, and plastic waste
		Fullerenes	~ 0.72 nm	Absorbs and adsorbs organic and organometallic compounds
Nanopolymers	Sorption, filtration encapsulation, chemical and biological catalysis	Chitosan NPs	10–1000 nm	Breakdown of organic substances (such as Bisphenol A, 2,6-Dimethoxyphenol, and various organic dyes) and the treatment of industrial wastewater
		CLEAs	1–100 μm	Organic dye degradation, degradation of acetaminophen and other organic drugs

(continued)

Table 15.2 (continued)

Type	Remediation process involved	Examples	Size	Application in remediation
		Poly (ethylene) glycol-modified urethane acrylate (PMUA)	~ 80 nm	Sorption of hydrophobic organic contaminants and increases solubility of PAHs
		Dendrimer	1.5–10 nm	Degrades organic compounds and arsenic pollutants
Other nanoparticles	Sorption, chemical catalysis	Nanoclay fluoranthene, phenanthrene	Nanoclay up to 1000 nm fluoranthene, phenanthrene	Sorption of PAHs (pyrene, anthracene, fluoranthene, phenanthrene)
		Nanosponge (highly cross-linked cyclodextrin-based polymers)	250–1000 nm	Removal of Triclopyr (3,5,6-trichloro-2-pyridinyloxyacetic acid)
		Anion-SAMMS (Self-Assembled Monolayer on Mesoporous Supports), Thiol-SAMMS and HOPO-SAMMS	2–20 nm	Remove inorganic ions, heavy metal ions, actinides, and lanthanides

Properties like soil fertility, enzymatic activity, and chlorophyll content in plants, seed germination can be increased by regulation of nutrient release in soil systems with NPs, including nano-fertilizers. It has also been examined that the application of Mn NPs increases nitrogen uptake, assimilation, and plant metabolism and also affects the elongation of roots and shoots (Pradhan et al. 2014).

15.7 Phytoremediation

Phytoremediation is a technique that uses plant science to enhance human living conditions. It makes use of plant physiology and rhizosphere organisms as cost-efficient and dependable methods for eliminating hazardous or persistent pollutants in areas where financial resources for pollution cleanup are limited. Most modern approaches for mitigating pollution are expensive and require penalties, transport to designated landfills, or manufacturing processes that either reduce or recycle waste. Phytoremediation represents a distinct form of bioremediation that uses the metabolic activities of plants and the rhizosphere to remove harmful substances from the environment (DeLorenzo 2018). Initially, bioremediation was created as an

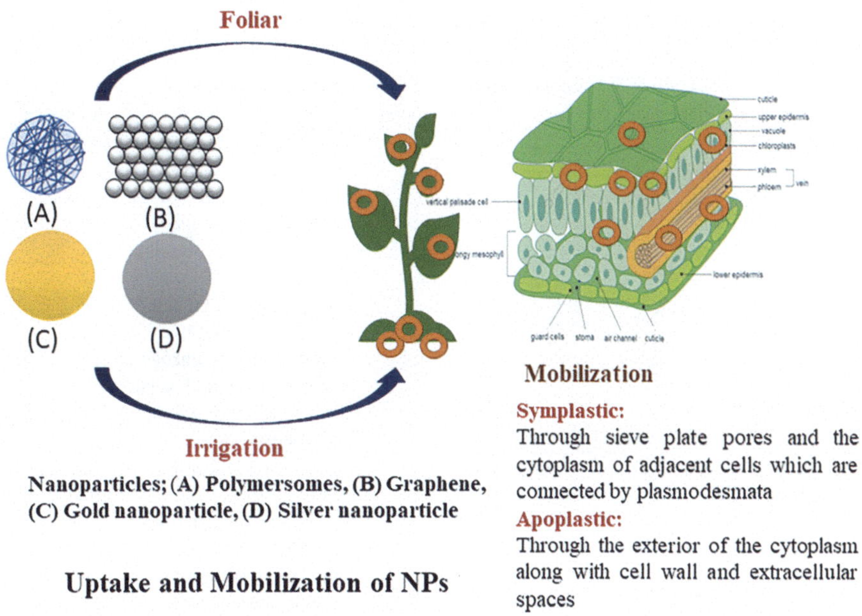

Fig. 15.4 Uptake and mobilization of NPs in plants

alternative to conventional chemical and physical methods for cleaning up pollution in soils and water, such as chemical neutralization or the removal of bulk soil (Conesa et al. 2012). It is a developing technology that utilizes plants and their associated rhizosphere microorganisms to eliminate and transform toxic substances from soils, sediments, groundwater, surface water, and even the air. Phytoremediation offers an efficient, non-invasive, and economical approach to soil remediation (Srivastav et al. 2018).

15.7.1 Types of Phytoremediation Techniques

Phytoremediation techniques can be several, based on the type of contaminant to be remediated, environmental conditions, and plant species used for the remediation (Fig. 15.5.). Contaminants can be extracted, volatilized, filtered, stabilized, and transformed owing to the different biological mechanisms involved in the process (Table 15.4).

15.7.1.1 Phytoextraction

Phytoextraction refers to the technique of absorption of contaminants from the soil and their transportation into the shoot tissues such that these are permanently

Table 15.3 Potential NPs in phytoremediation process

Nanoparticles	Plant species	Beneficiary concentration (s)	Heavy metal	Effects and mode of action	References
FeNPs, ZnO NPs	*Triticum aestivum*	5, 10, 15, and 20 mg/L	Cd	Triticum aestivum showed improved growth through enhanced photosynthesis, decreased activities of superoxide dismutase (SOD) and peroxidase (POD), minimized ion leakage, elevated nutrient levels, and reduced cadmium (Cd) absorption	Rizwan et al. (2019)
TiO_2 NPs	*Zea mays*	100, 250 mg/L	Cd	Decreased absorption of Cd toxicity, improved the activities of SOD and GST, and increased the regulation of various metabolic pathways	Lian et al. (2020)
ZnO	Fenugreek	1, 0.5, 0.25, and 0.10 g/L	Cr	In comparison to the control and zinc sulfate, the highest levels of seed germination, root development, and overall plant growth were noted with the application of zinc nanoparticles. The concentration of zinc in the root tissue was found to be elevated	Shaik et al. (2020)
Graphene oxide	Lettuce	30 mg/l		Increased net photosynthetic rate, stomatal conductance	Gao et al. (2020)
Mel-Au	Rice	50, 100 and 200 μM	Cd	Elevated internal accumulation of melatonin. Significant and intricate impacts on plant development. Synthesis of chlorophyll, aging of leaves, antioxidant capabilities, and the elimination of reactive oxygen species during cadmium stress	Jiang et al. (2021)
Ast-Au	Rice (*Oryza sativa* ssp.	100 μg/mL	Cd	Reestablishes the production of chlorophyll,	Dai et al. (2021)

(continued)

Table 15.3 (continued)

Nanoparticles	Plant species	Beneficiary concentration (s)	Heavy metal	Effects and mode of action	References
	Japonica cultivar Nipponbare)			enhances the elimination of reactive oxygen species (ROS), significantly boosts the activity of antioxidant enzymes, regulates the metabolism of free amino acids, and raises the concentration of non-enzymatic antioxidants	
Se	Coriander *(Coriandrum sativum)*	0, 5, 10, and 15 mg/l		Diminished Cd content, promote antioxidant response	Sardar et al. (2022)
CuO	*Brassica campestris*	20 mg/L	Cd	Enhanced oxidative damage and reduced nutritional contents in plants	Wang et al. (2022)
AgNPs	*Daucus carota*	50 mg/L	Cd	Reduced toxicity by neutralizing generated reactive oxygen species (ROS), lowered malondialdehyde (MDA) levels, and increased the activity of enzymes involved in phenol synthesis	Faiz et al. (2022)

removed from the polluted soil. The above-ground biomass, now rich in accumulated contaminants, needs to be safely disposed of. Phytoextraction is the most commonly used technique to remove heavy metal contamination from polluted soils (Sarret et al. 2002; Yan et al. 2020).

According to Ali et al. (2013) and Yan et al. (2020), the phytoextraction procedure of contaminants generally includes the following steps:

(I) Contaminant is immobilized in rhizosphere.
(II) Plant roots uptake the contaminant.
(III) Contaminant is translocated by the vascular system of the plant to the above-ground parts.
(IV) Contaminant is immobilized and stored in the aerial plant tissue.

Plants used for phytoextraction should have a high tolerance to toxic contaminants, extensive root systems for accelerated uptake, and high above-ground biomass (Yan et al. 2020). Generally, hyperaccumulator species are ideally suited for phytoextraction procedures.

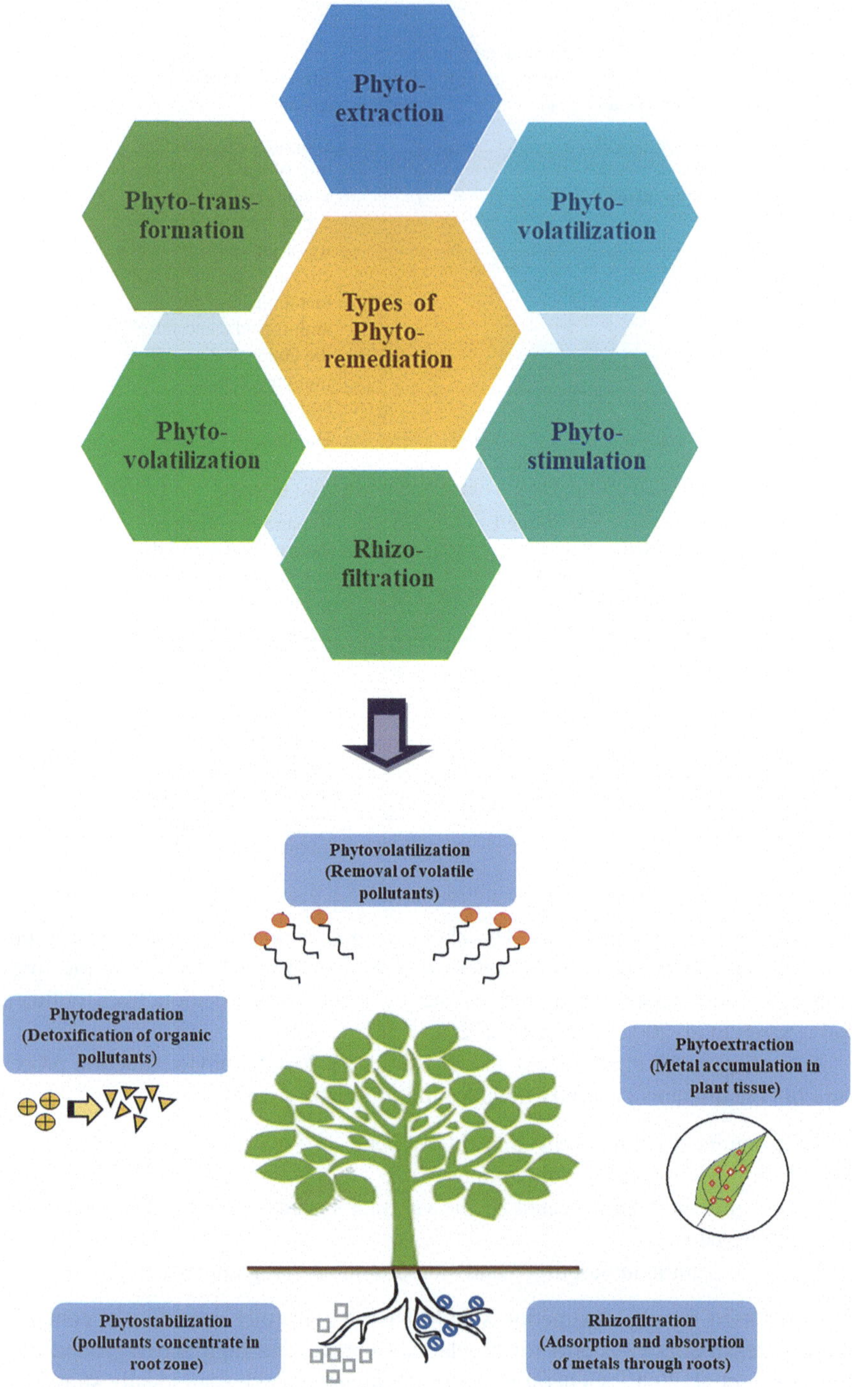

Fig. 15.5 Techniques of phytoremediation

Table 15.4 Inventory of plant species and nanoparticles employed for eliminating contaminants in nano-phytoremediation systems

Plant Species	Used NPs	Contaminants	Size of NP (nm)	References
Eucalyptus globulus (leaf)	Fe_3O_4	Arsenic	NA	Xiao et al. (2019)
Jatropha curcas (latex)	TiO_2	Chromium	10–120 75	Goutam et al. (2018)
Psidium guajava (leaves)	ZnO–$ZnFe_2O_4$	Chromium	10–20	Sahoo et al. (2021)
Citrus sinensis	Ag	Cadmium	44–50	Tavker et al. (2021)
Mentha piperita (leaf)	CuO	Cadmium	150	Mahmoud et al. (2021)
Thymus schimperi (leaf)	Fe_3O_4	Mercury	8.7	Geneti et al. (2022)
Carica papaya (leaf)	Ag	Blue CP	21.59	Bhuiyan et al. (2020)
Cinnamomum verum (bark)	Mn	Congo red	<100	Kamran et al. (2019)
Angelica gigas (stem)	Au	Eosin Y	NA	Chokkalingam et al. (2019)
Mentha piperita (leaf)	Ag	Crystal violet	15	Sharbaf Moghadas et al. (2020)
Aloe barbadensis (leaf)	Cu	Malachite green	60	Abbas et al. (2020a, b)
Carica papaya (leaf)	Fe_3O_4	Reactive yellow	21.59	Bhuiyan et al. (2020)
Nephelium lappaceum (seed)	Ag	Methylene blue	12.1	Perera et al. (2020)
Tephrosia purpurea (leaf), (stem), (inflorescences)	ZnO	Methylene blue	26–35	Wijesinghe et al. (2021)
Justicia gendarussa (leaf)	CuO	Methylene blue	10–15	Vasantharaj et al. (2023)
Crataegus pentagyn (leaf)	Ag	Rhodamine B	25–45	Ebrahimzadeh et al. (2020)
Alchemilla vulgaris (leaf)	ZnO	Rhodamine B	120	Rajendrachari et al. (2021)

15.7.1.2 Phytotransformation

Phytotransformation is the ability of certain plant species to convert a more toxic and complex organic contaminant to its simpler, less toxic version by employing certain metabolic activities occurring inside the plant tissue or via enzymes released by plant tissues. The breakdown products are usually non-toxic or less toxic compounds that help in plant growth. The rate of phytotransformation usually depends on the toxicant type, plant species used, and number of growing seasons the procedure is applied.

15.7.1.3 Phytovolatilization

Phytovolatilization works hand in hand with both phytoextraction and phytotransformation. It is the procedure where a plant body converts a contaminant to its volatile form and causes its diffusion into the atmosphere via its aerial tissues. Phytotransformed and phytoextracted organic compounds may at times be phytovolatilized through leaf stomata via gaseous exchange with the atmosphere. Photovolatilization can be applied to volatile metals like selenium, mercury, arsenic, etc. (Yan et al. 2020), and to chlorinated hydrocarbons, hydrophilic compounds, and BTEX compounds (Srivastav et al. 2018).

15.7.1.4 Rhizofiltration/Rhizodegradation

The rhizosphere is the area some millimeters from the soil surface where root-soil interaction is at its maximum. Rhizofiltration is, therefore, the use of plant roots to remove contaminants from polluted water/soil. Root exudates in the area, soil microflora including various bacterial, fungal, and mycorrhizal species, along with physical attributes like pH, high moisture availability, and redox conditions, favor degradation of organic contaminants, which are then easily taken up by the roots. For rhizofiltration, plant species with extensive root systems are generally preferred. The process includes growing plants till they acquire an extensive root system and then gradually shifting them to the site of degradation. Roots saturated with contaminants need to be disposed of (Wuana and Okieimann 2011; Yan et al. 2020).

15.7.1.5 Phytostimulation

Phytostimulation is the process of enhancing the degradation of contaminants in the soil vadose zone by increased microbial activity. Plants with the capability of increasing soil-microbe interaction via root exudation are preferred. At times phytostimulation can be used as a precursor/additive to rhizofiltration.

15.7.1.6 Phytostabilization

Phytostabilization is also known as in-place inactivation and is generally used for sediment and sludge remediation. It refers to the physical process that uses plants to contain contaminant migration to minimize soil erosion by wind/water. By using proper foliage cover coupled with other phytoremediation strategies, contaminant interaction with surrounding areas can be minimized. Plants with extensive root systems (especially grasses) are preferred for the process. The application of phytostabilization as a cleanup technology involves sorption, complexation, or precipitation mechanisms to immobilize contaminants present in the soil. This approach is beneficial because it avoids the need to manage waste materials or biomass, making it particularly effective when rapid immobilization is crucial to safeguard both ground and surface water sources. However, it has several notable disadvantages. For example, contaminants still persist in the soil, requiring significant fertilization or soil amendments. Additionally, continuous monitoring is essential to verify the treatment's success, and the stabilization of pollutants may largely be due to the soil amendments instead of the phytostabilization technique itself (Prakash 2023).

15.8 Phyto vs. Nano: The Need

Phytoremediation is a highly promising method for remediating soil and water pollution, both as a sole or supplemental approach, particularly for countries with emerging economies and economically advantaged countries. In comparison to traditional chemical and physical remediation methods, phytoremediation is a cost-effective alternative for economically emerging countries that require fewer engineered technologies. This method shows significant promise for application in a variety of climates and under severe environmental conditions. Invasive plants, when cultivated in a controlled environment, have proven to be particularly effective for countries lacking native phytoremediation options (Prabakaran et al. 2019). Every nation can promote opportunities for the transfer of green phytoremediation technologies, as recent evaluations have confirmed their effectiveness. Nevertheless, phytoremediation does have some technical constraints, as is the case with any remediation method. One strategic factor to consider is the extended timeframes necessary for plant growth or acclimation. Another tactical issue is that the contaminants must be located close to the plant roots, and the concentration of the contaminants influences the roots' ability to absorb or break down the pollutants.

Despite certain constraints, phytoremediation is comparable to traditional in situ remediation and may be more eco-friendly than conventional ex situ remediation methods. Efforts are underway to enhance the effectiveness of phytoremediation by finding ways to recycle or repurpose the plants once they have fulfilled their bioremediation role. Generally, after the completion of a phytoremediation process, the plants must be disposed of in some manner. Depending on the type of contaminant, the plants are either placed in a municipal landfill, incinerated, or treated as hazardous waste. It would be beneficial to explore methods for reusing or recycling these plants. Initial research has identified the potential to recover metals stored in the biomass of plants used in phytoextraction, which is particularly advantageous for reclaiming rare earth metals found in electronic waste sites where phytoremediation is occurring. Investigations into various phytodegradation and phytostabilization plants indicate that the harvested vegetation may also serve as animal feed. Likewise, there are efforts to explore the possibility of using harvested phytodegradation and phytostabilization plants for human consumption (Mitton et al. 2016). Research has also looked into utilizing energy crops for biofuel production or combustible biomass (Pandey et al. 2016). Spent phytoremediation plants have been found to be a valuable source for anaerobic digestion substrates (Cao et al. 2015).

Recent research has shown that various nanoparticles (NPs) and nanomaterials (NMs) are highly effective in removing organic, inorganic, and heavy metal contaminants from soils. Substances like chlorpyrifos, atrazine, lindane, pentachlorophenol, trichloroethylene (TCE), pyrene, polychlorinated biphenyls (PCBs), 2,4-dinitrotoluene, and ibuprofen, which can be present in the soil, can be quickly broken down by nano zerovalent iron (nZVI), magnetite nanoparticles (nFe3O4), and bimetallic nanoparticles (Pd/Fe). The application of nano-phytoremediation in the field of soil contaminant removal has become very prominent. A wide range of contaminants can be removed by this technique, from heavy metals to volatile organic compounds, with high efficiency. More effective removal of site contaminants like organic pollutants can be done with NM-assisted phytoremediation. Recent studies have shown that the application of nanoparticles (NPs) can enhance the stress tolerance of plants in both ex situ and in situ environments, thereby increasing their potential for phytoremediation.

As we have seen, for the environmental cleanup process nano-phytoremediation technology is a new field. Full-scale studies in this research field are very few; however, lab-scale and pot culture studies have already been done. Ji et al. (2011) conducted laboratory and pot culture experiments as well as field studies that demonstrate how different environmental factors such as temperature, pH, soil type, nutrient levels, and soil moisture significantly influence the real-world field environment. The reclamation of the environment by nanotechnology will offer green and eco-friendly options without harming nature. Plants having the ability to accumulate and translocate very large concentrations of metals are termed hyperaccumulators, consisting of several plants, fungi, and bacteria. Nanomaterials (NMs) derived from plants, fungi, and bacteria are crucial for the bioremediation of areas contaminated with heavy metals. These NMs can be employed in various forms to remove pollutants. Nanoparticles (NPs) from these plant, fungi, and bacteria species have been utilized for the extraction of certain heavy metals from contaminated locations.

NPs that are suitable for phytoremediation should have certain key characteristics. Firstly, they should be non-toxic to the plant, as any toxic effects could harm or even kill the plant, defeating the purpose of phytoremediation. The NPs should promote the growth and development of the plant. This includes an increase in germination, seedling growth, root-shoot elongation, plant height, and biomass. By enhancing these plant growth parameters, the plant becomes better equipped to take up and metabolize the contaminants from the soil. The NPs should significantly increase the production of phytoenzymes in plants. Phytoenzymes are enzymes produced by plants that are responsible for breaking down and metabolizing contaminants. By increasing their production, the plant becomes more efficient at removing contaminants from the soil. The NPs should be capable of enhancing plant growth hormones. This can lead to increased plant growth and development, further enhancing the plant's ability to remove contaminants from the soil. The NPs should be able to bind contaminants, making them more bioavailable for the plant.

This can help the plant to more effectively take up and metabolize the contaminants. The NPs should enhance the phytoremediation process as a whole. This can be achieved through a combination of the above characteristics, ensuring that the plant is better equipped to remove contaminants from the soil. Metallic-engineered nanomaterials (ENMs), consisting primarily of metals and metal oxides, are the most prevalent ENMs utilized for nano-remediation (Bardos et al. 2018). These materials are generally applied for in situ remediation through various injection techniques. Carbon-based materials rank as the second most frequently employed ENMs for environmental remediation, whereas polymers and nanocomposites, despite garnering considerable academic attention, still have limited practical uses (Mukhopadhyay et al. 2022).

Nevertheless, despite the growing interest in nano-remediation, there remains considerable debate within the scientific community regarding the risks to human health and the environment. Precursors frequently utilized in the synthesis of engineered nanomaterials (ENMs) can have detrimental effects on ecosystems when they are released into the environment (Abbas et al. 2020a, b). ENMs have the potential to disturb soil ecosystems by influencing the metabolic processes of soil microorganisms, hindering seed germination and plant development, adversely affecting invertebrate species, and being absorbed by crops, which could allow them to ascend the food chain to higher trophic levels (Qian et al. 2020). Therefore, the negative effects of ENMs on soil resulting from nano-remediation techniques should not be overlooked. Nanoparticles (NPs) have attracted considerable interest due to their remarkably small dimensions and large surface area relative to their volume, which present unique and extraordinary characteristics in comparison to larger bulk materials. Recently, NPs have become increasingly significant in the domain of environmental remediation (Zhang et al. 2020). Pollution of soil, water, and air has made ecosystems less viable, contaminating food and affecting biodiversity, which creates new risks for aquatic life and environments, water security, and public health and community welfare. To combat these challenges, the field of nanotechnology has introduced innovative nanoscale materials designed to detect and address persistent toxic pollutants that may be found in soil and water (Dörner et al. 2019).

While both phytoremediation and nano-phytoremediation are promising strategies for environmental cleanup, they have distinct advantages and limitations. Phytoremediation is a more established and widely used technique that has proven successful in many cases. It is a low-cost and sustainable method that can be implemented in large-scale projects. However, it may have limitations regarding its efficiency and the types of contaminants it can effectively treat. On the other hand, nano-phytoremediation has the potential to overcome some of the limitations of traditional phytoremediation. By increasing the absorption and conversion of pollutants, the overall effectiveness of the remediation process can be enhanced. Nonetheless, nano-phytoremediation remains a relatively emerging area of study, and further research is required to comprehend its long-term impacts on flora and ecosystems. There are also concerns regarding the potential risks associated with NPs, such as their potential toxicity and environmental impact. In conclusion, both phytoremediation and nano-phytoremediation are valuable tools for addressing environmental pollution. Phytoremediation provides a natural and sustainable approach that has been used successfully in many cases. On the other hand, nano-phytoremediation offers the potential for enhanced efficiency and targeted treatment of contaminants.

15.9 Environmental Concerns

Nano-phytoremediation, while holding promise as an innovative environmental remediation technique, raises significant concerns about potential adverse effects on the environment. The use of NPs in this process introduces the risk of toxicity to plants, microorganisms, and other organisms in the ecosystem. Prolonged exposure to certain NMs may disrupt cellular processes and have long-term implications for the health of both terrestrial and aquatic ecosystems. Furthermore, there is a potential for bioaccumulation and biomagnification of NPs, posing risks to organisms higher up the food chain, including animals and humans. The mobility of NPs in soil and the possibility of leaching into groundwater are additional concerns, threatening aquatic ecosystems and potentially impacting drinking water sources. The release of nanoparticles (NPs) into the environment endangers soil-microbe populations (Table 15.5). The effect of NPs on microbial function can be evaluated by measuring soil respiration and enzyme activity (Simonin and Richaume 2015). The unintended alteration of microbial communities in the rhizosphere and soil, as well as the potential development of resistance in plant species, could have cascading effects on ecosystem dynamics. Ethical and social considerations, including public acceptance and equitable access to benefits and risks, should also be carefully addressed.

The process of nano-remediation involves the use of zero-valent iron particles as NPs, which serve to facilitate the degradation of contaminants or immobilize them within soil matrices (Machado et al. 2017). Presently, zero-valent iron NPs are commonly utilized for the remediation of heavy metal contamination (Gil-Díaz et al. 2017) and can also be combined with electro-remediation to eliminate organic pollutants (Gomes et al. 2016). The practical implementation of nano-remediation in emerging economies faces challenges primarily due to the high costs involved in acquiring or producing the necessary amounts of zero-valent iron nanoparticles for extensive remediation projects (Shmaefsky 2020).

Recent research has shown that the rising application of nanoparticles in agriculture has led to negative consequences. Klaine et al. (2008) suggested that even though many natural nanoparticles exist in the environment, synthetic nanoparticles may exhibit different behaviors. Reports suggest that the presence of NPs in soils alters the pH of the soil, which is a crucial factor affecting various properties such as nutrient accessibility, microbial dynamics, plant growth and development, and overall soil ecosystem health. Higher concentrations of NPs are linked to reduced dehydrogenase activity, thus disrupting the balance of soil quality and nutrition (Jośko et al. 2014). The persistent release of NPs in the environment is likely to

Table 15.5 Application of NPs in different abiotic stresses

Abiotic stresses	Nanoparticles	Plant species	Physiological responses	References
Drought stress	Nano TiO_2	Basil (*Ocimum basilicum* L.)	Improving the toxic consequences of drought stress on basil plants	Kiapour et al. (2015)
	Nano SiO_2	*Crataegus* sp.	Rate of photosynthesis, stomatal conductance, and plant biomass all displayed a strong positive effect. Conversely, the levels of chlorophyll and carotenoids did not show significant results	Ashkavand et al. (2015)
Salinity stress	Nano ZnO	Sunflower (*Helianthus annuus* L.)	Enhanced growth, net rate of CO_2 absorption, concentration of sub-stomatal CO_2, chlorophyll levels, Fv/Fm ratio, and zinc levels, along with a reduction in sodium concentration in leaves	Torabian et al. (2016)
	Nano SiO_2	Faba bean (*Vicia faba* L.)	Improving seed sprouting, growth characteristics, antioxidant enzyme functionality, moisture content, and total production	Qados and Moftah (2015)
Flooding stress	Nano Ag	Saffron (*Crocus sativus*)	Blocking ethylene signals enhances root growth	Rezvani et al. (2012)
	Nano Al_2O_3	Soybean (*Glycine max* L.)	Rapid development, regulation of energy metabolism, and cellular death	Mustafa et al. (2015)
Heavy metal toxicity	ZnO NPs	*Leucaena leucocephala*	Reduced genotoxic effects of Cd and Pb, enhanced antioxidant defense mechanisms	Venkatachalam et al. (2017)
	TiO_2 NPs	Maize	Reduced Cd content	Lian et al. (2020)
	ZnO & Fe NPs	Wheat	Enhanced growth, decreased cadmium absorption, and lower oxidative stress	Rizwan et al. (2019)
	ZnO NPs	Spinach, parsley, cilantro	Reduced Pb and Cd toxicity	Sharifan et al. (2020)
	Mel-AuNPs Rice	Rice	Improved melatonin absorption and reduced cadmium uptake	Jiang et al. (2021)

result in their ubiquity, with NPs invading the food web at different trophic levels. This, in turn, can lead to toxicological impacts on various living organisms. Therefore, it is imperative to prioritize environmental concerns and develop NPs in a sustainable manner before utilizing them for bioremediation purposes (Prakash 2023; Ali et al. 2023). To ensure the responsible deployment of nano-phytoremediation, comprehensive research on the environmental fate and toxicity of NMs, coupled with robust risk assessments and regulatory frameworks, is imperative. Regular monitoring of sites where nano-phytoremediation is applied is essential to detect and mitigate any unforeseen environmental consequences.

15.10 Conclusion

Progression in research of nanotechnology requires the understanding of the transportation mechanism of NMs in the environment and also any toxicological effect on the plants or the environment. However, for a high-resolution clean-up process, a suitable selection of plants and NMs for pollutant removal is mandatory for the up-gradation of ecosphere management. Nanotechnology holds the promise of providing environmentally friendly solutions for cleaning and managing environmental issues without harming nature. Certain plants, fungi, and bacteria have been discovered to be hyperaccumulators, capable of absorbing significant amounts of metals, which makes them ideal for bioremediation in areas contaminated with heavy metals. Various forms of nanomaterials can be employed to eliminate environmental pollutants. Nanoparticles derived from these hyperaccumulator species have been used to extract heavy metals from contaminated locations. However, it is important to comprehend the mechanisms involved in the transport of nanomaterials to the environment to evaluate any possible toxic implications for the plants or surroundings. Selecting the right plant species and nanomaterials for the absorption of contaminants is essential, alongside optimizing agronomic management for an effective remediation process. Due to its interdisciplinary nature, capacity for innovation, proven track record, and potential benefits, nanotechnology is increasingly acknowledged as a significant field of research.

Over the last 10 years, scientists have increasingly turned to nanotechnology to investigate new methods for remediating heavy metals. Nano-remediation utilizes nanoparticles (NPs) to eliminate heavy metals from soil, air, vegetation, and water. Numerous studies have shown how NPs can help reduce the uptake of heavy metals by plants. This review emphasizes the expanding field of research focused on using NPs for the environmental cleanup of heavy metals. The effects of NPs introduced into the natural environment remain unclear. Although nanotechnology presents promising strategies for decreasing heavy metal pollution, it's vital to address the possible negative consequences for the ecosystem. Excessive use of specific antibacterial NPs, like graphene derivatives, might disrupt the ecological balance in soil and aquatic environments. More research is needed to comprehend the behavior and potential dangers of NPs in the environment. Processes such as aggregation are crucial to understanding NP outcomes, as they influence mobility, toxicity, and reactivity. Aggregation leads to a decrease in surface area and an increase in particle size, which results in NPs becoming less reactive and mobile, thereby reducing potential harmful effects in nature. Future studies should aim at elucidating the behavior and risks associated with NPs when used for environmental remediation. Furthermore, it is important to create more environmentally benign NPs that can decompose naturally, guaranteeing safe and sustainable remediation practices. However, nano-phytoremediation holds great potential; it's essential to proceed with caution and consider the ethical, social, and environmental implications of deploying NMs. Long-term studies on the ecological impact and safety of these technologies are crucial to ensuring their sustainability and minimizing unintended consequences. Responsible research and development, coupled with regulatory frameworks, will play a crucial role in shaping the future of nano-phytoremediation in the context of climate change.

References

Abbas S, Nasreen S, Haroon A, Ashraf MA (2020a) Synthesis of silver and copper nanoparticles from plants and application as adsorbents for naphthalene decontamination. Saudi J Biol Sci 27(4):1016–1023

Abbas Q, Yousaf B, Ali MU, Munir MAM et al (2020b) Transformation pathways and fate of engineered nanoparticles (ENPs) in distinct interactive environmental compartments: a review. Environ Int 138:105646

Afreen S, Omar RA, Talreja N, Chauhan D et al (2018) Carbon-based nanostructured materials for energy and environmental remediation applications. In: Prasad R, Aranda E (eds) Approaches in bioremediation: the new era of environmental microbiology and nanobiotechnology. Springer, Cham, pp 369–392

Alebachew N, Ananda Murthy HC, Abdisa B, Demissie TB (2023) Important features of nanomaterials for environmental remediation. In: Green nanoremediation: sustainable management of environmental pollution. Springer, Cham, pp 1–23

Ali H, Khan E (2018) What are heavy metals? Long-standing controversy over the scientific use of the term 'heavy metals'- proposal of a comprehensive definition. Toxicol Environ Chem 100(1): 6–19

Ali H, Naseer M, Sajad MA (2013) Phytoremediation of heavy metals by *Trifolium alexandrinum*. Int J Environ Sci 2:1459–1469

Ali S, Mehmood A, Khan N (2021) Uptake, translocation, and consequences of nanomaterials on plant growth and stress adaptation. J Nanomater 2021:1–17

Ali Q, Zia MA, Kamran M, Shabaan M et al (2023) Nanoremediation for heavy metal contamination: a review. Hybrid Adv 4:100091

Aragay, G., & Merkoçi, A. (2012). Nanomaterials application in electrochemical detection of heavy metals. Electrochimica Acta 84:49–61

Ashkavand P, Tabari M, Zarafshar M, Tomásková I et al (2015) Effect of SiO_2 nanoparticles on drought resistance in hawthorn seedlings. Leśne Prace Badawcze 76(4):350

Astefanei A, Núñez O, Galceran MT (2015) Characterisation and determination of fullerenes: a critical review. Anal Chim Acta 882:1–21

Bardos P, Merly C, Kvapil P, Koschitzky HP (2018) Status of nanoremediation and its potential for future deployment: risk-benefit and benchmarking appraisals. Remediat J 28(3):43–56

Bhuiyan MSH, Miah MY, Paul SC, Aka TD et al (2020) Green synthesis of iron oxide nanoparticle using *Carica papaya* leaf extract: application for photocatalytic degradation of remazol yellow RR dye and antibacterial activity. Heliyon 6(8):e04603

Cao Z, Wang S, Wang T, Chang Z et al (2015) Using contaminated plants involved in phytoremediation for anaerobic digestion. Int J Phytoremediation 17(3):201–207

Chokkalingam M, Rupa EJ, Huo Y, Mathiyalagan R et al (2019) Photocatalytic degradation of industrial dyes using Ag and Au nanoparticles synthesized from *Angelica gigas* ribbed stem extracts. Optik 185:1213–1219

Conesa HM, Evangelou MW, Robinson BH, Schulin R (2012) A critical view of current state of phytotechnologies to remediate soils: still a promising tool? Sci World J 2012:173829

Dai S, Wang B, Song Y, Xie Z et al (2021) Astaxanthin and its gold nanoparticles mitigate cadmium toxicity in rice by inhibiting cadmium translocation and uptake. Sci Total Environ 786:147496

Dejen KD, Sabir FK, Ananda Murthy HC, Ayanie GT et al (2023) Green synthesis of nanomaterials for environmental remediation. In: Green nanoremediation: sustainable management of environmental Pollution. Springer International Publishing, Cham, pp 27–65

DeLorenzo V (2018) Biodegradation and bioremediation: an introduction. Consequences of microbial interactions with hydrocarbons, oils, and lipids: biodegradation and bioremediation, Handbook of hydrocarbon and lipid microbiology. Springer, Cham, pp 1–21

Dhiman N, Kumar R, Kaushik A, Kumar R (2021) Nanotechnology for the remediation of heavy metals. In: New frontiers of nanomaterials in environmental science, pp 145–164

Dhir, B. (2021). Plant responses to heavy metals during cultivation in mining dump sites. In Phytorestoration of Abandoned Mining and Oil Drilling Sites (pp. 359–372). Elsevier

Dörner L, Cancellieri C, Rheingans B, Walter M et al (2019) Cost-effective sol-gel synthesis of porous CuO nanoparticle aggregates with tunable specific surface area. Sci Rep 9(1):11758

Ebrahimzadeh MA, Naghizadeh A, Amiri O, Shirzadi-Ahodashti M et al (2020) Green and facile synthesis of Ag nanoparticles using *Crataegus pentagyna* fruit extract (CP-AgNPs) for organic pollution dyes degradation and antibacterial application. Bioorg Chem 94:103425

El-Ramady, H., Alshaal, T., El-Henawy, A., Abdalla, N., Taha, H., Elmahrouk, M., ... & Domokos-Szabolcsy, É. (2017). Environmental nanoremediation under changing climate. Environment, Biodiversity and Soil Security, 1(2017), 109–128

Faiz S, Shah AA, Naveed NH, Nijabat A et al (2022) Synergistic application of silver nanoparticles and indole acetic acid alleviate cadmium-induced stress and improve growth of *Daucus carota* L. Chemosphere 290:133200

Gao M, Chang X, Yang Y, Song Z (2020) Foliar graphene oxide treatment increases photosynthetic capacity and reduces oxidative stress in cadmium-stressed lettuce. Plant Physiol Biochem 154: 287–294

Geneti ST, Mekonnen GA, Murthy HC, Mohammed ET et al (2022) Biogenic synthesis of magnetite nanoparticles using leaf extract of *Thymus schimperi* and their application for monocomponent removal of chromium and mercury ions from aqueous solution. J Nanomater 1:1–15

Gil-Díaz M, Pinilla P, Alonso J, Lobo MC (2017) Viability of a nanoremediation process in single or multi-metalloid contaminated soils. J Hazard Mater 321:812–819

Gomes HI, Fan G, Ottosen LM, Dias-Ferreira C et al (2016) Nanoremediation coupled to electrokinetics for PCB removal from soil. In: Ribeiro A, Mateus E, Couto N (eds) Electrokinetics across disciplines and continents: new strategies for sustainable development. Springer, Cham, pp 331–350

Goutam SP, Saxena G, Singh V, Yadav AK et al (2018) Green synthesis of TiO2 nanoparticles using leaf extract of *Jatropha curcas* L. for photocatalytic degradation of tannery wastewater. Chem Eng J 336:386–396

Guerra AM, Aimedieu P, Bornert M, Cui YJ et al (2018a) Analysis of the structural changes of a pellet/powder bentonite mixture upon wetting by X-ray computed microtomography. Appl Clay Sci 165:164–169

Guerra FD, Attia MF, Whitehead DC, Alexis F (2018b) Nanotechnology for environmental remediation: materials and applications. Molecules 23(7):1760

Islam F, Shohag S, Uddin MJ, Islam MR et al (2022) Exploring the journey of zinc oxide nanoparticles (ZnO-NPs) toward biomedical applications. Materials 15(6):2160

Jalilian F, Chahardoli A, Sadrjavadi K, Fattahi A et al (2020) Green synthesized silver nanoparticle from *Allium ampeloprasum* aqueous extract: characterization, antioxidant activities, antibacterial and cytotoxicity effects. Adv Powder Technol 31(3):1323–1332
Ji P, Sun T, Song Y, Ackland ML et al (2011) Strategies for enhancing the phytoremediation of cadmium-contaminated agricultural soils by *Solanum nigrum* L. Environ Pollut 159(3):762–768
Jiamjitrpanich W, Parkpian P, Polprasert C, Kosanlavit R (2012) The tolerance efficiency of *Panicum maximum* and *Helianthus annuus* in TNT-contaminated soil and nZVI-contaminated soil. J Environ Sci Health A 47:1506–1513
Jiang M, Dai S, Wang B, Xie Z et al (2021) Gold nanoparticles synthesized using melatonin suppress cadmium uptake and alleviate its toxicity in rice. Environ Sci Nano 8(4):1042–1056
Jośko I, Oleszczuk P, Futa B (2014) The effect of inorganic nanoparticles (ZnO, Cr2O3, CuO and Ni) and their bulk counterparts on enzyme activities in different soils. Geoderma 232:528–537
Kamran U, Bhatti HN, Iqbal M, Jamil S et al (2019) Biogenic synthesis, characterization and investigation of photocatalytic and antimicrobial activity of manganese nanoparticles synthesized from Cinnamomum verum bark extract. J Mol Struct 1179:532–539
Kanel SR, Manning B, Charlet L, Choi H (2005) Removal of arsenic (III) from groundwater by nanoscale zero-valent iron. Environ Sci Technol 39(5):1291–1298
Karn B, Kuiken T, Otto M (2009) Nanotechnology and in situ remediation: a review of the benefits and potential risks. Environ Health Perspect 117:1813–1831
Khan AG (2020) Promises and potential of in-situ nano-phytoremediation strategy to mycorrhizo-remediate heavy metal contaminated soils using non-food bioenergy crops (*Vetiver zizinoides* & *Cannabis sativa*). Int J Phytoremediation 22:900–915. https://doi.org/10.1080/15226514.2020.1774504
Khan MF, Murphy CD (2023) Environmental remediation by novel nanomaterials and fungi with high-degradation capacity of hazardous contaminants. In: Bio and Nanoremediation of hazardous environmental pollutants. CRC Press, pp 283–310
Kiapour H, Moaveni P, Habibi D (2015) Evaluation of the application of gibbrellic acid and titanium dioxide nanoparticles under drought stress on some traits of basil (*Ocimum basilicum* L.). Int. J. Agron. Agri. Res 6(4):138–150
Klaine SJ, Alvarez PJ, Batley GE, Fernandes TF et al (2008) Nanomaterials in the environment: behavior, fate, bioavailability, and effects. Environ Toxicol Chem An Int J 27(9):1825–1851
Kolahalam LA, Viswanath IK, Diwakar BS, Govindh B et al (2019) Review on nanomaterials: synthesis and applications. Mater Today Proc 18:2182–2190
Kumar P, Kumar A, Kumar R (2021) Phytoremediation and nanoremediation. In: Kumar R, Kumar R, Kaur G (eds) New frontiers of nanomaterials in environmental science. Springer, Singapore, pp 281–297
Li J, Wang X, Zhao G, Chen C et al (2018) Metal-organic framework-based materials: superior adsorbents for the capture of toxic and radioactive metal ions. Chem Soc Rev 47(7):2322–2356
Lian J, Zhao L, Wu J, Xiong H et al (2020) Foliar spray of TiO2 nanoparticles prevails over root application in reducing Cd accumulation and mitigating cd-induced phytotoxicity in maize (Zea mays L.). Chemosphere 239:124794
López-Villarreal J, Castro-Peña VC, Solis-Pomar F, Gutierrez-Lazos CD et al (2019) Thermal treatment of ZnO seed layer into atomic layer deposition to improve crystal quality of ZnO nanorod arrays. J Sci Technol Appl 6:108–117
Machado S, Pinto SL, Grosso JP, Nouws HP et al (2013) Green production of zero-valent iron nanoparticles using tree leaf extracts. Sci Total Environ 445–446:1–8
Machado S, Pacheco JG, Nouws HPA, Albergaria JT (2017) Green zero-valent iron nanoparticles for the degradation of amoxicillin. Int J Environ Sci Technol 14:1109–1118
Mahmood Q, Rashid A, Ahmad SS, Azim MR et al (2012) Current status of toxic metals addition to environment and its consequences. In: Anjum N, Ahmad I, Pereira M, Duarte A, Umar S, Khan N (eds) The plant family Brassicaceae: contribution towards phytoremediation. Springer, Dordrecht, pp 35–69

Mahmoud AED, Al-Qahtani KM, Alflaij SO, Al-Qahtani SF et al (2021) Green copper oxide nanoparticles for lead, nickel, and cadmium removal from contaminated water. Sci Rep 11(1): 12547

Masarovičová E, Kráľová K (2012) Plant-heavy metal interaction: phytoremediation, biofortification and nanoparticles. In: Advances in selected plant physiology aspects. InTech, Rijeka, pp 75–102

Mitton FM, Gonzalez M, Monserrat JM, Miglioranza KS (2016) Potential use of edible crops in the phytoremediation of endosulfan residues in soil. Chemosphere 148:300–306

Moghadas MRS, Motamedi E, Nasiri J, Naghavi MR, Sabokdast M (2020) Proficient dye removal from water using biogenic silver nanoparticles prepared through solid-state synthetic route. Heliyon 6(8):e04730

Mustafa, M. R., Mervat, S. S., Safaa, R. E. L., Ebtihal, M. A. E., & Magdi, T. A. (2015). Exogenous α-tocopherol has a beneficial effect on Glycine max (L.) plants irrigated with diluted sea water. The Journal of Horticultural Science and Biotechnology, 90(2):195–202

Mourdikoudis, S., Pallares, R. M., & Thanh, N. T. (2018). Characterization techniques for nanoparticles: comparison and complementarity upon studying nanoparticle properties. Nanoscale, 10(27), 12871–12934

Mukhopadhyay R, Sarkar B, Khan E, Alessi DS et al (2022) Nanomaterials for sustainable remediation of chemical contaminants in water and soil. Crit Rev Environ Sci Technol 52(15):2611–2660

Nwaadinigwe AO, Ugwu EC (2018) Overview of nano-phytoremediation applicants springer. Champions 34(6):377–382

Palencia M, García-Quintero A (2023) Green synthesis of nanomaterials and their use in bio-and nanoremediation. In: Bio and nanoremediation of hazardous environmental pollutants. CRC Press, pp 195–229

Pandey VC, Bajpai O, Singh N (2016) Energy crops in sustainable phytoremediation. Renew Sust Energ Rev 54:58–73

Pandit C, Roy A, Ghotekar S, Khusro A et al (2022) Biological agents for synthesis of nanoparticles and their applications. J King Saud Univ Sci 34(3):101869

Perera KMKG, Kuruppu KASS, Chamara AMR, Thiripuranathar G (2020) Characterization of spherical Ag nanoparticles synthesized from the agricultural wastes of *Garcinia mangostana* and *Nephelium lappaceum* and their applications as a photo catalyzer and fluorescence quencher. SN Appl Sci 2(12):1974

Pérez-de-Luque A (2017) Interaction of nanomaterials with plants: what do we need for real applications in agriculture? Front Environ Sci 5:12

Poguberovic SS, Krcmar DM, Maletic SP, Konya Z et al (2016) Removal of As(III) and Cr (VI) from aqueous solutions using green zero-valent iron nanoparticles produced by oak, mulberry and cherry leaf extracts. Ecol Eng 90:42–49

Prabakaran K, Li J, Anandkumar A, Leng Z et al (2019) Managing environmental contamination through phytoremediation by invasive plants: a review. Ecol Eng 138:28–37

Pradhan S, Patra P, Mitra S, Dey KK (2014) Manganese nanoparticles: impact on non-nodulated plant as a potent enhancer in nitrogen metabolism and toxicity study both *in vivo* and *in vitro*. J Agric Food Chem 62(35):8777–8785. https://doi.org/10.1021/jf502716c

Prakash P (2023) Nano-phytoremediation of heavy metals from soil: a critical review. Pollutants 3(3):360–380

Purushothaman B, Kannan S (2023) Green-based nanomaterials and plants in Nano-phytoremediation strategies. In: Green nanoremediation: sustainable management of environmental pollution. Springer, Cham, pp 277–288

Qados AA, Moftah AE (2015) Influence of silicon and nano-silicon on germination, growth and yield of faba bean (*Vicia faba* L.) under salt stress conditions. Am J Exp Agric 5(6):509–524

Qian Y, Qin C, Chen M, Lin S (2020) Nanotechnology in soil remediation-applications vs. implications. Ecotoxicol Environ Saf 201:110815

Qu X, Alvarez PJ, Li Q (2013) Applications of nanotechnology in water and wastewater treatment. Water Res 47(12):3931–3946

Rahman A, Kumar S, Nawaz T (2020) Biosynthesis of nanomaterials using algae. In: Microalgae cultivation for biofuels production. Academic Press, pp 265–279

Rai PK, Kumar V, Lee S, Raza N et al (2018) Nanoparticle-plant interaction: implications in energy, environment and agriculture. Environ Int 119:1–19

Rajendrachari S, Taslimi P, Karaoglanli AC, Uzun O et al (2021) Photocatalytic degradation of Rhodamine B (RhB) dye in waste water and enzymatic inhibition study using cauliflower shaped ZnO nanoparticles synthesized by a novel one-pot green synthesis method. Arab J Chem 14(6):103180

Rezvani N, Sorooshzadeh A, Farhadi N (2012) Effect of nano-silver on growth of saffron in flooding stress. Int J Agric Biosyst Eng 6(1):11–16

Rizwan M, Singh M, Mitra CK, Morve RK (2014) Ecofriendly application of nanomaterials: nanobioremediation. J Nanoparts 2014:1. https://doi.org/10.1155/2014/431787

Rizwan M, Ali S, Ali B, Adrees M et al (2019) Zinc and iron oxide nanoparticles improved the plant growth and reduced the oxidative stress and cadmium concentration in wheat. Chemosphere 214:269–277

Roberts, E. S., Richards, J. H., Jaskot, R., & Dreher, K. L. (2003). Oxidative stress mediates air pollution particleinduced acute lung injury and molecular pathology. Inhalation toxicology, 15 (13):1327–1346

Roy A, Singh V, Sharma S, Ali D et al (2022) Antibacterial and dye degradation activity of green synthesized iron nanoparticles. J Nanomater 2022:1–6

Ruffini Castiglione, M., Frediani, M., Ravalli, C., Venora, G., & Cremonini, R. (2009). Cytological characterization of Vicia oroboides Wulfen in Jacq. Protoplasma, 236, 21–27

Sahoo SK, Panigrahi GK, Sahoo A, Pradhan AK et al (2021) Bio-hydrothermal synthesis of ZnO-ZnFe2O4 nanoparticles using *Psidium guajava* leaf extract: role in waste water remediation and plant immunity. J Clean Prod 318:128522

Salve P, Vinchurkar A, Raut R, Chondekar R et al (2022) An evaluation of antimicrobial, anticancer, anti-inflammatory and antioxidant activities of silver nanoparticles synthesized from leaf extract of Madhuca longifolia utilizing quantitative and qualitative methods. Molecules 27(19):6404

Sardar R, Ahmed S, Shah AA, Yasin NA (2022) Selenium nanoparticles reduced cadmium uptake, regulated nutritional homeostasis and antioxidative system in Coriandrum sativum grown in cadmium toxic conditions. Chemosphere 287:132332

Sarret G, Saumitou-Laprade P, Bert V, Proux O et al (2002) Forma of Zinc accumulated in the hyperaccumulator *Arabidopsis halleri*. Plant Physiol 130:1815–1826

Sattelmatcher B (2001) The apoplast and its significance for plant mineral nutrition. New Phytol 149:167–192

Schwab F, Zhai G, Kern M, Turner A et al (2016) Barriers, pathways and processes for uptake, translocation and accumulation of nanomaterials in plants-critical review. Nanotoxicology 10(3):257–278

Shaik AM, David Raju M, Reddy RS, D. (2020) Green synthesis of zinc oxide nanoparticles using aqueous root extract of Sphagneticola trilobata Lin and investigate its role in toxic metal removal, sowing germination and fostering of plant growth. Inorg Nano-Met Chem 50(7): 569–579

Sharifan H, Moore J, Ma X (2020) Zinc oxide (ZnO) nanoparticles elevated iron and copper contents and mitigated the bioavailability of lead and cadmium in different leafy greens. Ecotoxicol Environ Saf 191:110177

Shmaefsky BR (2020) Principles of phytoremediation. In: Phytoremediation: In-situ applications. Springer, Cham, pp 1–26

Simonin M, Richaume A (2015) Impact of engineered nanoparticles on the activity, abundance, and diversity of soil microbial communities: a review. Environ Sci Pollut Res 22:13710–13723

Sohail MI, Waris AA, Ayub MA, Usman M et al (2019) Environmental application of nanomaterials: a promise to sustainable future. In: Comprehensive analytical chemistry, vol 87. Elsevier, pp 1–54
Srivastav A, Yadav KK, Yadav S, Gupta N et al (2018) Nano-phytoremediation of pollutants from contaminated soil environment: current scenario and future prospects. In: Ansari A, Gill S, Gill R, Lanza G, Newman L (eds) Phytoremediation: management of environmental contaminants, vol 6, pp 383–401
Tavker N, Yadav VK, Yadav KK, Cabral-Pinto MM et al (2021) Removal of cadmium and chromium by mixture of silver nanoparticles and nano-fibrillated cellulose isolated from waste peels of citrus sinensis. Polymers 13(2):234
Thangadurai D, Sangeetha J, Prasad R (2020) Nanotechnology for food, agriculture and environment. In: Nanotechnology in the life sciences. Springer, Cham, pp 378–393
Tripathi, A., & Shukla, S. (2018). Phytoremediation of lead and copper using ficus virens and azadirachta indica in Bareilly, Uttar Pradesh, India. Pollution Research, 37(4), 1109–1116
Torabian S, Zahedi M, Khoshgoftar AH (2016) Effects of foliar spray of two kinds of zinc oxide on the growth and ion concentration of sunflower cultivars under salt stress. J Plant Nutr 39(2): 172–180
Tripathi DK, Singh S, Singh VP, Prasad SM et al (2017) Silicon nanoparticles more effectively alleviated UV-B stress than silicon in wheat (Triticum aestivum) seedlings. Plant Physiol Biochem 110:70–81
Tsai CH, Chang WC, Saikia D, Wu CE et al (2016) Functionalization of cubic mesoporous silica SBA-16 with carboxylic acid via one-pot synthesis route for effective removal of cationic dyes. J Hazard Mater 309:236–248
Tungittiplakorn W, Cohen C, Lion LW (2005) Engineered polymeric nanoparticles for bioremediation of hydrophobic contaminants. IGI Global 39:202–219
Vasantharaj S, Shivakumar P, Sathiyavimal S, Senthilkumar P et al (2023) Antibacterial activity and photocatalytic dye degradation of copper oxide nanoparticles (CuONPs) using Justicia gendarussa. Appl Nanosci 13(3):2295–2302
Venkatachalam P, Jayaraj M, Manikandan R, Geetha N et al (2017) Zinc oxide nanoparticles (ZnONPs) alleviate heavy metal-induced toxicity in Leucaena leucocephala seedlings: a physiochemical analysis. Plant Physiol Biochem 110:59–69
Villalobos R, Medina C, Canales C, Melendrez M et al (2019a) Development of nanocomposite materials with thermostable matrix from nanoreinforcement of titanium dioxide (TNPS and TNTS) and zinc oxide (ZNPS and ZNBS) on epoxy resin matrix. J Sci Technol Appl 6:65–78
Villalobos RJ, Rojas D, Díaz A, Ramírez J et al (2019b) Synthesis and characterization of 1D metal oxide nano-structures: titanium dioxide nanotubes (TNTs) and zinc oxide nanobars (ZNBs). J Sci Technol Appl 6:79–95
Vyas S, Dhiman N, Negi S, Kumar R (2021) Management of waste using nanotechnology. In: New frontiers of nanomaterials in environmental science, pp 253–279
Wang CB, Zhang W (1997) Synthesizing nanoscale iron particles for rapid and complete dechlorination of TCE and PCBs. Environ Sci Technol 31:2154–2156
Wang S, Fu Y, Zheng S, Xu Y et al (2022) Phytotoxicity and accumulation of copper-based nanoparticles in brassica under cadmium stress. Nano 12(9):1497
Wijesinghe U, Thiripuranathar G, Iqbal H, Menaa F (2021) Biomimetic synthesis, characterization, and evaluation of fluorescence resonance energy transfer, photoluminescence, and photocatalytic activity of zinc oxide nanoparticles. Sustain For 13(4):2004
Wuana RA, Okieimann FE (2011) Heavy metal in contaminated soils: a review of sources, chemistry, risks and best available strategies for remediation. Commun Soil Sci Plant Anal 42:111–122
Xiao X, Wang Q, Owens G, Chiellini F et al (2019) Reduced graphene oxide/iron nanoparticles used for the removal of Pb (II) by one step green synthesis. J Colloid Interface Sci 557:598–607
Yan A, Wang Y, Tan S, Yusof N et al (2020) Phytoremediation: a promising approach for revegetation of heavy metal- polluted land. Front Plant Sci 11:359

Yang L, Ren Q, Ge S, Jiao Z et al (2022) Metalloids spatial distribution, accumulation, and potential health risk assessment in soil-wheat systems near a Pb/Zn smelter in Henan province, Central China. Int J Environ Res Public Health 19(5):2527

Zhang D, Ma XL, Gu Y, Huang H, Zhang GW (2020) Green synthesis of metallic nanoparticles and their potential applications to treat cancer. Front Chem 8:799

Chapter 16
An Overview of Current Application and Trends of Nanotechnological Approaches in the Dairy Industry: From Process to New Formulations

Patrícia Avelar Ribeiro, Jhonatan Rafael de Oliveira Bianchi, Marcos Silva de Sousa, Ana Luiza Schuenck Alberice Monteiro, Lucimara Gaziola de la Torre, Jane Sélia dos Reis Coimbra, and Igor José Boggione Santos

Abstract Nanotechnology deals with the class of materials that has been highlighted in materials science and engineering over the past decade. Many classes of nanomaterials (NMs) are classified as food grade and can potentially improve dairy products. The properties of NMs benefit dairy quality control, product properties, and nutritional values. Therefore, this review aims to introduce how nanotechnology is helpful in the dairy industry, showing the recent aspects of its application. For this purpose, we present the main technological approaches to nanofabrication using traditional bulk technologies and microfluidic approaches. Then, we discuss the recent aspects of NMs in the dairy industry, divided into two main approaches: the application of nanotechnology in the dairy industry process, i.e., the use of metallic nanoparticles (NPs) as nanosensors, biofilm control, and separation and purification processes. And second, the use of nanodelivery systems and emulsions in dairy products to improve texture, taste, and nutrition.

Keywords Nanomaterials · Nanosensors · Nutraceutical dairy products · Microbial contamination · Chemical adulterants

---

P. A. Ribeiro · M. S. de Sousa · A. L. S. A. Monteiro · I. J. B. Santos (✉)
Department of Chemistry, Biotechnology and Bioprocess Engineering, Federal University of São João del-Rei, Ouro Branco, MG, Brazil
e-mail: igorboggione@ufsj.edu.br

J. R. de Oliveira Bianchi · L. G. de la Torre
Department of Materials and Bioprocess Engineering, School of Chemical Engineering, University of Campinas, Campinas, Brazil
e-mail: ltorre@unicamp.br

J. S. dos Reis Coimbra
Department of Food Technology, Federal University of Viçosa (UFV), Viçosa, MG, Brazil
e-mail: jcoimbra@ufv.br

S. K. Yadav, A. Kumari (eds.), *Nanobiotechnology in Agri-food Sector*, Sustainability Sciences in Asia and Africa, https://doi.org/10.1007/978-981-96-7756-6_16

16.1 Introduction

The dairy industry has witnessed significant consumption growth globally, with market trends projecting a continuous increase in production by 35.0% by 2030. This surge in interest from consumers is driven by the nutritional and health benefits offered by dairy products, particularly their positive impact on the gut microbiota, which contributes to overall health and longevity (Chen et al. 2019a, b). However, the dairy industry faces numerous challenges in ensuring the production of high-quality products while adhering to regulatory standards (Kelly et al. 2020; Hernández-Castellano et al. 2019).

The dairy industry faces significant challenges in ensuring the supply of dairy products, which increases annually due to the growing global population (Hernández-Castellano et al. 2019). These challenges can be divided into two main areas: within the production process and in the formulation of new products. Throughout its production chain, various limitations exist in the control of packaging quality, separation, and filtration technology, primarily in microbiological control, which is essential to ensure food safety and product quality. Consequently, these limitations can result in losses in both the quality and safety of food products (Kuswandi and Jumina. 2020).

To address these challenges, new technologies are being adopted, with nanomaterial-derived technologies emerging as a promising solution. These technologies aim to enhance productivity within the production chain through the implementation of nanosensors, active and smart packaging, nanodisinfectants, and nanofiltration systems (Fanzo et al. 2023; Singh et al. 2021).

For the dairy products themselves, the challenges revolve around shelf life, sensory attributes, and nutritional content (Kuswandi and Jumina. 2020). Dairy products are susceptible to spoilage and microbial contamination, requiring effective preservation methods to extend their shelf life without compromising taste or texture (Silva et al. 2018). Furthermore, meeting consumer demand for healthier options while preserving the nutritional value of dairy products is a challenge, especially when considering factors such as fat content, sugar content, and fortification with essential nutrients (Fanzo et al. 2023).

Additionally, NMs are utilized in dairy product formulation to improve technical-functional properties, stability, and the controlled release of bioactive components, leading to enhanced flavor and texture (Dey et al. 2021; Hariyama et al. 2022; Liu et al. 2019). Furthermore, the use of NPs can positively impact microbiological control, improving safety and extending the shelf life of dairy products (Papadochristopoulos et al. 2021).

Nanotechnology has played a pivotal role in the development of new dairy products, offering versatility and a wide range of applications (Kim et al. 2018). The increasing interest in nanotechnology within the dairy industry is evident from the significant rise in publications and citations over the last 5 years (Fig. 16.1a). Figure 16.1b and c depict key authors and journals in the field, demonstrating their relevance and impact.

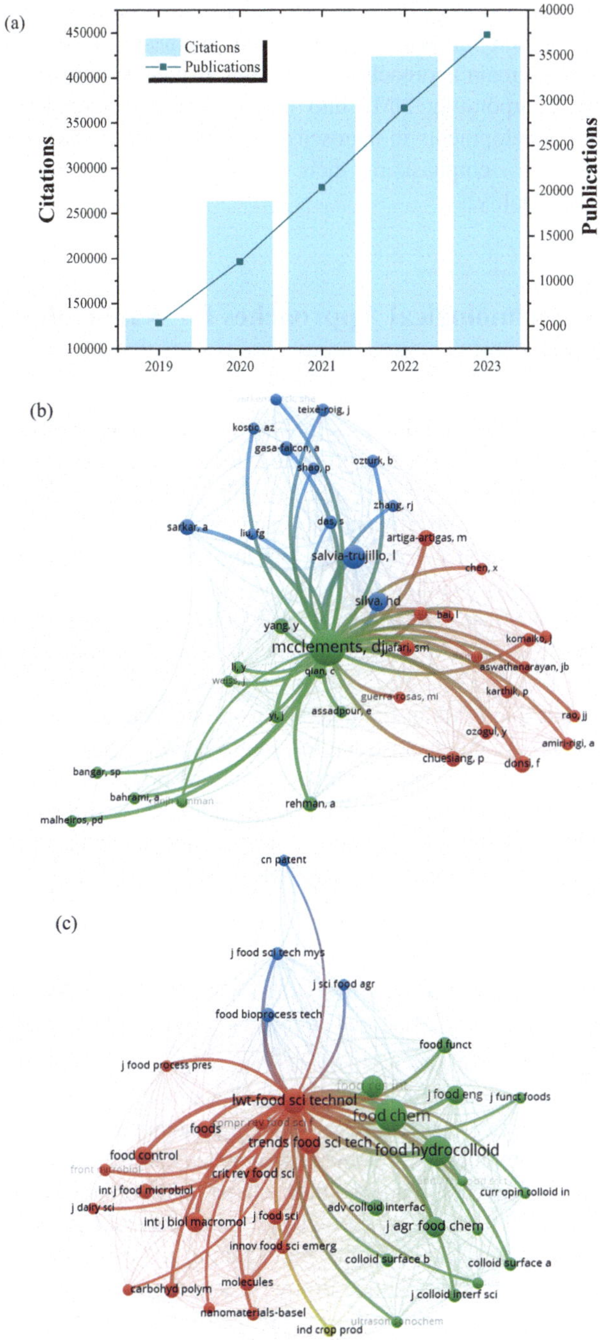

Fig. 16.1 (**a**) Cumulative number of publications and citations (2019–2023) on Web of Science using the keywords nanotechnology, dairy products, dairy industry, bioactive compounds, nanosensors for dairy products, and nanoemulsions. (**b**) Overlay visualization based on authorship distribution. (**c**) Visualization of overlap based on the distribution of journals

This review aims to provide a comprehensive overview of nanotechnology applications in the dairy industry. It covers the synthesis of NMs, technological approaches, and two main approaches: integrating nanotechnology into the production chain and incorporating NMs into dairy product formulation. The review addresses recent developments in nanosensors, nanosanitizers, nanoemulsions, and nanodelivery systems, emphasizing their significance in improving food safety, quality, and functionality.

16.2 Main Technological Approaches for Nanomaterials Synthesis

Over the years, different methodologies for NM synthesis have been developed. The researchers seek to understand the physical-chemical mechanisms, reduce costs, reduce the environmental impact, increase the yield, and develop scale-up approaches for industry applications. The main challenge is the reproducibility of synthesized NM once the size distribution variation rolls all the material's properties. Consequently, rigorous synthesis control is necessary to maintain regulatory approval and safe NM use. Therefore, several technological approaches have been studied. The synthesis of NM is done through bulk methodologies or microfluidic approaches.

The most applicable NMs in the dairy industry are inorganic NPs (silver, gold, and zinc oxide) and organic nanostructures (chitosan, whey protein, and liposome). This section summarizes the main synthesis route of principal NMs for the dairy industry. A rational discussion of recent methodologies is discussed, helping to choose the best methodology depending on NM and the final application. The methodologies are divided into two main topics: the bulk and microfluidics routes.

16.2.1 Current Approaches to Bulk Nanomaterial Synthesis

Inorganic NP synthesis follows a chemical route with a metal precursor, reducing agent, and colloid dispersion stabilizer. The bulk approach synthesis is a top-down process where the ionic metal reduces its size to a nanometric scale. For example, Raza et al. (2016) synthesized silver nanoparticles (AgNP) using silver nitrate ($AgNO_3$) as an Ag precursor and sodium citrate and sodium borohydride as a reducing agent. The reaction was carried out under constant agitation (400 rpm) and three different temperatures (room temperature, ice bath, and boiling conditions). The reducing agent was added dropwise into the silver nitrate solution. All samples reduced silver in AgNP with a size range from 15 to 150 nm and mostly a spherical shape. To avoid agglomeration of the AgNP colloidal system, PVP (polyvinylpyrrolidone) was added to guarantee stabilization (Raza et al. 2016).

As the temperature affects AgNP size, understanding the impact of operational conditions for AgNP synthesis was necessary. Therefore, De Oliveira Bianchi et al. (2023) studied the effects of temperature, stirring, and reducing agent drop flow (sodium citrate and sodium borohydride) on AgNP size. Using the design of experiments (DOE), the authors discovered that the temperature did not affect the particle size; the AgNP size was affected by stirring velocity. The particle size ranges from 14.08 to 157.60 nm with a PDI lower than 0.4. The TEM images show spherical-shape particles. So, the chemical synthesis of AgNP can be controlled by changing the stirring.

Moreover, another reduction of silver nitrate reaction synthesizes AgNP. Verma et al. 2021 synthesized AgNP via a single-step silver mirror reaction (SMR) on a filter paper substrate. The silver nitrate was pondered on the paper, and NH_3 solution was added; the reaction resulted in silver oxide with a brown color. After that, D-glucose was added to catalyze the reaction between silver and cellulose, and the paper was washed to keep only AgNP—the AgNP growth on the paper surface had an irregular shape and size. Another inorganic NP useful to the dairy industry is gold nanoparticles (AuNP). The fundamental methodology is reducing gold (III) chloride trihydrate ($HAuCl_3$). Zou et al. (2021a, b) synthesized AuNP via bulk wet chemical—the $HAuCl_3$ solution under heating and stirring with trisodium citrate. The authors found AuNP to be 32.6 nm in size. Following the same methodology synthesized spherical AuNP with a homogeneous size of around 20.13 nm (Jia et al. 2019).

For once, zinc oxide nanostructures also interest new technologies applied in the dairy industry. Jyoti et al. (2022) recently synthesized nano ZnO using a hydrothermal technique—a reaction between zinc acetate in ethanol and sodium hydroxide at room temperature results in a ZnO colloidal solution. A hexamethylenetetramine and zinc nitrate were prepared for ZnO deposition. Then, both solutions were mixed, allowing nanorod ZnO deposition to form a uniform coating at 150 °C in a tube furnace. The length of nanorods was 2.4–2.6 μm, and the diameter ranged from 190 to 205 nm (Jyoti et al. 2022). Differently, Barbosa Costa et al. (2023) synthesized nano ZnO via dry-method mixing $ZnCl_2$, NaCl, and Na_2CO_3 (1:1:8). The powder was mixed into the ball mill for 6 h at 12 rps, after that, it was heated in a muffle furnace at 400 °C for 30 min. The authors obtained spherical nano ZnO with a 175.15 nm diameter and PDI 0.601.

On the other hand, organic nanostructure follows different synthesis principles. The main methodologies focus on bottom-up self-assembly, creating nanostructures with different sizes and morphologies. The self-assembly technique depends on the molecule's nature and can modify the nanoparticles' properties. The chitosan nanoparticle synthesis follows mainly two principles: the deprotonation method and tripolyphosphate addition (Kahdestani et al. 2021; Ribeiro et al. 2020). The chitosan must be solubilized under agitation with 1% acetic acid. Deprotonation of amino groups and the D-glucosamine groups' units is conducted, changing the pH value from 3.5 to 6.7 using NaOH. For the ionic crosslink TPP solution, pH 8 is dropwise into the chitosan solution under stirring until pH 4.35 or 5.16. Both methods result in different NP properties; as shown by Ribeiro et al. (2020), the deprotonation method results in chitosan nanoparticles with a surface charge from 16.1 to 18.3 mV and PDI from 0.912 to 0.944. The TPP-add method results in a surface charge from 22 to 24.1 mV and PDI from 0.478 to 0.551.

Self-assembly methodologies also provide ways to nanostructure protein synthesis. The modification of ionic strength, temperature, and organic solvent denatures the protein. Then, the agitation process can reassemble the structures until the desired size. D'Onofre Couto et al. (2021) used DOE to optimize the synthesis of β-lactoglobulin nanostructures (nano β-lg). The authors studied the effects of ionic strength (NaCl concentration), temperature, and time on nanoprotein size and PDI. The nanoprotein size ranges from 105.00 ± 13.71 to 581.70 ± 59.98 nm, and the PDI from 0.297 ± 0.059 to 0.666 ± 0.293. The best parameters were 100 °C, 60 min, and 2 mg/ml of β-lg in 50 mmol/L of NaCl solution. The TEM images showed a nanofibril shape. A similar approach was used for α-lactalbumin nanostructures (nano α-la). Bianchi et al. 2023a synthesized nano α-la by self-assembly methodology. The authors used DOE to determine ionic strength, temperature, and time effects on nanoprotein size and PDI. The NaCl salt concentration and temperature were significant parameters affecting the nanostructure protein. The ionic strength with NaCl 100 mmol/L was the best condition for synthesizing 100–200 nm nanoprotein at room condition and 10 min. The TEM images showed spherical structures due to the use of monovalent salt (Bianchi et al. 2023a).

The self-assembly methodology also provides a way for liposome synthesis. The bulk synthesis is via the ethanol injection methodology. The principle is injecting a lipid-ethanolic solution into a large volume of the aqueous solution. The lipids exposed to an unfavorable aqueous medium self-assemble themselves and precipitate in a boundary phase between water and ethanol. The process creates a bilayer of phospholipid fragments (BPF) (Charcosset et al. 2015; Gouda et al. 2021). The ethanol must be removed, and the BPF must be resuspended to form spherical particles. The layers of each particle depend on the amount of energy. Multicellular vesicles are easier to form because they need less energy than unicellular vesicles (Charcosset et al. 2015; Gouda et al. 2021). Despite that, the lipid composition also affects the vesicle formation and characteristics.

16.2.2 Current Approaches to Nanomaterial Synthesis via Microfluidics Route

Box 16.1 Microfluidics Technology

Principle: based on fluid flow into micrometer channels in low Reynolds' number (laminar regime) but with the mass transfer phenomenon guided by molecular diffusion. The small channel sizes and the diffusion of the

(continued)

Box 16.1 (continued)
molecules decrease the mass transfer resistance, so inertial effects are not present (Bayareh et al. 2020). The diffusion of molecules guides the reactions inside the microfluidics device (Ismagilov et al. 2000).

The microfluidic device works as micromixers where a low amount of fluid flows into the channels with high mixing effectivity and low Reynolds number (Bayareh et al. 2020; Liu et al. 2019). Consequently, surface effects are more relevant than macroscale reactions. Most nanoparticle synthesis using microfluidics have high yield, shape control, and minimal reaction amount (using fewer reagents) (Khizar et al. 2022; Niculescu et al. 2021b).

Microfluidics technology (Box 16.1) is widely applicable nowadays for NM synthesis. The most common materials for microdevice fabrication are inert elastomers poly(dimethylsiloxane) (PDMS), poly(methyl methacrylate) (PMMA), and glass (Lei 2014). The material selection depends on the microfabrication technique; generally, soft lithography and 3D printing are widely used to fabricate different geometries of microdevices. There are reviews about materials and microfabrication techniques (Niculescu et al. 2021a; Ren et al. 2013; Scott and Ali 2021). We will focus on using these microdevices for inorganic and organic NM synthesis.

The microdevices to synthesize NMs demand a good relationship between channel length and size. The time that the reagents flow into the channels and the diffusion of the molecules affect the reaction efficiency due to the time of molecules' contact. So, different microdevice geometries have been developed to improve NMs synthesis. Two different microdevice geometries can synthesize NMs: capillary geometry and planar geometry. The capillary devices are commonly manufactured using glass, so they have high chemical resistance and are inert to most substances. However, they are complex to fabricate, have difficult reproducibility, and are high-cost (Bianchi et al. 2023a, b; Duncanson et al. 2012). The planar microdevices are usually based on PDMS or PMMA, which are more flexible materials. The design is fabricated using soft lithography, and the most common geometries are T- or Y-junctions; the material's flexibility allows the fabrication of more complex geometries. Nonetheless, the materials are not inert to the same organic solvents and are complex structures for industrial applications (Bianchi et al. 2023a, b; McDonald and Whitesides 2002).

Microfluidics devices are widely used for metal nanoparticles with different sizes and geometries. Using droplet-based microfluidics, Kašpar et al. (2019) synthesized spheric silver nanoparticles from 12 to 22.3 nm size range (81% yield). The aqueous phase of tannic acid and silver nitrate is pumped into two separate channels, and the oil phase meets both solutions, forming a droplet emulsion. Each emulsion works as a microreactor collected at the end of the process. The oil flow rate was kept constant at 80 μL/h, and the aqueous phase flow range was 10, 16, and 32 μL/h. The flow rate changes roles with the particle size and reaction time inside the droplet. Also, modify the droplet regime formation. Different size ranges and geometry were obtained by

Kwak et al. (2018); the author used droplet microfluidics with two T-junctions. Both aqueous solutions (silver nitrate and branched polyethyleneimine) flow and meet at the first T-junction, where the mixing and reaction begin. Then, the mixed solution meets the oil in the second T-junction to droplet emulsion formation. The AgNP presents cubic geometry with a 4.37 ± 0.73 to 11.45 ± 2.00 size range depending on flow rate and a 71% yield. Other inorganic nanoparticles can be synthesized using microfluidic approaches like AuNP (Abalde-Cela et al. 2018; Adamo et al. 2020; Ye et al. 2020) and ZnO nanoparticles (Baruah et al. 2017; Hao et al. 2019; Zukas and Gupta 2017).

Moreover, organic nanostructure synthesis via microfluidics has been studied. Droplet-based microfluidics improves the self-assembly of proteins inside microchannels to nanoaggregate formation. Recently, Zhang et al. (2023) used a microdevice with three flow channels: oil as the external phase, ethanol as the middle phase, and silk fibroin as the internal phase. The ethanol mixes with protein, reducing the solubility and resulting in protein nucleation. The oil separates the mixed protein/ethanol into droplets to control the nucleation, forming monodisperse protein nanoparticles (Zhang et al. 2023). Depending on silk fibroin's flow rate and concentration, nanoparticle sizes range from 80 to 250 nm. The mechanism is an interesting alternative to different protein nanoparticle formations.

Pessoa et al. (2017) tested regular flow-focusing devices and synthesized chitosan nanoparticles with ATP particles as crosslink agents. The regular flow-focusing device results work with the chitosan flow, which has been hydrodynamically focused by two ATP solution flows, resulting in microfiber-like nanostructures. The particle size ranges from 80 to 120 μm. Nanostructure formation happens during the mixing process in the out channel and is guided by molecular diffusion. The flow rate ratio and channel size can modify the synthesis's morphology, size, and yield (Pessoa et al. 2017).

For once, liposome synthesis by microfluidics is an alternative to ethanol injection. Several parameter considerations include lipid nature, flow rate ratio, and microdevice geometry. The basic principle follows the self-aggregation process but inside the microchannels. The lipid/ethanol solution flows into a central channel, and the aqueous solution flows into two outside channels, creating a three-phase flow on the out channel. Due to the diffusion of the molecules, the water is mixed with an ethanolic solution, which decreases the solubilization of lipids and consequently forms micelles (Balbino et al. 2013). The lipid nature, lipid concentration, and flow rate of all phases are fundamental parameters for optimizing liposome synthesis (Balbino et al. 2013). Microfluidic synthesis shows less PDI than ethanol injection methodology with different liposome size possibilities.

Balbino et al. (2013) produced cationic liposomes testing different synthesis conditions using a single flow-focusing microfluidic device. The lipid concentration ranges from 9 to 92 mM, surface velocity from 0.08 to 0.16 and from FRR 8 to 18. The liposome size results range from 85.8 nm to 508.6 nm. The highest liposome concentration obtained was 7 mM, with high and low FRR concentrations. On the other hand, the lowest liposome concentration was obtained with low lipid concentration and high FRR synthesis conditions (Balbino et al. 2013). Also, the authors discovered that using a double flow-focusing microfluidic device, the liposome size and PDI are higher, even at the lowest lipid concentration (Balbino et al. 2013). Despite the advantages of using microfluidics, liposome production is poor due to the predominance of laminar flow limiting scale-up applications. So, alternatives to promote a better mix inside of microchannels, reducing mass transfer phenomena (Eş et al. 2020; Firmino et al. 2021). Eş et al. (2020) developed a microdevice to improve mixing by chaotic advection in herringbone-like units. The herringbone unit creates a chaotic fluid flow in the microdevice, improving the mixture even with a low flow rate (μL/min). The cationic liposome productivity increased 70 times compared to the common flow-focusing device with a size of around 200 nm and low PDI (Eş et al. 2020).

Firmino et al. (2021) developed a different microfluidic device based on chaotic advection to explore liposome synthesis at a high flow rate. The mold of the high-throughput microfluidic device was fabricated using a PLA lateral shape sealed in a glass base. Treated nylon with 200 μm was put in the center of the mold. The microdevice was fabricated with PDMS inside the mold. After the PDMS curing and nylon were removed, the device was read. The central channel is a 3D-twisted cross channel that creates a chaotic advection to improve mixing. The liposome size ranges from 73 to 176.4 nm, and PDI is 0.193 to 0.394. The high flow rate was 20 mL/min (73.0 ± 3.93 nm and PDI 0.394 ± 0.048). Also, the final lipid concentration in this condition was 1.67 mM. Therefore, the microdevice is an alternative to scaling up liposome synthesis. However, until now, another NP synthesis has not been tested in these microdevices.

The benefits of using microfluid technologies to increase the yield and quality of reactions to produce different NPs are well known and have been increasing over the past decade. However, the scale-up still is a challenge that requires the attention of research groups to transfer the technology to industrial scale.

16.3 Current Applications of Nanotechnology in the Dairy Process Chain

Nanotechnology has several applications due to its significant benefits, especially in the dairy sector (Saini et al. 2023). The NMs have different characteristics helpful for controlling product quality, enhancing sensory attributes, targeted delivery of compounds, safety of dairy products, and stabilization of nutraceuticals (Smykov 2020). As part of the process chain, most NMs are useful for fabricating nanosensors, used as sanitizers for biofilm control in tubulations and reactors, and in the purification and separation process with nanofillers (Malik et al. 2023).

16.3.1 Nanosensors

Adulterating and contaminating dairy products pose significant problems that impact human health, the economy, and trade, especially in developing countries. To mitigate these effects, authorities worldwide have established strict food safety standards (Himshweta and Singh 2023). Using NM-based sensors offers a promising solution due to their high sensitivity, stability, and ease of synthesis, helping to ensure pure and healthy milk for consumers (Kuswandi et al. 2017).

Nanosensors are advanced devices that use NMs to detect and convey information about NPs, transforming unique nanoscale processes into signals for analysis. They are highly precise and sensitive, capable of detecting specific physical or chemical phenomena at the nanometer scale (Olsen 2006). NPs offer excellent sensing capabilities, such as strong reactivity, higher conductivity, strength, unique magnetic properties, and a considerable volume/area ratio. So, nanosensors require a smaller amount of analyte, reduce the complexity and cost of assays, and have the advantage of early detection of pathogens, toxins, and chemicals during the processing and storage of dairy products (Thakur et al. 2022). The main NMs used in the manufacture of sensors, due to their properties of versatility, simple synthesis, adjustable performance, and high load capacity of biomolecules, include noble metal nanostructures, magnetic nanoparticles, carbon quantum dots, and metal-organic frameworks (Pu et al. 2021).

Box 16.2 Classification of Nanosensors

Sensing Technique

Nanosensors are fabricated by combining a receptor and a transducer. Any organic or inorganic substance can be the receptor, interacting with the analyte or its derivatives. The transducer's role is to convert this interaction into a measurable signal, which can be mechanical, electrochemical, or optical (Yin et al. 2013).

- *Mechanical:* are based on the ultrasensitive detection of forces, displacements, and changes in the system's mass, as well as the measurement of changes in mechanical forces at the molecular level (Munawar et al. 2019).
- *Electrochemical:* occurs from chemical reactions in which electrons donate or receive ions, and consequently, proportional signals are generated with the concentration of the analyte of interest in the solution. These emitted signals can be based on potentiometry, voltammetry, amperometry, and electrochemical impedance spectroscopy (Munawar et al. 2019).
- *Optical:* uses the conversion of light rays into quantifiable signals, colorimetric sensors use the visible change from the interaction of the sensor with the analyte, and fluorescent sensors using luminescent materials to detect the analyte of interest (Jigyasa and Kaur Rajput 2023).

(continued)

Box 16.2 (continued)

- *Type of Interaction*
- Biological: use a biological material (bioreceptor) and a transducer to enable rapid, precise, and sensitive detection (Jafarizadeh-Malmiri et al. 2019).
- Chemical: generate signals by detecting the chemical nature or concentration of a molecule in the solution of interest (Kurbanoglu et al. 2019).
- Physical: monitor properties like force, pressure, and temperature, converting these variables into detectable and analyzable signals (Javaid et al. 2021).

Nanosensors in the dairy industry are classified (Box 16.2) based on the type of interaction (chemical, biological, or physical) or sensing technique (optical, mechanical, or electrical). They are primarily used to ensure product quality and safety by quantifying relevant analytes and detecting microorganisms (Table 16.2) (Javaid et al. 2021). The key requirements for these sensors include linearity of response, sensitivity, reproducibility of signal response, quick response and recovery times, selectivity, stability, and a long operating life (Jafarizadeh-Malmiri et al. 2019; Saleh Ahammad et al. 2009). Their efficiency is measured by the limit of detection (LOD), which indicates the lowest analyte concentration the sensor can detect in a given volume. Therefore, a lower LOD signifies a more efficient sensor (Jyoti et al. 2022).

16.3.2 Detection of Chemical Adulterants

Milk is a constant target for adulteration, leading not only to economic losses but also to significant health risks to consumers (Nascimento et al. 2017). Common milk adulteration practices involve dilution and/or the addition of cheap, low-quality, and sometimes hazardous substances to increase volume, mask inferior quality, or replace natural milk components for economic gain. Frequent adulterants include water, urea, formalin, hydrogen peroxide, and melamine, all of which have serious adverse health effects (Azad and Ahmed 2016). To address these practices, global regulatory agencies such as the World Health Organization (WHO) and the U.S. Food and Drug Administration (FDA) have established limits for chemical contaminants. Consequently, there is growing interest in techniques that can accurately and consistently detect adulterants in dairy products (Ceniti et al. 2023).

Melamine (2,4,6-triamino-1,3,5-triazine) is a nitrogen-rich compound widely utilized in the synthesis of melamine resins and fertilizers (Singh et al. 2021). Due to its high nitrogen content and cost-effectiveness, melamine is often subjected to illegal adulteration in dairy products to enhance their protein content. However, consuming melamine beyond the safe limit can trigger the formation of insoluble melamine cyanurate crystals in the kidneys, leading to kidney failure and potentially fatal outcomes (Liang et al. 2020). The FDA and WHO have set maximum melamine residue limits at 1 mg/kg for infant formula and 2.5 mg kg^{-1} for other dairy products. The tolerable daily intake limits are 0.2 mg kg^{-1} body weight by WHO and 0.063 mg kg^{-1} body weight by FDA (Nascimento et al. 2017).

Various metallic NPs have been used to develop optical sensors for detecting melamine in milk. Copper NPs (6.6 nm) with double-stranded DNA achieved a detection limit of 0.79 μM (Radhika and Gorthi 2019). Gold NPs (12 nm) reached a detection limit of 0.019 μM, and increasing the size of gold NPs from 13 to 42 nm reduced the limit from 1 μmol L^{-1} to 1 nmol L^{-1} (Bao et al. 2021; Chen et al. 2019a, b). Silver NPs (52 nm) on filter paper showed an even lower detection limit of 1.69×10^{-3} μM (Verma et al. 2021). The detection of melamine in milk using quantum dots has proven viable. Üzek et al. (2021) developed a fluorescence sensor with cadmium sulfide quantum dots (22 nm), achieving a detection limit of 1.11 μM. Similarly, studies using graphene quantum dots and carbon quantum dots with gold nanoparticles (5 nm) achieved a detection limit of 3.6×10^{-3} μM (Hu et al. 2019; Xue et al. 2020).

Hydrogen peroxide stands out as one of the major adulterants found in milk. This substance is often introduced to activate the natural enzyme lactoperoxidase, thereby imparting antimicrobial properties that extend the shelf life of dairy products. However, this practice comes at the expense of compromising the nutritional integrity of the food, leading to the degradation of essential vitamins such as A and E, which could potentially have adverse effects on public health (Ivanova et al. 2019; Lima et al. 2020a, b). In numerous countries, the use of hydrogen peroxide as a preservative in liquid milk is prohibited due to its associated toxic effects. For instance, in the United States, the Food and Drug Administration (FDA) permits the presence of hydrogen peroxide, albeit with a maximum limit of 0.05% by weight, exclusively in milk intended for cheese production (Lima et al. 2020a, b).

An electrochemical sensor using copper oxide NPs in carbon fabric (70 nm) detected H_2O_2 in milk with a limit of 3.4 10^{-2} μM (Amini et al. 2021). A more sensitive electrochemical nanosensor with l-proline-assisted silver NPs (7 nm) achieved a detection limit of 0.05 μM (Zhao et al. 2019). Colorimetric techniques showed that a hybrid sensor of silver NPs and cellulose nanofibers (15 nm) had a detection limit of 0.014 μM Teodoro et al. (2019). An electrochemiluminescence nanosensor with single-stranded DNA nanosheets (100 nm) reached a detection limit of 33 aM for H_2O_2 in milk (Liu et al. 2021).

Urea, a nitrogen-rich compound often added to milk to enhance its perceived protein content, is typically considered to pose low toxicity risks to human health (Nagraik et al. 2021; Shalileh et al. 2023). The Prevention of Food Adulteration Act (PFA) rules of 1955 set the upper allowable limit of urea in milk at 15.39 μM (Kutlu et al. 2022). Studies have shown that exceeding this limit, specifically at concentrations above 70 mg/dL, can cause adverse effects, including degenerative and necrotic damage to the liver and kidneys, even with short-term exposure (Kumar and Dash 2021).

A silver NP sensor, with a size of 52 nm, was developed on filter paper to detect urea in whole milk samples and obtained a detection limit of 480 μM (Verma et al.

2021). Similarly, a citrate silver nanoparticle optical sensor with a size of 40 nm achieved a detection limit of 349.65 μM for urea detection (Eid et al. 2022). Notably, fiber optic nanosensors utilizing zinc oxide nanobastons, approximately 197 nm in diameter, exhibited an outstanding detection limit of 1.185×10^{-3} μM for urea in milk samples compared to previous studies (Jyoti et al. 2022).

Formaldehyde, a chemical compound often added to milk to prevent spoilage and extend its shelf life, poses health risks when consumed in high doses (Himshweta and Singh 2023). Despite being naturally produced by the human body as part of metabolism, formaldehyde has been classified as carcinogenic by the International Agency for Research on Cancer (IARC) (IARC Working Group 2018). The World Health Organization—International Programme on Chemical Safety (WHO IPCS) has established a tolerable concentration (TC) of this compound of 2.6 mg L^{-1} in ingested products (World Health Organization 2017).

For the sensitive detection of formaldehyde in milk samples, a gold-based electrochemical sensor sized at 40 nm was developed, achieving a detection limit of 400 μM (Xi et al. 2020). Additionally, an electrochemical, non-enzymatic sensor constructed with copper oxide nanoparticles on carbon fabric, sized at 50 nm, attained a lower detection limit of 2.6 10^{-2} μM (Yang et al. 2021).

Most of the studies demonstrated detection limits significantly lower than the standards established by the FDA and WHO. Additionally, these methods are simpler and more sensitive compared to current analytical techniques for detecting these compounds, which include chromatographic approaches, spectroscopic techniques, enzyme-linked immunosorbent assays (ELISA), polymerase chain reaction (PCR), and electrophoresis. These traditional methods require complex and expensive instruments and specialized laboratories and make real-time detection of adulterants challenging (Nagraik et al. 2021).

16.3.3 Detection of Antibiotics and Antibiotic Residues

The utilization of antibiotics in livestock farming serves three primary purposes: therapeutic, preventive, and growth-promoting, offering practical and cost-effective methods for disease control and enhancing animal well-being (Pushparaj et al. 2022). However, the excessive or improper use of antibiotics can elevate the risk of antibiotic residues in milk and foster the proliferation and selection of antibiotic-resistant microorganisms (Ndahetuye et al. 2020). The consumption of milk containing antibiotic residues is emerging as a significant global concern for public health due to its potential to induce adverse effects, including damage to hearing and kidney function (ototoxicity and nephrotoxicity), disruption of endocrine systems, hypersensitivity reactions, and most critically, the development of bacterial resistance (Seth and Rathinasabapathi 2022).

To ensure food safety for consumers, regulatory authorities such as the FDA and Codex Alimentarius have established maximum residue limits (MRLs) for antibiotics in milk and other food products (Regulation (EU) No 37/2010). Beta-lactams,

Table 16.1 The antibiotics used in human medicine and/or dairy animals and their MRL values as established by the European regulation. (Adapted by Kemper 2008)

Chemical class	MRL (μg kg^{-1})
β-Lactams (penicillins)	4–30
Tetracyclines	100
Quinolones	30–100
Sulfonamides	100

tetracyclines, quinolones, and sulfonamides are the most commonly used antibiotics in dairy cattle, with specific MRLs detailed in Table 16.1 (Quintanilla et al. 2019). Methods for detecting veterinary drug residues in milk prioritize rapid results to facilitate timely processing. Control agencies employ reliable analytical techniques such as High-Performance Liquid Chromatography (HPLC), Gas Chromatography-Mass Spectrometry (GC-MS), and antibiotic residue tests for monitoring purposes (Teixeira et al. 2020). However, these methods have drawbacks, including sample preparation, the need for specialized labor, complex procedures, lengthy analysis times, high costs, and specificity for certain classes of antibiotics. Consequently, a sensitive, selective, and robust detection method, such as nanosensors, is desirable for determining antibiotics in milk samples to ensure food safety (Seth and Rathinasabapathi 2022).

Recognizing the critical need to monitor and manage antibiotic residues in food products, numerous nanosensors have been documented in the scientific literature as potential solutions. These nanosensors have demonstrated satisfactory results, with detection limits below the established MRL values.

Baezzat et al. (2021) demonstrated a novel sensor based on a Carbon Ionic Liquid Electrode (CILE) and TiO_2 NPs/Ionic Liquid (IL) (octylpyridinium iodide) for the determination of penicillin G potassium (PGK) in milk through Cyclic Voltammetry (CV) and Square-Wave Voltammetry (SWV) techniques. The TiO_2 nanostructures exhibited spherical, spongy shapes with diameters from 16 to 40 nm. The sensor achieved an LOD of 2.09 10^{-3} μM. In contrast, Jalili et al. (2020) reported a ratiometric fluorescent sensor comprising different colored carbon dots (CDs) as dual fluorophores and a mesoporous structured molecularly imprinted polymer as a receptor (B/YCDs@mMIP) for detecting penicillin G (PNG) in milk. The B/YCDs@mMIP nanoparticles presented a highly uniform spherical morphology with an average spherical size of 80 nm. A 3.4 10^{-4} μM detection limit and excellent recognition specificity for PNG over its analogs were observed. Safran et al. (2021) developed a selective and sensitive penicillin G (PEN−G) antibiotic detection method from milk samples using a molecular imprinting technique by surface plasmon resonance sensor based on gold nanoparticles (23.88 nm). The nanosensor exhibited 9.87 times more selectivity for the target PEN−G molecule than amoxicillin and 16.78 times more than ampicillin, with an LOD of 0.0017 ppb.

For the detection of oxytetracycline in spiked milk samples, sulfur- and nitrogen-doped carbon quantum dots (S, N-CQDs) with a size distribution smaller than 10 nm were prepared by a solid-phase hydrothermal method, achieving a detection limit of 2.5 10^{-2} μM (Amjadi et al. 2020). Conversely, Birader et al. (2021) presented a novel colorimetric lateral flow assay for the detection of antibiotic oxytetracycline

Table 16.2 Novel nanosensors for the detection of adulteration, antibiotics, and pathogens in milk and dairy product samples

Target element	Nanocomposite	Size (nm)	Detection method	Limit of detection—LOD	References
Adulteration				(μM)	
Melamine	Copper nanoparticle	6.6	Fluorescence	7.90×10^{-1}	Radhika and Gorthi (2019)
	Gold nanoparticle	12	Colorimetric	1.90×10^{-2}	Bao et al. (2021)
	Silver nanoparticle	52	Surface-Enhanced Raman Scattering (SERS)	1.69×10^{-3}	Verma et al. (2021)
	Cadmium sulfide quantum dots	22	Fluorescence	11.1×10^{-1}	Üzek et al. (2021)
	Gold nanoparticles @Carbon quantum dots nanocomposites	5	Fluorescence	3.6×10^{-3}	Hu et al. (2019)
	Graphene quantum dots (GQDs)	–	Fluorescence	3.6×10^{-3}	Xue et al. (2020)
Hydrogen peroxide	Copper oxide nanoparticle	70	Electrochemical	3.4×10^{-2}	Amini et al. (2021)
	Silver nanoparticle	7	Electrochemical	0.5×10^{-1}	Zhao et al. (2019)
	Silver nanoparticle	15	Colorimetric	1.4×10^{-2}	Teodoro et al. (2019)
Urea	Silver nanoparticle	52	Surface-Enhanced Raman Scattering (SERS)	4.8×10^{2}	Verma et al. (2021)
	Citrate silver nanoparticle	40	Colorimetric	3.5×10^{2}	Eid et al. (2022)
	Zinc oxide nanobastons	197	Colorimetric	1.18×10^{-3}	Jyoti et al. (2022)
Formaldehyde	Gold nanoparticle	40	Electrochemical	4.0×10^{2}	Xi et al. (2020)
	Copper oxide nanoparticle	50	Electrochemical	2.6×10^{-2}	Yang et al. (2021)
Antibiotic					
Penicillin G (PEN–G)	TiO_2 nanoparticles	16 and 40	Carbon ionic liquid electrode (CILE)	2.09×10^{-3}	Baezzat et al. (2021)
	Carbon dots	80	Fluorescence	3.4×10^{-4}	Jalili et al. (2020)
	Gold nanoparticles	23.88	Surface plasmon resonance	0.0017 ppb	Safran et al. (2021)

(continued)

Table 16.2 (continued)

Target element	Nanocomposite	Size (nm)	Detection method	Limit of detection—LOD	References
Oxytetracycline (OTC)	Sulfur- and nitrogen-doped carbon quantum dots	10	Chemiluminescence	2.5×10^{-2}	Amjadi et al. (2020)
	Gold nanoparticles	20	Colorimetric	5 ng mL^{-1}	Birader et al. (2021)
Kanamycin (KAN)	Magnetic molecularly imprinted particles	150	Electrochemical	4.88×10^{-3}	Zhang et al. (2022)
	Gold nanoparticles	32.6	Colorimetric	1.0×10^{-5}	Zou et al. (2021b)
Tobramycin (TOB)	Magnetic molecularly imprinted particles	150	Electrochemical	1.28×10^{-3}	Zhang et al. (2022)
	Nitrogen-doped carbon dots	4.25	Fluorescent	9.92×10^{-2}	Ren and Fan (2023)
Gentamicin (GEN)	Magnetic molecularly imprinted particles	150	Electrochemical	1.07×10^{-3}	Zhang et al. (2022)
	Graphene quantum dots Gold nanoparticles	3.37 15.44	Colorimetric Fluorometric	4.22×10^{-1} 4.93×10^{-1}	Sajwan and Solanki (2022)
Pathogens				CFU mL^{-1}	
S. aureus *E. coli*	Silver nanoparticles	37	Surface-enhanced Raman scattering	3.4×10^{1} 1.8×10^{1}	Mi et al. (2023)
S. aureus	Au/Pt nanoclusters	2–4	Colorimetric	8×10^{1}	Bagheri Pebdeni and Hosseini (2020)
S. aureus	Carbon nanotubes	10–20	Electrochemical	5×10^{0}	Farooq et al. (2020)
S. aureus	Carboxylate-modified magnetic nanobeads	50	Electrochemical	3×10^{0}	Eissa and Zourob (2020)
E. coli	Europium (III) chelated nanoparticles	114	Fluorescence	1×10^{4}	Wang et al. (2020)
E. coli	Gold nanoparticles	20.13	Electrochemical immunosensor	1.95×10^{4}	Jia et al. (2019)
E. coli	Gold nanoparticles	20	Chromatographic immunoassay	1×10^{3}	Luo et al. (2019)

Salmonella	Superparamagnetic nanoparticle	60	Time domain-nuclear magnetic resonance (TD-NMR)	2.3×10^3	Jin et al. (2020)
S. typhi	Nanozyme of zirconium and praseodymium	80	Colorimetric	3.7×10^1	Mirsadoughi et al. (2023)
S. ent *S. typhi*	Gold nanorods	50	Electrochemical immunosensor	1.35×10^2 6.8×10^1	Mahari et al. (2022)
B. cereus	Magnetic nanoparticle	–	Magnetic separation	1×10^3	Li et al. (2021)
B. cytotoxicus	Gold nanoparticles	10 and 20	Colorimetric	1×10^2	Rizzotto et al. (2023)

(OTC) in cow milk using a short anti-parallel G-quadruplex forming aptamer (27 nt) conjugated with gold NP, 20 nm in size. This assay enabled visual detection of 5 ng/mL.

Zhang et al. (2022) proposed a cost-effective, simple, and highly sensitive magnetic molecularly imprinted particles (MMIPs), with an average diameter of about 150 nm, electrochemical sensor to detect kanamycin (KAN), tobramycin (TOB), and gentamicin (GEN) in milk. The sensor achieved LOD values of 4.88, 1.28, and 1.07 nmol/L or KAN, TOB, and GEN, respectively. Zou et al. (2021a, b) reported a sensitive colorimetric aptasensor for kanamycin (KAN) detection in milk samples, employing target-triggered catalytic hairpin assembly amplification coupled with the assembly of AuNPs-DNA with a hydrodynamic diameter dispersed around 32.6 nm. This aptasensor exhibited high sensitivity with a detection limit of $1.0\ 10^{-5}$ μM for kanamycin in milk samples.

For the detection of tobramycin (TOB) in milk samples, Ren and Fan (2023) used a fluorescent probe based on molecularly imprinted polymers on nitrogen-doped carbon dots with a size of 4.25 nm. The probe met the requirements for detection in actual milk samples with high precision and accuracy, achieving an LOD of $9.92\ 10^{-2}$ μM. In the case of gentamicin (GEN) detection in milk samples, Sajwan and Solanki (2022) developed a rapid and sensitive hybrid nanosensor with a colorimetric and fluorometric readout system using graphene quantum dots (GQDs) with a size of 3.37 ± 0.61 nm and gold nanoparticles (AuNPs) with an average size of 15.44 ± 1.64 nm. This hybrid nanosensor exhibited a lower LOD of 0.422 μM and 0.493 μM in colorimetric and fluorimetric readout methods, respectively.

16.3.4 Detection of Microbial Contamination

Contamination by microorganisms in dairy products can occur at various processing stages, often due to unsanitary milking and handling practices. Such contamination poses risks of product spoilage and compromises its safety. Microbial proliferation can lead to the release of undesirable byproducts, such as biogenic amines, rendering the product unappetizing (Nyokabi et al. 2021; Pushparaj et al. 2022). Ensuring the quality and safety of milk globally is of paramount importance. Regulations aimed at safeguarding consumer health necessitate adherence to crucial guidelines for milk quality and safety, one of which is maintaining a low somatic cell count (SCC). This count reflects the total number of cells per milliliter of milk and is pivotal in ensuring product integrity (Ndahetuye et al. 2020).

Milk deterioration is characterized by the production of various microbial metabolites that cause undesirable changes in odor, taste, and texture. Bacteria are the primary microorganisms responsible for milk spoilage, as milk provides an excellent growth medium for these organisms (Martin et al. 2023). The main pathogens currently found in milk and dairy products, along with their respective maximum detection limits established by regulatory authorities, include *Escherichia coli* (20 to 100 CFU g^{-1}), *Salmonella* spp. (not detectable in 25 g), *Listeria monocytogenes*

(20 to 100 CFU g^{-1}), *Staphylococcus* spp. (20 to 100 CFU g^{-1}), and *Bacillus* spp. (10^3 to 10^4 CFU g^{-1}) (EFSA and ECDC 2017; FDA 2022).

Conventional methods for detecting microbial spoilage typically involve microbial growth on selective or non-selective media, biochemical assays, and serological confirmation. These approaches can be complex and time-consuming (Fusco et al. 2020). Thus, there is a critical need for alternative, rapid, and cost-effective techniques for real-time detection of milk spoilage. The dairy industry demands technologies that are not only fast and sensitive but also economical for assessing milk quality (Ziyaina et al. 2020). Concurrently, global food safety regulations are becoming more stringent. Recent developments have led to the creation of various nanosensors designed to monitor microbial contamination in dairy products effectively. These nanosensors have demonstrated satisfactory performance with low detection limits.

An ultra-small Au/Pt bi-metallic nanocluster (NC), with a size of about 2 to 4 nm, was prepared using DNA as a template in a one-pot chemical process and set as a bio-receptor conjugation for the colorimetric detection of *S. aureus*. A detection limit of 80 CFU mL^{-1} was obtained (Bagheri Pebdeni and Hosseini 2020). Another study reports the development of an electrochemical biosensor through the immobilization of bacteriophage in a surface-modified bacterial cellulose (BC) matrix with carboxylated multiwalled carbon nanotubes impregnated (a diameter ranging between 10 and 20 nm). The developed nanobiosensor effectively detected 5 CFU mL^{-1} of *S. aureus* in a milk sample within 30 min with high sensitivity, specificity, and accuracy (Farooq et al. 2020). On the other hand. A rapid, simple, and cost-effective biosensor for *S. aureus* using dual colorimetric and electrochemical detection on the same platform was developed. In this approach, gold screen-printed electrodes were used, on which specific peptide sequences coupled to carboxylate-modified magnetic nanobeads (around 50 nm in diameter). The developed peptide/magnetic nanoparticle biosensor was then examined in a milk sample, and the detection limit of the electrochemical assay was 3 CFU ml^{-1} after 1 min (Eissa and Zourob 2020).

Mi et al. (2023) report a low-cost and highly sensitive label-free SERS biosensor, 37 nm in size, produced from silver nanoparticles for the detection of *Staphylococcus aureus* (*S. aureus*) and *Escherichia coli* (*E. coli*), with a detection limit of 34 and 18 CFU mL^{-1}, respectively, in fresh milk samples. Wang et al. (2020) developed a fluorescence sensor utilizing europium (III) chelated NPs (EuNPs) measuring 114 nm in size for the determination of *E. coli* in milk samples, achieving a LOD of 10^4 CFU mL^{-1}. Jia et al. (2019) introduced an immunofiltration strip based on the photothermal effect of gold NPs for the sensitive, rapid, handheld, and low-cost detection of *E. coli* in skimmed milk samples. The gold NPs utilized were spherical and homogeneous with an average diameter of 20.13 $\pm$ 1.71 nm. This method achieved a detection limit of 1.95 10^4 CFU mL^{-1}, with sensitivity improved approximately tenfold compared to the conventional visual strip method. Luo et al. (2019) developed a paper-based radial flow chromatographic immunoassay employing gold NPs (AuNPs) with a mean particle diameter of 20 nm as chromatic agents for the detection of *E. coli*. This assay enabled the detection of the pathogen as low as 10^3 CFU mL^{-1} in whole milk within 10 min.

Jin et al. (2020) prepared a new time-domain nuclear magnetic resonance (TD-NMR) biosensor based on superparamagnetic nanoparticles (SMN) with a size of about 60 nm for detecting *Salmonella* in pasteurized milk samples. This biosensor achieved a detection limit of 2.3 10^3 CFU mL^{-1}. The total detection time, including sample preparation, membrane filtration, and NMR detection, was not more than 90 min. Mahari et al. (2022) fabricated and characterized an electrochemical immunosensor using gold nanorods (50 nm in size) as an electroactive NM for rapid detection of *Salmonella enteritidis (S. ent)* and *Salmonella typhimurium (S. typhi)*. For *S. ent* and *S. typhi*, the fabricated immunosensor achieved significant sensitivity, with an LOD of 135 and 68 CFU mL-1, respectively, in spiked milk samples. On the other hand, Mirsadoughi et al. (2023) presented a colorimetric biosensor, with a size of around 80 nm, based on a novel metal-organic framework (MOF) composed of zirconium (Zr) and praseodymium (Pr) as a nanozyme to detect *Salmonella Typhimurium* (*S. typhimurium)* in the spiked milk sample; the study obtained a LOD of 37 CFU mL^{-1}.

Li et al. (2021) reported a Van-modified poly-l-lysine (PLL)-coated magnetic nanoparticle for the detection of *Bacillus cereus* (*B. cereus*) in milk with a detection limit of 1 10^3 CFU mL^{-1}. Rizzotto et al. (2023) propose a colorimetric aptasensor containing gold nanoparticles, with a size of 10 and 20 nm, for the simple and rapid detection of *Bacillus cytotoxicus* (*B. cytotoxicus)* in milk samples. The aptasensor proved to be highly sensitive, achieving the naked-eye limit of detection as low as 10^2 CFU mL^{-1}.

A colorimetric nanosensor based on silicon dioxide (SiO_2) NPs was developed to detect volatile organic compounds and consequently, monitor the quality and shelf life of milk. The authors obtained a total color change (ΔE) of 5.77 in 60 h of storage (Ziyaina et al. 2019). Milk deterioration was detected through pH change using a colorimetric nanosensor using cellulose paper incorporated with nanostructures of zinc oxide, chitosan, and anthocyanins from black carrot, in which a color difference (ΔE) of 29.28 was obtained after 48 h (Ebrahimi Tirtashi et al. 2019).

16.4 New Sanitizers for the Industry

Food sanitization is a combination of procedures aimed at protecting against contamination, which need to be carried out frequently to ensure this protection (Singh 2020). It has an important role to perform in the health and well-being of consumers (Shineh et al. 2023), reducing the risk of disease; in the final quality of the product, extending shelf life, and increasing the integrity of the food (Okpala and Korzeniowska 2023), reducing economic losses caused by product spoilage; in addition to complying with the laws and standards established by health agencies, also contributing to the efficiency of the operation.

The legislation and regulatory requirements vary for each country, but all aim to reduce the incidents caused by a lack of food safety. However, there is the Codex Alimentarius Commission, which adopts a set of standards and guidelines aimed at protecting the health of consumers, established by the Food and Agriculture Organization of the United Nations (FAO) and the World Health Organization (WHO) (Okpala and Korzeniowska 2023; "Codex Alimentarius Commission Procedural Manual," 2023).

The cleaning-in-place (CIP) system, commonly used in the dairy industry, involves washing equipment with a sanitizing liquid without the need for disassembly associated with raising the temperature. It is used because it is a quick and simple process to carry out, in addition to reducing the risk of chemical exposure for those involved, and it can be applied several times (Meesilp and Mesil 2019). Sanitizing is mainly done with sanitizers based on chlorine, ammonium, acids, peroxides, and silver compounds, and their effectiveness depends on several factors such as pH, temperature, and contact time (Shineh et al. 2023), but even this method may not effectively eliminate contamination from all areas, including biofilms (Pant et al. 2023). Moreover, frequent sanitization is necessary to prevent biofilm formation, leading to increased water and energy consumption (Khade et al. 2018; Shemesh and Ostrov 2020).

Dairy biofilms (Box 16.3) pose a significant challenge for the industry, as they serve as a source of contamination for milk and contribute to various operational issues. Comprising a microbial community embedded within a polymeric extracellular matrix, biofilms have a remarkable ability to adhere to solid surfaces (Carrascosa et al. 2021). Several factors contribute to their formation, including prolonged processing times, delayed transportation and distribution, and the complexity of processing systems (Rather et al. 2021).

Box 16.3 Dairy Biofilms

In the dairy sector, biofilms commonly form on various equipment surfaces, including milk storage tanks, pump exteriors, silos, walls, heat exchangers, filling units, and pipes (Siqueira et al. 2021). Eliminating these biofilms presents a considerable challenge for the industry, as their presence can lead to the release of bacterial cells, posing risks to consumer health (Alonso and Kabuki 2019). Additionally, biofilms can impair operational efficiency by reducing heat exchange efficiency, narrowing tube diameters, increasing internal pressure, elevating energy consumption during processing, and causing surface damage due to corrosion (Jiménez-Pichardo et al. 2021). Therefore, effective strategies for biofilm control and elimination are essential for ensuring product safety, maintaining operational efficiency, and protecting consumer health in the dairy industry.

Different species of bacteria found in the dairy industry: Bacillus, Staphylococcus, Listeria, Streptococcus, Pediococcus, and *Leuconostoc* among

(continued)

Box 16.3 (continued)
Gram-positive bacteria, and *Klebsiella*, *Proteus*, *Alcaligenes*, *Enterobacter*, *Citrobacter*, *Moraxella*, *Aeromonas*, *Plesiomonas*, *Shigella*, *Escherichia*, and *Pseudomonas* species among Gram-negative bacteria (Rather et al. 2021). Additionally, *Salmonella* species are also found in the dairy industry, representing a significant health risk to consumers if they contaminate food products (O'Sullivan et al. 2019; Fusco et al. 2020).

The extracellular matrix of biofilms protects bacteria, making them resistant to unfavorable conditions and sanitizing agents (Fang et al. 2020). Despite rigorous sanitization processes, biofilms can persist due to the challenges associated with accessing certain areas, such as valve pipe connections, which are difficult to sanitize (Gonçalves et al. 2020). Therefore, effective strategies for biofilm control and elimination are essential for ensuring product safety, maintaining operational efficiency, and protecting consumer health in the dairy industry.

Nanoparticles have emerged as a promising technology for combating biofilms due to their high surface area-to-volume ratio, which allows them to penetrate the biofilm matrix and exert antimicrobial effects efficiently (Mohanta et al. 2022). NMs derived from metals and metal oxides are particularly noteworthy for their antimicrobial properties, stability, non-toxicity, and safety (Sharahi et al. 2019). Leveraging nanotechnology to address biofilm challenges is therefore crucial for enhancing sanitization processes and improving efficiency in the dairy industry.

Zinc oxide nanostructure (ZnONP) exhibits antimicrobial properties by interacting with bacterial cell surfaces, penetrating the cells, and inducing oxidation, thereby inhibiting bacterial growth and causing cell death (Mohd Yusof et al. 2019). Fontecha-Umaña et al. (2020) demonstrated the effectiveness of ZnO nanostructures incorporated into hard polyester surfaces, presenting rod-shaped structures with dimensions of 20 nm in width and 100 nm in length, against Escherichia coli biofilms. At a concentration of 850 ppm, these surfaces exhibited a remarkable reduction of 2.07 log CFU/cm^2 in *E. coli* biofilm formation on stainless steel disks. Similarly, Radwan et al. (2022) achieved satisfactory results using zinc oxide nanostructure (ZnO-NPs), with a spherical shape of 15 to 25 nm, combined with Proteinase K at a concentration of 50 μg/mL to combat *Bacillus*, using stainless steel coupons to form the biofilm, resulting in 99.19% efficiency. Moreover, Kaur et al. (2021) reported promising outcomes employing zinc oxide nanostructures, nearly spherical and ranging from 3 to 11 nm in size, as a sanitizer against bacterial biofilms commonly found in the dairy industry, including *Bacillus cereus*, *Streptococcus pyogenes*, *Escherichia coli*, and Pseudomonas aeruginosa. Tested on glass, plastic, and aluminum surfaces at concentrations of 50 μg/mL and 100 μg/mL, the nanostructures exhibited significant antibacterial activity against *P. aeruginosa* strains, followed by *S. aeruginosa*, *S. pyogenes*, *E. coli*, and *B. cereus*. The antibiofilm activity of these nanoparticles was notably pronounced on glass and aluminum surfaces.

Silver nanoparticles (AgNP) and gold nanoparticles (AuNP) are known for their ability to disrupt bacterial cell membranes, altering cell morphology and function. AgNP and AuNP can interact with lipids and proteins on the cell membrane, penetrate the cell as whole nanoparticles, or release Ag^+ and Au^+ ions, impacting DNA, respiratory processes, and cell division (Sharahi et al. 2019; Joshi et al. 2020).

Fontecha-Umaña et al. (2020) demonstrated the efficacy of silver nanostructures, spherical with a diameter of 20 nm, embedded in hard polyester surfaces on stainless steel discs, against *Staphylococcus aureus* and *Escherichia coli* biofilms. At a concentration of 400 ppm, the silver nanostructures significantly reduced *S. aureus* and *E. coli* biofilm formation by 4.90 log UFC/cm^2 and 3.84 log UFC/cm^2, respectively.

Aziz et al. (2022) used the silver nanoparticle together and separately from the Virokill disinfectant to control the biofilm production of *Pseudomonas aeruginosa*, isolated from a dairy farm. They found that Virokill/AgNPs at a concentration of 200/1000 mg/L after 24 h exhibited high sensitivity after 24 h, indicating increased sanitization efficiency when AgNPs were utilized. Gonzalez et al. (2021) utilized a coating of silver nanoparticles through electroless deposition on stainless steel to reduce biofilm formation by *E. coli*, *P. aeruginosa*, and *S. aureus*. This approach demonstrated a tendency to reduce cell viability in these bacteria, indicating its potential for biofilm control.

Similarly, gold NPs follow a similar mechanism in combating biofilms. Khan et al. (2019) used the gold nanostructure, having a spherical shape with a size of 56.01 nm, coated with chitosan oligosaccharide (COS-AuNP) to inhibit the biofilm of *Pseudomonas aeruginosa*, achieving efficient biofilm reduction of approximately 2.8 log UFC for the concentration of 512 μg/mL. Chlumsky et al. (2020) investigated the use of gold nanoparticles, spherical with sizes ranging from 10 to 12 nm, to reduce biofilm formation by various bacteria, including *S. aureus*, *L. monocytogenes*, *E. coli*, and *Salmonella enterica* Infantis. They observed up to 78% inhibition efficiency, preventing biofilm formation at a maximum concentration of 180 mg L^{-1}.

16.5 Nutraceutical Dairy Products

Box 16.4 Nutraceutical

Defined as a food or part of a food, such as a dietary supplement, that provides medical or health benefits, including the prevention and treatment of diseases. These products can include vitamins, minerals, amino acids, among other components, which contribute to health improvement and play a role in the prevention and treatment of specific health conditions (DeFelice 2014; Aronson and Aronson 2017). This trend has led the food industry to explore

(continued)

Box 16.4 (continued)
by-products generated during food processing, such as cereals, dairy, seafood, fruits, and vegetables, as they are abundant sources of these beneficial compounds (Tlais et al. 2020). Bioactive compounds, known for their antioxidant, antitumor, antimutagenic, antiproliferative, and anti-inflammatory properties, are often categorized as preventive disease ingredients (Soumya et al. 2021).

When it comes to applying nanotechnology to products, there is a wide range of possibilities. Nanotechnology can be used to improve the stability and bioavailability of bioactive compounds and increase the efficiency of encapsulation and controlled release of nutrients. The development of nutraceuticals (Box 16.4) has seen a remarkable rise in recent times, with a focus on foods rich in bioactive compounds that offer various health benefits (Carrizzo et al. 2020).

The development of nutraceuticals has seen a remarkable rise in recent times, with a focus on foods rich in bioactive compounds that offer various health benefits (Carrizzo et al. 2020). This trend has led the food industry to explore by-products generated during food processing, such as cereals, dairy, seafood, fruits, and vegetables, as they are abundant sources of these beneficial compounds (Tlais et al. 2020). Bioactive compounds, known for their antioxidant, antitumor, antimutagenic, antiproliferative, and anti-inflammatory properties, are often categorized as preventive disease ingredients (Soumya et al. 2021).

Bioaccessibility refers to the amount of a specific component released from the food matrix in the gastrointestinal tract, thereby becoming available for absorption (Shahidi and Pan 2022). On the other hand, bioavailability refers to the portion of nutrients or phytochemicals that are digested, absorbed, and metabolized through the body's normal pathways (Shahidi and Peng 2018). Therefore, for bioactive compounds to exert their beneficial effects, they must be bioavailable, as enhanced bioavailability directly correlates with the effectiveness of absorption and utilization of these compounds (Shahidi et al. 2019).

Nanotechnology can alter the properties of dairy products, offering improvements in terms of texture, stability, flavor, bioavailability, and even nutritional efficiency (Pateiro et al. 2021). For example, nanostructures derived from proteins and lipids are commonly used to encapsulate and release nutrients in a controlled manner, while polymer-based nanoparticles with antimicrobial properties contribute to improving the stability and preservation of these products (Namasivayam et al. 2022). Table 16.3 shows the advantages of potential nanostructures applied to dairy products.

Factors influencing bioavailability and bioaccessibility in dairy products include the presence of metabolizing enzymes like lactase and the food matrix affecting nutrient release and absorption (Szilagyi 2019) (Turgeon and Brisson 2020). Researchers have devised strategies to enhance the bioavailability of bioactives in dairy, such as developing delivery systems based on nanomaterials like protein or lipid nanoparticles. These systems can encapsulate and protect specific nutrients,

Table 16.3 Main advantages of nanoemulsions, solid lipid nanoparticles, nanostructured lipid carriers, liposomes, nanogels, polymeric nanoparticles, micelles, and polymeric micelles for foods (Jacobsen et al. 2018; Nasrollahzadeh et al. 2019)

	Nanostructure	Main advantages		Nanostructure	Main advantages
Nanoemulsionn	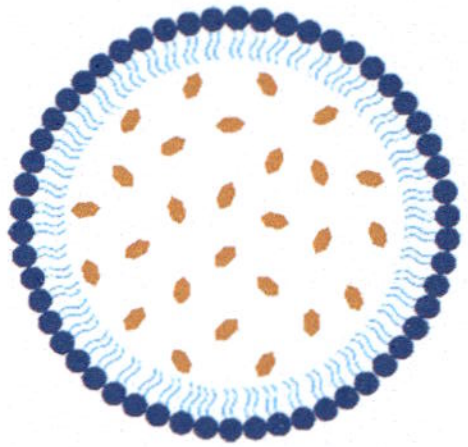	Improves the stability and bioavailability of sensitive ingredients Increases solubility in aqueous systems Facilitates the formulation of transparent emulsions.	Nanogel	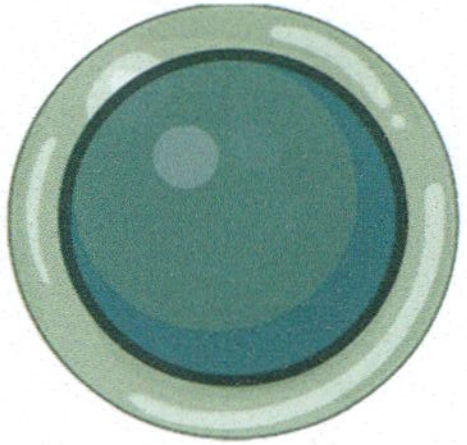	Stabilizes sensitive compounds Controls nutrient release Easy incorporation into foods.
Solid-lipid nanoparticles	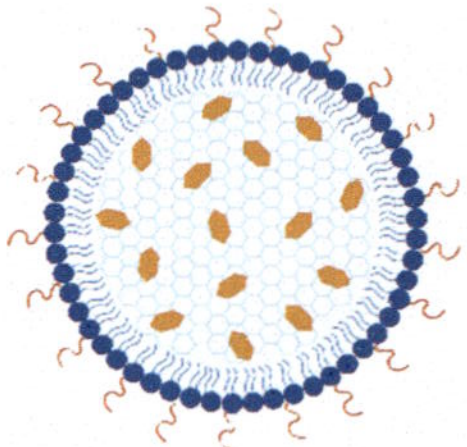	Protects ingredients from degradation Controls the release of flavors and nutrients Reduces migration of undesirable compounds in packaging.	Polymer nanoparticles	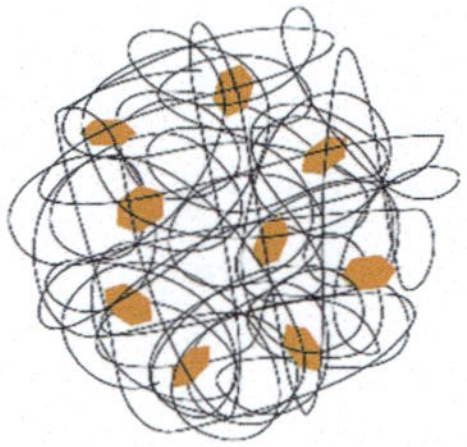	Increases stability of active ingredients Controls the release of functional compounds Protects against adverse conditions.
Nanostructured lipid carriers	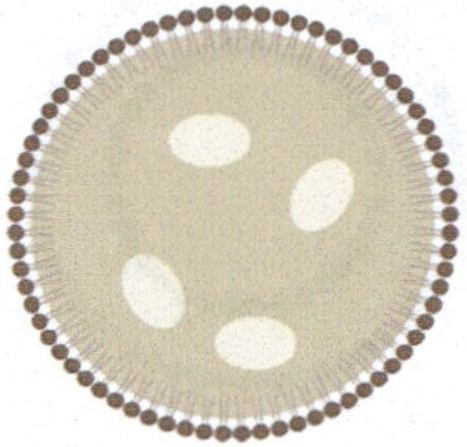	Improves solubility and stability of bioactive compounds Protects against degradation Easy incorporation into food formulations.	Micelles	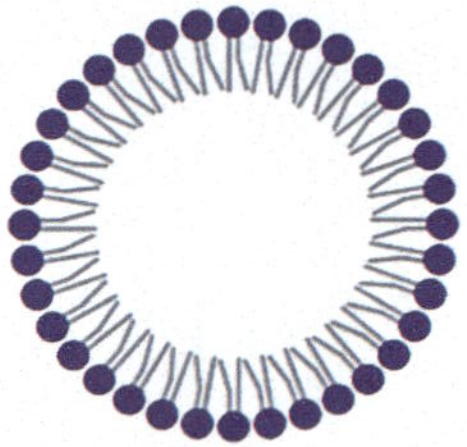	Improves solubility of hydrophobic compounds Facilitates the formulation of transparent emulsions Easy incorporation into foods.

(continued)

Table 16.3 (continued)

	Nanostructure	Main advantages		Nanostructure	Main advantages
Liposome	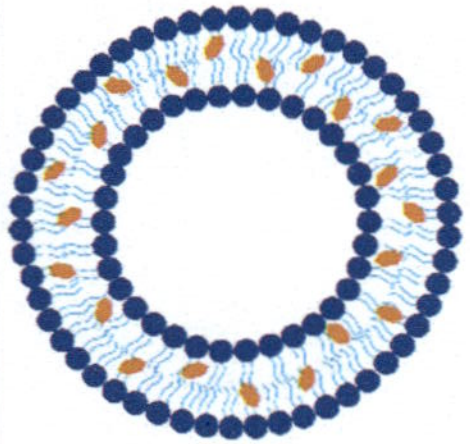	Improves delivery of nutrients and bioactive compounds Reduces rapid release of volatile compounds Directs nutrients to specific targets.	Polymeric micelle	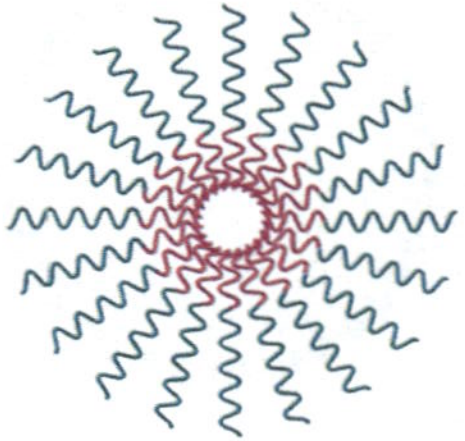	Stabilizes sensitive ingredients Controls the release of bioactive compounds Develops specific functional foods.

facilitating absorption by the body (Ali et al. 2021) (Jafari and McClements 2017). Such innovations aim to improve the absorption of key nutrients in dairy products, including calcium, fat-soluble vitamins, and proteins of high biological value (Verma and Pandey 2021).

In a study conducted by Zou et al. (2021a, b), the manufacturing of composite nanoparticles was carried out involving a coating of caseinate around zein NPs. These composite NPs, loaded with quercetin and featuring a size below 150 nm, demonstrated a high production efficiency (95.2%), a nutraceutical loading capacity of 5.6%, and a nutraceutical loading efficiency of 75.7%. The controlled release of quercetin primarily occurred in the simulated intestinal phase, resulting in a significant improvement in bioaccessibility and higher in vitro antioxidant activity. These findings indicate the potential of composite NPs for application in dietary supplements, functional foods, and pharmaceutical products.

Nanostructures of proteins/peptides offer unique properties for the encapsulation and controlled release of bioactive compounds, such as antioxidants, antimicrobials, and essential nutrients (Yusuf et al. 2017). In this context, the application of these nanoparticles stands out in enhancing the stability of ingredients in dairy products, such as vitamins and unsaturated fatty acids.

The ability of nanostructures, such as nanoparticles derived from proteins or peptides, to protect active ingredients from environmental factors like light and oxygen is crucial for preserving the quality of food products (Bajpai et al. 2018). Additionally, nanoparticles offer the potential to develop functional dairy products containing bioactive peptides sourced from milk proteins (Peighambardoust et al. 2021). These bioactive peptides, when encapsulated within nanocarriers, can be released in a controlled manner during digestion, providing health benefits such as anti-hypertensive and immunomodulatory properties (Alu'datt et al. 2022). Table 16.4 presents protein nanocarriers with potential applications in dairy products.

In a study conducted by Solghi et al. (2020), the encapsulation of curcumin in whey protein isolate nanoparticles (WPI-NPs) was investigated, which demonstrated a spherical structure with a smooth surface and a high encapsulation efficiency of 93.1%. Encapsulation significantly improved the oxidative stability of curcumin, increasing from 10 to 70%, and allowed controlled release, with minimal release in simulated gastric fluid and substantial release in simulated intestinal conditions.

The study by 'D'Onofre Couto et al. (2021) focused on the production of β-lg nanostructures through heat treatment at 100 °C in NaCl solution, resulting in particles with an average size of 109 nm and encapsulation efficiency of 96.50% for quercetin. In contrast, Simões et al. (2020) achieved particle sizes of 100 nm and an encapsulation efficiency of 67.9%, attributed to the use of sodium phosphate buffer solution and heat treatment at 70 °C. Both studies indicate stabilizing properties and effective release mechanisms, highlighting β-lg nanostructures as highly efficient in encapsulating hydrophobic molecules.

Jiménez-Cruz et al. (2022) investigated the effect of poly-L-lysine (PLL) coating on α-La nanoparticles against enzymatic degradation in the gastrointestinal tract, obtaining particles with a mean diameter of 151.2 nm, a PDI of 0.072, and a zeta

Table 16.4 Characteristics of nanostructures (proteins/peptides) for delivery of bioactive compounds with potential application in dairy products

Bioactive		Nanovehicle	Size (nm)	Encapsulation efficiency (EE) (%)	References
Carotenoids	β-Carotene	Whey protein	150.00	–	Mehrad et al. (2018)
	Lycopene	α-Lactalbumin	17.4 ± 1.3	70.00	Guo et al. (2023)
Polyphenols	Quercetin	β-Lactoglobulin (β-Lg)	87.8 ± 1.9	64.40	Simões et al. (2020)
			109.95	96.50	D'Onofre Couto et al. (2021)
	Curcumin	Whey protein	–	93.10	Solghi et al. (2020)
	Soy isoflavones	Whey protein	59.96	92.59	Liu et al. (2022)
	Olive leaf phenolics	Whey protein	232.3	–	Soleimanifar et al. (2020)
Minerals	Calcium	Whey protein hydrolysate	104.9 ± 0.9	–	Xixi et al. (2015)
	Calcium	Whey protein	150.00	–	Yi et al. (2020)
	Zinc	Whey protein	92.20	99.2	Shao et al. (2018)
Vitamins	Vitamin D_3	Sodium caseinate	120.16	96.14	Barri et al. (2023)
	Vitamin B_2	β-Lg	109. 95	36.39	D'Onofre Couto et al. (2021)
		α-Lactalbumin	163.84	98.50	Bianchi et al. (2023b)
	Vitamin D	β-Conglycinin	–	75.00	Levinson et al. (2014)
	Vitamin D3	β-Lg	130.00	–	Berino et al. (2019)
Polymer	Poly-L-lysine	α-Lactalbumin	151.2	95.00	Jiménez-Cruz et al. (2022)
	Chitosan	α-Lactalbumin	211.00	58.86	Liu et al. (2020)
	Chitosan	Whey protein	109.00	–	De Queiroz et al. (2018)
Other nutraceuticals	Chondroitin sulfate	Egg white protein	133.2 ± 3.4	–	Wang et al. (2021)
	Omega-3	Whey protein	196.00	–	Abbasi et al. (2019)
	Folic acid	β-Lg	109.77 ± 2.80	68.82 ± 1.76	Kayani et al. (2018)

potential of −29.7 mV. The coating efficiency increased with PLL concentration, regardless of molecular weight, reaching a maximum efficiency of 90–95% at a concentration of 1 mg/mL^{-1} of PLL.

Liu et al. (2022) developed whey protein nanoparticles to encapsulate soy isoflavones, with particle sizes ranging from 54.09 to 59.96 nm, high zeta potential (−35.63 to −31.03 mV), and high encapsulation efficiency (91.29%–92.59%). Increasing the ratio of isoflavones to soy proteins resulted in aggregated particles and reduced encapsulation efficiency. Encapsulation significantly improved the photochemical stability, antioxidant activity, and bioaccessibility of soy isoflavones, highlighting NPs as effective in improving the stability and bioactivity of these compounds in dairy products.

NPs, such as liposomes, giant liposomes, and lipid nanoparticles, represent advancements in delivering bioactive compounds in dairy products (Ayelén Vélez et al. 2016). Liposomes, with their bilayer structure, offer high biocompatibility and efficient encapsulation, ideal for the controlled release of sensitive nutrients (Subramani and Ganapathyswamy 2020). Giant liposomes, with larger capacity, are useful for transporting significant volumes of compounds, while lipid NPs, rich in diverse lipids, provide stability and protection against lipid oxidation (Fan et al. 2021). In dairy products, these nanoparticles stand out for nutritional preservation, controlled release, and stability, enhancing quality and extending the shelf life of foods (Sridhar et al. 2020). Table 16.5 presents studies on nanostructures, such as liposomes, giant liposomes, and lipid NPs, encapsulating various bioactive compounds.

Shtay et al. (2019) developed solid lipid nanoparticles (SLNs) loaded with EGCG (epigallocatechin gallate). Using hot homogenization and a pure cocoa butter lipid matrix, they achieved nanoparticles with an average size ranging from 108 to 122 nm and a maximum encapsulation efficiency of 68.5%. Hao et al. (2017) encapsulated quercetin in nanoliposomes via electrostatic deposition, resulting in homogeneous and spherical nanocapsules with a high encapsulation efficiency of 71.14%. These nanocapsules demonstrated better storage stability and antioxidant activity compared to free quercetin.

Jafarifar et al. (2022) produced lipid nanostructures (NLC) and nanoemulsions containing vitamin D for dairy product enrichment. The NLC ranged from 123.4 to 210.6 nm and the nanoemulsions from 137.6 to 171.6 nm, with vitamin D_3 release of 9.3 and 26.9% under simulated gastric conditions, respectively. Blanco-Llamero et al. (2022) used an extract from the microalga Nannochloropsis gaditana, rich in eicosapentaenoic acid and glycolipids, to develop an oral formulation of lipid NPs. The formulations, with a controlled particle size of 200 nm, showed an encapsulation efficiency exceeding 60%, highlighting their potential as effective vehicles for the controlled delivery of omega-3 in dairy products.

Table 16.5 Characteristics of nanostructures (liposomes/giant liposomes/lipid) for delivery of bioactive compounds with potential application in dairy products

Bioactive		Nanovehicle	Size (nm)	Encapsulation efficiency (%)	References
Carotenoids	β-carotene	Solid lipid nanoparticles	136.0	–	Helena de Abreu-Martins et al. (2020)
			130.0	90.00	Osanlou et al. (2022)
Polyphenols	Epigallocatechin-3-gallate (EGCG)	Solid lipid nanoparticles	108.00	68.50	Shtay et al. (2019)
	Curcuminoids	Liposome nanoparticles	120.16	96.14	Barri et al. (2023)
	3,5,7,2′,4′-pentahydroxyflavone	Lipid nanoparticles	200.00	82.00	Sunoqrot et al. (2023)
	Quercetin	Liposome nanoparticles	350.00	71.14	Hao et al. (2017)
Minerals	Calcium	Lipid nanoparticles	205.6 ± 3.9	75.60	Sapino et al. (2021)
Vitamins	Vitamin A	Lipid nanoparticles	100.00	99.10	Garcês et al. (2018)
			223.00 ± 10.0	90.00	Resende et al. (2020)
	Vitamin D3	Lipid nanoparticles	123.4	86.24	Jafarifar et al. (2022)
Other nutraceuticals	Coenzyme Q10	Lipid nanoparticles	275.00	–	Moschetti et al. (2020)
	Omega-3	Lipid nanoparticles	200.00	60.00	Blanco-Llamero et al. (2022)
	Omega-3	Liposome nanoparticles	73.20 ± 18.10	–	Rasti et al. (2017)

16.6 Conclusion and Future Perspective

The integration of nanotechnology into the dairy industry holds immense potential for addressing various challenges and enhancing product quality and safety. Understanding the synthesis process and choosing appropriate methods are crucial initial steps, as they directly influence the properties of NMs and their suitability for specific applications. While traditional bulk methods are well-established, microfluidic approaches offer advantages in reducing reagent volumes and addressing mass transfer issues, although scaling up remains a challenge.

Different NMs, including metallic NPs and organic NMs like carbohydrates, proteins, and lipids, show promising applications in the dairy industry. Metallic NPs are utilized for nanosensor fabrication, enabling rapid and specific detection of contaminants, while also offering potential in sanitizer formulations for biofilm removal. Organic NMs enhance dairy product quality and serve as effective nanocarriers for improving nutrient bioavailability.

Looking ahead, nanotechnology will continue to transform the dairy industry by enhancing product stability, controlled release of ingredients, and even real-time monitoring of milk quality through nanosensors. Intelligent packaging enabled by nanotechnology will extend product durability and preservation. However, ethical, regulatory, and safety considerations must be carefully addressed through ongoing dialogue among researchers, industry professionals, and regulatory bodies to ensure responsible implementation and public acceptance of nanotechnologies in the food industry. Through collaborative efforts and continued research, nanotechnology will drive innovation and advancements, contributing to the sustainable growth of the dairy sector while meeting consumer demands for safe and high-quality products.

References

Abalde-Cela S, Taladriz-Blanco P, De Oliveira MG, Abell C (2018) Droplet microfluidics for the highly controlled synthesis of branched gold nanoparticles. Sci Rep 8(1):1–6. https://doi.org/10.1038/s41598-018-20754-x

Abbasi F, Samadi F, Jafari SM, Ramezanpour S, Shams SM (2019) Ultrasound-assisted preparation of flaxseed oil nanoemulsions coated with alginate-whey protein for targeted delivery of omega-3 fatty acids into the lower sections of gastrointestinal tract to enrich broiler meat. Ultrason Sonochem 50:208–217. https://doi.org/10.1016/J.ULTSONCH.2018.09.014

Adamo CB, Junger AS, Bressan LP, da Silva JAF, Poppi RJ, de Jesus DP (2020) Fast and straightforward in-situ synthesis of gold nanoparticles on a thread-based microfluidic device for application in surface-enhanced Raman scattering detection. Microchem J 156:104985. https://doi.org/10.1016/J.MICROC.2020.104985

Ali MA, Kamal MM, Rahman MH, Siddiqui MN, Haque MA, Saha KK et al (2021) Functional dairy products as a source of bioactive peptides and probiotics: current trends and future prospectives. J Food Sci Technol 59(4):1263–1279. https://doi.org/10.1007/S13197-021-05091-8

Alonso VPP, Kabuki DY (2019) Formation and dispersal of biofilms in dairy substrates. Int J Dairy Technol 72(3):472–478. https://doi.org/10.1111/1471-0307.12587

Alu'datt MH, Alrosan M, Gammoh S, Tranchant CC, Alhamad MN, Rababah T et al (2022) Encapsulation-based technologies for bioactive compounds and their application in the food industry: a roadmap for food-derived functional and health-promoting ingredients. Food Biosci 50:101971. https://doi.org/10.1016/J.FBIO.2022.101971

Amini N, Rashidzadeh B, Amanollahi N, Maleki A, Yang JK, Lee SM (2021) Application of an electrochemical sensor using copper oxide nanoparticles/polyalizarin yellow R nanocomposite for hydrogen peroxide. Environ Sci Pollut Res 28(29):38809–38816. https://doi.org/10.1007/S11356-021-13299-6/TABLES/2

Amjadi M, Hallaj T, Mirbirang F (2020) A chemiluminescence reaction consisting of manganese (IV), sodium sulfite, and sulfur- and nitrogen-doped carbon quantum dots, and its application for the determination of oxytetracycline. Microchim Acta 187(3):1–7. https://doi.org/10.1007/S00604-020-4168-0/FIGURES/5

An Interview With Dr. Stephen DeFelice I Nutraceuticals world. (2014) Retrieved July 13, 2024, from https://www.nutraceuticalsworld.com/contents/view_health-e-insights/2011-10-28/an-interview-with-dr-stephen-defelice

Aronson JK, Aronson JK (2017) Defining 'nutraceuticals': neither nutritious nor pharmaceutical. Br J Clin Pharmacol 83(1):8–19. https://doi.org/10.1111/BCP.12935

Ayelén Vélez M, Cristina Perotti M, Santiago L, María Gennaro A, Hynes E (2016) Bioactive compounds delivery using nanotechnology: design and applications in dairy food. In: Nutrient Delivery, pp 221–250. https://doi.org/10.1016/B978-0-12-804304-2.00006-8

Azad T, Ahmed S (2016) Common milk adulteration and their detection techniques. Int J Food Contam 3(1):1–9. https://doi.org/10.1186/S40550-016-0045-3/TABLES/3

Aziz SAAA, Mahmoud R, Mohamed MBED (2022) Control of biofilm-producing Pseudomonas aeruginosa isolated from dairy farm using Virokill silver nano-based disinfectant as an alternative approach. Sci Rep 12(1):9452. https://doi.org/10.1038/s41598-022-13619-x

Baezzat MR, Pourghobadi Z, Pourghobadi R (2021) Nanomolar determination of penicillin G potassium (PGK) salt using a carbon paste electrode modified with TiO_2 nano particles/ionic liquids in real samples. Mater Chem Phys 270:124641. https://doi.org/10.1016/J.MATCHEMPHYS.2021.124641

Bagheri Pebdeni A, Hosseini M (2020) Fast and selective whole cell detection of Staphylococcus aureus bacteria in food samples by paper based colorimetric nanobiosensor using peroxidase-like catalytic activity of DNA-Au/Pt bimetallic nanoclusters. Microchem J 159:105475. https://doi.org/10.1016/J.MICROC.2020.105475

Bajpai VK, Kamle M, Shukla S, Mahato DK, Chandra P, Hwang SK, Kumar P, Huh YS, Han YK (2018) Prospects of using nanotechnology for food preservation, safety, and security. J Food Drug Anal 26(4):1201–1214. https://doi.org/10.1016/J.JFDA.2018.06.011

Balbino TA, Aoki NT, Gasperini AAM, Oliveira CLP et al (2013) Continuous flow production of cationic liposomes at high lipid concentration in microfluidic devices for gene delivery applications. Chem Eng J 226:423–433. https://doi.org/10.1016/J.CEJ.2013.04.053

Bao X, Liu J, Zheng Q, Duan L et al (2021) Colorimetric recognition of melamine in milk using novel pincer zinc complex stabilized gold nanoparticles. Chin Chem Lett 32(10):3023–3026. https://doi.org/10.1016/J.CCLET.2021.04.014

Barbosa Costa KC, Schlogl AE, Menezes De Souza S, Nazaré De Oliveira E et al (2023) Tara gum coating with embedded ZnO nanostructures for increased postharvest guava shelf life. ACS Food Sci Technol 3:738. https://doi.org/10.1021/ACSFOODSCITECH.3C00006

Barri A, Ghanbarzadeh B, Mohammadi M, Pezeshki A (2023) Application of sodium caseinate as protein-based coating/stabilizing agent in development and optimization of vitamin D3-loaded nanostructured lipid carriers (NLCs). Food Hydrocoll 144:108747. https://doi.org/10.1016/J.FOODHYD.2023.108747

Baruah A, Jindal A, Acharya C, Prakash B, Basu S, Ganguli AK (2017) Microfluidic reactors for the morphology-controlled synthesis and photocatalytic study of ZnO nanostructures. J Micromech Microeng 27(3):035013. https://doi.org/10.1088/1361-6439/AA5BC4

Bayareh M, Ashani MN, Usefian A (2020) Active and passive micromixers: a comprehensive review. Chem. Eng. Process. Process Intensif. 147:107771. https://doi.org/10.1016/J.CEP.2019.107771

Berino RP, Báez GD, Ballerini GA, Llopart EE et al (2019) Interaction of vitamin D3 with beta-lactoglobulin at high vitamin/protein ratios: characterization of size and surface charge of nanoparticles. Food Hydrocoll 90:182–188. https://doi.org/10.1016/J.FOODHYD.2018.11.027

Bianchi JRDO, de la Torre LG, Costa ALR (2023a) Droplet-based microfluidics as a platform to design food-grade delivery systems based on the entrapped compound type. Food Secur 12(18): 3385. https://doi.org/10.3390/FOODS12183385/S1

Bianchi JRDO, Paz BER, Fonseca LG, Carvalho RL et al (2023b) Cytotoxicity, structural, conformational, and techno-functional properties of α-lactalbumin nanostructured aggregates. J Food Sci 88(7):3049–3062. https://doi.org/10.1111/1750-3841.16622

Birader K, Kumar P, Tammineni Y, Barla JA et al (2021) Colorimetric aptasensor for on-site detection of oxytetracycline antibiotic in milk. Food Chem 356:129659. https://doi.org/10.1016/J.FOODCHEM.2021.129659

Blanco-Llamero C, Galindo-Camacho RM, Fonseca J, Santini A et al (2022) Development of lipid nanoparticles containing Omega-3-rich extract of microalga Nannochlorpsis gaditana. Food Secur 11(23):3749. https://doi.org/10.3390/FOODS11233749

Carrascosa C, Raheem D, Ramos F, Saraiva A et al (2021) Microbial biofilms in the food industry-a comprehensive review. Int J Environ Res Public Health 18(4):1–31. https://doi.org/10.3390/ijerph18042014

Carrizzo A, Izzo C, Forte M, Sommella E et al (2020) A novel promising frontier for human health: the beneficial effects of nutraceuticals in cardiovascular diseases. Int J Mol Sci 21(22):8706. https://doi.org/10.3390/IJMS21228706

Ceniti C, Spina AA, Piras C, Oppedisano F et al (2023) Recent advances in the determination of Milk adulterants and contaminants by mid-infrared spectroscopy. Food Secur 12(15):2917. https://doi.org/10.3390/FOODS12152917

Charcosset C, Juban A, Valour JP, Urbaniak S et al (2015) Preparation of liposomes at large scale using the ethanol injection method: effect of scale-up and injection devices. Chem Eng Res Des 94:508–515. https://doi.org/10.1016/J.CHERD.2014.09.008

Chen M, Ye X, Shen D, Ma C (2019a) Modulatory effects of gut microbiota on constipation: the commercial beverage Yakult shapes stool consistency. J Neurogastroenterol Motil 25(3): 475–477. https://doi.org/10.5056/JNM19048

Chen XY, Ha W, Shi YP (2019b) Sensitive colorimetric detection of melamine in processed raw milk using asymmetrically PEGylated gold nanoparticles. Talanta 194:475–484. https://doi.org/10.1016/J.TALANTA.2018.10.070

Chlumsky O, Purkrtova S, Michova H, Svarcova V et al (2020) The effect of gold and silver nanoparticles, chitosan and their combinations on bacterial biofilms of food-borne pathogens. Biofouling 36(2):222–233. https://doi.org/10.1080/08927014.2020.1751132

Codex Alimentarius Commission Procedural Manual (2023) In codex Alimentarius commission procedural manual. WHO, FAO. https://doi.org/10.4060/cc5042en

D'Onofre Couto B, Novaes da Costa R, Castro Laurindo W, Moraes da Silva H et al (2021) Characterization, techno-functional properties, and encapsulation efficiency of self-assembled β-lactoglobulin nanostructures. Food Chem 356:129719. https://doi.org/10.1016/J.FOODCHEM.2021.129719

De Oliveira Bianchi JR, De Souza SM, Santos IJB (2023) Post-harvest application of tara gum coating incorporated with silver nanoparticles for preservation of banana. Biointerface Res Appl Chem 13(1):1–15. https://doi.org/10.33263/BRIAC131.081

De Queiroz JLC, De Araújo Costa RO, Rodrigues Matias LL, De Medeiros AF et al (2018) Chitosan-whey protein nanoparticles improve encapsulation efficiency and stability of a trypsin inhibitor isolated from Tamarindus indica L. Food Hydrocoll 84:247–256. https://doi.org/10.1016/J.FOODHYD.2018.06.010

Dey S, Samadder A, Nandi S (2021) Exploring current role of nanotechnology used in food processing industry to control food additives and their biochemical mechanisms. Curr Drug Targets 23(5):513–539. https://doi.org/10.2174/1389450123666211216150355

Duncanson WJ, Lin T, Abate AR, Seiffert S et al (2012) Microfluidic synthesis of advanced microparticles for encapsulation and controlled release. Lab Chip 12(12):2135–2145. https://doi.org/10.1039/C2LC21164E

Ebrahimi Tirtashi F, Moradi M, Tajik H, Forough M et al (2019) Cellulose/chitosan pH-responsive indicator incorporated with carrot anthocyanins for intelligent food packaging. Int J Biol Macromol 136:920–926. https://doi.org/10.1016/J.IJBIOMAC.2019.06.148

EFSA & ECDC (2017) The European Union summary report on trends and sources of zoonoses, zoonotic agents and food-borne outbreaks in 2016. EFSA J 15(12). https://doi.org/10.2903/J.EFSA.2017.5077

Eid SM, el-Shamy S, Farag MA (2022) Identification of milk quality and adulteration by surface-enhanced infrared absorption spectroscopy coupled to artificial neural networks using citrate-capped silver nanoislands. Microchim Acta 189(8):1–16. https://doi.org/10.1007/S00604-022-05393-4/TABLES/4

Eissa S, Zourob M (2020) A dual electrochemical/colorimetric magnetic nanoparticle/peptide-based platform for the detection of Staphylococcus aureus. Analyst 145(13):4606–4614. https://doi.org/10.1039/D0AN00673D

Eş I, Montebugnoli LJ, Filippi MFP, Malfatti-Gasperini AA et al (2020) High-throughput conventional and stealth cationic liposome synthesis using a chaotic advection-based microfluidic device combined with a centrifugal vacuum concentrator. Chem Eng J 382:122821. https://doi.org/10.1016/J.CEJ.2019.122821

Fan Y, Marioli M, Zhang K (2021) Analytical characterization of liposomes and other lipid nanoparticles for drug delivery. J Pharm Biomed Anal 192:113642. https://doi.org/10.1016/J.JPBA.2020.113642

Fang K, Park OJ, Hong SH (2020) Controlling biofilms using synthetic biology approaches. Biotechnol Adv 40:107518. https://doi.org/10.1016/j.biotechadv.2020.107518

Fanzo J, McLaren R, Bellows A, Carducci B (2023) Challenges and opportunities for increasing the effectiveness of food reformulation and fortification to improve dietary and nutrition outcomes. Food Policy 119:102515. https://doi.org/10.1016/J.FOODPOL.2023.102515

Farooq U, Ullah MW, Yang Q, Aziz A, Xu J, Zhou L, Wang S et al (2020) High-density phage particles immobilization in surface-modified bacterial cellulose for ultra-sensitive and selective electrochemical detection of Staphylococcus aureus. Biosens Bioelectron 157:112163. https://doi.org/10.1016/J.BIOS.2020.112163

FDA, Food and Drug Administration. (2022) Food Code

Firmino PCOS, Vianna SSV, da Costa OMMM, Malfatti-Gasperini AA et al (2021) 3D micromixer for nanoliposome synthesis: a promising advance in high mass productivity. Lab Chip 21(15): 2971–2985. https://doi.org/10.1039/D1LC00232E

Fontecha-Umaña F, Ríos-Castillo AG, Ripolles-Avila C, Rodríguez-Jerez JJ (2020) Antimicrobial activity and prevention of bacterial biofilm formation of silver and zinc oxide nanoparticle-containing polyester surfaces at various concentrations for use. Food Secur 9(4):442. https://doi.org/10.3390/foods9040442

Fusco V, Chieffi D, Fanelli F, Logrieco AF et al (2020) Microbial quality and safety of milk and milk products in the 21st century. Compr Rev Food Sci Food Saf 19(4):2013–2049. https://doi.org/10.1111/1541-4337.12568

Garcês A, Amaral MH, Sousa Lobo JM, Silva AC (2018) Formulations based on solid lipid nanoparticles (SLN) and nanostructured lipid carriers (NLC) for cutaneous use: a review. Eur J Pharm Sci 112:159–167. https://doi.org/10.1016/J.EJPS.2017.11.023

Gonçalves SF, Lee SHI, Khaneghah AM, Oliveira CAF (2020) Enzyme-based approaches to control microbial biofilms in dairy processing environments: a review. Qual Assur Saf Crops Foods 12(1):50–58. https://doi.org/10.15586/qas.v12iSP1.828

Gonzalez AS, Riego Vega V, García J, Galié S et al (2021) Functional antimicrobial surface coatings deposited onto nanostructured 316l food-grade stainless steel. Nano 11(4):1055. https://doi.org/10.3390/nano11041055

Gouda A, Sakr OS, Nasr M, Sammour O (2021) Ethanol injection technique for liposomes formulation: an insight into development, influencing factors, challenges and applications. J. Drug Delivery Sci. Technol. 61:102174. https://doi.org/10.1016/J.JDDST.2020.102174

Guo S, Guo Q, Zhang Y, Peng X et al (2023) Preparation of enzymatically cross-linked α-lactalbumin nanoparticles and their application for encapsulating lycopene. Food Chem 429:136394. https://doi.org/10.1016/J.FOODCHEM.2023.136394

Hao J, Guo B, Yu S, Zhang W et al (2017) Encapsulation of the flavonoid quercetin with chitosan-coated nano-liposomes. LWT Food Sci Technol 85:37–44. https://doi.org/10.1016/J.LWT.2017.06.048

Hao N, Xu Z, Nie Y, Jin C (2019) Microfluidics-enabled rational design of ZnO micro−/nanoparticles with enhanced photocatalysis, cytotoxicity, and piezoelectric properties. Chem Eng J 378:122222. https://doi.org/10.1016/J.CEJ.2019.122222

Hariyama T, Takaku Y, Kawasaki H, Shimomura M, Senoh C, Yamahama Y, Hozumi A, Ito S, Matsuda N, Yamada S, Itoh T, Haseyama M, Ogawa T, Mori N, So S, Mitsuno H, Ohara M, Nomura S, Hirasaka M, Kimura Y et al (2022) In this issue. Microscopy 71(1):1–2. https://doi.org/10.1093/JMICRO/DFAB060

Helena de Abreu-Martins H, Artiga-Artigas M, Hilsdorf Piccoli R, Martín-Belloso O et al (2020) The lipid type affects the in vitro digestibility and β-carotene bioaccessibility of liquid or solid lipid nanoparticles. Food Chem 311:126024. https://doi.org/10.1016/J.FOODCHEM.2019.126024

Hernández-Castellano LE, Nally JE, Lindahl J, Wanapat M et al (2019) Dairy science and health in the tropics: challenges and opportunities for the next decades. Trop Anim Health Prod 51(5):1009–1017. https://doi.org/10.1007/S11250-019-01866-6/FIGURES/1

Himshweta, Singh M (2023) Nanosensor platforms for detection of milk adulterants. Sens. Actuators Rep 5:100159. https://doi.org/10.1016/J.SNR.2023.100159

Hu X, Shi J, Shi Y, Zou X et al (2019) Use of a smartphone for visual detection of melamine in milk based on Au@Carbon quantum dots nanocomposites. Food Chem 272:58–65. https://doi.org/10.1016/J.FOODCHEM.2018.08.021

IARC Working Group (2018) IARC monographs on the evaluation of carcinogenic risks to humans: Formaldehyde

Ismagilov RF, Stroock AD, Kenis PJA, Whitesides G et al (2000) Experimental and theoretical scaling laws for transverse diffusive broadening in two-phase laminar flows in microchannels. Appl Phys Lett 76(17):2376–2378. https://doi.org/10.1063/1.126351

Ivanova AS, Merkuleva AD, Andreev SV, Sakharov KA (2019) Method for determination of hydrogen peroxide in adulterated milk using high performance liquid chromatography. Food Chem 283:431–436. https://doi.org/10.1016/J.FOODCHEM.2019.01.051

Jacobsen C, García-Moreno PJ, Mendes AC, Mateiu RV, Chronakis IS (2018) Use of Electrohydrodynamic processing for encapsulation of sensitive bioactive compounds and applications in food. Annu Rev Food Sci Technol 9(9):525–549. https://doi.org/10.1146/ANNUREV-FOOD-030117-012348/1

Jafari, S. M., & McClements, D. J. (2017). Nanotechnology Approaches for Increasing Nutrient Bioavailability. Advances in Food and Nutrition Research, 81, 1–30. https://doi.org/10.1016/bs.afnr.2016.12.008

Jafarifar Z, Rezaie M, Sharifan P, Jahani V et al (2022) Preparation and characterization of nanostructured lipid carrier (NLC) and Nanoemulsion containing vitamin D3. Appl Biochem Biotechnol 194(2):914–929. https://doi.org/10.1007/S12010-021-03656-Z/FIGURES/9

Jafarizadeh-Malmiri H, Sayyar Z, Anarjan N, Berenjian A (2019) Nano-sensors in food Nanobiotechnology. In: Nanobiotechnology in food: concepts, applications and perspectives. Springer, Cham, pp 81–94. https://doi.org/10.1007/978-3-030-05846-3_6

Jalili R, Khataee A, Rashidi MR, Razmjou A (2020) Detection of penicillin G residues in milk based on dual-emission carbon dots and molecularly imprinted polymers. Food Chem 314: 126172. https://doi.org/10.1016/J.FOODCHEM.2020.126172

Javaid M, Haleem A, Singh RP, Rab S et al (2021) Exploring the potential of nanosensors: a brief overview. Sens Int 2:100130. https://doi.org/10.1016/J.SINTL.2021.100130

Jia M, Liu J, Zhang J, Zhang H (2019) An immunofiltration strip method based on the photothermal effect of gold nanoparticles for the detection of Escherichia coli O157:H7. Analyst 144(2): 573–578. https://doi.org/10.1039/C8AN01004H

Jigyasa, Kaur Rajput J (2023) Nanosensors: a smart remedy for early detection of clenbuterol contamination in food. Food Chem 426:136569. https://doi.org/10.1016/J.FOODCHEM.2023.136569

Jiménez-Cruz E, Cuevas-Gómez AP, Unsworth L, Cornejo-Mazón M et al (2022) Poly-L-lysine-coated α-lactalbumin nanoparticles: preparation, effect of pH, and stability under in vitro simulated gastrointestinal conditions. J Chem Technol Biotechnol 97(6):1597–1603. https://doi.org/10.1002/JCTB.6952

Jiménez-Pichardo R, Hernández-Martínez I, Regalado-González C, Santos-Cruz J et al (2021) Innovative control of biofilms on stainless steel surfaces using electrolyzed water in the dairy industry. Food Secur 10(1):103. https://doi.org/10.3390/foods10010103

Jin L, Li T, Wu B, Yang T et al (2020) Rapid detection of Salmonella in milk by nuclear magnetic resonance based on membrane filtration superparamagnetic nanobiosensor. Food Control 110: 107011. https://doi.org/10.1016/J.FOODCONT.2019.107011

Joshi AS, Singh P, Mijakovic I (2020) Interactions of gold and silver nanoparticles with bacterial biofilms: molecular interactions behind inhibition and resistance. Int J Mol Sci 21(20):1–24. https://doi.org/10.3390/ijms21207658

Kahdestani SA, Shahriari MH, Abdouss M (2021) Synthesis and characterization of chitosan nanoparticles containing teicoplanin using sol–gel. Polym Bull 78(2):1133–1148. https://doi.org/10.1007/S00289-020-03134-2/FIGURES/12

Kašpar, O., Koyuncu, A. H., Pittermannová, A., Ulbrich, P., & Tokárová, V. (2019). Governing factors for preparation of silver nanoparticles using droplet-based microfluidic device. Biomedical Microdevices, 21(4), 1–14. https://doi.org/10.1007/S10544-019-0435-4/FIGURES/11

Kaur T, Putatunda C, Vyas A, Kumar G (2021) Zinc oxide nanoparticles inhibit bacterial biofilm formation via altering cell membrane permeability. Prep Biochem Biotechnol 51(4):309–319. https://doi.org/10.1080/10826068.2020.1815057

Jyoti, Kavita, Verma RK (2022) Selective detection of urea as milk adulterant using LMR based fiber optic probe. J Food Compos Anal 114:104825. https://doi.org/10.1016/J.JFCA.2022.104825

Kayani Z, Bordbar AK, Firuzi O (2018) Novel folic acid-conjugated doxorubicin loaded β-lactoglobulin nanoparticles induce apoptosis in breast cancer cells. Biomed Pharmacother 107:945–956. https://doi.org/10.1016/J.BIOPHA.2018.08.047

Kelly P, Shalloo L, Wallace M, Dillon P (2020) The Irish dairy industry – recent history and strategy, current state and future challenges. Int J Dairy Technol 73(2):309–323. https://doi.org/10.1111/1471-0307.12682

Kemper N (2008) Veterinary antibiotics in the aquatic and terrestrial environment. Ecol Indic 8(1): 1–13. https://doi.org/10.1016/J.ECOLIND.2007.06.002

Khade PA, Patil SS, Takale MP (2018) CIP for dairy industry. J NX- A Multidiscip. Peer, Rev, J:250–253

Khan F, Lee JW, Manivasagan P, Pham DTN et al (2019) Synthesis and characterization of chitosan oligosaccharide-capped gold nanoparticles as an effective antibiofilm drug against the Pseudomonas aeruginosa PAO1. Microb Pathog 135:103623. https://doi.org/10.1016/j.micpath.2019.103623

Khizar S, Zine N, Errachid A, Jaffrezic-Renault N et al (2022) Microfluidic-based nanoparticle synthesis and their potential applications. Electrophoresis 43(7–8):819–838. https://doi.org/10.1002/ELPS.202100242

Kim DY, Kadam A, Shinde S, Saratale RG et al (2018) Recent developments in nanotechnology transforming the agricultural sector: a transition replete with opportunities. J Sci Food Agric 98(3):849–864. https://doi.org/10.1002/JSFA.8749

Kumar V, Dash S (2021) Evaporation-based low-cost method for the detection of adulterant in Milk. ACS Omega 6(41):27200–27207. https://doi.org/10.1021/ACSOMEGA.1C03887/ASSET/IMAGES/LARGE/AO1C03887_0007.JPEG

Kurbanoglu S, Bakirhan NK, Shah A, Ozkan SA (2019) Chemical nanosensors in pharmaceutical analysis. In: New developments in nanosensors for pharmaceutical analysis. Academic Press, pp 141–170. https://doi.org/10.1016/B978-0-12-816144-9.00005-5

Kuswandi B, Jumina (2020) Active and intelligent packaging, safety, and quality controls. In: Fresh-cut fruits and vegetables: technologies and mechanisms for safety control, pp 243–294. https://doi.org/10.1016/B978-0-12-816184-5.00012-4

Kuswandi B, Futra D, Heng LY (2017) Nanosensors for the detection of food contaminants. In: Nanotechnology applications in food: flavor, stability, nutrition and safety, pp 307–333. https://doi.org/10.1016/B978-0-12-811942-6.00015-7

Kutlu N, Meral R, Ekin MM, Erim Kose Y et al (2022) A new application for the valorisation of pomegranate seed oil: nanoencapsulation of pomegranate seed oil into electrospun nanomats for food preservation. Int J Food Sci Technol 57(2):1074–1082. https://doi.org/10.1111/IJFS.15472

Kwak CH, Kang SM, Jung E, Haldorai Y et al (2018) Customized microfluidic reactor based on droplet formation for the synthesis of monodispersed silver nanoparticles. J Ind Eng Chem 63: 405–410. https://doi.org/10.1016/J.JIEC.2018.02.040

Lei KF (2014) Materials and fabrication techniques for Nano- and Microfluidic devices. RSC Detect Sci 5:1–28. https://doi.org/10.1039/9781849737609-00001

Levinson Y, Israeli-Lev G, Livney YD (2014) Soybean β-Conglycinin nanoparticles for delivery of hydrophobic nutraceuticals. Food Biophys 9(4):332–340. https://doi.org/10.1007/S11483-014-9341-8/FIGURES/9

Li Q, Xie G, Wang Y, Aguilar ZP et al (2021) Vancomycin-modified poly-l-lysine magnetic separation combined with multiplex polymerase chain reaction assay for efficient detection of Bacillus cereus in milk. J Dairy Sci 104(2):1465–1473. https://doi.org/10.3168/JDS.2020-18962

Liang W, Wei Y, Gao M, Yan X et al (2020) Detection of melamine adulteration in milk powder by using optical spectroscopy Technologies in the Last Decade—a Review. Food Anal Methods 13(11):2059–2069. https://doi.org/10.1007/S12161-020-01822-3/TABLES/1

Lima LS, Rossini EL, Pezza L, Pezza HR (2020a) Bioactive paper platform for detection of hydrogen peroxide in milk. Spectrochim Acta A Mol Biomol Spectrosc 227:117774. https://doi.org/10.1016/J.SAA.2019.117774

Lima MJA, Sasaki MK, Marinho OR, Freitas TA et al (2020b) Spot test for fast determination of hydrogen peroxide as a milk adulterant by smartphone-based digital image colorimetry. Microchem J 157:105042. https://doi.org/10.1016/J.MICROC.2020.105042

Liu C, Li Y, Liu BF (2019) Micromixers and their applications in kinetic analysis of biochemical reactions. Talanta 205:120136. https://doi.org/10.1016/J.TALANTA.2019.120136

Liu Y, Gao L, Yi J, Fan Y et al (2020) A-Lactalbumin and chitosan core–shell nanoparticles: resveratrol loading, protection, and antioxidant activity. Food Funct 11(2):1525–1536. https://doi.org/10.1039/C9FO01998G

Liu Z, Wang L, Liu P, Zhao K et al (2021) Rapid, ultrasensitive and non-enzyme electrochemiluminescence detection of hydrogen peroxide in food based on the ssDNA/g-C3N4 nanosheets hybrid. Food Chem 357:129753. https://doi.org/10.1016/J.FOODCHEM.2021.129753

Liu Q, Sun Y, Cheng J, Guo M (2022) Development of whey protein nanoparticles as carriers to deliver soy isoflavones. LWT 155:112953. https://doi.org/10.1016/J.LWT.2021.112953

Luo K, Ryu J, Seol IH, Jeong KB et al (2019) Paper-based radial chromatographic immunoassay for the detection of pathogenic bacteria in milk. ACS Appl Mater Interfaces 11(50):46472–46478.

https://doi.org/10.1021/ACSAMI.9B16075/ASSET/IMAGES/LARGE/AM9B16075_0006.JPEG

Mahari S, Roberts A, Gandhi S (2022) Probe-free nanosensor for the detection of salmonella using gold nanorods as an electroactive modulator. Food Chem 390:133219. https://doi.org/10.1016/J.FOODCHEM.2022.133219

Malik S, Muhammad K, Waheed Y (2023) Nanotechnology: a revolution in modern industry. Molecules 28(2):661. https://doi.org/10.3390/MOLECULES28020661

Martin NH, Evanowski RL, Wiedmann M (2023) Invited review: redefining raw milk quality-evaluation of raw milk microbiological parameters to ensure high-quality processed dairy products. J Dairy Sci 106(3):1502–1517. https://doi.org/10.3168/JDS.2022-22416

McDonald JC, Whitesides GM (2002) Poly(dimethylsiloxane) as a material for fabricating microfluidic devices. Acc Chem Res 35(7):491–499. https://doi.org/10.1021/AR010110Q/ASSET/IMAGES/LARGE/AR010110QF00008.JPEG

Meesilp N, Mesil N (2019) Effect of microbial sanitizers for reducing biofilm formation of Staphylococcus aureus and Pseudomonas aeruginosa on stainless steel by cultivation with UHT milk. Food Sci Biotechnol 28(1):289–296. https://doi.org/10.1007/s10068-018-0448-4

Mehrad B, Ravanfar R, Licker J, Regenstein JM et al (2018) Enhancing the physicochemical stability of β-carotene solid lipid nanoparticle (SLNP) using whey protein isolate. Food Res Int 105:962–969. https://doi.org/10.1016/J.FOODRES.2017.12.036

Mi F, Guan M, Wang Y, Chen G et al (2023) FPBA-modified magnetic nanoparticles combined with Au@AgNPs for label-free SERS detection of foodborne pathogens in milk. Spectrochim Acta A Mol Biomol Spectrosc 302:123094. https://doi.org/10.1016/J.SAA.2023.123094

Mirsadoughi E, Pebdeni AB, Hosseini M (2023) Sensitive colorimetric aptasensor based on peroxidase-like activity of ZrPr-MOF to detect Salmonella Typhimurium in water and milk. Food Control 146:109500. https://doi.org/10.1016/J.FOODCONT.2022.109500

Mohanta YK, Chakrabartty I, Mishra AK, Chopra H et al (2022) Nanotechnology in combating biofilm: a smart and promising therapeutic strategy. Front Microbiol 13:1028086. https://doi.org/10.3389/fmicb.2022.1028086

Mohd Yusof H, Mohamad R, Zaidan UH, Abdul Rahman NA (2019) Microbial synthesis of zinc oxide nanoparticles and their potential application as an antimicrobial agent and a feed supplement in animal industry: a review. J Anim Sci Biotechnol 10(1):1–22. https://doi.org/10.1186/S40104-019-0368-Z

Moschetti A, Vine LN, Lethcoe K, Dagda RK et al (2020) Assembly and characterization of biocompatible coenzyme Q10-enriched lipid nanoparticles. Lipids 55(2):141–149. https://doi.org/10.1002/LIPD.12218

Munawar A, Ong Y, Schirhagl R, Tahir MA et al (2019) Nanosensors for diagnosis with optical, electric and mechanical transducers. RSC Adv 9(12):6793–6803. https://doi.org/10.1039/C8RA10144B

Nagraik R, Sharma A, Kumar D, Chawla P et al (2021) Milk adulterant detection: conventional and biosensor-based approaches: a review. Sens Bio-Sensing Res 33:100433. https://doi.org/10.1016/J.SBSR.2021.100433

Namasivayam SKR, John A, Bharani A, Kavisri RSM, Moovendhan M (2022) Biocompatible formulation of cationic antimicrobial peptide Polylysine (PL) through nanotechnology principles and its potential role in food preservation – a review. Int J Biol Macromol 222:1734–1746. https://doi.org/10.1016/J.IJBIOMAC.2022.09.238

Nascimento CF, Santos PM, Pereira-Filho ER, Rocha FRP (2017) Recent advances on determination of milk adulterants. Food Chem 221:1232–1244. https://doi.org/10.1016/J.FOODCHEM.2016.11.034

Nasrollahzadeh M, Issaabadi Z, Sajjadi M, Sajadi SM et al (2019) Types of nanostructures. Interface Sci Technol 28:29–80. https://doi.org/10.1016/B978-0-12-813586-0.00002-X

Ndahetuye JB, Artursson K, Båge R, Ingabire A et al (2020) MILK symposium review: microbiological quality and safety of milk from farm to milk collection centers in Rwanda. J Dairy Sci 103(11):9730–9739. https://doi.org/10.3168/JDS.2020-18302

Niculescu AG, Chircov C, Bîrcă AC, Grumezescu AM (2021a) Fabrication and applications of microfluidic devices: a review. Int J Mol Sci 22(4):2011. https://doi.org/10.3390/IJMS22042011

Niculescu AG, Chircov C, Bîrcă AC, Grumezescu AM (2021b) Nanomaterials synthesis through microfluidic methods: an updated overview. Nano 11(4):864. https://doi.org/10.3390/NANO11040864

Nyokabi SN, de Boer IJM, Luning PA, Korir L et al (2021) Milk quality along dairy farming systems and associated value chains in Kenya: an analysis of composition, contamination and adulteration. Food Control 119:107482. https://doi.org/10.1016/J.FOODCONT.2020.107482

O'Sullivan L, Bolton D, McAuliffe O, Coffey A (2019) The use of bacteriophages to control and detect pathogens in the dairy industry. Int J Dairy Technol 73(1):1–11

Okpala COR, Korzeniowska M (2023) Understanding the relevance of quality management in agro-food product industry: from ethical considerations to assuring food hygiene quality safety standards and its associated processes. Food Rev Intl 39(4):1879–1952. https://doi.org/10.1080/87559129.2021.1938600

Olsen DS (2006) Nanotechnology: science, innovation and opportunity. Sci Technol Stud 19(2):100–103. https://doi.org/10.23987/STS.55197

Osanlou R, Emtyazjoo M, Banaei A, Hesarinejad MA et al (2022) Preparation of solid lipid nanoparticles and nanostructured lipid carriers containing zeaxanthin and evaluation of physicochemical properties. Colloids Surf A Physicochem Eng Asp 641:128588. https://doi.org/10.1016/J.COLSURFA.2022.128588

Pant KJ, Cotter PD, Wilkinson MG, Sheehan JJ (2023) Towards sustainable cleaning-in-place (CIP) in dairy processing: exploring enzyme-based approaches to cleaning in the cheese industry. Compr Rev Food Sci Food Saf 22(5):3602–3619. https://doi.org/10.1111/1541-4337.13206

Papadochristopoulos, A., Kerry, J. P., Fegan, N., Burgess, C. M., & Duffy, G. (2021). Natural Anti-Microbials for Enhanced Microbial Safety and Shelf-Life of Processed Packaged Meat. Foods 2021, Vol. 10, Page 1598, 10(7), 1598. https://doi.org/10.3390/FOODS10071598

Pateiro M, Gómez B, Munekata PES, Barba FJ et al (2021) Nanoencapsulation of promising bioactive compounds to improve their absorption, stability, functionality and the appearance of the final food products. Molecules 26(6):1547. https://doi.org/10.3390/MOLECULES26061547

Peighambardoust S, Karami Z, Pateiro M, Lorentzo JM (2021) A review on health-promoting, biological, and functional aspects of bioactive peptides in food applications. Biomol Ther 11(5):631. https://doi.org/10.3390/biom11050631

Pessoa ACSN, Sipoli CC, De La Torre LG (2017) Effects of diffusion and mixing pattern on microfluidic-assisted synthesis of chitosan/ATP nanoparticles. Lab Chip 17(13):2281–2293. https://doi.org/10.1039/C7LC00291B

Pu H, Xu Y, Sun DW, Wei Q et al (2021) Optical nanosensors for biofilm detection in the food industry: principles, applications and challenges. Crit Rev Food Sci Nutr 61(13):2107–2124. https://doi.org/10.1080/10408398.2020.1808877

Pushparaj K, Liu WC, Meyyazhagan A, Orlacchio A et al (2022) Nano- from nature to nurture: a comprehensive review on facets, trends, perspectives and sustainability of nanotechnology in the food sector. Energy 240:122732. https://doi.org/10.1016/J.ENERGY.2021.122732

Quintanilla P, Doménech E, Escriche I, Carmen Beltrán M et al (2019) Food safety margin assessment of antibiotics: pasteurized goat's milk and fresh cheese. J Food Prot 82(9):1553–1559. https://doi.org/10.4315/0362-028X.JFP-18-434

Radhika NK, Gorthi SS (2019) Enhancement of the fluorescence properties of double stranded DNA templated copper nanoparticles. Mater Sci Eng C 98:1034–1042. https://doi.org/10.1016/J.MSEC.2019.01.042

Radwan AA, Darwesh OM, Emam MT, Mohamed KA et al (2022) A combined treatment of Proteinase K and biosynthesized ZnO-NPs for eradication of dairy biofilm of sporeformers. AIMS Microbiol 8(4):507–527

Rasti B, Erfanian A, Selamat J (2017) Novel nanoliposomal encapsulated omega-3 fatty acids and their applications in food. Food Chem 230:690–696. https://doi.org/10.1016/J.FOODCHEM.2017.03.089

Rather MA, Gupta K, Bardhan P, Borah M et al (2021) Microbial biofilm: a matter of grave concern for human health and food industry. J Basic Microbiol 61(5):380–395. https://doi.org/10.1002/jobm.202000678

Raza MA, Kanwal Z, Rauf A, Sabri AN et al (2016) Size- and shape-dependent antibacterial studies of silver nanoparticles synthesized by wet chemical routes. Nano 6(4):74. https://doi.org/10.3390/NANO6040074

Ren Y, Fan Z (2023) Synthesis of fluorescent probe based on molecularly imprinted polymers on nitrogen-doped carbon dots for determination of tobramycin in milk. Food Chem 416:135792. https://doi.org/10.1016/J.FOODCHEM.2023.135792

Ren K, Zhou J, Wu H (2013) Materials for microfluidic Chip fabrication. Acc Chem Res 46(11): 2396–2406. https://doi.org/10.1021/AR300314S

Resende D, Costa Lima SA, Reis S (2020) Nanoencapsulation approaches for oral delivery of vitamin A. Colloids Surf B: Biointerfaces 193:111121. https://doi.org/10.1016/J.COLSURFB.2020.111121

Ribeiro EF, de Barros-Alexandrino TT, Assis OBG, Junior AC et al (2020) Chitosan and crosslinked chitosan nanoparticles: synthesis, characterization and their role as Pickering emulsifiers. Carbohydr Polym 250:116878. https://doi.org/10.1016/J.CARBPOL.2020.116878

Rizzotto F, Marin M, Péchoux C, Auger S et al (2023) Colorimetric aptasensor for detection of bacillus cytotoxicus spores in milk and ready-to-use food. Heliyon 9(7):e17562. https://doi.org/10.1016/J.HELIYON.2023.E17562

Safran V, Göktürk I, Bakhshpour M, Yılmaz F et al (2021) Development of molecularly imprinted polymer-based optical sensor for the sensitive penicillin G detection in Milk. ChemistrySelect 6(43):11865–11875. https://doi.org/10.1002/SLCT.202103058

Saini P, Kamalesu, Lalita, Manikanika (2023) Review on nanotechnology impact on the food services industry. Mater Today: Proc 92(1):226–232. https://doi.org/10.1016/J.MATPR.2023.04.377

Sajwan RK, Solanki PR (2022) A hybrid optical strategy based on graphene quantum dots and gold nanoparticles for selective determination of gentamicin in the milk and egg samples. Food Chem 370:131312. https://doi.org/10.1016/J.FOODCHEM.2021.131312

Saleh Ahammad AJ, Lee JJ, Rahman MA (2009) Electrochemical sensors based on carbon nanotubes. Sensors 9(4):2289–2319. https://doi.org/10.3390/S90402289

Sapino S, Chindamo G, Chirio D, Manzoli M et al (2021) Calcium phosphate-coated lipid nanoparticles as a potential tool in bone diseases therapy. Nano 11(11):2983. https://doi.org/10.3390/NANO11112983/S1

Scott SM, Ali Z (2021) Fabrication methods for microfluidic devices: an overview. Micromachines 12(3):319. https://doi.org/10.3390/MI12030319

Seth S, Rathinasabapathi P (2022) A short review on detection of antibiotics in Milk using nanomaterial-based biosensor. Food Anal Methods 15(8):2181–2192. https://doi.org/10.1007/S12161-022-02291-6/METRICS

Shahidi F, Pan Y (2022) Influence of food matrix and food processing on the chemical interaction and bioaccessibility of dietary phytochemicals: a review. Crit Rev Food Sci Nutr 62(23): 6421–6445. https://doi.org/10.1080/10408398.2021.1901650

Shahidi F, Peng H (2018) Bioaccessibility and bioavailability of phenolic compounds. J Food Bioact 4:11–68. https://doi.org/10.31665/JFB.2018.4162

Shahidi F, Ramakrishnan VV, Oh WY (2019) Bioavailability and metabolism of food bioactives and their health effects: a review. J Food Bioact 8:6–41. https://doi.org/10.31665/JFB.2019.8204

Shalileh F, Sabahi H, Dadmehr M, Hosseini M (2023) Sensing approaches toward detection of urea adulteration in milk. Microchem J 193:108990. https://doi.org/10.1016/J.MICROC.2023.108990

Shao S, Shen X, Guo M (2018) Zinc-loaded whey protein nanoparticles prepared by enzymatic cross-linking and desolvation. Int J Food Sci Technol 53(9):2205–2211. https://doi.org/10.1111/IJFS.13809

Sharahi JY, Azimi T, Shariati A, Safari H et al (2019) Advanced strategies for combating bacterial biofilms. J Cell Physiol 234(9):14689–14708. https://doi.org/10.1002/jcp.28225

Shemesh M, Ostrov I (2020) Role of Bacillus species in biofilm persistence and emerging antibiofilm strategies in the dairy industry. J Sci Food Agric 100(6):2327–2336. https://doi.org/10.1002/jsfa.10285

Shineh G, Mobaraki M, Perves Bappy MJ, Mills DK (2023) Biofilm formation, and related impacts on healthcare, food processing and packaging, industrial manufacturing, marine industries, and sanitation-A review. Appl Microbiol 3(3):629–665. https://doi.org/10.3390/applmicrobiol3030044

Shtay R, Keppler JK, Schrader K, Schwarz K (2019) Encapsulation of (−)-epigallocatechin-3-gallate (EGCG) in solid lipid nanoparticles for food applications. J Food Eng 244:91–100. https://doi.org/10.1016/J.JFOODENG.2018.09.008

Silva CCG, Silva SPM, Ribeiro SC (2018) Application of bacteriocins and protective cultures in dairy food preservation. Front Microbiol 9:345996. https://doi.org/10.3389/FMICB.2018.00594/BIBTEX

Simões LS, Abrunhosa L, Vicente AA, Ramos OL (2020) Suitability of β-lactoglobulin micro- and nanostructures for loading and release of bioactive compounds. Food Hydrocoll 101:105492. https://doi.org/10.1016/J.FOODHYD.2019.105492

Singh A (2020) Sanitation and safety in food services. Res Rev Int J Multidiscip 5(8):205–210. https://doi.org/10.31305/rrijm.2020.v05.i08.048

Singh P, Singh S, Nara S (2021) Nanotechnology as a tool for contaminants detection in Milk or milk products. In: Nascimento RFd, Neto VdOS, Fechine PBA, Freire PdTC (eds) Materials horizons: from nature to nanomaterials. Springer, Singapore, pp 133–161. https://doi.org/10.1007/978-981-33-6056-3_5/COVER

Siqueira IND, Limeira CH, Cavalcanti AAC, Souza JGD et al (2021) Bactérias Formadoras de Biofilmes na Indústria de Laticínios: uma Breve Revisão. Ensaios e Ciência C Biológicas Agrárias e Da Saúde 25(4):491–500. https://doi.org/10.17921/1415-6938.2021v25n4p491-500

Smykov IT (2020) Nanotechnology in the dairy industry. In: The ELSI handbook of nanotechnology: risk, safety, ELSI and commercialization, pp 223–275. https://doi.org/10.1002/9781119592990.CH11

Soleimanifar M, Jafari SM, Assadpour E (2020) Encapsulation of olive leaf phenolics within electrosprayed whey protein nanoparticles; production and characterization. Food Hydrocoll 101:105572. https://doi.org/10.1016/J.FOODHYD.2019.105572

Solghi S, Emam-Djomeh Z, Fathi M, Farahani F (2020) The encapsulation of curcumin by whey protein: assessment of the stability and bioactivity. J Food Process Eng 43(6):e13403. https://doi.org/10.1111/JFPE.13403

Soumya NPP, Mini S, Sivan SK, Mondal S (2021) Bioactive compounds in functional food and their role as therapeutics. Bioact Compd Health Dis 4(3):24–39. https://doi.org/10.31989/BCHD.V4I3.786

Sridhar A, Ponnuchamy M, Kumar PS, Kapoor A (2020) Food preservation techniques and nanotechnology for increased shelf life of fruits, vegetables, beverages and spices: a review. Environ Chem Lett 19(2):1715–1735. https://doi.org/10.1007/S10311-020-01126-2

Subramani T, Ganapathyswamy H (2020) An overview of liposomal nano-encapsulation techniques and its applications in food and nutraceutical. J Food Sci Technol 57(10):3545–3555. https://doi.org/10.1007/S13197-020-04360-2/TABLES/2

Sunoqrot S, Alkurdi M, Al Bawab AQ, Hammad AM et al (2023) Encapsulation of morin in lipid core/PLGA shell nanoparticles significantly enhances its anti-inflammatory activity and oral bioavailability. Saudi Pharm J 31(6):845–853. https://doi.org/10.1016/J.JSPS.2023.04.010

Szilagyi A (2019) Digestion, absorption, metabolism, and physiological effects of lactose. In: Lactose: evolutionary role, health effects, and applications, pp 49–111. https://doi.org/10.1016/B978-0-12-811720-0.00002-7

Teixeira RC, Luiz LC, Junqueira GMA, Bell MJV et al (2020) Detection of antibiotic residues in cow's milk: a theoretical and experimental vibrational study. J Mol Struct 1215:128221. https://doi.org/10.1016/J.MOLSTRUC.2020.128221

Teodoro KBR, Migliorini FL, Christinelli WA, Correa DS (2019) Detection of hydrogen peroxide (H_2O_2) using a colorimetric sensor based on cellulose nanowhiskers and silver nanoparticles. Carbohydr Polym 212:235–241. https://doi.org/10.1016/J.CARBPOL.2019.02.053

Thakur M, Wang B, Verma ML (2022) Development and applications of nanobiosensors for sustainable agricultural and food industries: recent developments, challenges and perspectives. Environ Technol Innov 26:102371. https://doi.org/10.1016/J.ETI.2022.102371

Tlais AZA, Fiorino GM, Polo A, Filannino et al (2020) High-value compounds in fruit, vegetable and cereal byproducts: an overview of potential sustainable reuse and exploitation. Molecules 25(13):2987. https://doi.org/10.3390/MOLECULES25132987

Turgeon SL, Brisson G (2020) Symposium review: the dairy matrix—bioaccessibility and bioavailability of nutrients and physiological effects. J Dairy Sci 103(7):6727–6736. https://doi.org/10.3168/JDS.2019-17308

Üzek R, Sari E, Denizli A (2021) Detection of melamine by using fluorescent nanocomposites with specific recognition sites. ChemSelect 6(9):2149–2155. https://doi.org/10.1002/SLCT.202100130

Verma S, Pandey AK (2021) Improving bioavailability of nutrients through Nanotechnology. In: Maurya VK, Gothandam KM, Ranjan S, Dasgupta N, Lichtfouse E (eds) Sustainable agriculture reviews 55. Sustainable agriculture reviews, vol 55. Springer, Cham, pp 135–170. https://doi.org/10.1007/978-3-030-76813-3_6

Verma M, Naqvi TK, Tripathi SK, Kulkarni MM et al (2021) Paper based low-cost flexible SERS sensor for food adulterant detection. Environ Technol Innov 24:102033. https://doi.org/10.1016/J.ETI.2021.102033

Wang Q, Long MY, Lv CY, Xin SP et al (2020) Lanthanide-labeled fluorescent-nanoparticle immunochromatographic strips enable rapid and quantitative detection of Escherichia coli O157:H7 in food samples. Food Control 109:106894. https://doi.org/10.1016/J.FOODCONT.2019.106894

Wang Y, Tao H, Wang K, Li M et al (2021) Nanocapsules formed by interactions between chondroitin sulfate and egg white protein for encapsulating hydrophilic ingredients. Green Chem 23(19):7566–7575. https://doi.org/10.1039/D1GC01706C

World Health Organization (2017) Guidelines for drinking-water quality

Xi H, Chen X, Cao Y, Xu J et al (2020) Electrochemical determination of formaldehyde via reduced AuNPs@PPy composites modified electrode. Microchem J 156:104846. https://doi.org/10.1016/J.MICROC.2020.104846

Xixi C, Lina Z, Shaoyun W, Pingfan R (2015) Fabrication and characterization of the nanocomposite of whey protein hydrolysate chelated with calcium. Food Funct 6(3):816–823. https://doi.org/10.1039/C4FO00811A

Xue G, Zhiying M, Xiuying L, Lijun T et al (2020) A fluorescence resonance energy transfer biosensor based on graphene quantum dots and protoporphyrin IX for the detection of melamine. J Fluoresc 30(6):1463–1468. https://doi.org/10.1007/S10895-020-02524-Z/TABLES/2

Yang L, Yang J, Dong Q, Zhou F et al (2021) One-step synthesis of CuO nanoparticles based on flame synthesis: as a highly effective non-enzymatic sensor for glucose, hydrogen peroxide and formaldehyde. J Electroanal Chem 881:114965. https://doi.org/10.1016/J.JELECHEM.2020.114965

Ye Z, Wang K, Lou M, Jia X et al (2020) Consecutive synthesis of gold nanobipyramids with controllable morphologies using a microfluidic platform. Microfluid Nanofluid 24(5):1–8. https://doi.org/10.1007/S10404-020-02345-3/FIGURES/5

Yi J, Gao L, Zhong G, Fan Y (2020) Fabrication of high internal phase Pickering emulsions with calcium-crosslinked whey protein nanoparticles for β-carotene stabilization and delivery. Food Funct 11(1):768–778. https://doi.org/10.1039/C9FO02434D

Yin PT, Kim TH, Choi JW, Lee KB (2013) Prospects for graphene–nanoparticle-based hybrid sensors. Phys Chem Chem Phys 15(31):12785–12799. https://doi.org/10.1039/C3CP51901E

Yusuf CAGM, Demirci M, Bayramba SSK, Akır BC¸ Gü Lseren I (2017) Nanoencapsulation of enzymes, bioactive peptides, and biological molecules. Academic 297–332. https://doi.org/10.1016/B978-0-12-809740-3.00008-8

Zhang M, Zhang B, Li T, Zhu X et al (2022) Electrochemical detection of aminoglycoside antibiotics residuals in milk based on magnetic molecularly imprinted particles and metal ions. Food Chem 389:133120. https://doi.org/10.1016/J.FOODCHEM.2022.133120

Zhang Q, Toprakcioglu Z, Jayaram AK, Guo G et al (2023) Formation of protein nanoparticles in microdroplet flow reactors. ACS Nano 17(12):11335–11344. https://doi.org/10.1021/ACSNANO.3C00107/ASSET/IMAGES/LARGE/NN3C00107_0007.JPEG

Zhao F, Zhou M, Wang L, Huang Z et al (2019) One-step voltammetric deposition of l-proline assisted silver nanoparticles modified glassy carbon electrode for electrochemical detection of hydrogen peroxide. J Electroanal Chem 833:205–212. https://doi.org/10.1016/J.JELECHEM.2018.11.050

Ziyaina M, Rasco B, Coffey T, Ünlü G et al (2019) Colorimetric detection of volatile organic compounds for shelf-life monitoring of milk. Food Control 100:220–226. https://doi.org/10.1016/J.FOODCONT.2019.01.018

Ziyaina M, Rasco B, Sablani SS (2020) Rapid methods of microbial detection in dairy products. Food Control 110:107008. https://doi.org/10.1016/J.FOODCONT.2019.107008

Zou Y, Qian Y, Rong X, Cao K et al (2021a) Encapsulation of quercetin in biopolymer-coated zein nanoparticles: formation, stability, antioxidant capacity, and bioaccessibility. Food Hydrocoll 120:106980. https://doi.org/10.1016/J.FOODHYD.2021.106980

Zou L, Li X, Lai Y (2021b) Colorimetric aptasensor for sensitive detection of kanamycin based on target-triggered catalytic hairpin assembly amplification and DNA-gold nanoparticle probes. Microchem J 162:105858. https://doi.org/10.1016/J.MICROC.2020.105858

Zukas BG, Gupta NR (2017) Interphase synthesis of zinc oxide nanoparticles in a droplet flow reactor. Ind Eng Chem Res 56(25):7184–7191. https://doi.org/10.1021/ACS.IECR.7B00407/ASSET/IMAGES/LARGE/IE-2017-00407V_0010.J

The manufacturer's authorised representative in the EU is Springer Nature Customer Service Centre GmbH, Europaplatz 3, 69115 Heidelberg, Germany. If you have any concerns regarding our products, please contact ProductSafety@springernature.com

Printed and bound by CPI Group (UK) Ltd, Croydon, CR0 4YY
01/10/2026
02225922-0013